Lazarus J. Salop

# Geological Evolution of the Earth During the Precambrian

Translated by V. P. Grudina

With 78 Figures

Springer-Verlag
Berlin Heidelberg New York 1983

Professor Dr. Lazarus J. Salop
All-Union Geological Research Institute (VSEGEI)
Sredny pr. 74, 199026, Leningrad, USSR

Translation of: Geologicheskoe Razvitie Zemli V Dokembrii
© by Nedra Publishing House, Leningrad 1982

ISBN 3-540-11709-1 Springer-Verlag Berlin Heidelberg New York
ISBN 0-387-11709-1 Springer-Verlag New York Heidelberg Berlin

Library of Congress Cataloging in Publication Data. Geologicheskoe razvitie zemli v dokembrii.
English. Geological evolution of the earth during the Precambrian. Translation of: Geologiches-
koe razvitie zemli v dokembrii. Bibliography: p. Includes index. 1. Geology, Stratigraphic –
Pre-Cambrian. I. Salop, L.I. (Lazar̆ Iosifovich), 1926– . QE653.G4513. 1982. 551.7'1. 82–16985.

Typesetting: Daten- und Lichtsatz-Service, Würzburg
Printing and Bookbinding: Graphischer Betrieb Konrad Triltsch, Würzburg
2132/3130-543210

# Preface

Progress in Precambrian geology has been exceptionally great, indeed quite striking for geologists of the older generation; only some 30–40 years ago the Precambrian appeared as an uncertain and even mystic prelude to geologic evolution. Even the very name – Precambrian – means some indivisible unit in the early history of the Earth, the beginning of which is poorly known. At the same time it was obvious that the Precambrian formations are of extremely varied and complex composition and poor knowledge and lack of reliable methods of division and correlation were to blame for the lack of significant progress in studies of this early evolutionary stage of the planet. Certainly, even at the very start of Precambrian studies, the results obtained were quite promising, lifting as they did the mysterious veil over the regional Precambrian; but they presented no general realistic picture of this early stage in the Earth's evolution at that time. Recently, this situation has completely changed, due to new methods of study of the older formations, and due also to the refinement of some well-known methods, in particular of division, dating, and correlation of "silent" metamorphic strata. Application of different isotope methods of dating was most important in providing objective rock age and thereby the age of geologic events recorded in these rocks. Thus it became possible to reconstruct the oldest geologic period of our planet.

The title of this book, at first sight, represents a certain tautology, but geologic history can only be clarified by study of the rock records and the successions in the formation of these rocks. The oldest rocks cropping out in the Earth's surface are isotopically dated at 3.7–4.0 b.y. These rocks are, in fact, older in reality and these values characterize not the time of their formation, but the time of their transformation (metamorphism), and the oldest rocks formed during the time the Earth acquired the planet shape are not yet known for certain. The geologic evolution of the planet since ~ 4.3 m.y. approximately and up to 0.57 b.y. (the beginning of the Cambrian period of the Paleozoic Era) is the object of the present study. This interval of 3.7 b.y. embraces more than 85% of the total geologic evolution. It will be shown that certain traces of still earlier events are recorded in the rocks of the Earth, corresponding to the so-called "pre-geologic" stage in the Earth's evolution (4.5–4.2 b.y.). These events, however, can only be speculated upon by indirect evidence, by analogy based on data of planetary cosmology.

Naturally, the knowledge of a very long period in the Earth's history has exceptionally important scientific implications in its theoretical and philosophical aspects. Recent studies reveal that many of the regularities poorly based or uncertain for Phanerozoic time could become much clearer after analyzing data on the Precambrian time. The studies of the older formations revealed new, previously unknown general regularities and showed important evolutionary relations of different geologic phenomena and processes; in particular, these studies provided a possibility of determining certain trends in changes of lithogenesis, magmatism, and tectonics through time. As a result we can evaluate more exactly the validity in applying the concept of actualism in studies of geologic processes. Finally, Precambrian studies are of exceptionally great economic importance, for the oldest rocks are related to large, unique mineral deposits, many of which occur only in the Precambrian complexes of certain age and structure.

This book gives a brief outline of Precambrian history in the light of the data now available and an attempt is made here to determine certain general regularities in the evolution of the outer shells of the planet: the Earth's crust, hydrosphere, and atmosphere. Naturally, different aspects of this problem are not adequately discussed here; many phenomenae are still poorly known; the author's interests, scope of study, and knowledge must also be taken into account. This book is by no means a summary or a compilation; the problems discussed here are those with which the author was concerned personally. Primarily these are the problems of the Precambrian subdivision, the evolution of sedimentogenesis, of tectonic structures and of larger elements of the Earth's crust, the problems of periodicity of tectogenesis, the relations of tectonic processes and magmatism. Many more problems of Precambrian geology are also discussed, for instance the origin of the oldest astroblemes, their appearance on the Earth's surface, certain peculiar aspects of origin of glaciations, the possible causes of changes in organic evolution. In the final chapter an attempt is made to sum up all data on the Precambrian with a conclusion on a directed and irreversible geologic evolution of the planet.

Some other generalized works of the author have appeared earlier. Data on the older formations of the Northern continents were analyzed in the book *General Stratigraphic Scheme of the Precambrian* (Salop 1973a), of which a revised and enlarged version appeared in English (Salop 1977a). The Precambrian of Africa is discussed in detail in a separate book that was also recently published (Salop 1977b). Some aspects of the Precambrian of Australia and South America are examined in papers that are in preparation. Here are also included the results obtained in the works mentioned above and the data that have become available recently; in some cases they introduce appreciable changes to the earlier conceptions. At the same time, many concepts elaborated earlier were examined more fundamentally, and many new ideas are also suggested.

This book is addressed to Precambrian specialists, as well as to a wider readership among geologists, post-graduates, and students. The author hopes also that it will be of interest to scientists of different branches, to naturalists in particular with interest in the geologic evolution of our planet.

In the book the isotope datings are given in accordance with the decay constants $\lambda_{238U} = 1.5369 \cdot 10^{-10} \, y^{-1}$, $\lambda_{235U} = 9.7216 \cdot 10^{-10} \, y^{-1}$, $\lambda_{Rb} = 1.39 \cdot 10^{-11} \, y^{-1}$, $\lambda_K = 4.72 \cdot 10^{-10} \, y^{-1}$, $\lambda_{\beta K} = 0.557 \cdot 10^{-10} \, y^{-1}$. The U-Th-Pb or Pb-Pb datings given in the book are to be multiplied by a coefficient of 0.990, the Rb-Sr datings by 0.978, the K-Ar datings are to be related to the age interval: 0.4–1.0 b.y. by 0.975; 1–1.5 b.y. by 0.978; 1.5–3.0 b.y. by 0.968 and 3.0–5.0 b.y. by 0.963 (Zykov et al. 1979) in order to bring them into accord with the now recommended standard decay constants ($\lambda_{238U} = 1.55125 \cdot 10^{-10} \, y^{-1}$, $\lambda_{235U} = 9.8485 \cdot 10^{-10} \, y^{-1}$, $\lambda_{Rb} = 1.42 \cdot 10^{-11} \, y^{-1}$, $\lambda_K = 4.692 \cdot 10^{-11} \, y^{-1}$ and $\lambda_{\beta K} \, 0.581 \cdot 10^{-10} \, y^{-1}$). The approximate values thus obtained differ from the ones given in the book by insignificant figures (for Pb-methods ca. 1%) and they in no way affect the geochronologic boundaries accepted in the book, especially if a general scattering of values related to geologic factors is taken into consideration. The Rb-Sr dating is obtained by whole rock isochron analysis (except in cases where a special reservation is made).

Leningrad, August 1982                                                    L. J. SALOP

# Contents

## Chapter 6  The Epiprotozoic

## Chapter 7  The Eocambrian (Vendian sensu stricto)

## Chapter 8  Geologic Synthesis

# Chapter 1    General Problems in Division of the Precambrian

## I. Methods and Principles of the Precambrian Division

Rocks and their assemblages or rock records have long been the object of geologic studies. Many pages of this record have not been preserved, and many of those known have not been understood. Historical geology aims at deciphering these pages recorded in an unknown language, at reconstructing their general sequence and clarifying the contents of the buried pages. In reading these rock records geologists apply many different methods and theories their own science, as well as of many branches of the natural and physical sciences. These methods, refined with time, lead to progress in geologic science as a whole and in studies of Precambrian geology in particular. We face great difficulties in determining the succession of rock record studies or the succession of occurrence and origin of rocks even within a relatively small area. The determination of synchroneity of events imprinted in rocks of different remote areas becames much more difficult. How is this problem to be solved?

Firstly, some peculiarities exist in geologic science. Geologic study in any region starts with the accumulation of observation data which can conventionally be called facts; these data are then compared and analyzed, and result in a general picture of a certain process or event, for instance, the scheme of a normal sequence of rocks or of a geologic map of an area. Usually these reconstructions are called "factual material". In reality they represent results whose validity depends to a great extent on the natural characteristics of the area (its exposures, rock preservation, etc.), on study methods, on the capability, knowledge, and experience of the researcher, on up-to-date scientific data, and on many more factors. Thus, it can be regarded as "secondary" factual material, under constant revision and investigation. However it should be considered as factual, for new data enlarge rather than refute the old, as the history of research shows.

The next phase in geologic study is the elucidation of "empiric regularities", situations recurrent in different areas, revealed, for instance, by the existence of definite and specific paragenesis of rocks (formations) in succession of origin, by the presence of common tectonic patterns, by the manifestation of metamorphism etc. To establish the empiric regularities, which are of exceptional importance the worker must be able to compare at first sight uncoordinated data (the "primary" and "secondary" facts) and examine these data from different angles. This co-ordinated thinking is both necessary and valuable in geologic knowledge. Many geologic conceptions, including those presented here, are empiric generalizations.

The greatest achievement of geologic knowledge is to found empiric regularities on results from different natural and physical sciences. The theoretical understanding of factual data and empirical generalizations can be both inductive and deductive, but in both cases (in the latter in particular) it is the result of intuition, that is the insight of a researcher based on experience and characteristic thinking. At first sight it seems that empiric regularities based on theory lead to a law – the last and final phase of knowledge. Experience, however, proves that this is not the case, for many facts regarded as geologic "law" appeared to be a hypertrophic particular case, explained by erroneous and misinterpreted data of other sciences.

It seems, after presenting these reflections that we have no grounds to be optimistic. Nevertheless, it is impossible to deny a continuous and accelerated progress in geologic knowledge. Certainly we must never forget the relativity of our knowledge with its simultaneous expansion and investigation. Moreover, the possibility of errors is to some extent limited by the very existence of many facts (observations) which have a common or similar interpretation.

We now turn to the division and correlation of the events of the distant past and some peculiarities of methods applied in Precambrian global rock correlation will be briefly discussed.

## 1. Paleontological Methods

The problem of age and, thus, of stratigraphic correlation of the rocks of Phanerozoic strata is in most cases solved by paleontological (biostratigraphic) methods. In the case of Precambrian strata these methods are of limited application. The oldest strata generally lacks any determinable organic remains, and in the younger Precambrian sedimentary strata, where the microscopic and sub-microscopic remains of blue-green algae and bacteria are principally registered, their stratigraphic value is not quite clear.

In the Precambrian carbonate sediments with an age of 2400 m.y., phytolites are abundant, the products of the activity of blue-green algae and bacteria; sometimes they build extensive, thick horizons or large bioherms.

Single structures and forms of phytolites are also encountered in the strata older than 2800 m.y. (probably starting from 3400 m.y.). Phytolites, including stromatolites (columnar, tabular, nodular, and shelly forms), oncolites (concentrical, commonly small forms) and catagraphies (ornamented microstructures), do not represent the remains of organisms but rocks formed by carbonate precipitation on the algal mucus influenced by primitive microflora and partially by processes of metabolism. Thus, strictly speaking, the use of phytolites for stratigraphic correlation cannot be regarded as a paleontological method; that phytolites can be the subject of so-called parapaleontology and of lithology is more close to the truth. Studies reveal that the form of phytolites and, to some extent, their inner structure, strongly depend on the facial environment of sedimentation, but at the same time the changes in phytolites with time are likely to be due to evolution of the organisms themselves and also to the growth of population (biomass) accompanying it.

Soviet workers (Krylov, Korolyuk, Komar, Semikhatov, etc.) have elaborated the classification of stromatolite structures and have given Latin names to their different forms similar to the binar terms of biological and paleontological classification. They also outlined a vertical zonation in distribution of different forms of phytolites that is likely to be followed in many (even distant) areas. Thus, in the Upper Precambrian four stromatolite and microphytolite complexes were recognized, which are designated by ordinal numbers (starting with the oldest one) or by the names Lower Riphean, Middle Riphean, Upper Riphean and Vendian. Later, the still older pre-Riphean (Aphebian) stromatolites were also studied, but no complexes were determined among them.

At first it was suggested that the stromatolites and microphytolites typical of phytolite complexes occur only at definite levels, but further studies revealed that many of them (or even all of them) are of much wider vertical range of distribution. The regularity earlier established is only expressed statistically: certain phytolites mostly occur in rocks of definite age and composition; and also a vague tendency toward a change of complexes singly in the course of time. At the same time these complexes are characterized by a combination of stromatolites mainly differing from each other not by outer shape, i.e., not by group composition, but by the microstructural features which are taken as the base for determination of forms ("species") in the conventional classification of stromatolites. The microstructural features, however, do not necessarily reveal the primary structure of phytolites, but are in many cases due to the chemical and physical environment of sedimentation and to the epigenetic processes of recrystallization. Moreover, the systematics of formal classification of phytolites are diffuse and seldom applied, and thus a subjective determination becomes possible. These obstacles to the use of phytolites in stratigraphy lead to many errors in correlation. Recent works (Salop 1973 a, 1977 b, Hofmann 1977, Preiss 1977, Playford 1979, etc.) show that phytolites can mostly be used for correlation of monofacial deposits within relatively small areas (sedimentation basins). Global correlation is not possible with the help of phytolites.

In Precambrian rocks younger than 1600 m.y. various phytofossils, among them the acrytarchs – planktonic unicellular algae (?) – are of the greatest stratigraphic importance. In the youngest Precambrian rocks situated near the Cambrian boundary, well-preserved traces of various nonskeletal organisms and remains of multicellular algae are reported.

## 2. Isotope (Radiometric) Methods

These methods appear important for Precambrian stratigraphy, despite the fact that sometimes the values obtained admit different interpretation, as they can reflect the time of different events and even the time range between the last and previous events. In the case of recurrent superposition of thermal and other processes resulting in intensive isotope migration, the time of formation or early transformation of rock can only be determined approximately by the so-called relict datings obtained from detailed geochronologic studies.

In dating rocks the following methods seem to be the most valuable: lead isotope (U-Th-Pb), Rb-Sr isochron, and lead isochron (Pb-Pb), as there is a certain possibility of inner control of isotope migration with the above-mentioned methods. However, the general limitations of isotope methods also apply to these in many cases in determining the time of early events. If the rocks are of high grade and recurrent metamorphism the isotope methods only indicate the time of the last strong transformation of the rocks, when isotope equilibrium was established anew and homogenization in isotope distribution in cogenetic rocks (minerals) took place.

The K-Ar widely accepted method usually gives only the time of the latest thermal event. By this method it is possible to obtain the true age of the rocks (or minerals) which were formed on or shortly below the Earth's surface and were never later subjected to any (even slight) long heating or deformation. The K-Ar datings usually give that period of time when agents of the deeper parts of the Earth's crust (higher temperature and pressure) stop acting; they do not greatly change the rocks but initiate the loss of argon. This stopping of metamorphism (and cryptometamorphism) is mainly explained by uplifting of the crust blocks above a critical level that differs for dating of different minerals (for biotite, for instance, it coincides approximately with a geoisotherm of 300 °C). In cases when uplift of the Earth's crust terminates the tectono-plutonic cycle, the K-Ar method roughly shows the time of folding, metamorphism, and plutonism.

However, in cases where argon is lost from the crystalline lattice of minerals under extremely high pressures (that is at a great depth), it can be preserved for a long time in defects of crystals or in rocks. If there was no significant uplift of the Earth's crust at that time, then in later metamorphism argon could have been held again by the crystalline lattice and then the age of minerals (rocks) analyzed will not differ greatly from the time of their formation or their initial metamorphism. This environment seems to have been typical of some very old gneiss-granulite complexes, of those forming separate blocks of the Earth's crust within the Kola Peninsula in particular (certain relict datings of rocks from these blocks are given in Table 3).

Isotope methods are usually applied to obtain the time of formation (or transformation) of magmatic and metamorphic rocks. Dating of syntectonic and late-tectonic granitoids and of various metamorphic rocks provides a possibility to determine the time of tectonic-plutonic processes (i.e. the time of diastrophism). Dating of intrusive rocks of platform type or of other kinds of intrusive rocks not related to folding reveals the time of magmatic processes or of thermal activation of stable portions of the Earth's crust. Only rarely is it possible to determine the time of formation of sedimentary piles that originated in the intervals between the orogenic cycles and magma intrusions. It is possible to evaluate their age indirectly by dating volcanics emplaced in the sediments. But in the older complexes these rocks are commonly metamorphosed or altered under the influence of postmagmatic processes and thus the values obtained are not always adequate.

It is only possible to obtain the age of a sedimentary rock or the age of sedimentation directly by dating the unaltered or slightly altered rocks from platform strata and rarely from geosynclinal ones. Even then, however, we are not sure that true age values rejuvenated by later processes. Glauconite, a syngenetic mineral, is commonly used in the K-Ar dating of sedimentary deposits, it easily loses argon at low temperature and thus the values obtained are notably rejuvenated and give only the upper age limit of the wall rocks. It is known that the older the rock is, the greater the loss of argon observed in glauconite and the bigger scattering of the values obtained. Certain satisfactory results on glauconite are known, however, for the Precambrian rocks situated in the uppermost part of the sequence of platform strata. Glauconite dating by Rb-Sr analysis is free from defects either due to the unstable crystalline lattice of the mineral or to different secondary changes in composition. Many difficulties are also faced in interpreting datings obtained by K-Ar and Rb-Sr analyses on micaceous minerals of sediments or on clay shales, phyllites, and other similar specimens, for it is not usually possible to state whether these datings reveal an age of diagenesis (epigenesis) or an age of low-temperature metamorphism of primary or superimposed character. Some errors in age determinations in such cases can also be explained by the presence of clastic minerals in the rock. The age of carbonate rocks can be determined by the Pb-Pb method, but if the rocks are only slightly altered, the interpretation becomes inadequate.

Recently great progress has been made in the equipment of isotope dating and the accuracy of laboratory determinations has become very high and is being refined from year to year. However, we have to admit that deciphering the age values is less advantageous than determination. The difficulties in this sphere are primarily due to various considerable processes of migration of elements and their isotopes. In the older rocks these processes were recurrent and very intensive during orogenic cycles when a general rise of geoisotherms took place, and very extensive transportation of solutions and fluids was also recorded, together with different types of deformations. Incidentally, the isotope datings of such rocks are usually interpreted by geologists as an age of the rock (or geologic complexes,) and certain differences in datings, if they exceed the experimental error, are often regarded as a different age of the rock analyzed; in reality they simply exhibit their different geochemical evolution or different grade of preservation.

Isotope rejuvenation is mostly typical of the oldest polycyclic metamorphic rocks of the Precambrian. It is known that the age of their original metamorphism can be determined only by sophisticated studies using different methods (mostly isochron ones). However, even these methods do not always answer the purpose. Dating of the older gneisses of the Limpopo belt in Southern Africa by Rb-Sr analysis can be given as an example. At first their age was evaluated to be 1900–2000 m.y.; the same age was obtained by K-Ar dating of micas (Van Breemen et al. 1966). Further detailed studies (Van Breemen and Dodson 1972) gave older age values of gneisses (of granite-gneisses to be more exact): $2690 \pm 60$ m.y. ($^{87}Sr/^{86}Sr = 0.7038$). Recently (Barton et al. 1977) the same method gave an age of $3856 \pm 116$ m.y. ($^{87}Sr/^{86}Sr = 0.7014$) for the granite-gneiss of the Limpopo belt and an age of $3643 \pm 102$ m.y. ($^{87}Sr/^{86}Sr = 0.7014$) for the amphibolite dynes in it. Geologic material shows all three stages of metamorphism superimposed on the older rocks and revealed by isotope methods (see Salop 1977 b). Many examples of this kind can be given.

Geochronologists usually think that the low primary strontium isotope ratio, typical of the greatest portion of the old rocks, disagrees with the concept of isotope rejuvenation. Many cases can be cited when obviously rejuvenated datings were obtained for the oldest (Katarchean) rocks with low $^{87}Sr/^{86}Sr$ ratio. During metamorphism the loss of a number of elements from rocks is known to occur (of K, Na, Rb, Sr, U, Th, etc.) and, as a rule, the daughter elements (isotopes) of radioactive decay that have less stable bonds in the mineral crystal lattice are being evacuated more intensely. It is possible to suggest that the radiogenic $^{87}Sr$ will migrate during many thermal-metamorphic processes quicker than the stable $^{86}Sr$. If this is so, then the $^{87}Sr/^{86}Sr$ in the recurrently metamorphosed rock will not notably increase. Besides, during a long process (and the duration of thermal processes in the Early Precambrian was several million years), an intensive loss of Rb is to be registered that will also result in a decrease of radiogenic strontium in rocks. The most intense fractionation of isotopes propably happened during the ultrametamorphic processes typical of the Early Precambrian. It was shown (Heier 1964) that during ultrametamorphism the anatectic melts are enriched in radiogenic strontium and the residual rocks (gneisses) are being depleted in it and are thus characterized by a low $^{87}Sr/^{86}Sr$ ratio. Thus, it is not always possible to use the strontium isotope ratio for elucidating the pre-history of highly altered polymetamorphic rocks[1].

The "rejuvenation" phenomena of the older rocks can often be registered by coincidence or similarity of their isotope dating and the dating of rocks from a younger metamorphic complex unconformably overlying these former. For instance, the age values obtained by different methods for the rocks of the oldest gneiss (gneiss-granulite) complexes are commonly close to the values determined for the rocks from the greenstone strata resting on gneisses with a significant angular unconformity. In these cases the datings of both types of rocks fall within the 2600 to 2800 m.y. range; the same age is recorded for granitoids underlying the greenstone strata and cutting them. It is evidently suggestive of the older complex being rejuvenated by the late metamorphism. It will be shown below that the Precambrian orogenies (especially during the Early Precambrian) were separated by very long (several hundreds of millions of years) time intervals. It is known that some rare "relict" datings of the rocks of the gneiss-granulite complexes give values of the order of 3500–4000 m.y. This example and many more demonstrate that while interpreting the isotope datings it is necessary to take into consideration not only their analytical (geochemical) aspects but also geologic observation data as an obligatory element of it.

Still, in spite of many difficulties met in deciphering data of isotope analyses, the correctly interpreted radiometric ages are very important and constitute an objective base for correlation of Precambrian formations even in distant regions.

### 3. Geohistorical Methods

Geologic methods of division and correlation of rocks are many and various. To break up the older strata it is very important to establish and follow the angular and stratigraphic unconformities and to know the relations of bedded rocks and plutonic formations which serve the geologic and geochronologic markers denoting diastrophic events that separate the periods of sedimentation and volcanism. Correlation of the rocks distributed in the areas of relatively monotonous geologic structure that is within definite structural-formational zones or sedimentation basins can be achieved by comparing the strata by their lithology and by following their facial changes. This method, however, seems to be unsatisfactory for interregional correlation and especially in the case of very distant areas, for instance for the areas situated in different

---

1  The same can be stated about the application of strontium isotopes in studying the genesis of magmatic rocks; naturally, the high value of this ratio always points to the lithogenic (crustal) origin of magma but the low ratios are difficult to interpret without knowledge of the sample evolution.

Migration of radiogenic strontium is also likely to occur under the low-temperature processes. Thus, the loss of the $^{87}Sr$ isotope is reported under the effect of hydrothermal (metamorphic) solutions in the Tertiary basalts of zeolite facies in Iceland (Wood et al. 1976)

continents. Then the geohistorical (including the formational) methods become of exceptional importance. Many empiric generalizations are the basis of these methods; it is believed that certain irreversible changes in the tectonic evolution of the Earth exist, and that they are also known in the chemical composition and thermodynamic conditions of its outer shells which, when combined, govern the evolution of sedimentation and rock formation. The origin of specific, unique types of rocks and their associations (paragenesis, formations) is a result of this evolution and they build up definite levels in the normal Precambrian sequence and thus can be used for the global correlation.

In the first case these are the sedimentary rocks, as their origin greatly depends on the evolution of atmosphere and hydrosphere composition, and on the concentration of free oxygen and carbon dioxide in particular. These are various ferruginous, manganese formations, gold-uranium bearing conglomerates, red beds, carbonate rocks etc. Certain levels in the Precambrian are also characterized by supracrustal formations with some geochemical peculiarities, for instance, higher concentration of iron, manganese, copper, phosphorus, barium, uranium, gold, and some other elements that in some localities attain economic importance. Then the formations uprising under definite tectonic and climatic environment typical of certain stages in the Precambrian should be mentioned; these are, for example, crusts of weathering or different orogenic formations. Some peculiar magmatic formations should also be examined among the geogenerations under discussion, for instance, the komatiite, subaerial rhyolitic, or trachyte-rhyolitic lava; they are reported from strata of different age but they are mostly constant and characteristic in the Precambrian complexes of a definite stratigraphic level. Some specific intrusive formations also belong here, as they accompany some Precambrian orogenies, for example the rapakivi granites and anorthosites of different types. It seems reasonable to mention also the sediments originating from the activity of organisms and containing different organic remains.

The glacial formations are of exceptionally great significance for division and global correlation of the older strata, which were formed due to an abrupt fall of temperature registered in several periods during the geologic history when the typical hot Precambrian climate for a short period became cooler. Glaciations on the Earth were principally initiated by cosmic agents and in the Precambrian they embraced the entire surface of the planet irrespective of latitude (Salop 1973a, b, 1977c). The climate-stratigraphic criterion thus becomes of great significance for Precambrian stratigraphy if this special characteristic of glaciations in the far past is considered.

It is important also that some empirical regularities on the vertical distribution of specific formations and rock types, when repeatedly confirmed by isotope age determinations, acquire a separate and objective significance for the division and correlation of older strata. It is well known that if the "irreversible" formations are applied for correlation of Precambrian together with isotope method control the results obtained are very significant and revealing. On this basis it became possible to make a detailed subdivision of Precambrian into several lithostratigraphic complexes followed in different continents. Naturally, the formational (lithoparagenetical) method is much more important in Precambrian stratigraphy than in Phanerozoic. It was stated by the author in previous works and will be demonstrated here that certain characteristic aspects of early geologic evolution are responsible for this.

The principles of the Precambrian division will now be briefly explained. Division into geologic periods can be based on one or several important aspects typical of a definite time interval of geologic history. For instance, the Phanerozoic is mainly divided on the evidence of changes of organic remains in the rocks with time or, in other words, on the paleontological (biostratigraphic) principle. It was stated above that this principle is of little use for Precambrian stratigraphy, whose subdivision can be established on the basis of specific types of formations or on larger tectonic elements (structures) typical of definite stages of geologic evolution. This is then a lithostratigraphic and tectonic principle applied for division into periods. Some other criteria can also be used. For instance, it was suggested to divide the Precambrian into equal time intervals according to isotope dating. Then, however, it will be not the division into periods but simply a chronological scale ("calendar").

Here the so-called geohistorical principle for the Precambrian division is used; essentially it acknowledges that the existence of advancing physical and chemical evolution of the planet and the geologic processes that occurred were of cyclic and progressive nature. This principle of the Precambrian division was earlier applied by many workers but recently it has received a serious foundation and new contents due to the wide usage of results of radioactive dating methods and to the progress in lithoparagenetic (formational) analysis.

The analysis of geologic and geochronologic data on the Precambrian revealed that diastrophic events occurred at the same time intervals, and the supracrustals formed between the diastrophic cycles, and revealing the same age, are known to have great similarity in composition, sequence, in type of metamorphism, and tectonic pattern all over the world. Thus, it appeared possible to define in the Precambrian some natural stages of geologic evolution, that is, those intervals that are typified by common tectonic, geochemical, and physical environments affecting the origin of definite types of lithogenesis, of tectonic structures, of magmatism, and to a certain extent life evolution. Division into geologic periods must necessarily be based on a complex of data summarizing all information available from the rock record. This division, when based only on a single aspect, will seem to lack solid grounds for its application, for example the breaking of the Precambrian based only on phytolites that some geologists propose.

Some further comments serve to show this more accurately. The diastrophic (orogenic) cycles, due to their world-wide nature and synchroneity (within a definite time range), serve as good natural boundaries separating stages of the Earth's evolution. However, the geologic events are not always known to have an abrupt change in their course or to terminate at these boundaries. Some of the events continue their course from one evolutionary stage to another, and some notable changes sometimes occur at the beginning or even in the middle of the next stage. Life evolution is the least dependent on a change of environment, as it is governed by immanent biologic causes. Thus, when saying the "natural stages" we are to consider the most important and typical aspects of periodicity of geologic evolution.

# II. Division of the Precambrian

## 1. Stratigraphic Units

The essence of a general (unified) stratigraphic scale for the Precambrian is to break this geologic period into several stages that can be established only on analyzing data from every principal distribution area of the older strata in the world; this is the only possible way to elucidate general (global) regularities and to avoid the overstatement of a particular case. The analysis done showed that several large natural stages of geologic evolution can be recognized in the Precambrian. Conventionally they can be compared with the Phanerozoic eras; in duration they greatly exceed these latter and are, moreover, established on different principles. These units appeared to be more numerous than those accepted earlier and thus the author gave them new names. These larger units of the Precambrian were designated as follows (starting from the oldest): Archean, Paleoprotozoic, Mesoprotozoic, Neoprotozoic, and Epiprotozoic. These Precambrian eras are separated by global diastrophic cycles of the first order: 3750–3500 m.y. ago (the Saamian orogeny), 2800–2600 m.y. ago (the Kenoran orogeny), 2000–1900 m.y. ago (the Karelian orogeny), ~ 1000 m.y. ago (the Grenville orogeny) and 680–650 m.y. ago (the Katangan orogeny). The Neoprotozoic Era is additionally divided into four sub-eras by the diastrophic cycles of the second order: the Vyborgian orogeny (1750–1600 m.y.), the Kibaran orogeny (1400 to 1350 m.y.) and the Avzyan orogeny (~ 1200 m.y.), and the Epiprotozoic Era into two sub-eras by the Lufilian orogeny of the second order (approx. 780 m.y.) (Salop 1977 b). It will be shown that it is possible to divide the Paleoprotozoic Era into two sub-eras and the Mesoprotozoic Era into three sub-eras on the basis of some peculiarities in sequence and composition of the rocks and also thanks to manifestations of orogenies of the third order that delimit the sub-eras. Then at the end of the Precambrian it is possible to distinguish a small unit, Eocambrian (with a time range of 650–570 m.y.); in its duration it is comparable with the Phanerozoic periods and in geohistorical aspect it is closely related to the Cambrian period of the Paleozoic Era. Thus, 13 stratigraphic units of world-wide implication can be recognized in the Precambrian.

The most important turning point in geologic history that separates the Precambrian into the two largest time intervals (megachrone), conspicuously differing in thermal and tectonic regime and in many more aspects, is registered between the Archean and the Paleoprotozoic, and falls in the Saamian orogeny. Thus, it offers a possibility of dividing the Precambrian history into two eons: the Eozoic, embracing the Archean Era only, and the Protozoic, comprising all the remaining Precambrian Eras [2]. So the whole geologic history divides into three eons: the Eozoic ("dawn of life") Protozoic ("primitive life") and Phanerozoic ("true life"). Probably the Eocambrian is to be attributed to the last eon, but here the term Phanerozoic will be used as it is generally accepted.

---

2 The name "Protozoic" was proposed by Sedgwick in 1838 to designate all rocks older than Cambrian distributed in the British Isles. Later this name was forgotten, though it was the first among those proposed for Precambrian formations; it was only used by Kay in his well-known work on geosynclines in North America

It is clearly seen from the names of every large Precambrian unit that they accord well with the names of eras of the Phanerozoic and simply reflect general aspects of evolution of Life. Certainly it is merely a convention, as the Precambrian units are recognized on different principles. However, the use of these names does not contradict any factual data and it is convenient, for it keeps to the tradition and gives a clear idea of the sequence of the units.

In the new stratigraphic scale the Archean (Archeozoic) is the only name that the author uses here among those earlier used to designate the Precambrian Eras; this name was proposed for the first time by Dana in 1872 for the entire Precambrian, but since 1887 it has been applied only to the oldest Precambrian unit, as was suggested by Emmons. In the Soviet Union it was commonly accepted that the Archean comprises high-grade strata of gneisses and crystalline schists that build up the basement of all of the younger supracrustal formations. Later it was established that they are older than 3500 m.y. In many countries, however, including those with widespread Precambrian strata (as, for instance, Canada) these similar (synchronous) formations were not regarded as a single unit until lately, but were combined in one Archean Era together with the overlying low-grade strata (the "greenstone strata") with the upper age limit estimated to be of 2600–2800 m.y. According to new data these values characterize an age of metamorphism of the upper (Paleoprotozoic) supracrustal complex or, in other words, the time of tectonic-plutonic processes of the Kenoran orogenic cycle. The age of the oldest gneissic complex is regarded as being the same in the U.S.S.R. and elsewhere. When I (Salop 1973 a) proposed the old name "Archean" for it, that is accepted by Soviet geologists, I did not take into consideration the tradition of drawing the Archean-Proterozoic boundary at the 2600–2800 m.y. level (as is known to be commonly applied in the Precambrian division of Canada). Recently, many Soviet geologists have also adapted the 2500–2700 m.y. time interval as the Archean-Proterozoic boundary, and in the resolutions of the meeting in Ufa in 1976 it was proposed to accept the Archean-Proterozoic boundary at 2600 ± 100 m.y. In 1977 the Precambrian Subcommission of the IUGS recommended drawing this boundary at the 2500 m.y. level (James 1978). It should be noted that this latter value is an absolutely arbitrary geochrone and does not reflect any real boundary of natural geologic units (eras). It will be shown later that this value falls within the time interval of the Mesoprotozoic Era and does not denote any important event; its usage in drawing boundaries between eons and eras results in artificial division of single stratigraphic complexes.

Thus the usage of the Archean became ambiguous. To avoid it the author abandons the "Archean" and proposes the term "Katarchean" (Greek Kata-below, under; archean-old) to designate the oldest Precambrian unit; this term was once suggested by Sederholm for the oldest rocks of the Baltic shield at the end of the last century. Thus, the earliest Protozoic Era – Paleoprotozoic or Archeoprotozoic (the old primitive life) – corresponds to the Archean or Upper Archean as it is understood by many workers. It should be noted here that the term "Protozoic" is not a synonym of "Proterozoic" as is at times erroneously thought. This latter name was used to designate a unit combination of different stratigraphic range: starting from Mesoprotozoic (the Lower Proterozoic or Aphebian) and up to Eocambrian or Vendian inclusively, that is without the Paleoprotozoic in the lowermost part, but with the Eocambrian in the uppermost part. Generally the names "Archean" and "Proterozoic" are "the weeds in the rock garden of the Precambrian" as Rankama (1970) correctly thinks them to be. Now these names are known to have so many meanings that it is better to expurgate them completely.

It seems possible to avoid terminological difficulties if we call the large stratigraphic units of the Precambrian (eras, erathems) by the names of their most typical strata – stratotypes, or by the stratotypical locality, as was suggested by some workers. In general, in the Precambrian stratigraphy the significance of stratotypes is very great; they are much more important than in the stratigraphy of the Phanerozoic, that can rely on a widely applicable paleontological method. For the correlation of the older strata essentially based on comparison of formational composition of the strata, the standard or reference section is quite necessary. This stratotype should be the most complete, with definite boundaries, with available isotope data and it is to be fossiliferous in the case of the Upper Precambrian. A global stratotype is to be chosen for every erathem and sub-erathem of the Precambrian. Naturally, any larger region is to have its own local stratotypes including those of different formational types of the rocks.

If the "stratotypical principle" is accepted for the stratigraphic terminology of the Precambrian, then it is possible to use instead of Katarchean "Aldanian" (the Aldan Era, Erathem) according to the Aldan complex or the shield of the same name in Eastern Siberia; instead of Paleoprotozoic (or Archeoprotozoic), the "Ontarian", according to the Province of the same name in Canada where the stratotypical or parastratotypical strata of Keewatin and Timiskaming are distributed [3]; instead of Mesoprotozoic, the "Karelian" according to the Karelian complex in Karelia: instead of Neoprotozoic, the "Baikalian" according to the Baikal Highland, where the complete stratotypical and parastratotypical units of the erathem are represented; instead of "Epiprotozoic", the "Katangan" according to the Katanga Supergroup of Equatorial Africa (or "Adelaidean" according to the Adelaidean System of Southern Australia) [4]. It seems possible to use the name "Vendian" according to the Vendian complex (see below) of the Russian platform instead of Eocambrian. Almost all of the above – mentioned stratotypical names of eras have already been used as generalized or correlative units ("systems") in corresponding regions. However, these names possess certain defects: they do not give any idea of the age sequence and the local terms, when applied to different regions, will present certain difficulties and will at first appear unintelligible. For this reason terms of more general meaning are here given preference for a world-wide scale.

It was stated above that the Precambrian eras are delimited by the orogenies of the first order. The geologic formations (supracrustals and magmatic rocks) that originated after the preceding diastrophic cycle and up to the end of the cycle terminating the era form an erathem with the name of the era. In the Precambrian the orogenic cycles were of very long duration, in the Early Precambrian in particular: some hundreds of millions of years. Then during one cycle the tectonic-plutonic processes occurred recurrently, while in some areas they were very intensive only at

---

3 For the first time the "Ontarian" was proposed by Lawson (1913) for designation of the corresponding units in Canada
4 The widely used name "Riphean" is poor for the entire Neoprotozoic because in the Riphean complex of the Urals the lower part of the erathem seems to be lacking (see Chap. 5)

the beginning of a cycle. In the middle and at the end of it the sedimentary-volcanogenic strata, commonly of peculiar composition, could have accumulated there; in the formational type they are transitional between the strata of a given era and of the preceding one, but are usually more close to the later era rocks. Such strata and the stratigraphic units composed of them will here be called transitional. These piles as a rule rest unconformably on the older strata of the era and on the intrusions cutting them, and the piles are metamorphosed at the end of the next era during a new orogenic cycle. In such cases it seems reasonable to attribute these rocks to the lowermost part of a younger erathem. At the same time the synchronous strata of other regions are registered as being altered and cut by granites during certain later phases of a preceding orogenic cycle. These piles combined with other rocks under-lying them, will be assigned to a early erathem. Such cases will be examined later. Unfortunately, it is rarely possible to prove the existence of such cases, for we need the isotope age of sedimentation (or volcanism) to state the synchroneity of both types of transitional strata and these age data can rarely be obtained, as it can be done only by analysis of low-grade rocks. Lithological comparisons can easily give erroneous results.

The existence of transitional strata (units) is an absolutely undoubted fact; in such cases there is an element of uncertainty in the assignation of these strata to the boundary beds of one or the other erathem.

## 2. Lithostratigraphic Units (Complexes)

It was noted above that the Precambrian piles comprise some characteristic rocks or their associations, sometimes irreversible in geologic history, that were formed in response to a specific environment, for instance to the specific composition of the atmosphere-hydrosphere of that time. The presence of such rocks or formations permits us to divide many of the large Precambrian stratigraphic units (erathems and sub-erathems) into lithostratigraphic complexes that can be traced in many regions of the world by isotope methods.

Naturally, the co-eval strata developed in different tectonic regions, for instance, in geosynclinal belts and on platforms, are conspicuously different, but even then the strata contain similar formations typical of a given level and it is possible, in combination with other peculiarities of the strata, to correlate the rocks and distinguish separate lithostratigraphic complexes. It should be noted also that the complexes younger than 2600 m.y. are mostly recognizable in the miogeosynclinal and platform types of strata: among the eugeosynclinal strata their lithological and geochemical aspects are masked by intensive volcanism and quick burial of sediments.

At present in the Precambrian it is possible to distinguish 23 non-coeval correlative lithostratigraphic complexes. Many of the complexes are developed on every continent and occur in the most complete sequences of erathems and sub-erathems. However, some of them are absent in the sequences of certain regions or their nature is not obviously expressed under the influence of different agents. The stratigraphic range of the complexes recognized is unequal, the implication in correlation is not at all adequate; the vertical range of some complexes is very narrow and at the same time there are complexes of very wide stratigraphic range. The lithologic-geochemical characteristic of some complexes is not sufficiently specified, nevertheless they are easily recognizable, as they are situated between evidently expressed markers or delineated by dated intrusive rocks related to orogenic cycles. Generally the 23 complexes above mentioned embrace almost all supracrustal formations of the Precambrian (except the Neoprotozoic and Eocambrian). Naturally, further studies will result in breaking up many of the complexes and their characteristics will become more sophisticated. For mere convenience the lithostratigraphic complexes are given proper names according to their most typical local units in different regions of the world and in some cases according to stratotypical locality. The Epiprotozoic complexes form an exception, for they are named according to their position in relation to non-coeval horizons of glacial deposits.

Recognition of lithostratigraphic complexes is a typical example of empiric generalization. In some cases the occurrence of certain complexes at definite stratigraphic levels is sufficiently theoretically explained, in other cases this theoretic explanation is lacking at the moment. This latter fact, however, does not diminish the role of empiric generalization. It will be shown below that recognition of lithostratigraphic complexes is of great significance not only for Precambrian stratigraphy, but also for the scientific prediction of mineral deposits, as many of the complexes are related to definite types of exogenic and metamorphogenic mineral deposits.

In addition to lithostratigraphic complexes some special plutonic complexes (associations) can also be established in the Precambrian; they are typical of some older diastrophic cycles, but we are at the initial stage of study of the problem and many specialized petrological and geochemical studies must be carried out to solve it. It is possible to suggest that the number of "irreversible" plutonic complexes or formations will not be great, for the evolution of matter of the interior part of the Earth was going on at a much lower speed and not so noticeably as in the case of the surface shells of the planet. In the Precambrian 16 orogenic cycles of different order are recognized and some of the plutonic formations are related to them. Not all of the plutonic associations are sufficiently specified, but they are easily distinguishable according to their relation to stratified beds and due to the radiometric dating that is mostly done on plutonic rocks. Plutonic complexes are not examined in this book, but they are sometimes mentioned to characterize events during orogenic cycles.

## 3. The Global Stratigraphic Scheme of the Precambrian

Three tables are given here for better understanding of the material dealt with; Table 1 represents the global stratigraphic scheme of the Precambrian accepted here (Salop 1973a, 1979a). Table 2 gives the major lithostratigraphic complexes of the Precambrian and the last table (Fig. 1) shows a comparison of the scheme accepted here with other well-known general schemes of the Precambrian according to isotope datings of the boundaries of the units established. Advantages and disadvantages of some of those schemes were discussed earlier in my previous works (Salop 1973a, 1979a, 1980). It is to be noted that when comparing schemes by age boundaries of general stratigraphic units there is no harmony apparent with correlation of concrete (regional) stratigraphic units commonly included in the general units. This can be

**Table 1.** General stratigraphic scheme of the Precambrian

| Eon | Era | Sub-era and "complex" | Age boundaries of units in m.y. | Diastrophic cycles and their age ranges in m.y. (principal cycles are shown in bold print) |
|---|---|---|---|---|
| Phanerozoic? | Paleozoic (?) | | 570 | |
| | | Eocambrian (Vendian s.str.) "complex" | | |
| | | | 650 | |
| | Epiprotozoic (Katangian, Adelaidean) | Upper Epiprotozoic | | **Katangan of the first order, 680–650** |
| | | Lower Epiprotozoic | 780 | Lufilian of the second order, ~ 780 |
| | | | 1000 | |
| Protozoic | Neoprotozoic (Baikalian) | Terminal Neoprotozoic | | **Grenville of the first order, 1100–1000** |
| | | Upper Neoprotozoic | 1200 | Avzyan of the third order, 1250–1200 |
| | | Middle Neoprotozoic | 1300 | Kibaran of the second order, 1400–1300 |
| | | Lower Neoprotozoic | 1600 | Vyborgian of the second order, 1750–1600 |
| | | | 1900 | |
| | Mesoproto-zoic (Karelian) | Upper Mesoprotozoic | | **Karelian of the first order, 2000–1900** |
| | | Middle Mesoprotozoic | 2200 | Ladogan of the third order, 2200 ($\pm$) |
| | | Lower Mesoprotozoic | 2400 | Seletskian of the third order, 2400 ($\pm$) |
| | | | 2600 | |
| | Paleoproto-zoic or Ar-cheoproto-zoic (Ontarian) | Upper Paleoprotozoic | | **Kenoran of the first order, 2800–2600** |
| | | Lower Paleoprotozoic | 3200 | Swazilandian of the second order, 3200 ($\pm$) |
| | | | 3500 | |
| Eozoic | Katarchean (Aldanian) | | | **Saamian of the first order, 3750–3500** |

**Table 2.** Stratigraphic and lithostratigraphic units of the Precambrian

| Eras | Suberas | Lithostratigraphic complexes. Age range of the units in m.y. in brackets | Concise characteristics of the units |
|------|---------|----------------------------------------------------------------------------|--------------------------------------|
| 1 | 2 | 3 | 4 |
| Paleozoic (?) | Eocambrian (Vendian) | Lithostratigraphic complexes are not recognized (670–570) | Platform, rare geosynclinal terrigenous and (or) carbonate sediments containing Ediacara fauna remains, calcareous algae, worm imprints, various phytolites etc. |
| Epiprotozoic (Katangan, Adelaidean) | Upper Epiprotozoic | *Katangan diastrophism of the first order, 680–650 m.y.* | |
| | | Superglacial (750–670) | Molasse-like terrigenous sediments, commonly red-colored |
| | | Upper Glacial (780–750) | Glacial deposits (tillite etc.); usually overlain by sediments with a higher concentration of barium |
| | Lower Epiprotozoic | *Lufilian diastrophism of the second order, ~ 780 m.y.* | |
| | | Interglacial (820–780) | Terrigenous and carbonate sediments; rare basic and acid volcanics. Various phytolites; *Chuaria* acrytarchs |
| | | Lower Glacial (850–820) | Glacial deposits (tillite etc.); commonly associate with sheet deposits of iron ores and basic volcanics |
| | | Subglacial (1000–850) | Platform and geosynclinal (mainly miogeosynclinal), terrigenous and carbonate sediments, locally substituted by abundant volcanics of acid, rarely basic composition. Stratiform copper deposits in terrigenous-carbonate strata. Numerous phytolites in carbonate rocks. *Chuaria* is registered among acrytarchs; eukaryota remains in microbiota |

**Table 2** (continued)

| Eras | Suberas | Lithostratigraphic complexes. Age range of the units in m.y. in brackets | Concise characteristics of the units |
|---|---|---|---|
| 1 | 2 | 3 | 4 |
| Neoprotozoic (Baikalian) | Terminal Neoprotozoic (Upper Riphean) | Lithostratigraphic complexes are not recognized (1200–1000) | *Grenville diastrophism of the first order, ~ 1000 m.y.*<br>Various platform and geosynclinal strata. Red beds are abundant among platform strata. Numerous, different phytolites in carbonate rocks. *Chuaria* is typical among acrytarchs; eukaryota remains |
|  | Upper Neoprotozoic (Middle Riphean) | Lithostratigraphic complexes are not recognized (1350–1200) | *Avzyan diastrophism of the second–third order, 1250–1200 m.y.*<br>Various platform and geosynclinal strata. Abundant basalt flows and dolerite sills among platform strata. Red beds are widespread. Phytolites: in carbonate rocks; acrytarchs: in terrigenous rocks; rare eukaryota remains among microbiota. Sedimentary iron ores of the sheet (Urals) type |
|  | Middle Neoprotozoic (Lower Riphean) | Lithostratigraphic complexes are not recognized (1600–1350) | *Kibaran diastrophism of the second order, 1400–1300 m.y.*<br>Various platform and geosynclinal strata. Abundant red terrigenous rocks among platform strata. Phytolites: in carbonate rocks; acrytarchs: in terrigenous rocks. Sedimentary iron ores of the sheet (Urals) type |

| | | | |
|---|---|---|---|
| Neoprotozoic (Baikalian) | Lower Neoprotozoic (Akitkanian) | *Vyborgian diastrophism of the second order, 1600 m.y.* | |
| | | Chayan (1750–1600) | Sedimentary essentially terrigenous strata, continental red beds with local acid volcanics in the lower part; littoral-marine rocks with basalt in the upper part |
| | | Khibelenian (1900–1750) | Acid subaerial volcanics alternating with subordinate continental terrigenous strata. Comagmatic subvolcanic and hypabyssal intrusions of granite and syenite |
| Mesoprotozoic (Karelian) | Upper Mesoprotozoic | *Karelian diastrophism of the first order, 2000–1900 m.y.* | |
| | | Vepsian (∼ 2000) | Terrigenous, molasse-like strata, local red beds. Commonly occur with a break on the underlying rocks |
| | | Ladogan (or Transvaalian) (2200–2000) | Flyschoid terrigenous or terrigenous-carbonate strata with a tillite horizon in the lower part |
| | Middle Mesoprotozoic | *Ladogan diastrophic episode (diastrophism of the third order), 2200–2160 m.y.* | |
| | | Upper Jatulian (or Animikian) (2350–2200) | Quartzite-slate and dolomite strata with local thick horizons of jaspilite of the Superior type. Common stromatolitic bioherms in dolomites; abundant microscopic remains of prokaryotes and rare eukaryotes (?) in cherty interbeds among jaspilites |

**Table 2** (continued)

| Eras | Suberas | Lithostratigraphic complexes. Age range of the units in m.y. in brackets | Concise characteristics of the units |
|---|---|---|---|
| 1 | 2 | 3 | 4 |
| Mesoprotozoic (Karelian) | Middle Mesoprotozoic | Lower Jatulian (2450–2350) | Terrigenous, partly terrigenous carbonate strata with abundant quartzite. Carbonate rocks contain stromatolites that sometimes make up large bioherms: first rock-forming stromatolitic horizons. Jaspilite of the Krivoy Rog type |
| | | *Seletskian diastrophic episode (diastrophism of the third order), 2450–2400 m.y.* | |
| | Lower Mesoprotozoic | Witwatersrand (2750–2450) | Terrigenous strata enclosing up to three horizons of tillite. Gold-uranium conglomerates of great economic importance mostly occur in the lower part. Volcanics (from basic up to acid composition) are in the upper part of the complex in many regions |
| | | Dominion-Reef (2800–2750) | Basic, rare acid volcanics alternating with terrigenous rocks. Gold-uranium conglomerates (of no economic importance) |
| | | *Kenoran diastrophism of the first order, 2800–2600 m.y.* | |
| Paleoprotozoic or Archeoprotozoic (Ontarian) | Upper Paleoprotozoic | Moodies (3000–2800) | Conglomerates, subgreywackes, quartzites, carbonate rocks, rare jaspilite of the Krivoy Rog type. Rest on the underlying rock with a break |
| | | Timiskamingian (3200–3000) | Terrigenous rocks (mainly greywackes) with subordinate volcanics and tuffs. Jaspilite of the Krivoy Rog and Algoma types |
| | | *Swazilandian diastrophic episode (diastrophism of the second–third order), 3200–3150 m.y.* | |

| | | Unit | Description |
|---|---|---|---|
| Paleoprotozoic or Archeoprotozoic (Ontarian) | Lower Paleoprotozoic | Keewatinian (3400–3200) | Basic volcanics are dominant in the lower part, alternation of acid and basic volcanics and their tuffs in the upper part, subordinate tuffites, clastic rocks, cherts, jaspilite of the Algome type. The lowermost part of the complex is built up of strata of cherty, dolomite and clastic rocks (the Steep Rock subcomplex); first rare phytolites are reported there. The complex rests with a break on the underlying rocks |
| | | Komatian (3550–3400) | Peridotite and basalt komatiites. Quite rare first microscopic remains of prokaryotes in cherty interbeds. Local terrigenous rocks (the Pontiac subcomplex) in the lower part of the complex |
| | | *Saamian diastrophism of the first order, 3750–3500 m.y.* | |
| Katarchean (Aldanian) | | Isuan (4000–3750) | Amphibolite, para- and orthoschists, carbonate and carbonate-silicate rocks, jaspilite (of the Algome type ?), acid metavolcanics, metaconglomerates |
| | | *Godthaabian diastrophism of the second (?) order, ~ 4000 m.y.* | |
| | | Slyudyankan | Carbonate rocks (marble and calciphyre) alternating with various crystalline paraschists, paragneiss and quartzite; subordinate metabasite |
| | | Sutamian | Garnet gneiss, leucocratic granulite; subordinate interbeds of different gneisses, metabasites and carbonate rocks |
| | | Fedorovian | Basic pyroxenic crystalline schists and amphibolites (metabasites) with subordinate horizons of carbonate rocks (marble, calcareous-silicate schists). Interbeds of various gneisses (garnet-bearing in particular), quartzite, magnetite rocks of the Azovian type |
| | | Ungran | Basic pyroxenic (commonly bi-pyroxenic) crystalline schists and amphibolites (metabasites) |
| | | Iyengran | Basic crystalline schists and amphibolites (metabasites) with thick horizons of quartzite and interbeds of high-alumina gneiss. Basement is not known |

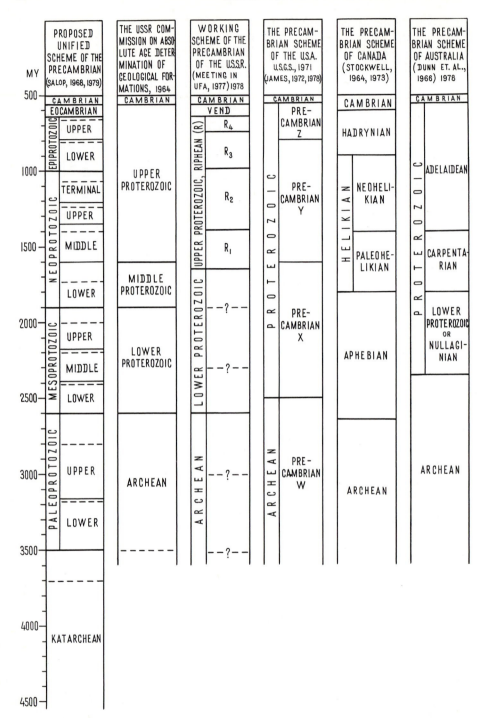

**Fig. 1.** General stratigraphic schemes of Precambrian compiled by different authors and scientific institutions

explained by inadequate estimation of an age of regional units by different workers and by different local traditions that sometimes contradict data of new studies [5].

A short characteristic of principal stratigraphic units of the Precambrian and major aspects of geologic environment of formation of the rocks will be discussed in the following chapters of this part of the book. Description of every unit will consist of two parts: (1) rock records (stratigraphy, matter composition, tectonic structures etc. and (2) geologic interpretation of rock records.

---

5 Age boundaries of the units shown on the Precambrian scheme of Australia were given by Dunn et al. (1966), then these data appeared again on the geologic map of Australia published in 1976 for the 25th IGC in Sydney. They are out-of-date and are to be revised, as the Australian geologists informed the session of IGC

# Chapter 2   The Katarchean

## A. Rock Records

### 1. General Characteristics of the Katarchean

**Definition.**  The oldest (older than 3500 m.y.) supracrustals and plutonic rocks building up the basement of all of the remaining geologic formations belong to the Katarchean or Aldanian. This erathem is characterized by: (a) high-grade metamorphism of granulite and amphibolite facies (in most cases the rocks are the result of polymetamorphism); (b) ubiquitous, though irregular granitization and migmatization; (c) the supracrustals are principally represented by metavolcanics and metasedimentary rocks of less abundance, psephites are rare or absent; (d) absence of determinable organic remains; (e) absence of structural-facial and metamorphic zonation (in place to place it is faintly expressed); (f) grouping of linear and dome-shaped folds into large isometric or irregularly shaped systems; (g) dominance of plastic deformation. The above mentioned features of the erathem and some more will be examined below. The basement of the erathem is unknown. In many regions of the world the supracrustals and plutonic rocks are unconformably overlain by the Paleoprotozoic (Archeoprotozoic) greenstone strata, but their original relations are commonly masked by later thermal-tectonic processes.

**Isotope Age.**  The superimposed processes are responsible for the recurrently polymetamorphosed Katarchean rocks. That is why the isotope datings of the oldest rocks usually reveal rejuvenated values corresponding to the time of late metamorphism. Locally the Katarchean rock dating gives several groups of age values characterizing phases of thermal activity occurring at different periods of time. It is only rarely possible to reconstruct the time of the earliest thermal events. But even these relict datings are also commonly rejuvenated and approximately give the time of the early events. Only very detailed geochronological studies are known to make a breach in the thick wall of rejuvenated values when applying different isotope methods, the isochron (Rb-Sr and Pb-Pb) ones in particular. We have the most reliable datings of the Katarchean rocks within the 3750–3500 m.y. range, though there are some older values of the order of 4000–4100 m.y.

Up to now the age of the oldest rocks was mostly evaluated according to the rejuvenated datings and it seems reasonable here to give the oldest age values available at present for the Katarchean complexes from different regions of the world (Table 3). In this table are given all datings exceeding 3000 m.y., that means several

**Table 3.** Isotope "relict" datings of Katarchean rocks

| NN | Region | Objects of dating | Method | Age in m.y. | References |
|----|--------|-------------------|--------|-------------|------------|
| 1 | 2 | 3 | 4 | 5 | 6 |

**Europe**

| NN | Region | Objects of dating | Method | Age in m.y. | References |
|----|--------|-------------------|--------|-------------|------------|
| 1 | Kola Peninsula, Monchegorsk region | Gneiss of the Kola Group, rocks | Pb-Pb | 3150 ± 50 | Sobotovich et al. (1963) |
| 2 | Kola Peninsula, Central part | Gneiss of the Kola Group, amphibole (several samples) | K-Ar | 3250–3500 | Lobach-Zhuchenko et al. (1972) |
| 3 | Kola Peninsula, Chuna-Volch'y tundra | Enderbite, amphibole | K-Ar | 3620 | Maslenikov (1968) |
| 4 | Kola Peninsula, Pinkeljavr Lake | Oligoclase gneiss-granite (tonalite), biotite | K-Ar | 3400 | Lobach-Zhuchenko et al. (1972) |
| 5 | Kola Peninsula, Voronya Mt. | Gneiss-migmatite and gneiss-granite (tonalite), biotites (several samples) | K-Ar | 3300–3580 | Polkanov and Gerling (1961) |
| 6 | Kola Peninsula, Voronya Mt. | The same rocks, zircon | U-Th-Pb | 3300 (concordia) | Gerling and Lobach-Zhuchenko (1967) |
| 7 | Kola Peninsula, Central part | Gneiss-granite and migmatite, biotite (several samples) | K-Ar | 3580–3590 | Bondarenko and Dagelaysky (1968) |
| 8 | Kola Peninsula, Kola Bay | Charnockite, biotite | K-Ar | 3200 | Polkanov and Gerling (1961) |
| 9 | Karelia, White Sea shore | Metabasite in the Belomorian Group, amphibole | K-Ar | 3300 | Maslenikov (1968) |
| 10 | Ukraine, Bug River | Marble of the Bug Group, rocks | Pb-Pb | 3600 | Yeliseeva et al. (1973) |
| 11 | Ukraine, Dnieper River left bank | Amphibolite of the Orekhovo-Pavlograd complex, amphiboles (several samples) | K-Ar | 3400–3560 | Ladieva (1965), Salop (1970, unpubl.) |
| 12 | Ukraine, Dnieper River, Jamburg town | Gneissoid tonalite-quartz diorite, amphibole | K-Ar | 3420 | Ivantishin and Orsa (1965) |
| 13 | Northern Norway, Vesterålen Islands | Gneiss of granulite facies of the Lofoten complex, rocks | Pb-Pb | 3460 ± 60 rejuvenated values: 2700 and 1800 | Taylor (1975), Griffin et al. (1978) |
| 14 | Scotland, Rona (Inner Hebrides) | Quartzite of the Lewisian complex, zircon(clastic?) | U-Th-Pb | 3250 | Bowes et al. (1976) |

**Asia**

| NN | Region | Objects of dating | Method | Age in m.y. | References |
|----|--------|-------------------|--------|-------------|------------|
| 15 | Yenisei Ridge | Charnockite of the Kansk Group, rocks | Pb-Pb | 4000 ± 200 | Volobuev et al. (1977b) |

**Table 3** (continued)

| NN | Region | Objects of dating | Method | Age in m.y. | References |
|----|--------|-------------------|--------|-------------|------------|
| 1 | 2 | 3 | 4 | 5 | 6 |
| 16 | Yenisei Ridge | Charnockite of the Kansk Group, rocks | Pb-Pb | 3400 | Rudnik (pers. comm. 1978) |
| 17 | Eastern Sayan | Marble of the Biryusa Group, rocks | Pb-Pb | 3700 | Volobuyev et al. (1977) |
| 18 | Baikal Region | Gneiss and crystalline schist of the Sharyzhalgay Group, rocks | Rb-Sr | 3750 | Sandimirova et al. (1979) |
| 19 | Anabar massif | Pegmatite cutting gneiss of the Anabar Group, monazite | U-Th-Pb | 3500 | Manuylova (ed.) (1968) |
| 20 | Aldan shield, Timpton River | Crystalline schist and amphibolite of the Aldan Group, Timpton Subgroup, rocks | Pb-Pb | 3500 | Rudnik et al. (1969) |
| 21 | Aldan shield, Timpton River | Marble of the Aldan Group, Timpton Subgroup, rocks | Pb-Pb | 3200 | Gerling et al. (1970) |
| 22 | Aldan shield, Uchur River | Metasomatite in gneiss of the Aldan Group, phlogopite | K-Ar | 3800 | Vinogradov et al. (1976 b) |
| 23 | Okhotsk massif (Far East USSR) | Gneiss of granulite facies, Okhotsk complex, rocks | Pb-Pb | 3500, 4100 | Sobotovich et al. (1973) |
| 24 | Omolon massif (North-East of the USSR) | Basic crystalline schist of granulite facies, Avekov complex, rocks | Pb-Pb | ~ 3250 | Sobotovich et al. (1973) |
| 25 | Omolon massif (North-East of the USSR) | Plagiogneiss among granulites of the Avekov complex, zircon | U-Th-Pb | 3400 | Bibikova et al. (1978) |
| 26 | Taygonos Peninsula (North-East of the USSR) | Basic crystalline schist of granulite facies, Purnoss Group, rocks | Pb-Pb | ~ 3100 | Sobotovich et al. (1977) |
| 27 | Stanovoy Range, Western part, the "Mogocha block" | Gneiss and crystalloschist of the Mogocha Group, rocks | Pb-Pb | 3400–3500 | Iskanderova et al. (1977) |
| 28 | Stanovoy Range, Western part | Amphibolite of the Nikitkino Group, rocks | Pb-Pb | 3700 ± 200 | Neimark (1981) |
| 29 | Stanovoy Range, central part | Gneiss of the Zverevo Group, micas (several samples) | K-Ar | 3400–3500 | Knorre et al. (1970) |
| 30 | Stanovoy Range, central part, Tynda River | Granite-gneiss, zircon, orthite | U-Th-Pb | 3200 | Tugarinov et al. (1977) |

**Table 3** (continued)

| NN | Region | Objects of dating | Method | Age in m.y. | References |
|---|---|---|---|---|---|
| 1 | 2 | 3 | 4 | 5 | 6 |
| 31 | Far East USSR, Dzhugdzhur Range | Anorthosite among gneisses and pyroxenic schists, rocks (several determinations) | Pb-Pb | up to 3200 | Moshkin et al. (1977) |
| 32 | India, Karnataca | Cordierite gneiss ("Peninsular gneiss"), rocks | Rb-Sr | > 3100 | Jayaram et al. (1976) |
| 33 | India, Karnataca | "Peninsular gneiss", rocks | Rb-Sr | 3065 | Crawford (1969) |
| 34 | India, Karnataca, Nilgiri Mt. | Charnockite ("Peninsular gneiss"), rocks. | Rb-Sr | ~ 3200 | Crawford (1969) |
| 35 | India, Karnataca | Tonalite from pebbles in the Kaldooga conglomerate, Dharwar Supergroup | Rb-Sr | 3250 ± 150 | Venkatasubramanian and Narayanaswamy (1974) |
| 36 | India, Orissa | Gneiss underlying the "Ferruginous Formation", mica and amphibole | K-Ar | 3450 | Sarkar et al. (1967) |
| 37 | India, Orissa | Tonalite gneiss underlying the "Ferruginous Formation", rocks | Rb-Sr | 3200 | Sarkar et al. (1979) |
| 38 | India, Rajasthan | Clastic zircon from the Aravalli metasandstone that overlies the Bandelkand gneiss | U-Th-Pb | 3500 | Tugarinov and Voytkevich (1970) |
| **North America** | | | | | |
| 39 | Western Greenland, Godthaab region | The Amitsoq tonalite gneiss (amphibolite facies), rocks | Rb-Sr | 3980 ± 175 | Black et al. (1971) |
| 40 | Western Greenland, Godthaab region | The Amitsoq tonalite gneiss (amphibolite facies), rocks | Rb-Sr | 3700–3750 | Moorbath et al. (1972) |
| 41 | Western Greenland, Godthaab region | The Amitsoq tonalite gneiss (amphibolite facies), rocks | Rb-Sr | 4065 (4155?) | Baadsgaard et al. (1976) |
| 42 | Western Greenland, Godthaab region | The same rocks, zircon | U-Th-Pb | 3600 | Baadsgaard et al. (1976) |
| 43 | Western Greenland, Godthaab region | The same rocks, zircon | U-Th-Pb | 3650 | Michard-Vitrac et al. (1977) |
| 44 | Western Greenland, Godthaab region | The same rocks, zircon | U-Th-Pb | ~ 4000 | Lovering (1979) |

**Table 3** (continued)

| NN | Region | Objects of dating | Method | Age in m.y. | References |
|----|--------|-------------------|--------|-------------|------------|
| 1 | 2 | 3 | 4 | 5 | 6 |
| 45 | Western Greenland, Godthaab region | The Amitsoq tonalite gneiss (granulite facies), rocks | Rb-Sr | 3600 ± 140 | Griffin et al. (1980) |
| 46 | Western Greenland, Godthaab region | The Amitsoq tonalite gneiss (granulite facies), rocks | Pb-Pb | 3625 ± 125 | Griffin et al. (1980) |
| 47 | Western Greenland, Godthaab region | Supracrustals of the Isua Group, rocks | Pb-Pb | 3760 | Moorbath et al. (1973) |
| 48 | Western Greenland, Godthaab region | Quartz-feldspar rocks from the Isua pebbles in conglomerate, rocks | Rb-Sr | 3860 ± 240 | Moorbath et al. (1975) |
| 49 | Western Greenland, Godthaab region | The same rocks, zircon | U-Th-Pb | 3820 | Michard-Vitrac et al. (1977) |
| 50 | Western Greenland, Godthaab region | Vein granite from the Isua Group | Rb-Sr | 3740 | Moorbath et al. (1977a) |
| 51 | Canada, Labrador Peninsula | The Hebron gneiss, rocks | Rb-Sr | 3600 | Barton (1975) |
| 52 | Canada, Labrador Peninsula | The Lost Channel migmatite, zircon | U-Th-Pb | 3640 | Barton (1975) |
| 53 | Canada, Labrador Peninsula | The Uivak gneiss, rocks | Rb-Sr | 3620 | Hurst et al. (1975) |
| 54 | Canada, Grenville tectonic province | Tonalite gneiss, rocks | Rb-Sr | > 3000 | Frith and Doig (1975) |
| 55 | Canada, Slave tectonic province | Gneiss underlying the Yellowknife Group, rocks | Rb-Sr | > 3000 | Frith et al. (1974) |
| 56 | Canada, SW Manitoba | English River gneiss, zircon | U-Th-Pb | > 3000 | Krogh et al. (1975) |
| 57 | Canada, British Columbia | Malton gneiss, rocks | Rb-Sr | 3235 ± 258 | Chamberlain et al. (1979) |
| 58 | U.S.A., Michigan | Gneiss from granitegneiss basement, zircon | U-Th-Pb | > 3400 | Peterman et al. (1978) |
| 59 | U.S.A., SW Minnesota | Morton (Montevideo) gneiss, zircon | U-Th-Pb | 3550 | Catanzaro (1963) |
| 60 | U.S.A., SW Minnesota | Morton (Montevideo) tonalite gneiss, rocks | Rb-Sr | 3800 | Goldich and Hedge (1974) |
| 61 | U.S.A., SW Minnesota | The same rocks, zircon | U-Th-Pb | 3300 | Michard-Vitrac et al. (1977) |
| 62 | U.S.A., SW Montana | Jardine-Crevice gneiss (Pony complex), rocks | Rb-Sr | 3300 | Brookins (1968) |

**Table 3** (continued)

| NN | Region | Objects of dating | Method | Age in m.y. | References |
|---|---|---|---|---|---|
| 1 | 2 | 3 | 4 | 5 | 6 |
| 63 | U.S.A., Montana Beartooth Mts | Gneiss from gneiss-granulite complex, zircon | U-Th-Pb | 3500 | Catanzaro and Kulp (1964) |
| 64 | U.S.A., Wyoming, Laramie Range | Gneiss-migmatite, rocks | Rb-Sr | 3020 ± 200 | Johnson and Hills (1976) |
| 65 | U.S.A., Idaho Raft River Range | Granite-gneiss, rocks | Rb-Sr | 3730 | Sayyah (1965) |
| **South America** | | | | | |
| 66 | Venezuela | Granite-gneiss in Itamaca gneiss-granulite complex, rocks | Rb-Sr | 3200 | Posadas and Kalliokoski (1967) |
| 67 | Venezuela | Gneiss of Itamaca complex | Rb-Sr | 3200–3400 | Hurley et al. (1972a) |
| 68 | Venezuela | Gneiss of Itamaca complex | Rb-Sr | 3200–3700 | Hurley et al. (1976) |
| 69 | Venezuela | Trondhjemite (tonalite) gneiss, rocks | Rb-Sr | 3000 | Dougan (1976) |
| 70 | Guiana | Amphibolite from the Ile-de-Cayenna Group, amphibole | K-Ar | ~ 4000 | Choubert (1966) |
| 71 | Brazil | Amphibolite from Xingu gneiss-granulite complex | K-Ar | 3420 | Amaral (1969) |
| 72 | Brazil | Granulite from gneiss-granulite complex of Bahia state | Rb-Sr | 3220 | Gordani and Iyer (1979) |
| **Africa** | | | | | |
| 73 | Swaziland, Barberton | Tonalite gneiss of the "Ancient gneiss complex", rocks | Rb-Sr | 3440 ± 400 | Allsopp et al. (1962) |
| 74 | Swaziland, Barberton | Tonalite gneiss of the "Ancient gneiss complex", rocks | Rb-Sr | 3340 | Allsopp et al. (1969) |
| 75 | Swaziland, Barberton | Tonalite gneiss in the area surrounding the Swaziland synclinorium, several determinations from different massifs, zircon | U-Th-Pb | 3170–3310 | Oosthuyzen (see Anhaeusser 1973) |
| 76 | Limpopo belt, Sandriver basin | Tonalite gneiss of Limpopo complex, rocks | Rb-Sr | 3856 ± 116 | Barton et al. (1977) |
| 77 | Limpopo belt, Sandriver basin | Amphibolite from dykes cutting tonalite gneiss of Limpopo complex, rocks | Rb-Sr | 3643 ± 102; 3128 ± 84 | Barton et al. (1977) |

**Table 3** (continued)

| NN | Region | Objects of dating | Method | Age in m.y. | References |
|----|--------|-------------------|--------|-------------|------------|
| 1 | 2 | 3 | 4 | 5 | 6 |
| 78 | Limpopo belt, Sandriver basin | Diorite (tonalite) gneiss of Limpopo complex, rocks | Rb-Sr | $3727 \pm 57$ | Barton et al. (1978) |
| 79 | Limpopo belt, Sandriver basin | Diorite (tonalite) gneiss of Limpopo complex, zircon | U-Th-Pb | $3190 \pm 34$ | Barton et al. (1978) |
| 80 | Limpopo belt, Sandriver basin | Leucocratic grano-diorite gneiss of Limpopo complex, rocks | Rb-Sr | $3776 \pm 97$ | Barton et al. (1978) |
| 81 | Zimbabwe, Fort-Victoria | Mushendyke tona-lite gneiss, rocks | Rb-Sr | $3520 \pm 260$ | Hickman (1974) |
| 82 | Zimbabwe, town of Mashaba | Mashaba tonalite banded gneiss, rocks | Rb-Sr | 3600 | Hawkesworth et al. (1975) |
| 83 | Zimbabwe, town of Mashaba | Mashaba tonalite banded gneiss, rocks | Rb-Sr | 3570 | Moorbath et al. (1977b) |
| 84 | Zimbabwe, environs of town of Selukwe | Tonalite granite | Rb-Sr | $3420 \pm 60$ | Moorbath et al. (1976) |
| 85 | Madagascar, Behar | Granite-gneiss from Androyan complex, rocks | Rb-Sr | 3200 | Hottin (1972, 1976) |
| 86 | Madagascar, Eastern shore | Migmatite, zircon | $\alpha$ Pb | 3200 | Hottin (1972, 1976) |
| 87 | Tanzania, Tan-ganyika craton | Gneiss from gneiss-granulite complex, biotites | K-Ar | 3100–3400 | Kulp and Engels (1963) |
| 88 | Tanzania, Tan-ganyika craton | Msagali charnockite, biotite | K-Ar | $3700 \pm 300$ | Kulp and Engels (1963) |
| 89 | Malawi | Metagabbro among gneiss, amphibole | K-Ar | 3550 | Cahen and Snelling (1966) |
| 90 | Zair, Katanga | Pegmatite cutting Lu-anyi gneiss-granulite complex, microcline | Rb-Sr | $3470 \pm 160$ | Delhal and Ledent (1971) |
| 91 | Zair, Katanga | Charnockite from Kasai complex, rocks | Rb-Sr | 3100 | Delhal and Ledent (1971) |
| 92 | Cameroun, southern part | Enderbite from Ebo-lowa complex, rocks | Rb-Sr | 3440 | Delhal and Ledent (1975) |
| 93 | Liberia, Bong area | Gneiss of Kasila complex, rocks | Rb-Sr | 3280 | Leo (1967) |
| 94 | Liberia, Nbomi Hills | Gneiss of Kasila complex, rocks | Rb-Sr | 3020 | Leo (1967) |
| 95 | Liberia, Nbomi Hills | Gneiss-granite from Kasila complex, rocks | Rb-Sr | 3315 | Leo (1967) |
| 96 | Liberia, south-western part | Gneiss of Kasila complex, rocks | Rb-Sr | 3470 | Hurley et al. (1970) |

**Table 3** (continued)

| NN | Region | Objects of dating | Method | Age in m.y. | References |
|----|--------|-------------------|--------|-------------|------------|
| 1 | 2 | 3 | 4 | 5 | 6 |
| 97 | Liberia, south-eastern part | Gneiss from Man and Kasila complexes, rocks | Rb-Sr | 3450 | Hurley et al. (1970) |
| 98 | Liberia and Sierra Leone | Gneiss from Man and Kasila complexes, rocks | Rb-Sr | 3300 | Hurley et al. (1976) |
| 99 | Algeria, Ahaggar | Charnockite from in Ouzzal complex, rocks | Rb-Sr | 3300 | Allègre and Caby (1972) |
| **Australia** | | | | | |
| 100 | Western Australia, southwestern part | Gneiss from gneiss-granulite complex of Wheat belt, rocks | Rb-Sr | 2900–3300 | Compston and Arriens (1968) |
| 101 | Western Australia, Yilgarn block | Gneiss Mt. Narryer, rocks | Rb-Sr | 3350 | De Laeter et al. (1981) |
| 102 | North-Western Australia, Pilbara block | Shaw migmatite, zircon | U-Th-Pb | 3417 ± 40 | Pidgeon (1978a) |
| **Antarctic** | | | | | |
| 103 | Enderby Land | Pyroxenic crystalline schist, rocks | Pb-Pb | 3800 ± 300 | Sobotovich et al. (1973) |
| 104 | Enderby Land | Enderbite, rocks | Pb-Pb | 4000 ± 200 | Sobotovich et al. (1973) |
| 105 | Enderby Land | Charnockite (enderbite?), zircon | U-Th-Pb | ~ 4000 | Lovering (1979) |
| 106 | Queen Maud Land | Leucocratic granite, rocks | Rb-Sr | 3060 ± 80 | Solov'yev and Halpern (1975) |

of them are rejuvenated. This value of 3000 (or 3000–3100) m.y. is certainly conventional, but it is important from the point of view that the younger Paleoprotozoic (Archeoprotozoic) rocks were usually metamorphosed much later, commonly the values 2800–2600 m.y. are referred to in this case. The datings shown in the table belong to highly metamorphosed supracrustals and plutonic rocks of the Katarchean, thus, to a certain degree, they reflect the time of the early metamorphic and plutonic processes.

**Distribution.** The Katarchean rocks are exceptionally widespread on every continent. However, their significance in the geologic structure of many regions is still underestimated, for they are usually and erroneously combined with the younger greenstone strata and granitoids of the Paleoprotozoic. Certain larger regional units of different continents can be given as examples of the Katarchean formations: in Europe these are the Kola Group of the Kola Peninsula, the Belomorian (White Sea) Group of Karelia, the Bug and Priazov Groups of the Ukraine, the gneiss-granulite complexes of Southern Sweden and Norway, the Lewisian complex of Scotland; in Asia, the Aldan Group of Eastern Siberia, the Anabar Group of Northern Siberia, the Kansk Group of the Yenisei Range, the Zerenda Group of Kazakhstan, the Hindustan complex (the "Peninsular gneisses") of Hindustan, the Tai Shan and Liangtou complexes of China; in Africa, the Malagasy complex of Madagascar, the pre-Swaziland complex (the "Older Gneisses") of Swaziland, the Kakamas complex of Namaqualand, the gneiss-granulite complexes of Zimbabwe, Malawi, and Tanzania, the Watian complex of Uganda and Kenya, the Dibaya and Bomu complexes of Zair, the Dahomey complex of Togo and Benin, the Man and Kasila complexes of Liberia and Sierra-Leone, the Amsaga, Ghallaman, Arechchoum and others gneiss-granulite complexes of North-Western Africa, the Tibesti, Western Nile and Meatiq complexes of North Eastern Africa; in North America, the pre-Keewatin and pre-Yellowknife gneiss-granulite complexes of the Superior and Slave Provinces of Canada, the Ungava and Uivak of Labrador Peninsula, the Grenville complex (s. str.) of South-Eastern Ontario, the gneiss-granulite and gneiss complexes of the Arctic shore-line of Canada and Greenland, the Montevideo (Morton) complex of Minnesota, U.S.A., the gneiss-granulite complexes of Montana, Nevada, Idaho, Arizona and of other central and southern states of the U.S.A.; in South America, the Itamaca complex of Venezuela, the Canucu complex of Gayana, the Ampada-Falavatra complex of Surinam and the Xingu complex of Brazil; in Australia, the Wheat Range, Fraser, Albany-Esperance, Musgrave-Mann, Goler, Willyama (?), Tikalara, Mirarmina gneiss-granulite complexes, the pre-Pilbara gneisses; in the Antarctic, the gneiss and gneiss-granulite complexes of the Queen Maude Land, the Enderby Land (the Ragat and Tula Groups).

**Supracrustals.** Among the widespread and typical rocks of Katarchean are the melanocratic amphibole, amphibole-pyroxene and pyroxene (including hypersthene and bipyroxene) plagiogneisses, and crystalline schists, and also amphibolites. Judging from their chemical and mineralogical composition and also from their mode of occurrence they represent high-grade basic, locally ultrabasic lavas and probably tuffs, and in some cases the sheet or sheet-cutting intrusive bodies of gabbroids and rare ultrabasites. The rocks close in composition to komatiites (Suslova 1976, McGregor and Mason 1977 etc.) are recognized among the basic and ultrabasic metabasites of Western Greenland (the Akilia association), of Kola Peninsula (the granulite complex of the Kola Group) and of Eastern Siberia (the Aldan and Anabar Groups). In many localities the metabasites are strongly granitized and transformed into different biotite-amphibole plagiogneisses (migmatites), enderbites, and charnokites. Sometimes they are encountered only as inclusions or ghost-relicts among plagiogneisses or diorite gneisses.

The metamorphosed acid volcanics in the Katarchean are not known for certain, but that does not mean that these rocks are absent. Up to the present we do not have any reliable criteria to distinguish these rocks when they are so strongly altered. Many workers reasonably suggest that certain leucocratic biotite-bearing leptite-like gneisses distributed in many Katarchean strata are the altered acid volcanics. It is possible that the so-called acid granulites – the leucocratic garnet-bearing quartzfeldspathic rocks with low concentration of pyroxene and biotite widely developed in Katarchean – are also highly altered acid volcanics and tuffs. However, their sedimentary origin is equally possible; it is not to be excluded that these "acid granulites" can reveal different modes of origin.

Metavolcanics associate with biotite, garnet-biotite, sillimanite- and cordierite-bearing gneisses and also quartzites. These rocks build up separate, sometimes thick horizons or strata. Locally these rocks contain varieties with abundant sillimanite, with corund and spinel, sometimes enriched with magnetite. By their composition these rocks are to be regarded as metamorphosed high-alumina clay rocks. Quartzites are likely to be formed also at the expense of sedimentary rocks, but their genesis is not as simple as it seems at first sight, as will be shown below.

Marbles and graphite gneisses or crystallo-schists are undoubtedly of sedimentary origin. The marbles of dolomite and calcite composition commonly associate with calciphyres, diopside-plagioclase, diopside-quartz-plagioclase, diopside-scapolite and other types of crystallo-schists or calc-silicate rocks that are likely to be formed at the expense of marly sediments. Marbles and calc-silicate rocks locally contain a high concentration of phosphorus as apatite; the banded sheet bodies of diopside-quartz-apatite rocks are also reported. Some marbles are known to contain native sulfur and sulfates (anhydrite). These latter locally form thin interlayers and reveal a stable equilibrium association with minerals typical of granulite facies of metamorphism; sometimes the traces of past salinity are registered in marbles (Brown 1973, Karagat'yev 1970, Leake et al. 1979, Litsarev et al. 1977, Vinogradov 1977).

Finally, the much rarer but typical Katarchean formations are to be mentioned; these are metamorphosed ferruginous rocks that build up interlayers in the older strata. Their typical and constant feature is their close relation to metabasites (amphibolites and basic crystalloschists) in particular if these are situated among supracrustals with abundant marble interbeds. Two principal varieties of such rocks are known to exist. To the first belong the banded, as a rule, roughly banded or faintly stratified rocks with interlayers rich in magnetite alternating with layers depleted of this mineral or lacking it completely, but all layers contain abundant amphibole, pyroxene (monoclinic and rhombic) and also garnet, biotite, quartz, and feldspar. Quartz is registered as mainly occurring in the ore-free layers. Generally such rocks are close to jaspilites but differ, however, as a rule from the typical Paleo-Meso-protozoic jaspilites by faintly expressed banding, by the presence of abundant silicates and alumosilicates, by commonly low abundance of quartz and absence of carbonates. The alumosilicate and silicate-magnetite ores occurring as lenses and irregularly shaped portions and also as impregnation of magnetite among metabasites are the other still more typical Katarchean ore variety. Both varieties of these ferruginous rocks are encountered in the same strata and form a typical association. It was suggested that this should be regarded as a special Azovian (or Pri-Azovian) type of syngenetic iron ore characteristic of Katarchean (Salop 1973a).

The primary structures of lavas and sediments of their relicts are not usually known among the Katarchean supracrustals. However, the large structural aspects of original rocks are conspicuously observed as, for instance, horizontal bedding. In a few cases it is possible to see different cross-bedding, traces of washout, and even ripple marks (Travin 1974).

**Metamorphism.** It was stated that the Katarchean rocks are typical of alteration under granulite and amphibolite facies. The rocks of these facies are related by way of intermediate amphibole-granulite sub-facies and commonly occur separately, in any case the observed zonation is of irregular (nonlinear) character. The amphibolite facies, as well as the granulite one, is progressive, but more commonly it is superimposed on the granulite facies, which is regressive. Generally the regressive alterations of the Katarchean rocks are quite common and in some places, in particular in the cataclase zones, they are accompanied by the origin of mineral associations of greenschist facies. At the same time, the cases of superposition of granulite facies on amphibolite facies, as suggested by some workers, is not confirmed by any geologic observations or petrographic studies.

It seems important that the granulite facies of regional metamorphism (in contrast to the amphibolite one) is exceptionally characteristic of Katarchean. This statement is sometimes disputed but up to the present there are no available data on the younger rocks regionally altered under this facies. Usually the examples given that are taken as proof of an opposite concept are based on misinterpreted rejuvenated isotope datings. The metamorphic zonation from greenschist facies to granulite facies sometimes observed in Protozoic rocks appears to be erroneous after more detailed studies; in reality the granulite facies rocks represented tectonic blocks among younger rocks of lower alteration (see, e.g., Salop 1979 b), and the statement above concerns the exposures of rocks of granulite facies in the Phanerozoic fold belts in particular, their basement having suffered multifold thermal-tectonic activation. If this statement seems too categorical, it, at any rate, brings no essential change, for there is no doubt that in major cases the granulite facies is typical of the oldest rocks of the Precambrian.

**Plutonic Formations.** The Saamian orogeny terminating the Katarchean Era was accompanied by intensive intrusive magmatism, anatectic and metasomatic granitization and migmatization. The Katarchean supracrustals commonly occur as ghost inclusions among granite-gneisses. Widely distributed are the grey gneissoid, essentially plagioclase granitoids of granodiorite composition (tonalite gneisses); potassic-microcline and microcline-plagioclase granites are noticeably less abundant and usually form small cutting bodies emplaced at the end of the cycle. Sometimes their assignation to the Katarchean is doubtful. The complex examined is also typified by metasomatic (or anatectic) enderbites and charnockites formed under granulite facies (the intrusive charnockitoids formed as a result of rheomorphism of the Katarchean granulite rocks are also reported from the younger Precambrian complexes). Intrusions of basic rocks preceded the origin of Katarchean granitoids; now they are represented by massifs of gabbroids and by folded sheet and layered bodies of meta-anorthosites with local chromite mineralization.

**Tectonic Structures.** The tectonic pattern of Katarchean formations is quite unique. It was stated above that it is typified by isometrical, round or oval, locally irregularly shaped fold systems; in plan they resemble at first sight gneiss domes, but they differ completely by their inner structure and by their much greater size – from 100 to 800 km across. Such fold systems (the "gneiss fold ovals") represent the concentric grouping of folds of different sizes and shapes, mostly of linear or isoclinal type, with obviously expressed centripetal vergence (mass transportation toward the system center). Large folds are complicated by folds of smaller orders up to microplication.

**Fig. 2.** Tectonic scheme of Katarchean of the Aldan Shield.
1 gneiss fold ovals, 2 intra-oval fields, 3 platform cover, 4 strike of fold axes in fold ovals, 5 strike of old axes and partly of schistosity in intra-oval fields, 6 synclinal zones, 7 gneiss domes, 8 northern boundary of the Stanovoy Range foldbelt. The fold ovals: I Chara, II Nelyuka, III Verkhnealdan, IV Verkhnetimpton, V Nizhnetimpton, VI Sunnagin

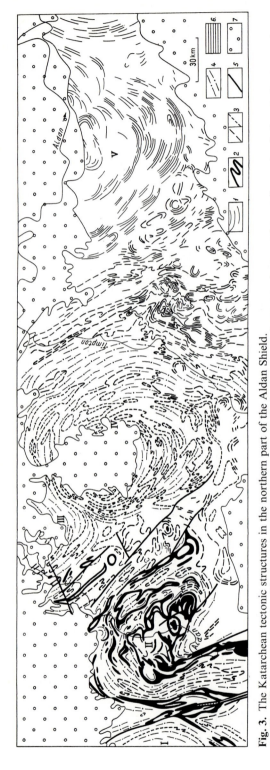

**Fig. 3.** The Katarchean tectonic structures in the northern part of the Aldan Shield.
*1* strike of fold axes and strike of bed seen from aerophotos, *2* quartzite horizons seen from aerophotos and according to detailed geologic observations, *3* boundaries of stratigraphic units traced according to geologic surveys of different scale, *4* strike of minor fold axes, *5* Precambrian fractures, *6* Mesoprotozoic strata, *7* platform cover (Eocambrian, Cambrian, Jurassic). Fold ovals: *I* Nelyuka, *II* Verkhnealdan ("amoeboid"), *III* Nimnyrykan, *IV* Nizhnetimpton, *V* Sunnagin

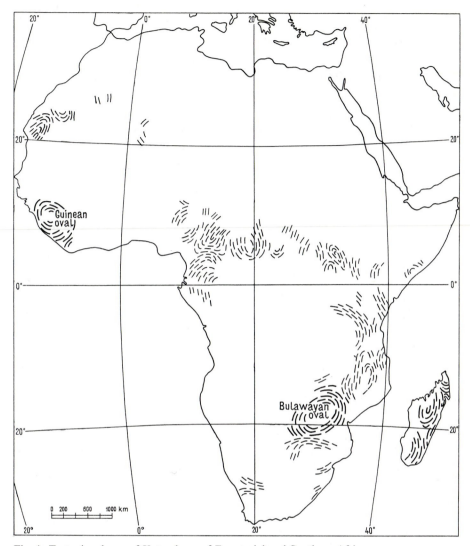

**Fig. 4.** Tectonic scheme of Katarchean of Equatorial and Southern Africa

The axial surfaces of the folds are commonly strongly curved in a horizontal plane and the detailed structural studies prove that these curves were formed during folding and not as a result of later deformations, as is sometimes suggested. In the center of the fold systems granite massifs are commonly observed, formed as a result of ana-texis of the gneiss substratum at a late stage of the tectono-plutonic cycle. Some earlier autochtonic gneissoid plagiogranites (tonalites) are ubiquitously developed and build up this complex folding together with gneisses, but their largest masses also occur in the inner part of the "gneiss ovals". Areas of different extension characterized by chaotic orientation of folds, by lacking of vergence and by wide development of the brachyform syncline; small dome structures are situated in between the "gneiss

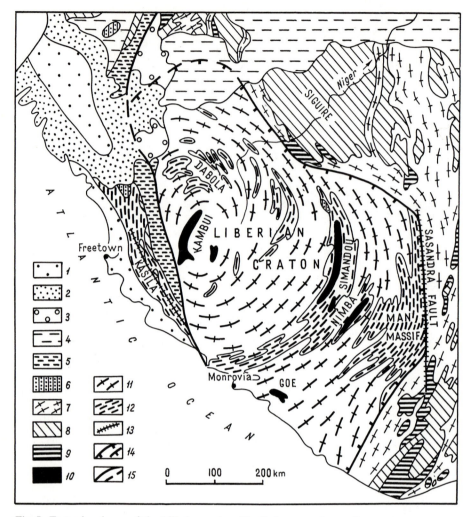

**Fig. 5.** Tectonic scheme of the Liberian craton (the Guinean gneiss oval).
*1–2* Phanerozoic (Eocambrian including) strata, *3* Epiprotozoic strata, *4–6* Neoprotozoic
strata, *7* Mesoprotozoic granite, *8* Mesoprotozoic strata (the Birrimian Group), *9* Meso-
Paleoprotozoic undivided strata, *10* Paleoprotozoic strata (the Simandou, Nimba, Kambui
Groups), *11* Katarchean gneiss-granite (tonalite, quartz diorite), *12* Katarchean supracrustal
gneiss-granulite complex, *13* abyssal fracture, *14* boundary of the Mesoprotozoic Liberian craton, *15* western boundary of the Epiprotozoic West-African platform

ovals". The genesis of all these structures was examined in the specialized work (Salop
1971 b) and here it will be briefly discussed at the end of the chapter.

The tectonic patterns of the Katarchean mentioned above were established for the
first time by the present author on the Aldan shield; now these structures are known
in many regions of almost every continent where the wide-spread oldest strata are
being studied (Salop 1971b, 1973a, 1977b, Salop and Sheinmann 1969, Salop and
Travin 1974), (Figs. 2–6). Structures of the gneiss oval type are also reported from

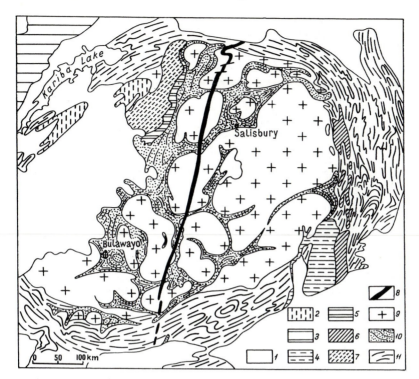

**Fig. 6.** Tectonic scheme of the Rhodesian craton (the Bulawayan gneiss oval).
*1* Phanerozoic platform strata, *2, 3* Epiprotozoic (*2* the Sijarira Group, *3* the Katangan Supergroup), *4* Neoprotozoic or Mesoprotozoic (the Umkondo Group), *5–7* Mesoprotozoic (*5* the
Lomagundi Group, *6* the Frontier and Gairezi Groups, *7* the Piriwiri Group), *8* the ultrabasic
and basic rocks of the Great Dyke, *9* Paleoprotozoic and Katarchean granite (tonalite),
*10* Paleoprotozoic greenstone strata (the Bulawayan and Shamvaian Groups), *11* Katarchean
gneiss-granulite complex

some portions of the ocean bottom when explored by geophysical methods. Large
ring structures (up to 300 km across) conspicuously seen according to magnetic fields
in North Atlantic in the area of the Rockall underwater plateau (Fig. 7). The granulite
rocks typical of Katarchean, reported from sea-drilling in this area, are suggestive of
a submerged continental crust probably related with the granulite complexes of
Lofoten Islands, Scotland (the Eria Massif) and Greenland.

## 2. Stratotype of the Katarchean (the Aldan Group)

Stratigraphy of the oldest supracrustals in many regions of the world has been poorly
known up to now because of the very complex tectonic structures of the Katarchean
that are difficult to interpret and because of the limited methods of establishing the
normal bed sequence. However, dataileded geologic mapping offers a possibility of
overcoming these difficulties; the tectonic structures can be recognized with the help
of aerophoto-surveying and by tracing marker horizons along the strike; the normal

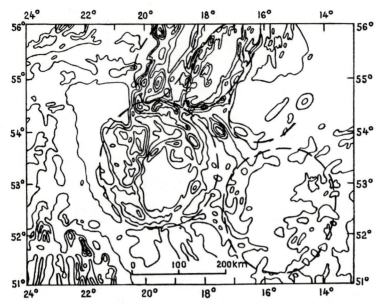

**Fig. 7.** The magnetic field chart in Northern Atlantic in the area of the Rockall underwater plateau. (After Vogt and Every 1974). Ring magnetic anomalies reflect the fold systems of the "gneiss oval" in the gneiss-granulite complex of Katarchean, that builds up the bed in this region

sequence of beds can be distinguished from the overturned type by studying the rhythmically bedded rocks and also cross-bedding and erosion pockets that are sometimes observed in quartzites.

The Katarchean formations are best known within the limits of the Aldan shield, which is a favorable object for studying the oldest supracrustals. These occupy there a very extensive territory (about 300,000 km$^2$), are well-exposed and are not very strongly reworked by post-Katarchean thermal-tectonic processes, this last being very important. The gneiss complex of the shield under the name the "Aldan Group" is the most complete among the known regional units of the Katarchean. It permits us to regard it as a world stratotype of the erathem.

The Aldan Group occupies the lowest position in the Precambrian of Eastern Siberia. The rocks of the group and the granite-gneisses piercing it are overlain transgressively and with an angular unconformity by low-grade sedimentary-volcanogenic strata of the Paleoprotozoic (the Sugan and Olonda Groups) and Meso-protozoic (the Udokan Group and its equivalents). The age of early metamorphism of rocks of the Aldan Group and of its certain equivalents is 3500–3800 m.y. (see Table 3, numbers 20–27, 29). Further data evidence that the rocks of the Aldan Group were under the influence of still earlier ($\sim$ 4000 m.y.) metamorphic processes.

Many workers over many years have studied the stratigraphy of the Aldan Group and recently many details have come to light due to very sophisticated mapping and tectonic structure studies in the central area of the shield that are very favorable for deciphering from aerophoto-survey (Salop and Travin 1974). The lower members of the group crop out in the core of a large unique structure situated in the Upper Aldan River, it is distinguished as the Verkhnealdan fold oval or "amoeboid" (Fig. 8, see

**Fig. 8.** Geologic map of the Verkhnealdan fold oval ("amoeboid"), Central-Aldan region. *1* alluvial deposits, *2* Jurassic strata, *3* Lower Cambrian strata, *4* Eocambrian strata (the Yudoma Formation), *5* diabase dykes (Neoprotozoic?), *6–15* Katarchean: *6* late tectonic alaskite granite and granite-gneiss (conformable bodies of syntectonic tonalite gneisses are ubiquitous and are not shown in the map), *7–15* the Aldan Group: *7* the Fedorov Mines Formation, *8* marble horizons in the Fedorov Mines Formation, *9* the Ungra Formation, *10* the Nimgerkan Formation, *11* the upper subformation of the Suontit Formation (quartzite), *12* the middle subformation of the Suontit Formation, *13* the lower subformation of the lower Formation (quartzite), *14* the Ayanakh Formation, *15* the Kurumkan Formation (quartzite), *16* bed orientation seen from aerophotos, *17* old fractures, *18* young (post-Jurassic) fractures

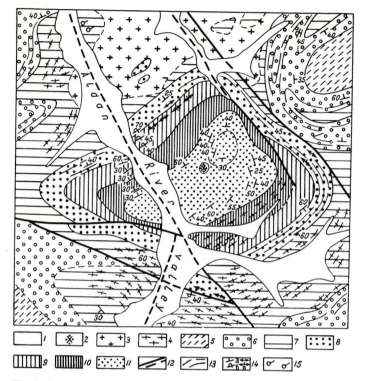

**Fig. 9.** Exposures of the oldest rocks of Katarchean. Geologic map of the Upper Aldan River in the area of the Kurumkan dome.
*1* modern alluvium, *2* Mesoprotozoic granite, *3–11* Katarchean: *3* alaskite granite, *4* ghost to-nalite gneiss, *5* the Nimgerkan Formation (strata of "variegated composition"), *6–8* the Suontit Formation (*6* the upper subformation-quartzite, *7* the middle subformation: amphibolite alter-nating with quartzite, *8* the lower subformation – quartzite), *9–10* the Ayanakh Formation (*9* the upper subformation: strata of garnetbiotite gneiss and some other gneiss types, *10* the lower subformation: strata of pyroxene-amphibole crystalline schist and amphibolite with quartz-zite interbeds), *11* the Kurumkan Formation: quartzite, locally it is sillimanite-bearing, *12* fracture, *13* schistosity elements seen from aerophotos, *14* bed position (*a* normal position, *b* overturned position, *c* vertical and subvertical position), *15* submergence of minor fold bends

also Figs. 2 and 3). In the apical part of it there are several domes with the oldest rocks of the Aldan shield in their cores, and probably these rocks are the oldest in the world (see below), they are represented by quartzites (Fig. 9). Limbs of the domes build up the higher units of the group typified by alternation of bands of pyroxene crystalline schists, amphibolites, and gneisses with thick horizons (strata) of quartzites. All these units belong to the Iyengra Subgroup. The quartzite horizons resistant to denudation are easily traced as ridges and clearly demonstrated in relief and on aerophotos of the tectonic structure of the area. The apical part of the "amoeboid" is surrounded by strata made up of different melanocratic pyroxenic crystalline schists, amphibolites and gneisses, and the upper parts of these strata comprise bands of marbles and calc-silicate rocks. This portion of the Aldan Group is divided as the Timpton Subgroup; in the basin of the Upper Aldan River it builds up the synclinorium structures between great anticlines formed by the rocks of the Iyengra Subgroup. It is more extensive in other areas of the shield and in the Timpton River basin in particular. There it makes up the Nizhnetimpton fold oval. The highest units of the group made up of different gneisses, mostly garnet-biotite, partly pyroxene, and also of granulites and marbles, are principally developed in the periphery of the fold ovals and in the inter-oval fields; they are divided as the Dzheltula Subgroup; in the Timpton basin they surround the Nizhnetimpton oval.

Roughly the normal sequence of the Aldan Group rocks in the Central Aldan area can be described as follows (up-section):

## Iyengra Subgroup

1. The Kurumkan Formation: quartzites, locally sillimanite-bearing, with rare thin interbeds of sillimanite gneisses. The known thickness is 1000 m.
2. The Ayanakh Formation: in the bottom the pyroxene-amphibole and amphibole crystalline schists and gneisses with quartzite interbeds; in the top alternating garnet-biotite, biotite-sillimanite, cordierite-biotite, and hypersthene-bearing gneisses with subordinate interbeds of bipyroxene schists, amphibolites, and quartzites. Thickness is 1500 m.
3. The Suontit Formation comprises three subformations, the lower and the upper ones are made up of quartzites with interbeds of sillimanite gneisses, the middle one (the thickest, 850–1000 m) is characterized by intercalation of amphibolites and quartzites. Thickness is 1600–2000 m.

## The Timpton Subgroup

4. The Nimgerkan Formation is of varying composition: amphibolites (commonly pyroxenic), amphibole, hypersthene, and bipyroxene gneisses and schists with thin interbeds of quartzites and garnet-bearing gneisses. Thickness is 1200–1300 m.
5. The Ungra Formation: amphibolites, pyroxene amphibolites, bipyroxenic schists or gneisses and enderbites and charnockites closely related to the former. Thickness is 1200–2500 m.
6. The Fedorov Mines Formation: basic crystalline schists and gneisses (the same as in the Ungra Formation) with several horizons of marbles, calciphyres, calcic diopside-bearing crystalline schists, diopside metasomatites, and local quartzites. Thickness is 2300–3100 m.
7. The Seim Formation: basic crystalline schists and gneisses with the thick (up to 600 m) horizon of garnet-biotite gneisses in the bottom. Thickness is 1500–1800 m.

## The Dzheltula Subgroup

8. The Kyurikan Formation: rhythmically alternating garnet-biotite, garnet-pyroxene, biotite, bipyroxene, diopside gneisses and other gneiss types with interbeds of marbles and calciphyres and also of graphite gneisses. Thickness is 1700–2100 m.

9. The Sutam Formation: monotonous garnet-biotite and biotite gneisses and leucocratic garnet-bearing granulites, the gneisses locally contain graphite, sillimanite,
and cordierite. Thickness is more than 2000 m.

Total thickness of the group is of the order of 15,000 m.

## 3. The Lithostratigraphic Complexes of the Katarchean

In different regions of the world the Katarchean supracrustals, complete enough and
well studied, resemble each other greatly when compared in composition and in
sequence structure, they also exhibit similarity with the Aldan stratotype. This striking aspect of the Katarchean formations permits us to distinguish four successively
occurring lithostratigraphic complexes within this oldest erathem, and gives a preliminary supposition as to the existence of two more complexes of a higher level, but their
singularity must still be investigated more thoroughly. The first four complexes were
examined in the previous works of the author (Salop 1968, 1973a, 1977b), the other
two are described for the first time here [6].

The oldest Katarchean strata that make up the *Iyengran or metabasite-quartzite
lithostratigraphic complex* (stratotype: the Iyengra Subgroup of Aldan) are composed
of basic pyroxene crystalloschists or (and) amphibolites alternating with thick quartzite horizons, with local interlayers of sillimanite gneisses and rare thin interlayers of
magnetite rocks. On the Aldan shield this complex is related to the epigenetic lateral-
secretional veins with rock crystal (among quartzites) and small deposits of corundum
(in the aluminous gneisses). Usually the rocks of the Iyengra complex in many regions
of the world are not eroded or they crop out only partially. In addition to the
stratotype, the lower parts of the Kurulta and Zverevo strata ("groups") of the
Stanovoy Ridge area also belong to this complex, as also the Daldyn Subgroup of the
Anabar shield and possibly the Ranomena Group of Madagascar to a certain extent
(Fig. 10) [7].

The overlying *Ungran (metabasite) complex* is principally composed of melanocratic pyroxene and amphibole crystalline schists and (or) amphibolites formed at the
expense of basic and partly ultrabasic volcanics. The amphibole, pyroxene, and other
types of gneisses are evidently of subordinate importance; thin interbeds and bands
of stratified silicate-magnetite and quartz-magnetite rocks are rare. This complex is
widespread all over the world and usually composes the major part of the erathem.
In the Aldan stratotype the Nimgerkan and the Ungra Formations correspond to it.

The Ungran complex can be demonstrated by many examples, several of which
are given here. In Asia there is the Verkneanabar Subgroup of the Anabar shield, the
major part of the Kansk Group of the Yenisei Ridge, the lower part of the Sharyzhalgay Group (the Zhidoy and Zoga Formations of the Sayan area), the Talanchan
Subgroup of the Baikal area, the Ileir Subgroup of the Middle Vitim River; in Europe

---

6 Characteristics of the Katarchean units developed in the Northern Hemisphere and Africa are
given in the works mentioned above

7 Here the stratigraphic division of the Katarchean of Madagascar is principally accepted
according to Besairie (1973). Some later schemes proposed by Hottin (1976) and Jourde (1972)
and by some other authors are mainly based on isotope datings of rocks and not on geologic
data, but the ages vary greatly under the influence of Pan-African activation (Salop 1977b)

there is the Khetalambino Formation of the White Sea, the Pobugian and Dniester-Bug Formations of the Bug area, the Volodarsk Formation of the central part of the Ukrainian shield, the West-Priazov Subgroup of the Azov area; in North America, the lower part of the Grenville complex ("Pyroxene Gneiss"); the major lower parts of all of the Katarchean complexes of Equatorial, Western and North-Western Africa, the lower part of the Androyan Group of Madagascar and probably the lower parts of many gneiss-granulite complexes of Australia that are conventionally attributed to the Katarchean here.

The third complex up-section is the *Fedorovian* (*metabasite-carbonate*) complex typified by the presence of carbonate horizons among metabasites; they are represented by marbles and calc-silicate rocks. Metabasites as a rule are dominant, carbonates are subordinate, but their abundance varies from place to place. This complex is also characterized by the presence of graphite-bearing gneisses, relatively thin interbeds of quartzites, aluminous gneisses and banded silicate-magnetite or quartz-magnetite rocks that occur together with metabasites and belong to the so-called Azovian type of ferruginous formations of the Precambrian. Evaporites, the oldest in the Earth's history occur in carbonate strata in some regions of the world, and represent anhydrite-bearing marbles and calcic crystalline schists (the Aldan shield, Pamir the Grenville Province of Canada, the Caraiba area of Brazil); rocks rich in phosphorus also occur in the carbonate rocks. This complex associates with different sedimentary-metamorphogenic deposits and occurrences of graphite, iron, manganese, and phosphorus and also with the contact-metasomatic deposits of phlogopite, iron, and boron.

The Fedorovian complex is widely developed in many regions of the world. The Fedorov Mines, Seim and Kyurikan Formations of the Aldan Group are stratotypes. Their analogs in other parts of the world are numerous. Several examples are given here from different continents: in Asia the Khaptasynnakh Formation of Anabar, the Kalantat Formation of the Yenisei Ridge, the upper part of the Sharyzhalgay Group (the Cheremsha Formation) of the Sayan area; in Europe the Teterev-Bug Formation of the Bug area, the Belotserkovsk Formation of the central part of the Ukrainian shield, the Cental Priazov (Korsak-Shovkay) Subgroup of the Azov area, the Chudzyavr Formation of the Kola Peninsula; in North America the upper part of the Grenville complex (?); in Africa the upper parts of many Katarchean complexes of different parts of the continent (the Masai, Karasuk, Menera-Roguga, Toufourfuse, Serchouf, and other formations), the upper part of the Androyan Group and major part of the Graphite Group of Madagascar.

The fourth *Sutamian complex* is principally made up of fine-bedded garnet-bearing biotite gneisses, coarse-bedded or massive leucocratic garnet granulites. Different other types of rocks (metabasites, marbles, high-aluminous gneisses) are reported to occur in subordinate abundance. The primary formations of which the mentioned gneisses were formed were different pelitic sediments and probably the acid volcanics, the part of the latter, however, is not sufficiently understood because the criteria of their recognition are not established. As to the mineral deposits, this complex associates only with garnet deposits (a raw material in abrasive manufacturing) and of sillimanite in part.

The Sutamian complex is less estended than the two previous but is still well known in many regions of the world. Its stratotype is the uppermost Sutam Forma-

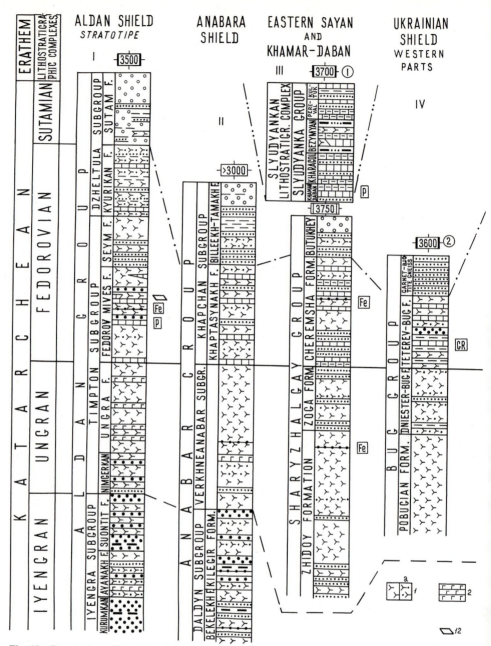

**Fig. 10.** Correlation of the Katarchean lithostratigraphic complexes in different parts of the world.

Stratigraphic columns: *I* after Salop and Travin (1974), *II* after Rabkin and Lopatin (1966), *III* after Prokof'yev (1971), Konikov et al. (1974), *IV* after Polovinkina (1960) and Drevin (1967), *V* after Polunovsky (1969 etc.), *VI* after Misharev et al. (1960) and Bondarenko and Dagelaysky (1968), *VII* after Wynne-Edwards (1967), *VIII* after Walsh (1966), *IX* after Besairie (1973 etc.). Only fundamental works are referred to, additional information (isotope datings etc.) is taken from different sources.

Symbols: *1* basic (pyroxene and amphibole) crystalline orthoschists or gneiss, amphibolite and other metabasites and also associated enderbite and charnockite (*a* crystalline ortho- and paraschist that is likely tuffaceous in part), *2* ultrabasic crystalline schist and metapyroxenite, *3* highly granitized gneiss with relics of amphibole-pyroxene rocks, *4* leucocratic granulite,

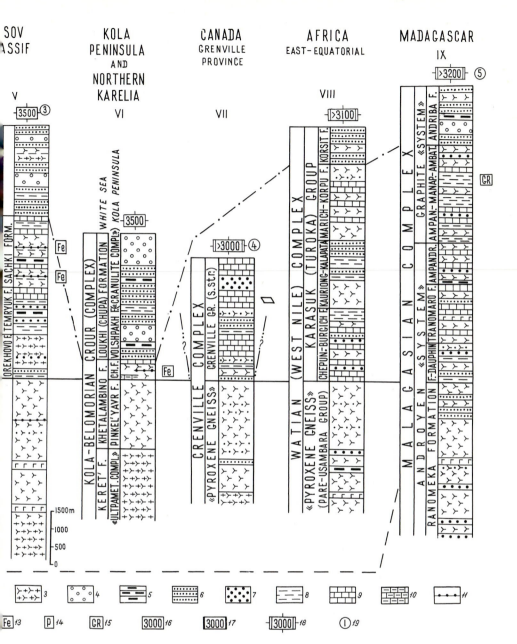

5 high-alumina gneiss, 6 garnet-bearing biotite gneiss, 7 quartzite, 8 graphitic gneiss, 9 marble, calciphyre, and calc-silicate rocks, 10 diopside-plagioclase and diopside-scapolite crystalline schist, 11 ferruginous rocks, the banded type including, 12 metamorphosed evaporite (marble with anhydrite), 13–16 stratiform deposits: 13 iron, 14 phosphorus (apatite), 15 graphite, 16–18 isotope datings of metamorphic and plutonic rocks: 16 K-Ar method, 17 Pb-Pb method, 18 Rb-Sr method, 19 number of a footnote.

Foot-notes (figures in circles): 1 dating of marble of the Biryusa Group – analog of the Slyudyanka Group (Volobuev et al. 1977a), 2 dating obtained on marble of the Teterevo-Bug Formation, 3 dating of granite of the Aul complex – analog of the Priazov Group (only the rejuvenated K-Ar datings of 2900 m.y. are available for the latter), 4 dating of tonalite gneiss cutting the Grenville complex close to the Grenville front (Frith and Doig 1975), 5 dating of granite cutting the Androyen rocks (major portion of datings of the Katarchean rocks of Madagascar is "rejuvenated" under the influence of later thermal processes)

tion of the Aldan Group. Examples are as follows: in Asia the Bileekh-Tamakh Formation of Anabar, the Butukhey Formation of the Sharyzhalgay Group of the Eastern Sayan area, the Chilcha and Kudulikan Formations of the Stanovoy Ridge; in Europe the Loukhi Formation of the White Sea, the Volshpakh Formation of the Kola Peninsula, the garnet-biotite gneiss strata of the Bug area, the Karatysh Subgroup of the Azov area; in Africa the Djougou-Agbandi strata of Benin, the Andriba Formation of Madagascar.

Thus, the four lithostratigraphic complexes of Katarchean examined are situated in the same sequence in different parts of the world. The synchroneity of these complexes is impossible to establish at present, as the isotope methods give only the time of the final metamorphism or plutonism and reliable marker horizons are lacking there. However, it will be shown below that under a specific Katarchean environment the successive formation of piles of different composition could have occurred simultaneously all over the world, certainly only with rough approximation. The suggested combination of the Katarchean formations into four lithostratigraphic complexes permits us, however, to forecast certain mineral deposits. In this respect it is well substantiated; further basing is needed for the stratigraphic significance of these complexes.

The four complexes examined seem to comprise the major part of the oldest formations in the world, but they still do not comprise the whole sequence of Katarchean supracrustals. In some places, for instance in the south of Siberia (in the fold belts surrounding the Siberian platform) thick gneissic strata are also recognized, containing abundant carbonate rocks and subordinate metabasites; the strata notably differ from the complexes described above. The relations of these strata and the complexes are not known, but it is very likely that they occur above the highest among them – the Sutamian complex. Some workers (Shafeyev 1970, Konikov et al. 1974) suggest an unconformity between them, but this has never been observed. The Slyudyanka group developed in the Khamar-Daban Range (the South Baikal area) is the most typical of these strata; it can be taken as a stratotype of the *fifth lithostratigraphic complex – the Slyudyankan*. This group is adjacent to the Sharyzhalgay Group of the Eastern Sayan, which is a certain equivalent of the Aldan Group (see Fig. 10), however, both groups are separated by a great and wide fracture zone. Earlier many workers, including the present author compared the Slyudyanka Group and the Cheremsha Formation (the Fedorovian complex) of the Sharyzhalgay Group. Now Konikov et al. (1975) have shown that they differ so much that it is impossible to explain these differences by facial changes that in the Katarchean strata are, as a rule faintly expressed.

The stratigraphic division of the Slyudyanka group proposed by different authors is also different (in Fig. 10, column III the formations are shown according to Konikov), its thickness is also estimated to be from 2500 to 8000 m. Generally the group is characterized by alternation of horizons or bands of carbonate and silicate-carbonate rocks: marbles (locally of dolomite composition), calciphyres, amphibole-diopside and calcite-amphibole-diopside crystalline schists and bands of garnet-biotite, biotite, sillimanite-cordierite gneisses (the hypersthene-bearing and graphite gneisses rarely). Marbles and calciphyres make not less than 30% of the total thickness of the group. The occurrence of carbonate rocks with a high concentration of phosphorus (apatite) and local vanadium is typical. In the lower part of the group horizons of bedded quartz-apatite-diopside schists or apatite- and diopside-bearing quartzites are reported. Rhythmicity of different orders is observed in alternation of carbonate and silicate rocks.

Metabasites (amphibolites and pyroxene amphibolites) are subordinate and are reported in greater abundance only in the middle part of the group (the Bezymyan Formation). In this lies the great difference of the Slyudyanka group (complex) and the Katarchean units belonging to the Fedorovian complex. The supracrustals of the Slyadyanka Group (complex) are metamorphosed under amphibolite facies or amphibole-granulite sub-facies; the rocks with mineral associations of pyroxene-granulite sub-facies are of much smaller distribution.

The mineral deposits associated with the Slyudyankan complex are apatite occurrences of little practical importance, then metasomatic deposits of lazurite and phlogopite, and also marbles of different types. The Biryusa and Derba Groups of the Eastern Sayan in Eastern Siberia and the Priolkhon Group of the Western Baikal area and probably the Vakhan Group of Pamir are analogs of the Slyudyanka Group. It cannot be excluded that the Grenville Group (s. str.) of Canada and the U.S.A. is also to be attributed to the Slyudyankan complex, as it is typified by abundant marbles, quartzites, and gneisses mainly garnet-bearing. Some workers (De Waard and Walton 1967, Wynne-Edwards 1967 etc.) suggest an unconformity between the Grenville Group (s. str.) and the underlying "Pyroxene gneiss". At the same time this group also resembles the Fedorovian complex (in particular, evaporites occur in carbonate rocks in Canada and on Aldan; the Vakhan Group also comparises evaporites, but its belonging to the Slyudyankan complex is not certain).

Up to now there are no available isotope data on the age of the earliest metamorphism of the Slyudyanka Group rocks; the later thermal events are dated at 2800 m.y. However, the marbles from the Biryusa Group (the closest and most reliable equivalent of the Slyudyanka Group) are dated at 3700 m.y. by Pb-Pb isochron method (see Table 3, N 17).

The five complexes described embrace the entire Katarchean in essence, but in the western part of Greenland very-specific supracrustals are developed that belong to the Katarchean by isotope dating though differ from any other formations of the era-them. These rocks, divided as the Isua Group, occur among the Amitsoq extensive tonalite augen gneisses with dark-colored minerals of granulite and amphibolite facies; they enclose xenoliths (or skialithes) and large portions of supracrustal metabasites and gneisses under granulite and amphibolite facies (the so-called Akilia supracrustal association; McGregor and Mason 1977).

Arguments exist as to the relations of the Isua Group rocks and the Amitsoq gneisses. Now many workers (Moorbath et al. 1973a, 1975, 1977, Bridgwater et al. 1979 etc.) think them to be older than the granite-gneisses. This concept is usually supported by data on the granite-tonalite veins resembling the surrounding Amitsoq tonalite gneisses cutting the supracrustals of the group; it is also explained by the presence of xenoliths of the rocks similar to the Isua Group rocks among the latter. Finally, the isotope datings of supracrustal metamorphic rocks and of tonalite veins are very similar and close to the dating of the Amitsoq gneisses (see Table 3, N 37 and 50).

However, a general structural position of the Isua Group makes us doubt the concept of many workers and makes us comply with the opinion of James (1976), who thinks that the Amitsoq gneisses originally served as basement of the supracrustal rocks, but later suffered strong thermal-tectonic mobilization and were deformed, together with the covering piles. The geologic map of the Isua area (Fig. 11) shows that the supracrustals form an arch-like belt that, previous to formation of a fault delimiting it from N−W had probably been an almost closed ring structure (in the north-east the structure is buried under the glacial cap). In the Isua structure a bilateral symmetry is clearly seen, expressed in repetition of the same strata from both sides of its central axis. This symmetry seems to be a result of compressed folding and the structure of the supracrustals proper can reasonably be interpreted as composite folds curved in plan, mostly synclines squeezed between the granite-gneisses. Generally it resembles structure of the mantled gneiss domes made up of remobilized basement and deformed supracrustal cover. In this case and in many other similar cases the cover rocks are preserved only as tectonic outliers – the keeled "roots" of the folds (cf. Figs. 16, 17 etc.). It is remarkable that the Amitsoq gneisses never cross the foldbelt, but it would have to be so if they were younger than the supracrustals. The granite veins cutting the supracrustal rocks at the contact with the gneisses are very likely to represent the rheomorphosed mobilizate of the latter, being the products of the late phase of magmatism of the Saamian tectono-plutonic cycle. Xenoliths in the Amitsoq granite-gneisses partially belong to an older Akilia gneiss-granulite substratum (they are ubiquitous there) and may probably belong to the Isua Group rocks separated in the contact zone as a result of tectonic-magmatic activation of the basement and then emplaced in melted or semi-melted substratum.

The Amitsoq gneisses and the Isua Group rocks are cut by dykes of amphibolites or metadiabases that are altered and dislocated at a greatly differing rate. They are supposed to belong to different generations (Bridgwater et al. 1979); it cannot be ruled out that there are dykes that are older and younger than the Isua Group; at least, many dykes cutting the Amitsoq gneisses do not pierce the supracrustal rocks.

The inner structure of the Isua Group presents many problems and is interpreted differently because of its complex tectonic pattern. According to data of Allaart (1976) the lower part of the group is built up of amphibolites with thin interbeds of garnet-biotite schists and banded magnetite quartzites: up-section occur quartz- and carbonate-bearing biotite-muscovite schist tectonites (?) with local lens-like fragments of acid metavolcanics (?); and still upward the carbonate and carbonate-silicate rocks, garnet-biotite and garnet-staurolite schists and quartzites with interbeds of banded magnetite cummingtonite rocks (jaspilites)[8] are encountered; These strata comprise the thickest horizons of schistose conglomerates with cigar-like pebbles and boulders of quartzites and quartz-feldspathic rocks that are commonly regarded as acid metavolcanics. In early publications (Bridgwater 1974, Bridgwater and McGregor 1974), the coarse-grained granite-gneisses and gneisses were reported to exist among the pebbles, but now the presence of such rocks in conglomerates is doubted or even denied. Others think that rocks very similar to the Amitsoq granite-gneisses surrounding the Isua Group are quite common among the pebbles (B. Windley 1979, pers. comm.). However, it is difficult to judge the primary nature of the rocks in clasts

---

8 From the descriptions, available jaspilites from the Isua Group are notably different from the ferruginous rocks of Katarchean of the Azov type and they are similar to the ferruginous-siliceous formation of the Algoma type characteristic of the Paleoprotozoic

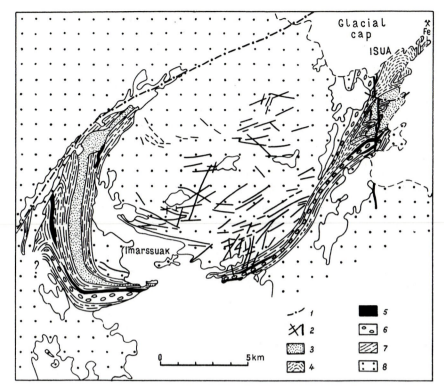

**Fig. 11.** Geologic map of the Isua region (Western Greenland). (After Allaart 1976).
*1* fault, *2* metadiabase dykes, *3* amphibole-chlorite-albite rocks, *4* metasedimentary rocks, *5* talc schist, dunite, *6* carbonate-bearing cherty slate with conglomerate interbeds, *7* amphibolite, *8* the Amitsoq gneiss

because of strong tectonic schistosity and recrystallization. In the uppermost part of the group the strata of bi-micaceous schists with a horizon of crushed and boudinaged acid metavolcanics (?) are likely to occur; it cannot be excluded that these strata present an analog of the second group of strata repeated in the sequence as a result of folding. Total thickness of the group is of the order of 2000 m.

It is clearly seen that the Isua Group differs sharply from the remaining supra-crustals of the Katarchean by its composition (including the presence of conglo-merates) and by its lower grade of alteration (amphibolite and epidote-amphibolite facies up to greenschist facies). It will also be shown that it greatly resembles certain "greenstone strata" of Paleoprotozoic (Archeoprotozoic). At the same time radiomet-ric datings evidence its Katarchean age. The metamorphism age of ferruginous rocks from the Isua group is 3760 m.y. by the Pb-Pb method; a close value was obtained for amphibolite of this group by the Rb-Sr method. Quartz-feldspathic rocks (acid meta-volcanics ?) from conglomerate pebbles are dated in the range 3710–3860 m.y. by Rb-Sr and U-Th-Pb methods. Granitoids from the veins cutting the Isua rocks yielded an age of 3720 m.y. according to Rb-Sr analysis (Moorbath et al. 1977a).

Radiometric ages of the Amitsoq gneisses usually reveal the same age as metamor-phic rocks of the Isua Group and granite veins cutting them: values in the range of

3650–3750 m.y. are usually accepted (different datings by Rb-Sr, U-Th-Pb, and Pb-Pb methods; see Table 3, N 39–46). At first sight the supposition that the Amitsoq gneisses form the basis of the group seems not valid in connection with the values cited; but still there are forcible arguments for considering the gneiss datings rejuvenated under the influence of the Saamian and later thermal events. In addition to the above-mentioned values certain older ages – 3980 ± 175 m.y. (Rb-Sr method) were obtained earlier (Black et al. 1971), but later they were acknowledged to be insignificant due to poor sample choice (Moorbath et al. 1972). Baadsgaard et al. (1976) obtained the Rb-Sr age of 4065 ± 30 m.y. for the Amitsoq granite-gneisses from the typical Godthaab area. The same authors demonstrated that the Amitsoq gneisses are polymetamorphic formations that suffered changes during two or three stages of thermal activity. However, these workers suggest that the age of the Amitsoq gneisses is closer to the true one when determined by the U-Th-Pb method on zircons, and the zircon isochron on the Wetherill diagram crosses the concordia in the point corresponding to 3600 m.y. This conclusion, however, does not seem quite satisfactory, for this isochron does not answer the usual case of a two-stage model: its lower crossing with concordia in the point of 1000 m.y. has no geologic meaning, as the real geologic events fixed by isotope methods had taken place 1550 and 2500 m.y. ago. It is possible that the zircon isochron reflects a mixture of the processes of different periods of time; that is of the period older than 3600 m.y. and of the younger periods.

An age of 3625 m.y. was obtained by Griffin et al. (1980) for the Amitsoq tonalite gneisses (of granulite facies) by the Pb-Pb method, and is interpreted as an age of regional metamorphism, however, the value of the $\mu_1$ parameter $(^{238}U/^{204}Pb) = 9.96$ known from the isochron is very low and differs greatly from a value of $\mu_1 \sim 7.5$ for other acid and intermediate gneisses of Greenland. The authors reasonably think that this apparent low value of $\mu_1$ in granulites is a result of "stopping" in isotope evolution of lead. In the opinion of L. Neymark (1981, pers. comm.) this effect is very likely to be related to the fact that this value probably gives the time of the superimposed metamorphism of amphibolite facies and not the time of granulite metamorphism, having occurred in the rocks depleted in uranium against lead (with the lower value of $\mu_1$) during the earlier granulite metamorphism (this assumption is based on the well-known fact of depletion of uranium during metamorphic transformations under granulite facies).

Lovering (1979) obtained an age of about 4000 m.y. for the Amitsoq gneisses by ion-microprobe determinations of $^{207}Pb/^{206}Pb$ in single grains of zircon.

Thus, the age of the Isua Group and its complicated relations with the Amitsoq gneisses can easily be explained if it is assumed that it was formed between two periods of tectono-plutonic activity.

It can reasonably be supposed that the tectonic-plutonic processes that preceded deposition of the Isua Group were of a separate orogeny that can be called Godthaabian, and this cycle is likely to be related to the formation of the Amitsoq tonalite gneisses and to the early metamorphism of granulite facies. It is possible that the Isua Group may be regarded as a peculiar late formation of the Katarchean. Its stratigraphic position accords well with the "transit formations" that are here given this name, this kind of strata constitutes the transitional beds that in formational aspect connect the Precambrian erathems close in age. It is to be noted that in the same area of south-western Greenland the Malene sedimentary-volcanogenic strata are locally developed, but these are younger and belong to the Paleoprotozoic (they are cut only by the Nuk granite-gneisses dated at 2800 m.y.).

Here a comparatively detailed description of the Isua Group was given which, though brief is of great scientific significance. It is possible that further studies will result in finding this type of old orogenic and transit formations in other parts of the world, though the chance of success is very small, as these rocks occupied the highest stratigraphic position and were the first to be affected by erosion.

It is possible that the Nikitkino and Ilikan Groups of gneiss (schist)-amphibolite in the Stanovoy belt of Eastern Siberia belong to this type of formation; according to data of Shuldiner and Ozersky (1967 etc.) and Mironyuk (1981, pers. comm.) these groups are separated by a structural unconformity from the underlying rocks of the Mogocha and Kurulta Groups altered under granulite facies. Meanwhile, the rocks of the Nikitkino and Ilikan Groups were metamorphosed under amphibolite facies approximately 3700 m.y. ago (Pb-Pb and U-Pb methods; Neymark 1981); at that time the rocks of the gneiss-granulite basement suffered recurrent and regressive metamorphism. The Mogocha and Kurulta Groups are very close analogs of the Aldan Group and, thus the latter, together with the former ones, is much older that 3700 m.y.

It seems reasonable to recognize a single sixth lithostratigraphic complex of the Katarchean – the *Isuan complex* that would embrace the Isua Group and its equivalents. It seems possible that the Godthaabian orogeny embraced many regions and not only Western Greenland; however, in most cases its traces are obliterated or masked by strong structural and metamorphic reworking of the Katarchean rocks under the influence of later tectonic-plutonic processes (of the Saamian cycle in particular) and these are responsible for the strong isotope rejuvenation of the rocks. Nevertheless, certain (though few) relict datings of the Katarchean gneisses are known from different regions of the world (the Yenisei Ridge, the Okhotsk massif, Guiana, the Limpopo belt of South Africa, Queen Maude Land in Antarctic; see Table 3), these values fall closely to the time of manifestation of this orogeny, that is about 4000 m.y.

# B. Geologic Interpretation of Rock Record

High-grade metamorphism destroyed the textural aspects of the Katarchean supracrustals almost everywhere, in elucidating their genesis we can mainly rely on their modern mineralogical and chemical composition and on some very conservative structural peculiarities that are well-preserved, that is bedding or rhythmicity of stratified rocks. It was noted above, however, that there is a large group of quartz-feldspathic rocks whose origin is uncertain. Some rocks, for instance psephytes, are lacking in Katarchean (the uppermost Isuan complex is an exception). On the contrary, there are rocks or rock associations, as an association of quartzites and basites or ultrabasites typical of Katarchean that occur among formations resembling the eugeosynclinal type at first sight, but in the younger Precambrian strata-since Mesoprotozoic this association known to occur only among the platform or subplatform type, partially among miogeosynclinal formations and among the Phanerozoic deposits it is absolutely lacking and more than, that it becomes a "forbidden" type. The tectonic pattern of the Katarchean is also specific: the fold systems of the "gneiss oval" type have no equivalents in some younger formations, not only Phanerozoic but also Protozoic. Many examples are possible of the peculiarities of Katarchean, demonstrating the limitations in apllying the actualistic method in the reconstruction of geologic environment during the oldest stage of geologic history.

Some conclusions at the end of this section will therefore be based not only on rock record data, but also on logical deductions resulting from physical, chemical, and cosmochemical studies and on analogies with other planets of the Earth's group.

## 1. Age of the Beginning of the Katarchean Era

A hypothesis of the Earth formation from the protoplanetary nebula under the influence of accumulation of particles in the gravitational field of the Sun seems to be quite reasonable. It adequately explains all the principal peculiarities in structure and evolution of the Sun system, considering the effects due to the existence of the magnetic field and of corpuscular emission of the proto-Sun. According to meteorite datings and to determinations of lead isotope composition, the age of the Earth is 4.5–5 b.y., the first value seeming to be the closest to the true age; at least it cannot be older than the minimal age of its chemical elements, that is 5.5 b.y.

It seems very important that with any accepted variant of the hypothesis (the original "cold" or "hot" formation of the Earth), its intensive heating or even melting is absolutely unavoidable due to huge energy released during chemical and gravitational differentiation of the planet (Birch 1965). Thus, the primary differentiation of the Earth's interior into shells of different composition, including formation of lithosphere with the granite crust, must have occurred over several hundreds or the first tens of millions of years.

Isotope methods are not suitable to determine the age of the beginning of the Katarchean supracrustal formation. All the available radiometric datings of the oldest rocks only give the age of the Saamian or Godthaabian thermal events which could have been separated from the unknown beginning by a significant period of time. It will be shown later that the duration of large geologic cycles corresponding to eras increases with their age: the older the eras, the longer they are (see Table 1). This problem will be examined later, here it will be pointed out that the duration of the Mesoprotozoic Era is 800 m.y. and of the Paleoprotozoic is 900 m.y. Considering this regularity, it is possible to assume that the Katarchean Era lasted not less than 900 m.y. and that its beginning thus falls quite close to the time of formation of the Earth. Therefore, the Katarchean Era is really the first and oldest geologic era in the history of the planet.

## 2. Physical Conditions on the Earth's Surface

Naturally, during this milliard of years that the Katarchean Era lasted, these conditions could not be kept invariable. Below it will be shown that evolution of the environment is evidenced by the rock record. It seems likely that at the beginning of the Katarchean the differentiation of the Earth's interior was principally completed and was the main source of endogenic energy at the very start of the planet's existence; however, due to low heat conductivity of the lithosphere rocks it could have happened that the planet did not become quite cool at that time. Besides, during the Katarchean the input of radioactive heat into the total energy budget of the Earth was much more significant than later because some radioactive isotopes with a relatively short decay period ($^{40}$K and $^{235}$U mainly) were quite common (Fig. 12). A relatively short period of time, however, elapsed and endogenic heat sources no longer greatly affected the temperature regime of the planet surface because of low heat conductivity of rocks; and solar emission must have replaced that source.

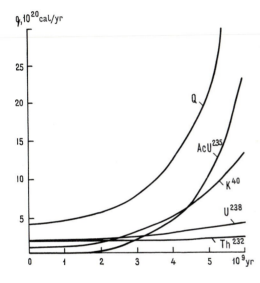

**Fig. 12.** Heat emission by major radioactive elements during geologic time. (After Lyubimova 1960). $Q$ curve of total heat emission by all major radioactive elements

It will be discussed below that it is easy to explain many peculiarities of the Katarchean metasedimentary rocks if we admit that they had been formed under the influence of a hot hydrosphere. Knaut and Epstein (1976), by studying isotopes of unaltered and slightly altered cherty rocks of different age, proved that deuterium-hydrogen ($\delta$D) and $^{18}$O/$^{16}$O ($\delta^{18}$O) ratios of silicium hydroxyl that depend on temperature, change with rock age. According to their data some 3.2 b.y. ago the average annual temperature of the Earth surface (the time of formation of the cherty rocks of the Paleoprotozoic Fig-Tree Group) was about 70 °C, then in the Late Precambrian (1200 m.y. ago) it lowered down to 52 °C and at the end of the Palezoic down to 20 °C, after a slight increase at the beginning of the Triassic up to 35 °C, it gradually decreased again down to 15 °C, which we witness now. Proceeding from the regularity established by the authors it is possible to assume that at any rate the temperature of the Earth's surface in the Katarchean was not less than 70 °C. The paleotemperature of formation of siliceous rocks of the Late Katarchean Isua Group was determined by Oskwarek and Perry (1976) according to the oxygen isotope ratio. These analyses showed that the sea temperature at that time was probably in the range of 90° – 150 °C, but the result obtained cannot be regarded as reliable, due to secondary transformations of rocks.

There are no data available on the temperature of the Earth's surface in the Early Katarchean and it seems hardly possible to obtain these data by isotope methods because of high-grade metamorphism of rocks. As is known, the oldest supracrustals were deposited in seawater (see below), the maximal temperature limit is determined to be 370 °C – the critical temperature of water boiling. But this temperature is likely to be realisitic for the "pre-geologic" (pre-Katarchean) time; undoubtedly it was lower during the Katarchean, but still it had been higher than 100 °C. This high temperature on the Earth's surface can be explained only by the hothouse effect created by the thick atmosphere, assuming the Sun's luminescence (Sagan and Mullen 1972) to have been lower in the distant past than it is now.

Usually the analogy with modern Venus is made in reconstructing physical conditions on the Earth during the Early Precambrian, the temperature of Venus' surface reaching 480 °C and the carbon dioxide atmosphere pressure about 90 bar. It is to be kept in mind, however, that water is absent on Venus' surface (probably it is present in very low abundance in its cloud layer, but it is not known for certain at present). Still, this analogy is true in the respect that this temperature is mainly due to the hothouse effect. If the extreme variations of temperature on the Earth are assumed: from 370 °C to 150 °C from the beginning of the Katarchean to its end, then the minimal atmospheric pressure must have changed from 220 to 5 bar. It seems, however, very likely that at the beginning of the Katarchean the temperature did not exceed 200° – 250 °C, and minimal pressure was 15 – 40 bar. Certainly this is a very rough approximation.

### 3. Composition of Atmosphere and Hydrosphere

It is widely accepted that atmosphere and hydrosphere are mainly products of magma differentiation: its degassing and separation of liquid components (Rubey 1951, Vinogradov 1967). At the very beginning of the Katarchean, the primary atmosphere of the Earth no longer existed: during a general heating of the planet the volatile components which had been quite abundant were dissipated in space or appeared to be strongly fixed in rocks or the hydrosphere; only the inert gases, crypton, xenon, and some argon isotopes, seem to be preserved in the modern atmosphere. Hydrogen is being lost from atmosphere now, helium is at the limit of its capacity. It cannot be excluded that the oldest – Katarchean – atmosphere was supplied with hydrogen from the interior of the planet, thus compensating its constant loss.

The secondary primitive (and primary in the geologic sense) atmosphere could have originated only after the temperature of the Earth's surface decreased so that gases could not overcome the attraction of gravity of the planet. Different conceptions are offered concerning the origin of atmosphere. According to one the major mass of atmosphere was extracted at the initial stages of the Earth's existence and then some insignificant addition of gases took place. According to another, the atmosphere was formed more or less evenly during the whole geologic history. Depending on which of the conceptions is accepted, the evaluations of changes in the atmosphere composition with time and of its parameters give results differing to some degree (see, e.g., Neruchev 1977). Instead of these extreme opinions it seems more reasonable to assume that the major portion of atmosphere and hydrosphere (though it is difficult to estimate accurately) appeared soon after the principal stage in the Earth's differentiation and then certain additions or losses of some components from its composition were always taking place, but irregularly depending on dominant processes of volcanism, sedimentation, and biologic activity (mostly photosynthesis). In particular, Ronov (1976) gives data on the interrelations of intensity of volcanism and carbon dioxide concentration in atmosphere and hydrosphere.

It is admitted that the primitive atmosphere composition was close to the composition of gas products of volcanic eruptions. It is known that intensive volcanic activity was typical of the Katarchean Era, of its beginning and middle stages in particular. It is natural to assume that the primitive atmosphere largely consisted of water vapor and carbon dioxide, then of nitrogen, acid smokes (HCl, HF, $H_2S$), ammonia, and possibly of methane. The following facts show the dominance of water and carbon dioxide in the atmosphere. The water content in the Earth's mantle is three times that of the water mass in modern oceans. Chemically fixed water also occurs in abundance in rocks of the Earth's crust. Estimates reveal that much more water that is now observed in the oceans had to be extracted from the upper mantle and crust during formation of basalt and andesite lavas. Carbon dioxide was deposited in carbonates during the whole Earth's history at a concentration 10,000 times greater than in the atmosphere now; carbon dioxide assimilated by plants and buried in sediments was 1000 times more than in the modern atmosphere. If it is acknowledged that the atmosphere was principally formed during the initial stage of the Earth's evolution, then the major portion of lithified carbon dioxide was present.

According to estimates of Schidlowski (1972), based on composition of modern volcanic exhalations, the older atmosphere contained approximately 80% water vapor, 10% carbon dioxide 5%–7% sulphur compounds (mainly $H_2S$); 0.5%–1% CO, $H_2$, and $N_2$, $CH_4$, $NH_3$, HF, and HCl of insignificant quantity. According to estimates of Neruchev (1977) the primitive atmosphere consisted of $CO_2$ at 98.8% and of $N_2$ at 1.1% (water vapor is not considered); this is based on the assumption that the atmosphere was almost fully extracted soon after the Earth's origin. The pressure of that atmosphere had to be about 70 bar, and when considering the solution of abundant carbon dioxide in the hydrosphere it had to be in the order of 50–60 bar. The temperature of water boiling at that pressure had to be 260°–285 °C.

Free oxygen was lacking in the primitive atmosphere. Biogenic photosynthesis is known to be the principal source of $O_2$ in the atmosphere. During the Katarchean the photosynthesizing organisms did not exist at all or were very few. At the same time, free oxygen was not registered in volcanic emanations and if it is recorded in magmas, it was formed as a result of thermic dissociation of water and carbon dioxide and then became fully fixed in the oxidation of bivalent ferrum. In the uppermost layers of the atmosphere an insignificant proportion of oxygen could have been produced under cosmic photosynthesis, but simultaneously it had to be consumed during oxidation of ammonia, sulfur, methane, etc. The absence of free oxygen in the atmosphere of the Early Precambrian (Katarchean, Paleoprotozoic, and the first half of Mesoprotozoic up to a 2300–2400 m.y. limit) is also evidenced by data on isotope composition of sulfur in low-grade sedimentary rocks of corresponding age and data on environments of formation of certain specific sedimentary formations. These factors will be discussed in the following chapters. Data on sulfur isotopes in the Katarchean formations give discrepant results, for the rocks studied were strongly altered and usually subjected to metasomatic transformations (Grinenko and Grinenko 1974, Kazansky 1978, Ohmoto 1972, Vinogradov et al. 1976 b).

The concepts of the composition of the oldest atmosphere are based on indirect data and logical conclusions; they are cogently argued, but still they need corroboration. Travin and I made an attempt to determine the Katarchean atmosphere composition by analyzing gas inclusions (bubbles) in quartzites of the Iyengra Group of the Aldan shield proceeding from an assumption that originally quartzites had been chemogenic sediments (see below) and the capture of gas from solution had occurred during precipitation of silica. It was also suggested that during later metamorphism the bubbles could be redistributed in a certain volume of rock, but that generally the system was closed. The analysis of inclusions of several samples was done by a group of workers headed by Kazansky, who proved that the gas component in the bubbles is composed of $CO_2$ – about 60%, $H_2S$, $SO_2$, $NH_3$, HCl and HF up to 35% and $N_2$ and rare gases from 1% to 8%. Oxygen, hydrocarbons, CO and $H_2$ were not detected (details of analyses are included in the work of Salop (1973a)). Thus, if the above suggestions on the genesis of gas inclusions in the Iyengra quartzites are correct, then the analytical data can be ragarded as evidence of the suggested composition of the primitive atmosphere. It should be noted here that analysis of gas bubbles from the Katarchean quartzites of the gneiss-granulite complex from Southern Norway also revealed that they are mainly composed of carbon dioxide with traces of methane and hydrocarbons sometimes (Al-Khatib and Touret 1973, Touret 1971 etc.).

Kazansky and his co-authors (1973) showed that regular increase of oxygen concentration is observed in gas inclusions of the younger chemogenic siliceous rocks (cherts) from 5.5% in Paleoprotozoic rocks up to 12% in Upper Precambrian-Lower Paleozoic rocks and up to 18% in Middle Paleozoic-Cenozoic rocks. A decrease of concentration of carbon dioxide is simultaneously noted: from 42% in the Paleoprotozoic rocks down to the level close to the modern concentration in the Middle Paleozoic-Cenozoic rocks. This regularity certainly suggests that gas inclusion in cherts to some extent truly reflects the atmosphere composition and its change with time.

Thus, the Katarchean atmosphere was extremely dense, lacking oxygen, hot, and mainly composed of water vapor, carbon dioxide, and the admixture of other components with very typical acid smokes. Such an atmosphere must have created a strong hothouse effect on the Earth's surface, for it is known that it was caused by molecules of $H_2O$ and $CO_2$; water vapor can shield 60% of the Earth's emission and carbon dioxide 18%. Hot rains are supposed to have been common at that time, they represented solutions of carbonic acids and strong acids. The atmosphere was likely to have been poorly transparent for the Sun's rays, but the earlier concept of its being quite opaque seems to be wrong in the light of new data on the surface of Venus being illuminated. In this case, however, we can judge the composition of the Katarchean ocean only by logical thinking, for the results of analysis of liquid inclusions from the older quartzites of the Aldan and Baikal areas and Norway are unstable and contradictory; probably it can be explained by the presence of metamorphic solutions in the bubbles.

We can judge the gas composition of the ocean more adequately, as it is known that a constant exchange exists between atmosphere and hydrosphere and that the ocean is the principal agent regulating $CO_2$ concentration in the atmosphere; now $CO_2$ concentration in the ocean is approximately 100 times greater than in the atmosphere ($3 \, cm^3/l$). The temperature of the planet's surface, atmospheric pressure, and $CO_2$ concentration in it during the Katarchean were much higher than now and so the quantity of $CO_2$ dissolved in the ocean was different. Temperature and pressure affect the solubility of $CO_2$ in opposite directions, but experiments show that the pressure is of greater importance within the limits of T and P values we are discussing. Thus, at $T = 120\,°C$ and $P_{CO_2} = 20$ bar, $CO_2$ concentration in water will be $5000 \, cm^3/l$ (but at the same temperature and $P = 70$ bar it will be $13,000 \, cm^3/l$). Thus, the hydrosphere of Katarchean seems to have been characterized by a rather high $CO_2$ concentration. Then, the original ocean is supposed to have contained many strong acids ($HCl$, $HF$, $H_2SO_4$). Thus the older hydrosphere was an aggressive agent affecting rocks and sediments and must have extracted from them the equivalent quantity of alkali and silica.

This original ocean must have been notably mineralized and salty from the very beginning; this is why natural evaporites are encountered in the oldest rocks of the Earth.

The problem of salt concentration in the Katarchean ocean is not clear at the moment. In opinion of Vinogradov "it did not differ much from the modern one, as the juvenile water was accompanied by volatiles at approximately the same rate as it is now" (1967, p. 30). This reasoning could have been correct if we admit that water mass has been gradually increasing during the whole evolution of the planet. However, the major proportion of water mass seems to have been extracted, together with the atmosphere, at the very start of the Earth's existence, and the quantitative ratio of other volatiles could then have been different from that now observed in volcanic exhalations [9]. Then, the salt concentration had to be affected also by conspiciously different physical conditions at that time.

The water in the original ocean is likely to have been acid, but later, after interaction with abundant alkali and alkaline earths extracted from rocks, it soon approximated the neutral state with pH $\sim$ 7 (Vinogradov 1967). It is then, suggested that at the beginning of Katarchean the oceanic water had been evidently chloridic and lacking any sulfates, for there had been no free oxygen necessary for oxidation of sulfur.

The presence of low abundance of sedimentary sulfates (anhydrite) in carbonate rocks of the Fedorovian and Slyudyankan complexes probably points to the origin of local low oxidation environment ("oxygen oases") in the Late Katarchean; it was due possibly to biogenic photosynthesis. At the same time the sulfur from sulfates of the Fedorov Mine Formation of Aldan is characterized by a great abundance of light isotope: $\delta^{34}S = 0 - + 8\%_0$ (Vinogradov et al. 1976a) that evidences the absence of

---

9 Chase and Perry (1972) analyzed different models explaining the increase of $^{18}O$ concentration in sea carbonates within the limits of $15\%_0$ during the last 3 b.y.; they conclude that the notable increase of water volume in the ocean had stopped since the Early Precambrian. Nevertheless, it is possible to suggest that the original ocean was not deep, for much water had been in the atmosphere and the ocean covered the whole territory of the Earth or almost all of it (see below)

intensive processes of oxidation and the low effect of the supposed photosynthesis in an atmosphere lacking oxygen.

## 4. Sedimentation Environment

A striking consistency in composition and sequence of the extensive Katarchean supracrustals shows a uniform environment for their formation. The lack of any evident facial zonation (at least of linear type) and any traces of existence of older land (areas of erosion) is suggestive of the fact that sedimentation and lava outflow occurred in a huge ocean, Panthalassa, that at times covered the major part of the Earth's surface or the whole planet. It was noted that this original ocean was not deep and its bottom, even emerged, was not a land for a long time, but banks and low islands; in any case the contrasting forms of relief did not exist. For this reason the psephytic rocks are absent in the Katarchean strata (the uppermost Isuan complex is an exception).

It appeared possible to establish several lithostratigraphic complexes vertically substitutable for each other due to similarity in rock sequence in many regions of the world, and this latter can reasonably be explained by the peculiarities of the older atmosphere and hydrosphere discussed above. The problem of quartzite genesis is of special significance for elucidating the environment of Katarchean sedimentation; these quartzites are abundant in the lowermost, Iyengran complex, but they are also known from the overlying units, though less abundantly. It is known that quartzites, Phanerozoic and Late Precambrian quartz sandstones formed from the latter are the extreme and the most mature products of differentiation of sediments. In these strata they are typical of platform and miogeosynclinal formations. What is the genesis of the Katarchean quartzites?

The oldest quartzites are always recrystallized; there are no clastic textures preserved. However, the commonly observed conspicuous bedding and presence of sillimanite-bearing varieties and alternation with high-aluminous rocks: sillimanite or corund-bearing gneisses (crystalline schists) definitely evidence their sedimentary origin. Taking a common association of Katarchean quartzites and metavolcanics into consideration, it is possible to assume that silica is a product of volcanic activity and the result of precipitation as gel. Thus the quartzite-amphibolite association of Katarchean could have been regarded as an analog of siliceous-volcanogenic formations of certain younger piles.

However, this concept appears to be in contradiction with an exceptionally great thickness (locally more than 1000 m) of quartzites, the uniformity of strata, association with high-aluminous rocks and presence of rare small (0.1–0.01 mm) rounded zircon grains that are very likely to be of clastic origin. Thus, the only assumption left is that the principal source of material for quartzites was certain rocks still older, rich in silica, that is some granitoids or gneisses. These could have been the rocks of original sialic crust formed during a general planet differentiation. Such rocks are not (or not yet) exposed on the Earth's surface (it is quite a problem to find them because of many methodical difficulties). The oldest rocks on the Aldan shield where the lower part of Katarchean is exposed are the quartzites of the Kurumkan Formation and thus it is suggested that the "initial" granitoids were not eroded there.

It could have been suggested that silica for quartzites had appeared as a result of chemical weathering of basic rocks, but the balance of the material that could have originated this way contradicts the real distribution of the corresponding types of metasedimentary rocks in the Iyengra complex and in many other Katarchean complexes. It can easily be estimated that reworking an exceptionally great volume of rocks must have occurred for such thick strata of quartzites to be formed (as for instance in the lowermost part of the Aldan Group) in the case of chemical decomposition of basites, and then huge thickness of iron ores, magnesites and many other carbonates must have appeared as complementary products, but these are not known in the oldest complexes. This idea also fails if the presence of clastic zircon grains in quartzites and certain chemical features of quartzites and of aluminous rocks established by Travin are considered (see Konikov et al. 1975). Thus we face the necessity of acknowledging the existence of pre-Katarchean granitoids and consequently, the "initial" Earth's crust with sialic material.

If the Katarchean quartzites represented the clastic rocks previous to metamorphism (quartz sands), then why do we find no facial changes in quartzites and why are there no quartz conglomerates among the quartzites? In spite of the fact that many regions have been studied in detail, up to now there are no data which permit us to establish the areas of supply of clastic material.

This contradiction and many more are avoided if the following assumption is accepted: the Katarchean quartz sediments had principally been formed as a result of precipitation from silica solutions extracted from the crust of weathering on the sialic rocks. It can be stated that the available data and the logical conclusions based on them combined support a concept of the oldest (initial) hydrosphere as having been very hot and composed of diluted solutions of carbonic acid and strong acids. It is known that silica is easily dissolved in warm water if not even hot and mineralized water. Khitarov and other workers established that quartz could have been completely dissolved from granites in such water and the residium would consist of feldspar and dark-colored minerals. The experiments done by Balitsky and his colleagues (1971) show that quartz solubility greatly increases with increase of temperature, in the case of the solutions containing sulfureous sodium in particular. It is reasonably assumed that this latter had to be quite abundant in a primary ocean, for the oldest atmosphere was lacking in any free oxygen, and compounds of the $Na_2S$ type had to originate instead of sulfureous sodium. It is evidenced in particular by a common occurrence of lazurite deposits (the mineral contains $N_2S$) in Katarchean marbles. Previous to metamorphism such marbles are likely to have represented sulfureous limestones or dolomites (even now they locally contain sulfur).

It is therefore assumed that silica (quartz) from crust of weathering partially or fully dissolved in hot and mineralized water of the old ocean and became a chemical precipitate in the process of redeposition. This probably explains the fact that there is no connection between exposures of quartzite strata and the areas of erosion. In some cases the quartz rocks could be of mixed chemogenic-clastic origin, as some small fragmental particles of quartz and of indissoluble minerals of heavy fraction (of zircon and monazite?) could have been transported for a long distance as suspensions in the colloidal and true solutions. Then this concept of the origin of the Archean quartzites from redeposited silica of the oldest subaerial and subaqueous crusts of chemical weathering is also favored by the association of these rocks with high-aluminous rocks and by common admixture of sillimanite (and rare magnetite) in quartzites.

The specific features of temperature and composition of the Katarchean initial ocean must have resulted in its aggressive activity not only in sialic but also in any rock, the subaqueous basic and acid volcanics included. If the rocks were rich in silica, the solution of quartz must have led to the destruction of their skeleton and favored the current process of halmyrolysis. The notable abundance of high-aluminous sedimentary rocks in the Katarchean strata is probably related to a high reactional activity of the oldest oceanic water. From these facts it is also concluded that the oldest ocean had a high concentration of dissolved silicic acid, as was stated by many workers though from slightly different view-point (see, e.g., Holland 1972).

It can be assumed that many banded garnet-biotite and biotite gneisses commonly interpreted as metamorphosed pelitic sediments were formed as a result of deposition of chemical and mechanical products of subaqueous weathering of lava and tuff, and not as a result of supply of terrigene material in the oceanic basin from land.

The major part of the Katarchean is known to consist of basic (partly ultrabasic) metavolcanics that dominate in the extensive Ungran complex situated above the Iyengran complex. But up-section, in the Fedorovian complex, their proportion becomes obviously less; various metasedimentary rocks are widespread there, first carbonates appear and then these are reported from every overlying lithostratigraphic complex. The appearance of carbonate rocks is an event of exceptional stratigraphic significance. It seems that this world-wide regularity is to be explained by some common cause. Precipitation of carbonates is assumed to be related to decrease of concentration of carbon dioxide and strong acids (acid smokes) in atmosphere and hydrosphere due to lessening of volcanic activity and to lower consumption of $CO_2$ in different mineral-forming and cosmic photochemical processes. An exceptionally high concentration of carbon dioxide in the still earlier Katarchean hydrosphere must have been retained to decompose carbonates and to extract $CO_2$ back into the atmosphere. The critical level was reached and precipitation of carbonates became possible, then the concentration of carbon dioxide in the Katarchean atmosphere must have decreased sharply as a result of being fixed in the carbonate strata. Certainly, the carbon dioxide reserves were being constantly replenished during the volcanic eruptions, but its concentration decreased notably after the deposition of the fourth Sutamian complex, essentially of sedimentary nature, and a formation of the fifth Slyudyankan complex became possible, where carbonates play an important part. Thus, when studying the evolution of the older atmosphere composition we find an explanation of a striking and at first sight hardly understandable, ubiquitous uniform structure of Katarchean supracrustal formations.

Finally, the genesis of ferruginous, carbon-bearing (graphite) and phosphate-bearing stratified rocks of the Katarchean must also be discussed in brief. The stratified ferruginous rocks of the Azov type, constantly occurring among metabasalts, are very likely to be related to volcanic activity, but the sedimentary chemogenic processes also played an important part in their formation: iron seems to have been supplied from the volcanic outflows and silica and dissolved silicates were in excess in seawater.

The genesis of the Katarchean graphite-bearing gneisses and marbles is not certain. There are two alternative opinions concerning it, according to the first of which graphite is organic, to the second inorganic in its origin. A detailed discussion of the problem is beyond the scope of this work. Still, it seems that the Katarchean biomass was not extensive enough to provide formation of abnormal abundance of graphite-bearing rocks. It is known that such rocks build up extremely thick horizons and strata, in the Fedorovian and partly in the Slyudyankan complexes in particular. In some regions even single formations are recognized built up entirely of such rocks, and on Madagascar the "Graphite system" is divided as the upper part of the Katarchean sequence. In certain parts of the world rather large (on a world-wide scale) deposits of graphite occur. It was noted that the organic remains in the Katarchean are not known, they are reported as quite rare even in the younger Paleoprotozoic Era where the graphite-bearing rocks are of much more limited distribution than in Katarchean.

Experiments in the field of inorganic synthesis of hydrocarbons revealed that certain organic matter can easily originate under the influence of ultraviolet Sun rays on water containing ammonia and hydrogen gases, affecting also the composition of organic (biologic) matter (experiments of Miller and Ponnamperuma); the simple hydrocarbons thus formed can be synthetized into large "organic" molecules simply in hot water that does not contain any oxygen (experiments of Oro, see Rutten 1971). Environments equivalent to those created in the laboratory were certainly typical of the Katarchean Era. According to experiments done by Salotti et al. (1971), the pyrolytic dissociation of methane with formation of free oxygen is possible at $200°-600°C$. Methane is formed in the reaction of hydrogen and carbonates. It should be kept in mind that methane could have been an important initial component of the Katarchean atmosphere. Pyrolysis of methane is likely to have occurred during metamorphism of sedimentary strata. The workers mentioned above suggest that this is the way the nonbiogenic carbon of graphite deposits originated.

The organogenic nature of graphite in the Katarchean gneisses is sometimes explained by using on the $^{13}C/^{12}C$ isotope ratio. However, there are discrepancies and the value of this ratio varies greatly for graphites from different localities. It is also known that this ratio varies in the carbon minerals obtained in experimental synthesis, which seems to be explained by the exchange reaction between isotopes and thus it appears impossible to separate the natural carbon compounds into biogenic and nonbiogenic according to the $^{13}C/^{12}C$ ratio (Hahn-Weinheimer 1971).

The problem of genesis of the Katarchean phosphate- and apatite-bearing rocks is no better known than graphite rock origin. In the younger strata (Protozoic and Phanerozoic) the formation and accumulation of phosphates is commonly due to life activity of organisms. It cannot be excluded that certain of the oldest initially sedimentary phosphate-bearing rocks, for instance, fine-bedded apatite-diopside-quartz schists or marbles rich in apatite also underwent the biogenic cycle. In this case the existence of relatively highly developed life is to be assumed in the Katarchean (in the Late Katarchean, to be more exact). But these rare and generally insignificant occurrences of phosphates did not support as much biomass as was the case with graphite rocks.

## 5. Life in the Katarchean

The absence of any determinable organic remains (biogenic structures) by no means signifies that no life (even primitive) existed then. It would have been striking if certain visual evidence of that life could have been preserved in highly altered older rocks. However, the presence of microscopic remains of prokaryota (algae and bacteria) in the lowermost parts of the Paleoprotozoic (Archeoprotozoic) strata and of products of their life activity (stromatolites) up-section is suggestive of life having occurred much earlier – during the Katarchean – as the evolutionary span from the pre-biologic "organic" molecules to the eobionts could not have been a short one. At the same time, the physical and chemical environments of the Katarchean seem to have been favorable for an origin of the pre-biologic compounds according to modern concepts of the chemical evolution of organic matter: a hot atmosphere and a hydrosphere with abundant $CO_2$, methane and ammonia, reducing medium, effect of UV emission (due to lack of thick ozone shield), probable occurrence of intensive electrical discharges (thunderstorms) in the atmosphere, intense volcanism [according to data of Markhinin and Podkletnov (1978) synthesis of "organic" molecules occurs in the volcanic gas outbursts].

In the opinion of some workers the eobionts seem to have been the anaerobic heterotrophic uni-cellular organisms inhabiting seawater that protected them from the pernicious UV emission of the Sun. It is possible that in the second half of the Katarchean the photosynthesizing organisms (blue-green algae) could have originated [10]. The high temperature of the old ocean could not prevent their evolution, for the adaptability of these plants to extreme conditions is striking. Many modern blue-green algae successfully reproduce in water of hot (more than 85 °C) springs. They are also adapted to inhabit water of different salinity and composition, they are registered in sulfate, chloric, and sodic lakes and in waters of different gas regime, for instance in water almost completely saturated with carbon dioxide. Some forms are anaerobic, others develop under abundant $O_2$ (aerophyles). The blue-green algae also live in the sulfurous and sulfuric water of volcanoes. Still, the Katarchean environment does not seem to have been favorable for the existence and evolution of such relatively highly organized life, and it had to suppress its expansion and to prevent the growth of biomass.

## 6. Genesis of Magmatic and Metamorphic Rocks

The reconstruction of the environment of the Katarchean magmatic and metamorphic rock formation is as complicated as for the oldest sedimentary rocks, and different data are needed to elucidate it, both geologic and petrologic and experimental (physico-chemical) data. Here it is impossible to examine all the information available and only some geologic aspects of the problem will be discussed in brief.

It is known that the major portion of the Katarchean supracrustals is made up of high-grade metamorphic volcanics of basic and partly ultrabasic composition. The altered acid lavas also possibly occur, though it is not certain, in particular in the upper lithostratigraphic complexes, from the Sutamian to the Isuan. Pyroclastic products are commonly known to accompany acid lavas. In the Katarchean there are

---

10 The microscopic spherical and filamentous structures found recently in cherts of the Isua Group are probably remains of blue-green algae (Pflug and Jaeschke-Boyer 1979), but their organogenic nature is doubtful (see Bridgwater et al. 1981)

rocks similar to tuffs of acid lavas in composition and mode of occurrence, but the psephytic-agglomerate welded tuffs are not known; their structural aspects, however, could help in determining them (similar formations are reported only from the Isua Group). The agglomerate rocks mainly originate under terrestrial explosions and in the Katarchean the lava outflows seem to have been underwater. Then, the extremely high atmospheric pressure in the Katarchean could have prevented explosive eruptions.

There exists a widely accepted and well-grounded idea on the isochemical character of metamorphism of the Katarchean metavolcanics or of their varieties, to be more exact, that were not subjected to metasomatic transformations; if we proceed from this idea we can obtain valuable information not only on their initial composition, but also on their evolution, producing chemical analyses. The most detailed study of chemism of basic crystalline schists and amphibolites belonging to different stratigraphic levels of the Katarchean was done for the stratotype of the erathem, the Aldan Group of Eastern Siberia.

With this aim, more than 400 complete silicate analyses and several hundreds of approximate-quantitative spectral analyses were done. According to data of Velikoslavinsky (1978), metabasites of the Iyengra Subgroup and of the lower part of the Timpton Group (the Ungran complex) are almost identical with the tholeiite basalts, those of the Fedorov Mines Formation with the alkaline basalts, and of the overlying units of the Aldan Group with the volcanics of tholeiite and alkaline-basalt series with the analogs of basanites and nephelinites present among the rocks of the latter. In the opinion of Travin (see Konikov et al. 1974) metabasites of the Aldan Group and of its equivalents in other parts of Eastern Siberia (the Anabar, Kurulta, and Zverevo Groups) are very close to the traps of the older platform in petrochemistry, but in some other aspects (higher concentration of magnesium and lower content of ferrum, absence of positive connection in concentration of potassium and silicium) they are similar to the oceanic tholeiite basalts. Metabasites of the younger Katarchean Slyudyankan complex are conspicuously different from the Aldan metabasites. According to Konikov et al. (1974) they are close to basalts of the andesite-basalt formation of the island arches (after Kutolin) and partly to the basalts of geosynclinal formation of andesite-basalt porphyrites and they resemble basalt of the trappean formation only in some parameters.

Further studies will show if the evolution of volcanism in the Aldan and Slyudyanka Groups can serve as a model for Katarchean complexes of other parts of the world.

It is to be noted that highly altered ultrabasic volcanics corresponding in composition to komatiites occur in certain Katarchean complexes. These rocks, wide-spread in the younger Paleoprotozoic Erathem are supposed to be the products of the "primitive" basalt magma that is very likely to have generated at a shallow depth.

The plutonic processes of the Katarchean were extremely intensive and went on for an exceptionally long period of time, judging from a great scattering of isotope age values of corresponding rocks. They are also evidenced by a multifold intersection of different granitoids that reveal a complicated pulsationary character of Katarchean plutonism. In some cases the dykes of deformed and granitized amphibolites are established, they separate single periods of emplacement of granitoids (that are Katarchean by isotope datings and geologic data). The most intensive plutonic processes were in the range of 3750–3500 m.y., that is they lasted for 250 m.y. and correspond to the Saamian orogenic cycle. However, in some places the granitoid intrusions occurred even earlier; reliable datings give values of the order of 3850 m.y. and more (see Table 3).

The interrelations of plutonic rocks of different genetic types give the impression that the formation of the magmatic granitoids proper, of juvenile as well as of anatectic origin, and also of the metasomatic rocks and migmatites had been related in time and space. In many concrete conditions it is possible to state only the principal process; anatectic and metasomatic processes are specially difficult to disguise in the widerspread phenomena of granitization.

The most widely distributed gneissoid granodiorites or tonalites (the "grey granites") and the associated diorite gneisses of the Katarchean are very likely of juvenile (hypogenic) nature. Locally they are quite extensive and do not comprise significant inclusions of supracrustals. This fact makes it too reasonable to regard these granitoids as the "initial" formations with the oldest sedimentary strata accumulated on them later. Detailed field studies, however, reveal that in these particular regions the granitoids examined almost everywhere contain xenolithes or skialithes of bedded melanocratic rocks and in the adjacent areas they pierce thick supracrustal piles. Extensive fields of tonalite gneisses usually occupy the central parts of the oval fold systems and in the periphery of the latter they fill in the bedded strata and are folded together. In many cases the sheet and cutting bodies of such are so abundant that they cannot be separated even in large-scale mapping.

In some regions small bodies of pink alaskitoid microcline granites are rarely encountered in addition to the widespread "grey granites" (gneisses). Commonly they reveal the cutting contacts with the emplacing rocks, but sometimes they appear to have gradual transitions to the "grey granites". These alaskitoid granites are usually associated with the exposures of quartzites in the cores of domes (see Fig. 8). Their common occurrence with charnockites is also known, these latter being mainly formed as a result of microclinization of hypersthene plagiogranites (enderbites). It is most probable that the alaskitoid granites were products of anatectic processes that were accompanied by a supply of potassium. It was noted above that the older age of the potassic granites situated in the field of the Katarchean rocks is doubted. Some finds of these rocks in the Paleoprotozoic basal conglomerates are known, however.

Intrusions of basic rocks preceded formation of granitoids. The intrusive bodies of anorthosites are most facinating in this case, for they shed light on the tectonic environment in the period of their formation. A large mass of anorthosites could have been formed only during long quiet crystallization of magma under its slow movement by the laminar stream, for high temperature and absence of tectonic activity during the whole period of formation of the intrusive body is needed for the plagioclase crystals to grow large and for their fractionation to continue successfully. Quick tectonic pressures had to destroy the process of fractionation and to call out homogenization of magma. It will be shown below that some special aspects of the Earth's evolution evidence a relatively low grade of differentiation and amplitude of tectonic movements during the Precambrian (the Early Precambrian in particular) and their gradual aggravation with time. This probably explains the occurrence of anorthosites in the older complexes (from Archean to the Lower Neoprotozoic) and in the relatively stable areas. It is also possible to assume that the wide distribution of anorthosites on the Moon is related to the quiet tectonic environment reigning on this planet.

The metamorphic processes of the Katarchean were inseparably linked with other plutonic phenomena. It was pointed out that in the Katarchean the ultrametamorphic rocks were widespread, the rocks of granulite facies occurred exceptionally in the older complexes, the metamorphic zonation was rather faint in these complexes and the linear type of this zonation was completely lacking. These and many more aspects of the Katarchean endogenic processes are the real result of a special thermal regime of the interior of the older Earth that still had huge resources of heat energy inherited from the primary gravitational differentiation of the planet, and from unexpended radioactive resources. The latter decreased essentially only at the very end of the Katarchean (see Fig. 12).

The problem of geothermic gradient in the Katarchean seems to be the most complicated. The opinions of many workers vary greatly in this aspect. Here the idea is accepted that this gradient was much higher than later on, especially compared with the modern epoch. It is possible to give some facts and speculations in favor of this opinion. Firstly, it is necessary to keep in mind the wide distribution of cordierite granulite in the Katarchean, these rocks being formed under high temperature and relatively low pressure. It is still more important that the low-grade greenstone strata of Paleoprotozoic overlie the high grade metamorphic supracrustals of the Katarchean and more often the Katarchean granitoids. If the geothermic gradient in the Early Precambrian had equaled, for instance, the Phanerozoic one, then a huge thickness of clastic rocks would have to have been accumulated previous to the deposition of greenstone rocks, due to the extremely deep erosion of the oldest formations. Meanwhile, such detractive formations are of limited distribution and are commonly not very thick. The Katarchean granitoids often directly underlie the Paleoprotozoic volcanics. The important moment in the problem discussed is the existence of the Late Katarchean sedimentary-volcanogenic strata that were formed in the interval between orogenic events. The Isua Group is meant in this case; this group supposedly rests unconformably on the Katarchean Amitsoq granite-gneisses and is cut by later (but still Katarchean) granitoids. These relationships are only possible in the case of the granites having been formed close to the surface.

In the opinion of Fyfe (1973) and Saxena (1977) the geothermic gradient of the Katarchean could not have been less than 100 °C for a kilometer and the granulite rocks could have originated at a relatively shallow depth, of the order of 6–8 km. If modern gradients (20°–30 °C) are accepted, than the corresponding depth of granulite formation would appear to have been 50–30 km and thus the probability of their exposure the Earth's surface is close to zero. In my opinion the geothermic gradient in the Katarchean could have been still higher than these authors suppose. Considering all the above aspects of the thermal regime of the "primitive" Earth, it seems permissible to suggest that in the Katarchean the large granitoid could have originated at a depth of only 2–4 km.

## 7. Tectonic Regime

It was noted that the most typical elements of the Katarchean tectonic structure are the large concentric (rounded or irregularly shaped) fold systems, here separated as "gneiss fold ovals", and also the inter-oval fields with their peculiar combination of domes and extensive brachy-form synclines situated between them. It is possible to state that the vertical movements were of principal significance in the erection of these structures, while the horizontal or any other oblique directions in transportation of mass were secondary or derivative. This is definitely proved by the isometric closed shape of the structures, the occurrence of vast fields of granitoids in their central part, and the centripetal vergence of folds on the limbs of the "ovals". Structure analysis also reveals that the source of tectonic activity during deformation was inside the oval systems and that the inter-oval fields were relatively passive and were characterized by weaker movement gradients and by a different style of tectonics. Judging from the isometric shape of the "fold ovals" and their general unregulated situation it is possible to suggest that the fold field had no frame, that is, it was not surrounded by any cratonic blocks (platforms). This conclusions seems to be of exceptional importance in understanding the tectonic environment of the Katarchean.

It is also evident that during formation of the infrastructures in question the processes of folding, metamorphism, granitization (s. lato) were related in time, space, and genesis. The following evidence is available: (1) a complete conformity of tectonic structures of supracrustal rocks (gneisses) and conformable granitoid bodies (granite-gneisses) situated among them; (2) dependence of tectonic forms (of style and complexity) on grade of granitization and migmatization of supracrustals; (3) a complete coincidence of crystallizational schistosity and tectonic pattern (the conformity of shistosity and bedding); (4) the isofacial character of metamorphism and granitization (the mineral paragenesis in crystalline schists or gneisses and granite gneisses belong to a single facies). Finally, it is absolutely true that during the formation of the folded structures deformation had been going on under great, though irregular plasticity of the matter.

The concentric fold systems ("ovals") are likely to have originated as a result of the rising of a great mass of mobilized and partially rheomorphic matter (rheon) of the Earth's crust. The great size (up to 800 km across) of the structures is suggestive of a source of energy at significant depth, probably in the upper or even in the middle mantle. Many gneiss ovals are accompanied by negative gravity anomalies (Fig. 13), indicating that the depth of occurrence of the light granitoid mass in the upper part of an infrastructure reaches 6–10 km.

The cause of the rheon uplift lies in the irregular movement of the heat from the interior of the planet toward its surface. The rheomorphic matter was moving upward because of its higher plasticity and lower density in comparison with the surrounding portions of the crust and also due to expansion in heating, to the supply of juvenile matter and to the presence of pore solutions and gases. An advance of the heat front in the areas of the gneiss ovals could have been caused not only by conductive transfer of heat, but also by diffusion and filtration of abyssal ("through-magmatic") solutions. A high partial pressure of water (and carbon dioxide?) in rheon is evidenced by the fact that the rocks are commonly altered under amphibole-granulite sub-facies or even amphibolite facies, as in the inter-oval fields the rocks are principally of pyroxene-granulite sub-facies.

It is obvious that with the advance of the heat and emanation front toward the surface (in the areas of low temperatures) the rheon grows narrower, or in other words its diameter shortens, and as result its surface acquires a dome-like shape. Deforma-

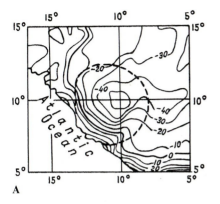

A

**Fig. 13A, B.** Gravity anomalies (Bouguer's reduction) over the Katarchean "gneiss ovals" in Southern and Western Africa. Parts of the map of gravity field of Africa. (After Slettene et al. 1973). **A** Guinean oval (Western Africa, see Fig. 5), **B** Bulawayan oval (Southern Africa, see Fig. 6). *Dotted lines* are the contours of the ovals. In **B** a contour between isoanomalies (− 100 mgal) is shown in *dots*

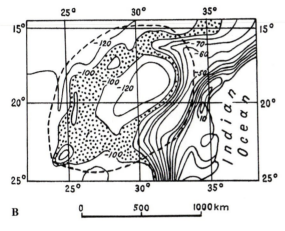

B

tion of the rocks in rheon can be regarded as a result of laminar flow of plastic matter in direction of the least resistivity. Because of the complicated character of strain distribution in rheon and due to the different plasticity of rocks that depend on their physical properties, heating degree, saturation with pore liquid, and many more factors, different portions (streams) of the flow move with different speed, creating a folded pattern in the bedded strata. The geometry of deformation of rheomorphic mass was examined in the work of Carey (1954). According to this author the thickness of layers in the laminar flow that move in the parallel streams will change in relation to the velocity of the streams. The thickness will also change when the streams approach each other, when they diverge or curve. Because of the dome-shaped surface of the rheon front, the streams of the laminar flow in its upper part had to curve and thus the flow of the matter will be directed parallel to the "top" of the rheon toward its center, causing the centripetal vergence of folds (Fig. 14).

Summarizing the above facts on the physical environment and structural forms of the Katarchean, we reach a conclusion as to a specific tectonic environment that governed the early stage of evolution of the Earth. It is stated firstly that differentiation of the Earth's crust into platforms and geosynclines did not exist in the Katarchean. This is evidenced by intensive and ubiquitous deformation and metamorphism of the rocks, by exceptional distribution of concentrical fold systems, by the absence of structural-facial zonation and many other specific features of this era. At the same time, there were in the Katarchean no high risings or deep subsidences; the morpho-structural forms did not exhibit a contrasting nature and the high gradients were not typical of relief-forming movements. That is why psephytic deposits are lacking in the oldest supracrustal complexes (they appear only at the very end of the era before the Saamian orogeny). On the contrary, the Katarchean supracrustal complexes are typified by the presence of such mature sedimentary rocks as quartzites and high-aluminous schists. Many specific features of sequence and composition of the supra-crustal strata and their striking uniformity all over the world are suggestive of the existence of a vast and even all-planetary ocean (the "Panthalassa") in Katarchean.

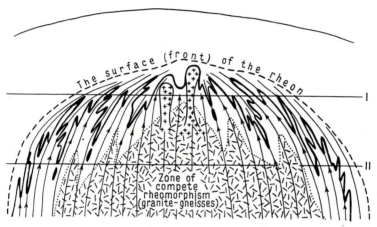

**Fig. 14.** Schematic vertical section through the fold ovals, showing character of material flow in rheon. *Horizontal lines* show different levels of denudation that correspond to those observed in the Aldan Shield (*I*) and in Southern Africa (*II*)

The similarity in composition of many of the Katarchean metavolcanics (meta-basites) and of basalts of the trappean formation of the Phanerozoic platforms and also a wide distribution of anorthosites among the Katarchean intrusive basic rocks evidence quiet conditions of magma differentiation in the abyssal and intermediate magmatic chambers, which means a relatively stable tectonic regime for the major part of the era. It is only in the second half of the Katarchean Era, during the formation of the Sutamian complex, that the tectonic regime seems to have changed a little, as is seen from the formation of thick fine-clastic (pelitic) strata accompanied by lava outflow in the nearby geosynclinal areas and even the lava of acid composition typical of the late stages of mobile zones seems to have flowed.

The primary Earth's crust on which the oldest sedimentary-volcanogenic strata were laid was very probably thin and plastic enough. Thus the high arched uplifts and large abyssal fractures could not have originated during the Katarchean. The absence of linear structural and metamorphic zonation is to be explained by this peculiarity.

The Saamian diastrophism, that lasted for a long time, is known for its intensive plutonic processes that resulted in injection of a huge mass of granitoids. To the end of the Katarchean the Earth's crust became much thicker and more enriched in sialic matter. In the opinion of Condie (1973), the Earth's crust of that time was almost as thick as it is now, of the order of 25–30 km; since that time the thickness of the crust has not much increased. If the lithogenic nature of the major portion of the post-Katarchean (the Late Precambrian and Phanerozoic) granitoids is taken into consideration, then this idea seems quite reasonable.

By nature of the tectonic regime it was proposed (Salop and Scheinmann 1969) to assign the Katarchean Era to a special permobile stage in the evolution of the Earth's crust typical of the earliest stage in geologic history. Other workers call this stage nuclear or pan-geosynclinal. It is obvious from the above discussion that these names are not semantically justified.

## 8. Certain Features of the Earliest Stage in the Evolution of the Earth (the Stage of Cosmic Bombardment)

It is known that the planets of the Earth's group were subjected to intensive bombardment by large meteoritic bodies at an early stage of its development; as a result the ring (the "little marine") and caldera-like morphostructures up to several hundreds of kilometers across on the surface of the planets appeared. According to data of radiometric age determinations of the Moon rocks, the large cosmic bodies (planetosimals) impacted the Earth mostly in the period of 4.5–3.9 (4.0) b.y., that is soon after the formation of the planets when the major portion of the protoplanet nebula matter was not yet fully exhausted. Smaller meteorites have subsequently been falling constantly on the Earth.

The Earth was certainly affected by this collision with large cosmic bodies. Moreover, due to its large mass, it must have exhausted from the surrounding space much more meteoritic matter than many other planets of the Earth's group. According to Urey (1973), the collision of large meteorites with the Earth must have had significant geologic aftereffects, for instance, the temperature of the Earth's atmosphere could have risen by 190 °C or of the oceanic water down to a depth of 100 m by 5 °C, or the water could have evaporated in the volume of $4.10^{20}$ g (corresponding to a cube with a side of 74 km or to an area of the ocean down to a depth of 3 km on a square of $1.33 \cdot 10^5$ km$^2$) or the Earth's mass of $3.24 \cdot 10^{19}$ g could have been ejected in the vicinity of the planet; intensive earthquakes could have caused a common outflow of lava along fractures.

In the opinion of Grieve (1980), the impact effect of large cosmic bodies had to consist of fracturing of the oldest Earth's crust and of formation of deep ($> 3$ km) multi-ring basins on its surface filled in with volcanic crustal-mantle material, regolith and sedimentary rocks – products of their decomposition. Dealing with data of the Moon, studies and results of modeling of the eruptive impact processes, this author suggests that for a time range of 4.5–3.9 b.y. about 3000 impactite structures with a diameter of more that 100 km must have appeared on the Earth's surface and about 25 structures among these must have been about 1000 km in diameter.

A number of traces of meteorite impact, the so-called astroblemes, are known on the Earth's surface. However, the oldest among them are not older than 2.3 b.y. (the Sudbury astrobleme in Canada and the Vredefort astrobleme in Southern Africa).

There are no direct geologic data that prove the existence of a still older bombardment of the Earth by meteorites, and at the early stage of the Earth's evolution they would have had to be manifested at an incomparably more intense rate. In the oldest, Katarchean and Paleoprotozoic (Archeoprotozoic), rocks there are no structures created by impact, such as collapse cones (shatter-cones), impactites or minerals forming during the meteorite impact (coesite, lechatelierite, maskelynite etc.). Certainly, here is meant the impact effects characterizing the Early Precambrian events; the traces of younger falls of meteorites in the areas of the older strata distribution are well known. Glikson (1976a) proposes looking for traces of old collision in the agmatite breccias and in the detritic matter of sedimentary rocks of the corresponding age, but up to now such evidence is lacking (the agmatites extensively developed in gneiss complexes as a rule are evidently related to the processes of migmatization). The stage of intensive cosmic bombardment is beyond doubt, according to Glikson (1976a) and Goodwin (1976), who suggest that the impact of gigantic meteorites could have been manifested to a certain extent in the outflow of basic and utlrabasic lavas along fractures at the earliest stage of geologic evolution of the planet.

In my opinion, however, the oldest tectonic structures observed now can evidence those cosmic events. In shape, size, and situation the concentrical fold systems (the "gneiss ovals") of the Katarchean (see Figs. 2–6) greatly resemble the huge meteoritic craters of the Moon, Mars, and Mercury and the so-called "little lunar maria" in particular, the Mare Orientale for instance (see Lipsky 1969). Incidentally, it is known that these Katarchean structures are of certain endogenic origin and no features of impact phenomena are impressed on them. To solve this contradiction it is necessary to suggest that those structures are simply the reflection of cosmogenic structures of an early "pre-geologic" stage in the evolution of the planet in the later formations. It is reasonably admitted that during this stage abundant cosmic matter fell on the Earth's surface from the protoplanet nebula that was then rather dense; it could have been very large bodies with a diameter of up to 100 km in the opinon of some workers (see, e.g., Smith 1976). The gigantic falling cosmic bodies, touching the thin primary crust, could pierce it and penetrate to certain levels of its mantle through the interior that had not cooled down completely at that time, thus providing an additional heating of the mantle. On the crust surface and inside it, the falling of large meteorites caused crushing and partial dusting of rocks and the formation of the thick piles of semiloose clastic matter over the impact depression, of the type of the lunar regoliths. Due to low heat conductivity this matter must have represented a very good shield that would protect the exceptionally high accumulation of heat during the following endogenic thermal events [11].

During the Katarchean, and in particular, during the Godthaabian and Saamian orogenic cycles the extensive centers of melting must have appeared as a result of uplift of the thermal front under the portions of thick accumulations of meteoritic outbursts and so the diapir columns (asthenoliths) were formed, composed of rheomorphic crust or mantle-crust matter of rheon and finally the structures of the "gneiss

---

11 This regolith must have been easily weathered and partially dissolved due to its low density and great specific surface of the particles. Could it not have been the principal source of the chemogenic-clastic strata formed in the lowermost part of the Katarchean supracrustal complex?

ovals" type appeared. Thus, the oldest astroblemes become visible through the thick strata of intensely dislocated and metamorphosed supracrustals of Katarchean overlying them.

The absence (or rarity?) of the Katarchean astroblemes preserved seems to be caused by decrease in size and abundance of meteorite bodies in the protoplanetary nebula and finally by the most important factor: high alteration and granitization of the older rocks and their extremely deep erosion.

## 9. Principal Stages of Geologic Evolution

It is thus possible to present the following succession of the principal events that took place during the early stage of the Earth's evolution (starting from the oldest "pre-geologic" events):

1. 4500 (5000?) − ≈ 4400 m.y. Formation of the Earth and of the Solar system from the proto-planetary nebula. Intensive differentiation of the matter of the planet interior. Origin of the concentric Earth's shells differing in composition, of the Earth's crust, including the local layer of granitoid composition (formation of the "primary granites"), of the primary essentially hydrogenic atmosphere that was unstable due to dissipation. Extremely intensive bombardment of the newly formed Earth's crust by asteroids and meteorites, including by very large cosmic bodies (scooping out the matter of the protoplanetary nebula); the appearance of certain heterogeneity in the structure of the Earth's crust and mantle as a result of this process.

2. Later than ≈ 4400 m.y. Almost complete dissipation of the primary atmosphere and appearance of thick, dense essentially water-vapor carbon-dioxide atmosphere lacking oxygen; the formation of the primary hot ocean ("Panthalassa") of chloride composition mainly with the temperature lowering below 370 °C on the Earth's surface. Formation of the ozone shield in the upper layers of the atmosphere. Strong hothouse effect on the planet surface.

3. Intense chemical decomposition ("weathering") of the primitive crust. Deposition of the oldest, mainly chemogenic siliceous and high-alumina sediments; outflow of basic lavas – formation of the Iyengran (metabasite-quartzite) lithostratigraphic complex

4. Intensive volcanic activity: subaqueous outflow of lavas of basic and ultrabasic composition mainly; local accumulation of tuffogenic and chemogenic sediments – formation of the Ungran (metabasite) lithostratigraphic complex.

5. Certain lessening of volcanism accompanied by slight decrease of $CO_2$ concentration in atmosphere and hydrosphere. For the first time in the Earth's history accumulation of carbonate strata, evaporites, chemogenic-silliceous sediments and also sediments rich in carbon; outflows of lavas of mainly basic composition – formation of the Fedorovian (metabasite-carbonate) lithostratigraphic complex. This stage or probably the previous one is characterized by the appearance of the primitive forms of life from organic (pre-biologic) matter; of certain anaerobic heterotrophic eobiontes, including prokaryotic organisms and probably uni-cellular blue-green algae and bacteria.

6. Further lessening of the volcanic processes accompanied by a change in composition of the products of eruption: the acid lavas are likely to outflow together with

basalts. Possible formation of isolated extensive areas of land. Deposition of the chemogenic and also of tuffogenic pelitic, fine-grained psammitic and carbonate sediments – formation of the Sutamian (granite-gneiss) lithostratigraphic complex.

7. Later than ~ 4400 m.y. and earlier than 4060 m.y. Slight orogeny (?). Formation of uplifts, extension of land.

8. Later than ~ 4400 m.y. and earlier than 4060 m.y. Deposition of carbonate, pelitic, and psammitic (mainly quartz) sediments, of local phosphate-siliceous sediments and evaporites – formation of the Slyudyankan lithostratigraphic complex. Certain increase of biomass (of prokaryotic algae and bacteria).

9. 4060–4000 m.y. ago, approximately. The Godthaabian orogeny (of the second order?). Folding, metamorphism of the granulite facies, injection of tonaltic gneisses (of the Amitsoq type), migmatization, granitization. Late-tectonic or post-tectonic intrusions of the basic magma along fissures (formation of metabasite dykes?).

10. 4000–3750 m.y. Erosion of the older rocks and local deposition of clastic and chemogenic (carbonate and ferruginous-siliceous) sediments, outflow of basic and acid lavas – formation of the Isuan "transit" lithostratigraphic complex. A gradual lessening of bombardement of the Earth by exceptionally large cosmic bodies (planetosimals).

11. 3750–3500 m.y. The Saamian orogeny of the first order. Intensive folding (formation of concentric fold systems), the early-tectonic intrusions of basic and ultrabasic magma, the syntectonic intrusions of large mass of granitoid (granodioritic-quartz-dioritic) magma, high-grade metamorphism under granulite and amphibolite facies, ultrametamorphic processes (migmatization and granitization). Late-tectonic (?) small intrusions of anatectic potassic granites and alaskites. Post-tectonic intrusions of basic magma (dykes) along fissures.

# Chapter 3   The Paleoprotozoic (Archeoprotozoic)

## A. Rock Records

### 1. General Characteristics of the Paleoprotozoic

**Definition.** The Paleoprotozoic (Archeoprotozoic or Ontarian) Erathem comprises the supracrustal and plutonic rocks that were formed in the time interval of 3600–3500 m.y. and 2800–2600 m.y., that is after (or at the end) of the Saamian orogeny of the Katarchean Era and before the end of the Kenoran orogeny that completed the Paleoprotozoic Era. It was stated in the first chapter of this book that the rocks here attributed to the Paleoprotozoic are assigned by many to the Archean Era, together with the older Katarchean formations, or that they are sometimes divided as the "Upper Archean". I think it reasonable and necessary to give them a status of a single erathem – the lower one in the Protozoic Eon, for they are conspicuously different from the oldest formations and signify a new large stage in the history of the planet; the platform-geosynclinal stage (see below). The term "Archeoprotozoic" here proposed for the era (erathem) seems to be more suitable, for it has a pronunciation similar to the "Archean", a widely accepted name.

**A Concise Characterization of the Era.** The Paleoprotozoic supracrustal complexes are typified by sedimentary-volcanogenic strata close in their formational aspect to eugeosynclinal types of rocks. The miogeosynclinal and platform strata are lacking or their distribution is absolutely insignificant (some essentially carbonate or terrigene-carbonate strata of Eastern Siberia that recently were regarded as Paleoprotozoic are to be assigned to the Fedorovian and Slyudyankan complexes of the Katarchean according to new information). The sedimentary-volcanogenic strata are varied and reval a notable facial variability. Their thickness also varies strongly: in the most complete sequences it reaches 14–15 thousands of meters (and even up to 20,000 m?). The primary nature of the rocks is recognizable from visual observation; under the microscope the textural aspects are easily identified. Conglomerates are common. Some strata are characterized by jaspilites usually associated with volcanics; they belong to a special type of ferruginous-silicious formation typical only of the era examined. The Paleoprotozoic rocks are metamorphosed under greenschist and amphibolite facies; the regionally altered rocks of granulite strata are absent, granitization is local. Organic remains are quite rare.

**Distribution.** The Paleoprotozoic sedimentary-volcanogenic piles and the intrusive rocks cutting them are widespread enough all over the world. Here only some of the large stratigraphic units of the erathem will be enumerated. In Europe: the Gimol Group and its analogs (the Lopian complex) of Karelia, the Tundra and Kolmozero-Voronya Groups of the Kola Peninsula, the Suomussalmi Group of Eastern Finland, the Bua Group and the "Leptite formation" of Sweden, the Pentevrian complex of Britain, the Teterevo and Konko-Verkhovstevo Groups of the Ukraine, the Mikhay-lovsk Group of the Kursk Magnetic Anomaly area; in Asia: the Yenisei Group of the Yenisei Ridge, the Muya Group of the Baikal Mountain Land, the Olonda and Subgan Groups of the Aldan shield, the Aralbay and Karsakpay Groups (Ulutau), the Borovsk Group of Kazakhstan, the Amur Group of the Far East of the U.S.S.R., the Wutai and Anshun Groups of China, the Lower Dharwar Group of India; in Africa: the Swaziland Supergroup of the Republic of South Africa, the Sebakwian II (s. str.), Bulawayan and Shamvaian Groups of Zimbabwe, the Nyanza and Kawi-rondo Groups of Tanzania, Uganda and Kenya, the Lufubu Group of Zambia, the Palabala Group (and its analogues) of Zair and Angola, the Kambui Group of Sierra-Leone, the Simandou Group of Guinea, the Nimba Group of Liberia; in North America: the Yellowknife Group in the north of Canada, the Keewatin and Timis-kaming Group and their numerous analogs in the south-east of Canada, the Tartoq and Malene Groups in South Western Greenland, the Sherry Creek Group in the U.S.A. (Montana); in South America: the Carichapo Group of Venezuela, the Para-maka Group of the Guinean shield, the Rio das Velhas of Brazil; in Australia: the Pilbara Group of the block bearing the same name, the Kalgoorlie Supergroup of the Yilgarn block, the Rum-Jungle complex of the Pine Creek belt.

**Relations with the Underlying and Overlying Rocks.** In all of the above-mentioned areas and in many more, Paleoprotozoic greenstone strata are distributed as relatively narrow and commonly irregularly shaped portions representing the synclinorium structures separated by extensive granitoids, granite-gneisses and gneisses, and other kinds of high-grade supracrustals of the Katarchean that generally occur in the anticlinal zones and dome-like structures. Granitoids separating the synclinal zones (the "greenstone belts") or cropping out in the cores of domes are heterogenous and are of different age: some belong to the rocks of the Katarchean basement that suffered the structural-thermal reworking at the end of the Paleoprotozoic (and even later sometimes), others belong to different plutonic formations of the Paleoproto-zoic; the latter comprise abundant granitoids of the Kenoran cycle; they cut all the greenstone strata and usually occur as large discordant massifs among them.

The relations of the Paleoprotozoic greenstone strata and the older granitoids and gneisses are not quite simple and evident, they are very often masked by remobilization of the basement. At the contact then an abrupt intensification of metamorphism is observed in the rocks of sedimentary-volcanogenic cover, till formation of migmatites and veins of anatectic granites or pegmatites. If the basal formations of the Paleopro-tozoic are the conglomerates and with clasts of the underlying rocks, then the rela-tions of the complexes are easily identified, but if the metavolcanics directly contact the basement rocks (as is quite common), then the interpretation is not as simple, and data from other areas are necessary to solve the problem and a sophisticated analysis of different geologic and geochronologic material must also be done Thus, it is quite natural that debates are continuing on the relations of the greenstone strata and the Katarchean underlying rocks even in the well-studied areas.

The problem becomes still more complicated because of a certain similarity in composition of the Katarchean and Paleoprotozoic granitoids: both contain grano-diorite series rocks (tonalites) and quartz diorites. However, the important differences between them are always determined by the mineralogical and chemical composition and structural and textural aspects, if these rocks are studied in detail in a particular area. The presence of xenoliths and ghost inclusions of supracrustal rocks of granulite and amphibolite facies (amphibolites, pyroxene crystalline schists, ferruginous rocks etc.) the equivalents of which are found in the greenstone strata in the Katarchean granite-gneisses is often regarded as evidence of a younger age of the granite-gneisses. Detailed studies of inclusions in the granite-gneisses reveal, however, that they are conspicuously different in many aspects (petrochemical, structural, and others) from the Paleoprotozoic rocks and are identical with the rocks of the gneiss complexes of Katarchean. Moreover, the Paleoprotozoic rocks under granulite facies are lacking or at least not known (they are not registered in xenoliths among the Paleoprotozoic granitoids cutting the greenstone rocks).

The rejuvenated isotope datings are often used to establish the Paleoprotozoic age of the older granitoids of the basement and an argument of low $^{87}Sr/^{86}Sr$ ratio is used to reject the isotope rejuvenation of the rocks. It was shown earlier, however (see Chap. 1), that this ratio cannot be used in reconstructing pre-history of these rocks. The following example illustrates the case: in Nothern Canada (the Slave tectonic province) the isotopically rejuvenated Katarchean granitoids dated at 2600–2700 m.y. by Rb-Sr isochron method are unconformably overlain by the Yellowknife greenstone group cut by granitoids, that are of the same isotope age as underlying rocks and have the same primary strontium ratio (Baragar and McGlynn 1976). A direct transgressive occurrence of the Group on the older granitoids with conglo-merates at the base has been established by Henderson and Easton (1977). There is an excellent photo of this unconformity in their work.

The primary stratigraphic relationships of the Paleoprotozoic strata and the Katarchean basement rocks have now been established in most locations in the world. If the contacts are masked by tectonic displacement or thermal reactivation of the basement, the younger age of the greenstone strata can easily be determined by the presence of the older rocks in the clastic matter of intraformational conglomerates and greywackes and also by the much lower grade of the rocks, the different compo-sition and sequence, the different style of tectonics, the presence of organic remains, etc. The isotope datings also prove it.

The upper geologic boundary of the Paleoprotozoic is drawn according to a sharply unconformable occurrence (usually of angular or structural type) of Meso-protozoic rocks on the supracrustal and plutonic rocks comprising the erathem. Such relations are almost ubiquitously observed and are in most cases certain. It is only in some locations of Eastern Siberia (for instance in the inner parts of the Baikal Mountain Land) that the Paleoprotozoic strata reveal a comparatively close relation with the overlying Mesoprotozoic piles, and even there local stratigraphic gaps and insignificant angular unconformities are know to exist. On the whole the Paleoproto-zoic formations make up a single large structural stage sharply different from the underlying Katarchean and the overlying Mesoprotozoic.

**Isotope Age Boundaries of the Erathem.** The lower age boundary of the Paleoprotozoic corresponds to the time of termination of the Saamian orogenic cycle. This boundary in most cases is dated at 3500 m.y., but in some localities the last intensive manifestations of the Saamian orogeny occurred earlier and thus some older supracrustals that originated at the end of the Saamian cycle of Katarchean, that is 3600–3500 m.y. ago, can also be attributed to this erathem, in particular, the so-called "transit formations" belong there. In Southern Africa (Swaziland) and Western Australia, the slightly altered basic volcanics occurring at the base of thick Paleoprotozoic complex are dated by Rb-Sr isochron and U-Th-Pb methods at 3500 and 3450 m.y., respectively (Jahn and Shih 1974, Pidgeon 1978 b). These values are interpreted as the time of lava outflow or, in other words, beginning of the Paleoprotozoic Era.

The upper age limit of the erathem is determined according to datings of metamorphic and plutonic rocks that were transformed or originated during the Kenoran orogenic cycle. The datings obtained by different isotope methods are numerous now, amounting to many hundreds, and they give the time interval of 2800–2600 m.y. for the Kenoran diastrophism; the most intensive thermal events are likely to have taken place at the beginning of the cycle. These values (2800–2600 m.y.) are usually obtained when dating (even by the K-Ar method that is known for a high sensibility to the processes of rejuvenation) in the areas consolidated by the Kenoran orogeny, for instance in the Superior tectonic province of Canada, in the Swaziland, Rhodesian and Tanganyika cratons of Africa, in the Yilgarn and Pilbara blocks of Australia; as in the areas where the Paleoprotozoic formations were reworked for the second and many more times by the later tectono-thermal processes; these values are simply the relict datings commonly obtained by Rb-Sr and Pb-Pb isochron methods or on application of the U-Th-Pb method.

It will be shown later, however, that there are data revealing that the thermal processes resulting in granitoid emplacement also occurred previously to the Kenoran cycle during two or three diastrophic episodes (the cycles of the second–third order) that are registered as having place during the Paleoprotozoic Era.

**The Supracrustal Rocks.** Different volcanics extruding in the submarine environment are most typical of the Paleoprotozoic. The subaerial volcanics (mainly of acid composition) are only registered in the upper part of the erathem. A definite regularity is known in the distribution of different types of volcanics in the stratigraphical vertical range; this will be examined while describing the lithostratigraphic complexes. Here we will only mention that the basic and ultrabasic volcanics mainly occur in the lower and middle parts of supracrustal complexes; in the middle and upper parts they alternate with acid and intermediate volcanics. The erupted rocks are accompanied by pyroclastic formations: tuffs, lava-breccias, and agglomerates which are widespread in the upper part of the sequence in association with acid lavas. In most cases the basic lavas dominate over the acid lavas. In the Abitibi greenstone belt typical of Paleoprotozoic (the area of Timmins-Kirkland-Noranda, Canada) the ratio of basalts, andesites, dacites, and rhyolites is 55:30:10:5 (Goodwin 1977)

Among the basic volcanics the tholeiitic basalts are mostly common; the basalt komatiites, diabases, andesite-basalts and spilites are slightly less abundant. All these rocks contain globular (pillow) structure, amygdaloidal and variolite varieties are locally known. Ultrabasic volcanics are represented by picrites with typical varieties rich in magnesium and characterized by $CaO/Al_2O_3$ ratio > 1; these rocks are known to be recognized as peridotite komatiites. Different green schists (chloritic, actinolitic, epidote-bearing etc.) are formed at the expense of basic volcanics in the process of metamorphism and also amphibolites and talc or carbonate-talc schists originate at the expense of ultrabasic lavas. Among the acid volcanics the varieties from rhyolites (porphyry) to dacites are known, but the quartz albitophyres rich in sodium are very typical, in the zones of high-grade metamorphism they are transformed into porphyroids, quartz-chloritic, sericite-albite schists, and other types of schists.

The sedimentary rocks either alternate with volcanics or form separate horizons and strata of great local thickness among them, the latter being a more common mode of occurrence than the former. They are particularly abundant in the upper part of the erathem, but in some places they are encountered in quite sufficient volume in its lowermost parts. The clastic rocks are the dominant type: greywackes, arkoses, siltstones, and different pelites. The polymictic conglomerates are know to occur at several levels, but mainly in the upper part of the erathem and in its base. Quartzites are of limited distribution, and mostly occur in the uppermost part of the sedimentary-volcanogenic complexes, where they associate with slates rich in alumina. The latter are locally known from the lower terrigene complex. Various volcanics and associated hypabyssal rocks are mainly registered among the clastic material of greywakes and conglomerates; the rocks of the Katarchean basement are more rarely recorded. Many clastic rocks associate closely with tuffites and locally exhibit lateral transition into tuffs and agglomerates; the facial substitution of clastic rocks by volcanics is also known.

The carbonate rocks are of notably limited extension; usually they build up separate bands (strata) in certain parts of the Paleoprotozoic sequence and permit a regional (and interregional?) correlation of strata. Among the chemogenic rocks of the Paleoprotozoic the siliceous (cherts and schists) and ferruginous-siliceous (jaspilites) types are best known. The latter are valuable not only from the point of view of economics, but also as typomorphic rocks of the erathem in question.

The Paleoprotozoic jaspilites quite commonly associate with volcanogenic strata, but in contrast to the banded iron ores of the Katarchean usually occurring directly in metabasites (basic volcanics), these do not always exhibit such close paragenesis with volcanics. Jaspilites are commonly situated in bands of sedimentary rocks among volcanics or in the sedimentary-tuffogenic strata substituting lavas laterally; however, the ferruginous rocks display a constant and stable association with volcanics. Canadian geologists distinguish these banded iron ores as a special Algoma type of the banded iron formations. Different subtypes are distinguished within this type, sometimes representing facies. Goodwin (1966) established three main facies in the sedimentary-volcanogenic complex of Canada (the Keewatin Group): oxide, carbonate, and sulfide. The oxide and partially carbonate facies are widely distributed in many other parts of the world. These facies are represented by thin-banded silicamagnetite and magnitite-siderite rocks – jaspilites with ore bands alternating with essentially quartz (chert) bands, the ore components of which are of extremely low rank. Pyrite and pyrrhotite constitute a common admixture. The ore minerals locally contain hematite or hematite mixed with goethite, but in most cases these minerals are secondary and originated as a result of oxidation and hydration of magnetite. Jaspilites of high grade contain various silicates, mostly amphiboles from the grünerite and cummingtonite group. The sulfide facies is represented by banded carbonate-sulfide ores that are laterally substituted by poorly banded and massive sulfide ores. The sulfides are pyrite and (or) pyrrhotite.

In the Michipikoten district (Ontario, Canada) the ferruginous formation of all three facies (subtypes) mentioned is situated at the contact of acid volcanics of rhyolite-dacite composition and the overlying basic volcanics of andesite-basalt type; one facies passes into another along the strike; the oxide facies is the most distant from the volcanic outflow centers, it penetrates in the sedimentary deposits of slate-greywacke composition that substitute volcanics along the strike. This zonation seems to be typical of other regions of development of the Algoma-type jaspilites.

**Organic Remains.** The first determinable organic remains are registered in the Paleoprotozoic, and are phytolites (stromatolites and oncolites) and also microphytofossils. Phytolites are known only from single localities; they occur in corbonate rocks resting among greenstone strata that in Zimbabwe belong to the Bulawayan Group, in Natal to the Pongola Supergroup, in Canada to the Steep Rock and Yellowknife Groups, in Australia to the Warrawoona Group (Macgregor 1951, Hofmann 1971, Bickle et al. 1975, Mason and Von Brunn 1977, Henderson 1975, Walter et al. 1980). Stromatolites are represented by rather small festoons, low dome-shaped and tabular forms; they associate locally with nodular structures of oncolite type.

The microscopic remains of prokaryota (blue-green algae and bacteria) are certainly of wider distribution. They are recognized in many locations in Southern Africa and Canada and occur in low-grade siliceous rocks (cherts) with an age of 3400 m.y. and younger. They are mainly represented by very small (4–8 mcm, rarely up to 20 mcm) unicellular spherical forms with certain signs of binary fission (Knoll and Barghorn 1977). Microbiota are recorded in the organic compounds of kerogen and polymere-like type (porphyrines etc.). In the cherts of Southern Africa (the Fig Tree Group) the optically active amino acids are established, possibly indicating significant biologic activity during the accumulation of sediments. These same sediments are locally known to contain filament structures, they are also likely to be the remains of the blue-green algae.

**Metamorphism.** The Paleoprotozoic supracrustals are altered to a different extent. Metamorphism is distinctly of zonal and linear-zonal character; the zones of higher-grade metamorphism are commonly related to belts of abyssal fractures. The vertical metamorphic zonation is noted near the mantled gneiss domes and the high-grade rocks in this case (up to amphibolite facies) occur in the base of the supracrustal cover at its contact with the remobilized gneiss basement. However, up-section the alteration grade of rocks quickly decreases down to greenschist facies; this kind of zonation is locally known in the range of only several tens of meters or the first hundreds of meters. The rocks regionally altered under greenschist and epidote-amphibolite facies with conspicuous relicts of different structures and textures are fairly widespread. In some places the rocks of the lowest grades of the greenschists facies are developed with very fine textural selvages that are sometimes poorly preserved even in the young Phanerozoic rocks. The highly altered rocks of amphibolite facies are less extended; they occur in the root zones of large fold belts and as a rule are related to wide contact aureoles surrounding large plutonic massifs of granitoids. It was stated above that the

rocks of regional metamorphism under granulite facies are lacking, but local alterations close to the character of facies of the contact pyroxene hornfels seem to occur.
Metasomatic granitization, anatexis and migmatization are limited to definite narrow
zones and commonly registered in the cores of mantled domes.

**Plutonic Formations.** The Paleoprotozoic instrusive rocks are varied and belong to
different stages of magmatism development. The earliest phases of plutonic activity
are known in the period of formation of volcanogenic strata. They are represented by
small sub-volcanic hypabyssal instrusions comagmatic to volcanics of albite granite-
porphyries (and granophyres), plagiogranites (trondhjemite), granodiorite, gabbro-
diabases, gabbro and hyperbasites. These formations are usually located among
volcanogenic strata as swarms of dykes and stocks, and during geologic mapping it
is difficult to separate them from the wall-rocks. In composition and chemism in
particular they are fully identic with volcanics from which they are derived. Relatively
large mesoabyssal granitoid bodies are also known, mostly of granodiorite series, very
close in composition to the corresponding subvolcanic rocks, but these were formed,
however, after intensive lava outflow during the diastrophic events separating different stages in formation of sedimentary-volcanogenic strata of the Paleoprotozoic
(see below).

In many regions of the world the largest widespread massifs of the Paleoprotozoic
instrusive rocks belong to the Kenoran orogenic cycle of the first order. They cut all
the supracrustal strata of the erathem and are principally made up of granodiorites
and quartz diorites of massive as well as of gneissoid (protoclastic) type; the bodies
of microcline-plagiolase granites are much rarer. Granitoids mostly occur within the
anticlinorium zones and in the dome cores. The injection of basic and ultrabasic
magma occurring in the zones of abyssal fractures preceded the granitoid instrusions.
The basic and ultrabasic rocks in combination with lavas of corresponding composition make up a characteristic association resembling the ophiolitic association in the
Phanerozoic fold belts.

**Tectonic Structures.** Paleoprotozoic structures of exceptionally wide distribution are
the so-called mantled gneiss domes and the narrow and irregularly shaped keeled
synclines or composite synclinoria squeezed between the former. The remobilized
Katarchean rocks of the basement accompanied by younger Paleoprotozoic granitoids of anatectic and hypogenic (juvenile) origin crop out in the cores of domes or
anticlines.

The mantled domes differ sharply from the Katarchean "gneiss ovals" not only
by size, which is much smaller and does not exceed 100 km across (commonly it is
10–40 km), but also by structure. The Katarchean concentric structures are known
to represent the foldbelts (fold ensembles) of uniform structure, as the mantled domes
are built up of basement rocks as well as rocks of sedimentary-volcanogenic cover. It
will be shown later that the genesis of mantled domes is also different.

The mantled domes as a rule are not isolated, they form groups that Macgregor
(1951) figuratively called swarms. In combination with linear folds they build up
rather extensive fold-domal greenstone belts that represent one of the major and
typical elements of Paleoprotozoic tectonics (Figs. 15–18). The anticlinorium zones
are situated between the greenstone belts; they are built up of the reworked rocks of

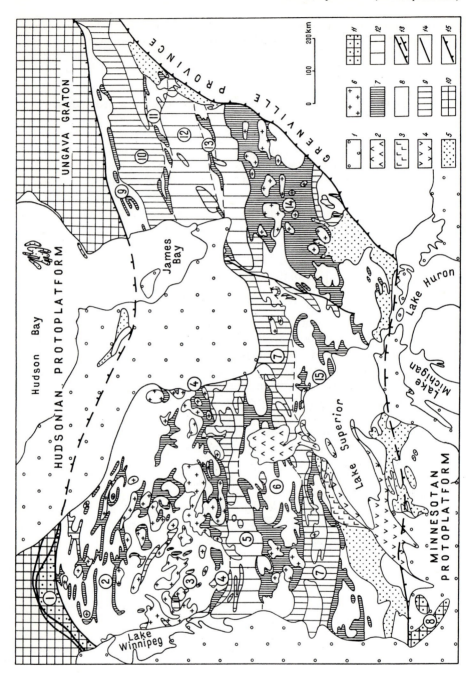

**Fig. 15.** Greenstone belts and protoplatforms of the Canadian Shield within the limits of the Superior tectonic province and the areas surrounding it.
*1* Paleozoic and Mesozoic (platform cover), *2–4* Neoprotozoic (*2* the Nipigon volcanics, *3* gabbroid of the Duluth massif, *4* the Keweenawan Supergroup), *5* Mesoprotozoic (the Huronian and Animikie Supergroups), *6* Paleoprotozoic granitoid of the Kenoran cycle, *7* Paleoprotozoic greenstone volcanics, *8* Katarchean gneissoid granitoid (mainly tonalite gneiss) strongly reworked in the Paleoprotozoic, *9* Katarchean (partially Paleoprotozoic) paragneiss, strongly reworked in the Paleoprotozoic, *10* Katarchean para- and orthogneiss in the areas of protoplatforms, *11* Katarchean grunulite, *12* anticlinorium zones with widespread paragneiss of Katarchean, *13* boundaries of Paleoprotozoic protoplatforms, *14* large fractures, including abyssal type, *15* fracture zone of the Grenville front.
Numbers in circles: *1* the Pikwitonei block, *2* the Sachigo greenstone belt, *3* the Berens zone, *4* the Uchi greenstone belt, *5* the English River paragneiss belt, *6* the Wabigoon greenstone belt, *7* the Coohetico paragneiss belt, *8* the Montevideo (Minnesota River) granulite complex, *9* the Fort George greenstone belt, *10* the Sakami paragneiss belt, *11* the Eastmain greenstone belt, *12* the Rupert paragneiss belt, *13* the Evans greenstone belt, *14* the Abitibi greenstone belt, *15* the Wawa greenstone belt

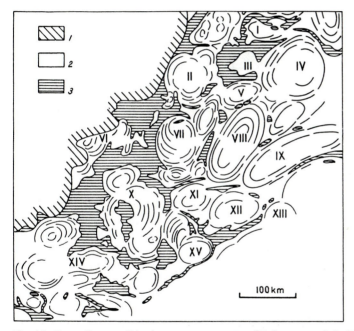

**Fig. 16.** Dome "swarms" in the greenstone strata of Paleoprotozoic in Zimbabwe. (After Macgregor 1951).
*1* young strata, *2* granite, *3* greenstone rocks. Domes: *I* Madziwa, *II* Zimba, *III* Chindamora, *IV* Mtoko, *V* Goromonzi, *VI* Sesombi, *VII* Rhodesdale, *VIII* Charter, *IX* Manika, *X* Shangani, *XI* Chilimanzi, *XII* Gutu, *XIII* Bikita, *XIV* Matopo, *XV* Chibi

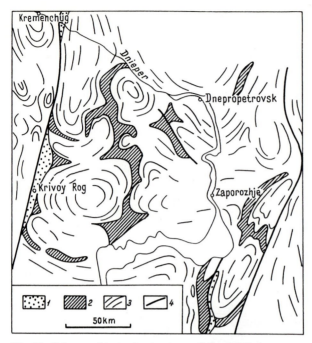

**Fig. 17.** Scheme of tectonic structure of the Pri-Dnieper area (the Ukrainian Shield). (After Kalyaev 1965 with some revision).
*1* Mesoprotozoic strata, *2* greenstone strata of the Paleoprotozoic, *3* Katarchean granite-gneiss mobilized at the end of the Paleoprotozoic, *4* large dislocations

the basement (gneisses and granite-gneisses) with abundant massifs of "young" granitoids. In certain regions the greenstone belts and the anticlinorium gneiss zones separating them are situated between large relatively stable massifs or protoplatforms. Thus, in the south-east of the Canadian shield, in the Superior tectonic province, there is a wide Ontarian fold (geosynclinal) system of Paleoprotozoic comprising greenstone belts and gneiss zones or the "paragneiss belts" according to Young's (1978) term; it is surrounded by protoplatforms made up of the Katarchean gneisses from the north and south (here we suggested calling them the "Gudsonian and Minnesotian protoplatforms"; see Fig. 15).

## 2. The Paleoprotozoic Stratotype and Its Correlatives in Different Parts of the World

Knowledge of Paleoprotozoic stratigraphy is incomparably greater than of the Katarchean. The most complete and well-studied sequences of the erathem are situated in Southern Africa, in Canada, and in Western Australia. Less complete but sufficiently typical sequences are also known in the east of the Baltic shield (in Karelia and the Kola Peninsula), in the Baikal Mountain Land, in the south of India and in many more regions. Some will be examined below.

**The Swaziland Supergroup is a Stratotype of the Paleoprotozoic.** The Swaziland Supergroup ("System") is developed in Southern Africa in the territory of the Republic of

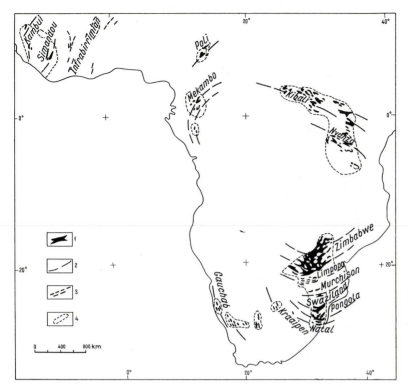

**Fig. 18.** Paleoprotozoic greenstone belts of Equatorial and Southern Africa.
*1* Paleoprotozoic greenstone strata, *2* principal strike of greenstone belts, *3* strike of Paleoprotozoic greenstone strata in Mesoprotozoic Eburnean foldbelt, *4* boundaris of the Lower Precambrian exposures

South Africa and in Swaziland. It is chosen as a stratotype of the erathem, for it seems there are no other large units of the erathem that could be compared with it for completeness of sequence, and for grade of knowledge, though some other strata in the greenstone belts of Canada and Australia can compete with it in some other aspects. In Southern Africa, as in other regions, the greenstone strata form several narrow synclinal or synclinorium structures ("belts") separated by fields of granites and gneisses (see Fig. 18). The largest field of this supergroup is in the Barberton Mountain Land and there it represents the Swaziland synclinorium (Fig. 19). Many workers have studied the stratigraphy of the sedimentary-volcanogenic strata making up this synclinorium; the works of Anhaeusser et al. (1968, 1969) and of Viljoen and Viljoen (1969) are of special value.

The Swaziland Supergroup is divided into three groups (upsection): the Onverwacht, Fig Tree, and Moodies Groups.

The Onverwacht Group is made up of two subgroups: the lower ultrabasic one and the upper one embracing rocks from basic up to acid composition, they in turn are divided into a number of formations and bands.

The Tjakasted or ultrabasic Subgroup comprises three formations: Sandspruit, Theespruit, and Komati. The lowest Sandspruit Formation (up to 2130 m thick) is

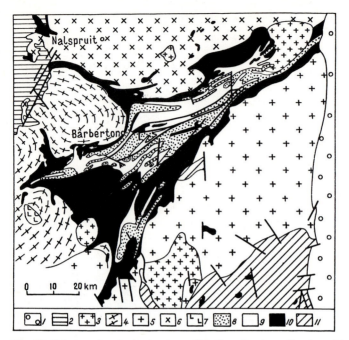

**Fig. 19.** Scheme of geologic structure of the Swaziland synclinorium. [According to the geologic map of Southern Africa (1970), with additional data after Anhaeusser et al. (1968), and Davies and Allsopp (1976)].
*1* Phanerozoic strata, *2* Mesoprotozoic strata, *3–7* Paleoprotozoic: *3* "young" granite and syenite, *4* gneissoid tonalite (this seems have been formed as a result of partial or complete remobilization of the Katarchean tonalite gneiss), *5* granite palingenic in part, formed at the expense of the older, Katarchean granitoid, *6* the Nalspruit migmatite and granite (probably formed at the expense of the older substratum), *7* gabbroid, *8* the Moodies Group, *9* the Fig Tree Group, *10* the Onverwacht Group, *11* Katarchean gneiss and granulite (the "Ancient gneiss complex")

locally developed and is composed of metamorphosed peridotite komatiites alternating with basalt komatiites and metadolerites represented in subordinate abundance. Pillow structures are sometimes observed in lavas. The Theespruit Formation (1890 m) is composed of altered basic lavas of basalt composition intercalating with tuffs of acid and basic volcanics. The lens-like bodies of ultrabasic rocks are also encountered, and are represented by talc and carbonate schists. The felsite tuffs locally transformed into secondary quartzites with high-alumina minerals are typical. In the interbeds of cherts among tuffs the microscopic spherical structures are reported, and somehow resemble the remains of prokaryota (their biogenic origin is doubted). The Komati Formation (3500 m) is characterized by interbedding of pillow and massive basalts and ultrabasic komatiites; acid volcanics and sedimentary rocks are lacking.

The Geluk Subgroup of basic to acid volcanics is principally built up of low-grade tholeiite basalts and acid lavas becoming more abundant up-section. The subgroup is divided into three formations: Hooggenoeg, Kromberg, and Swartkoppie. The Hooggenoeg Formation (4850) is composed of several rhythmical volcanogenic strata, each starting with tholeiite basalts, commonly pillow and amygdaloidal, and terminating with dacites or rhyolite-dacites. The acid volcanics comprise 10% of all types of the rocks of this formation and commonly they alternate with siliceous interbeds. The so-called "Middle Marker" is situated at the base of the formation, and is used for division of the lower and upper subgroups. This horizon is followed for tens of kilometers though its thickness is not more than 6–9 m. It is composed of white and black banded siliceous rocks (cherts) with subordinate interbeds of siliceous-carbonate slates and tuffs. In the upper part of the formation (up to 1500 m thick) the acid lavas are dominant, and contain bands of chert and jaspilites. At different levels the conformable bodies of peridotites and pyroxenites are reported in the formation. The Kromberg Formation (1400 m) is generally made up of the same rocks as the underlying one, but the cyclicity is not so obvious here. The dark cherts and carbonate-siliceous rocks with microscopic remains of prokaryota – the oldest determinable organogenic remains (their origin is certain) occur among acid lavas. Some siliceous rocks display micro-oolitic structure. The Swartkoppie Formation (900 m) is composed of green and grey slates formed at the expense of acid and basic lavas, and subordinate tuffs, banded siliceous rocks (cherts) with microbiota, grey-wackes, and talc schists.

The Fig Tree Group overlies conformably or with an insignificant local break the Onverwacht Group and is made up of three formations (upward): Sheba, Belvue Road, and Schoongesicht. The Sheba Formation (1000 m) comprises greywackes and schists with rare interbeds of cherts and ferruginous-siliceous (jaspilite) rocks. The Belvue Road Formation (600 m) is made up of siltstones and trachyte tuffs; in its bottom there is a thick band of cherts, and thin bands of jaspilites occur through the whole sequence of the formation. The Schoongesicht Formation (up to 550 m) starts with trachyte and trachyandesite tuffs that pass upward into agglomerates and tuff-conglomerates and trachyte lava. Generally the group is typified by rhythmical, graded bedding due to alternation of greywackes and slates; syngenetic sliding and current marks and other aspects of shallow-water environment of sedimentation are quite common. The clastic material of greywackes was supplied by the rocks of the Onverwacht Group and the Katarchean granite-gneiss basement (Condie et al. 1970).

The Moodies Group (3100 m) overlies the Fig Tree Group sometimes conformably, sometimes with a significant break; locally it occurs directly on the Onverwacht Group. It is divided into three formations (Clutha, Joe's Luck, and Bavianskop), they differ slightly in composition and structure. On the whole the group is composed of subgrey-wackes, feldspathic and calcareous quartzites, sandstones, gritstones, conglomerates, and slates. The interbeds and bands of magnetite slates, jaspilites, and horizons of amygdaloidal lavas also occur. The quartzites with cross-bedding, rupple marks and sun cracks are known. The calcareous quartzites pass into dolomites and limestones. At the base of the group the conglomerates with pebbles and boulders of various rocks of the underlying groups and of the granite-gneiss basement commonly occur. Conglomerates (or quartzites) occur also in the base of the overlying formations.

The total thickness of the supergroup is estimated to exceed 20,000 m, but possibly it is overstated due to fold structures and to problems in determination of bed position of volcanogenic rocks.

The Swaziland Supergroup rocks are generally altered under greenschist facies and locally they are of such low-grade metamorphism as to resemble diagenesis. It increases notably, however, toward the base of the supergroup, close to its contact with the reactivated basement in particular and also in the aureoles surrounding the post-Swaziland granites.

The relations of the greenstone rocks of the supergroup and the gneiss and granites cropping out in the environs of the Swaziland synclinorium are being debated. Anhaeusser and Viljoen and Viljoen also think the greenstone rocks to be the oldest formations and all the granitoids, the granite-gneiss including, to be younger. As to the banded supracrustal gneisses extensively exposed to the south-east of the Swaziland synclinorium, it is admitted that they are high-grade metamorphosed equivalents of the greenstone rocks. Hunter and other workers (Hunter 1974a, Davies and Allsopp 1976, Hunter et al. 1978) suggest that the greenstone rocks were deposited on the crystalline basement made up of various supracrustal gneisses, granulites, and granite-gneisses that are regarded as a separate "Ancient Gneiss Complex". The age of original metamorphism of the rocks of this complex is supposed to be 3500–3600 m.y. All information now available on the problem speaks in favor of the conceptions of the last workers.

Many workers who study the area think that almost all of the granitoids immediately surrounding the Swaziland synclinorium are younger than the Swaziland Supergroup, and three groups are distinguished among them differing in composition and structure and also in isotopic age. The first group is represented by the gneissoid tonalites developed in the western margin of the synclinorium (the Kaap-Valley, Stolzburg, Theespruit massifs, etc.). These rocks form dome-shaped massifs partially framed with ultrbasic rocks of the lower subgroup of the Onverwacht Group. Their isotopic age falls within the range of 3130–3310 m.y. (U-Th-Pb method). The second granitoid group is represented by the so-called Hood homogenic granodiorites, the Nalspruit migmatites, the Salisbury-Kop and Dalmein granodiorites and others; its age is about 3000 m.y. by Rb-Sr analysis (Davies and Allsopp 1976). Some massifs of this group cut the strata of the Fig Tree Group and probably of the Moodies Group. The Moodies Group is certainly cut by the Consort pegmatites, their Rb-Sr age is 3030 m.y., but this value is not reliable as the analyzed rocks are characterized by exceptionally high $^{87}Sr/^{86}Sr = 0.770$, which indicates excess radiogenic strontium supplied from the older rocks. Finally, the so-called "young" porphyraceous granites with local abundant kalispar (the Mpageni, Mbabane granites etc.) belong to the group. Most of these granite massifs are situated in the environs of the Swaziland synclinorium but away from the fields of the greenstone rocks. Thus, in the Paleoprotozoic three stages of intrusive granitoid magmatism began to show, and only the last one corresponds to the time of the Kenoran orogeny.

Still, the genesis of a part of the granitoids mentioned is not clear. In my opinion, some gneissoid tonalites of the first group and granitoids of the second group (the Nalspruit migmatites and partially the Hood "homogenic granodiorites") are the Katarchean granodiorite gneisses mobilized (rheomorphosed?) to a different extent. Certainly, it does not rule out the existence of separate hypogenic granodiorite-tonalites that had injected during different periods of magmatic activity, and probably they were emplaced simultaneously with activation of the basement[12].

12 Earlier, the present author (Salop 1977b) admitted the possibility of extensive manifestation of rheomorphism of the basement rocks in the area of Barberton Mountain Land due to the instrusion of "younger" granites of the Kennoran cycle, but now this concept seems to me not quite correct, if all of the available isotope datings and geologic data are considered

Datings of low-grade supracrustals of the Swaziland Supergroup reveal different values depending on their stratigraphic position. The oldest value is $3500 \pm 200$ m.y. and is obtained for the basalt komatiites of the Komati Formation (Rb-Sr method). The method of obtaining this dating seems dubious because it was done on different granulometric fraction of a single sample. Such "inner" isochron is usually obtained on different mineral fractions and one can hardly expect that in this case the granulometric fractions correspond to the mineral ones. The authors (Jahn and Shih 1974) undertook this kind of radiometric study because they failed to determine the age of different rocks of the Onverwacht Group by common Rb-Sr isochron whole-rock method, as the rocks lack isotope homogenization. The Rb-Sr age of rocks from the "Middle Marker" resting over the Komati Formation is $3370 \pm 20$ m.y. (Hurley et al. 1972 b). The age of sulfide concentrate and ziron from quartz porphyry of the Hooggenoeg Formation is 3360 m.y. (U-Th-Pb method, Van Niekerk and Burger 1969 a). The pillow tholeiite basalts from the higher formations of the Onverwacht Group give an age of about 3300 m.y. (Pb-Pb method and U-Pb concordia, Sinha 1972). All the datings mentioned were obtained for rocks of extremely low alteration, the age gradually decreases up-section and so it is possible to state that these datings reflect the time of lava outflow or the time of sedimentation.

At the same time, Rb-Sr datings of altered volcanics from the Onverwacht Group and of phyllitized schists from the Fig Tree Group give an age of 2980 and 2620–2540 m.y., respectively (Allsopp et al. 1968). These values are to be interpreted as revealing the time of the two last stages of plutonic activity (or in other words, rock metamorphism). The first stage of plutonic activity (3200–3150 m.y.) related to emplacement of tonalite massifs and early activation of the gneiss basement is very likely to have taken place at the end of the Onverwacht Group formation before the Fig Tree Group started to accumulate. It is also possible that many tonalites are comagmatic to acid lavas occurring in the upper part of the Onverwacht Group (in the top of the Geluk Subgroup). The second and the third stages (3000 m.y. and 2800–2600 m.y.) took place after the formation of the whole Swaziland Supergroup. It cannot be excluded, however, that the second stage preceded accumulation of the Moodies Group.

Within this portion of Southern Africa the Pongola Supergroup also belongs to the Paleoprotozoic in addition to the Swaziland Supergroup; it is developed some 50–60 km to the south of the Barberton Mountain Land, in the adjacent areas of Swaziland, Natal, and Transvaal (see Fig. 18). This supergroup is divided into two groups: Insusi, the lower one and Mozaan, the upper one.

The Insusi Group is made of volcanics alternating with impure quartzites and in part with slates. Rare horizons of dolomites are also reported, with small stromatolites in places (Mason and von Brunn 1977). Amygdaloidal andesites are dominant among volcanics; basic and acid lavas are of less abundance. These latter are mainly known from the upper part of the group. Cross-bedding is commonly observed in quartzite-sanstones. In Natal thickness of the group does not exceed 1800 m, but in northern areas, in Transvaal, it is much greater, up to 6000 m. The group rests unconformably on a weathered surface of gneissoid plagiogranites with porphyroblasts of microcline; regoliths passing into conglomerates are known from the base of the group (Matthews and Scharrer 1968).

The Mozaan Group overlies the Insusi volcanics with a stratigraphic or slight angular unconformity. Locally it rests directly on granites that are correlated with the Lochiel granites with an Rb-Sr age of 3060 m.y. (Hunter 1974, Mason and von Brunn 1977). The group is built up of quartzites and dark slates with lava interbeds near its top: additionally, bands of jaspilites are also present. The quartzites display cross-bedding and ripple marks. In Natal the thickness of the group does not exceed 700 m, but in Transvaal it probably reaches 5000 m.

Both groups are cut by gabbroids and granites that belong to the Usushwana composite intrusive complex; the granophyre granites from this complex are dated at 2875 m.y. by the Rb-Sr method (Davies and Allsopp 1976). In the contact aureoles of granites the sedimentary rocks are transformed into various micaceous schists, phyllites, hornfels, etc. Away from the zones of effect of granites the rocks of the group are slightly altered, they are even of a lower grade than in the Swaziland Supergroup.

The Problem of the placing of the Pongola Supergroup in the Paleoprotozoic sequence of Southern Africa is a constant subject of debates, but lately many workers have suggested that this supergroup belongs to the uppermost part of the Paleoprotozoic ("Archean") that occurs over the Moodies Group of Swaziland, and that its age is in the 3050–2870 m.y. range. However, the age boundaries mentioned were determined only for the Mozaan Group, and it will be shown later that its lower boundary is still doubtful.

The Rb-Sr age of the gneissoid granites underlying the Insusi Group is 3230 ± 80 m.y. (determinations done by Allsopp, see Burger and Coertze 1973). Juding from their description, they are likely to belong to the gneiss complex of the basement mobilized during injection of Paleoprotozoic granitoids of the first group ($\sim$ 3200 m.y.). Zircon from the Insusi lavas is dated at 3090 m.y. and the age of acid lavas from the same group is 3150 ± 15 m.y. by the Rb-Sr method (see Mason and von Brunn 1977). The second dating is much higher than the lower boundary of the group, as suggested by African workers (3050 m.y.). Then, for the abyssal rocks underlying the group to become exposed, a long period of denudation is needed, particularly because of the low geothermic gradient of the Late Paleoprotozoic (see below).

It is observed that a certain similarity exists between the Insusi Group and the upper part of the Fig Tree Group and partially the Geluk Subgroup of the Onverwacht Group. Their correlation, however, seems not to be justified, as the Mozaan Group overlies the Lochiel granites; but probably this is just an apparent contradiction. Do we know definitely that the Moodies Group is cut by the granites of the second group (the age of the Consort pegmatites is not certain, as was stated earlier)? On the other hand: is the age of the granites underlying the Mozaan Group quite reliable? It is possible that the so-called Lochiel granites are heterogenous and the dating of 3060 m.y. corresponds not to the granites that the Mozaan Group directly overlies, but to their supposed analogs. There are many fields of gneiss-granites of the Katarchean basement among the Lochiel-type granites (with an age of about 3400 m.y.), and it is quite possible that the granites are the product of mobilization of older rocks.

It seems that the doubts cast are quite justified, and so the position of the Pongola Supergroup among the Paleoprotozoic formations of the region cannot be taken for granted. At the same time, it is impossible to refute the conceptions of South African geologists as poorly based, particularly as concerns the Mozaan Group. Thus the problem remains open for discussion.

**Paleoprotozoic of Zimbabwe.** The greenstone strata of Paleoprotozoic are widespread in Southern Zimbabwe in addition to the Republic of South Africa, Swaziland, and Natal; there they were called the "primary systems". These are: the Sebakwian, Bulawayan and Shamvaian "systems" or groups; the latter seems more correct.

The Sebakwian Group is often regarded as comprising noncoeval formations: the Sebakwian metavolcanics and metasedimentary rocks proper and also the highly altered rocks of the Archean basement. That is why some workers, for instance Bliss and Stidolph (1969), distinguish two Sebakwian systems: (Sebakwian I and Sebakwian II) that are separated by an unconformity, the lower one being assigned to the basement. The relations of the basement and the Sebakwian rocks (s. str.) appear to be known from the Selukwe area, where the greenstone rocks overlie the Rhodesdale granite gneisses or migmatites enclosing xenoliths or skialithes of supracrustals metamorphosed under granulite facies. The Sebakwian Group (s. str.) is typified by ultrabasic and basic volcanics that possibly contain varieties that correspond to komatiites. The ferruginous-siliceous rocks of jaspilite type, phyllites and micaceous schists are subordinate to volcanics. The thickness of the group is not known, but it is likely to amount to some thousands of meters. In many localities the group is absolutely lacking, in which case the Bulawayan Group rests directly on the basement.

The Bulawayan Group overlies conformably or with a break the Sebakwian Group or the basement. The dominant rocks are the volcanics of basic, intermediate, and partly of acid composition and also the related pyroclastic rocks. In the basic volcanics (basalt komatiite an tholeiite basalt) the pillow lavas are commonly observed. The acid volcanics and pyroclasts are dominant in the upper part of the group together with volcanics. The sedimentary rocks: quartzites, siliceous and graphitic slates, and jaspilites occur in subordinate abundance. The siliceous rocks (cherts) are known to contain microphytofossils (prokaryota). In the lower part of the group occur one to two thin horizons of sandy dolomites with small dome-shaped stromatolites. In the base of the group there are conglomerates with pebbles and boulders of granites, gneisses, and greenstone rocks of the Sebakwian Group. The thickness of the group varies, but it is estimated to be several thousands of meters: commonly it is of the order of 6–8 thousands of meters and can reach 13,500 m.

The Bulawayan Group is known to rest on the tonalite gneisses of the Katarchean basement in the area of the towns of Mashaba, Fort-Victoria, and Belingwe. The gneisses of the first locality are dated at 3600 and 3570 m.y., of the second at 3520 m.y. (see Table 3, NN 81–83). In the area to the east of the town of Belingwe the sequence of the group is well known, its lower part in particular (Bickle et al. 1975). The lower part of the group there is characterized by a band of sedimentary rocks with thickness up to 120 m, starting with thin conglomerates, upward following siltstones with a horizon of sandy stromatolite dolomites (1–3 m), the first band of jaspilites, a band of cross-bedded sandstones with interbeds of conglomerates, the second band of jaspilites and a band of rhythmically bedded sandstones with interbeds of siltstones. The sedimentary strata are overlain by thick (up to 1000 m) of basic and ultrabasic komatiites, these are in turn overlain by a still thicker pile of basalts and andesites of acid lavas in part, up-section (in the core of syncline) occur sedimentary rocks belonging to the upper-Shamvaian Group (Fig. 20).

The Rb-Sr age of volcanics from different stratigraphic levels of the Bulawayan Group is similar and falls within the interval of 2540–2720 m.y. (Hawkesworth et al. 1975). This fact and the fact of the granites cutting the group (see below) permit us to regard these datings as reflecting the time of metamorphism of the rocks.

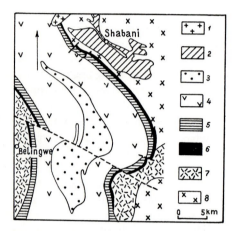

**Fig. 20.** Geologic chart of a part of the Belingwe greenstone belt. (After Bickle et al. 1975). *1* "young" Paleoprotozoic granite (2500–2600 m.y.), *2* ultrabasic intrusive rocks, *3* the upper sedimentary strata (the Shamvaian Group), *4–6* Bulawayan Group: *4* basalt and andesite, *5* komatiite, *6* the lower sedimentary strata, *7* the Sebakwian (s. str.) Group, lower greenstone rocks, *8* tonalite gneiss of the Katarchean basement

The Shamvaian Group unconformably overlies the Bulawayan Group and is folded similarly and in company with it. It is principally composed of arkoses, greywackes, and conglomerates; quartzites, phyllites, schists, and felsites are subordinate. In the area of Fort Victoria the lower part of the group is made up of gritstones and comglomerates interbedding with phyllites and several beds of jaspilites; the upper part is composed of conglomerates and felsites (porphyroids) with bands of jaspilites, crystalline limestones and quartzites. Locally thickness of the group reaches 3500 m.

The Paleoprotozoic of Zimbabwe is specially characterized by structures of the mantled gneiss-dome type. This peculiar pattern of tectonics of the region was first recognized in the 1930's by Macgregor (1937, 1951). The greenstone rocks make up synclines between the domes composed of Katarchean gneisses partially reactivated and cut by younger granitoids.

The Paleoprotozoic granites and tonalites give an age of 2680 (2600) up to 3055 (2960) m.y. by the Rb-Sr method (Bond et al. 1973). According to other datings their age is of the order of 2700 m.y. (Hawkesworth et al. 1975). It is possible that still older Paleoprotozoic granitoids are developed in Zimbabwe, probably corresponding to granitoids of the first group of the Barberton area, but they are not known for certain as yet[13].

Correlations of the Paleoprotozoic strata of Zimbabwe and the Republic of South Africa-Swaziland reveal their common features. The Sebakwian Group (Sebakwian II, see above) corresponds to the Tjakasted Subgroup of the Onverwacht Group by its composition and the situation in the sequence, and the Bulawayan Group corresponds fairly well to the overlying Geluk Subgroup. In the area examined, however, there is a significant break between both groups. It is assumed that the sedimentary strata with a horizon of dolomites can be compared with the "Middle Marker" that separates the Onverwacht Group into two subgroups. In the opinion of Glikson (1976b, 1978) this horizon gives a picture of a nonevident disconformity. Thick strata of komatiites in the lower part of the Bulawayan Group made it similar to the Tjakasted Subgroup, but komatiites are also known from the lowermost part of the Geluk Subgroup. Notable is the presence of dolomites with stromatolites in the lowermost Bulawayan Group, also known to occur in the lowermost Insusi Group of Natal. These two groups seem to be of different stratigraphic position and so their correlation is reasonably to be doubted, at least until the true situation of the Insusi Group in the Paleoprotozoic sequence is known.

**Paleoprotozoic of Equatorial Africa.** The Paleoprotozoic sedimentary-volcanogenic strata are widespread in the east of Equatorial Africa, in the territory of Uganda, Tanzania, and Kenya in particular, there they are represented by two groups – Nyanza and Kawirondo (Bulugwe). The Nyanza Group rests on the Katarchean gneiss-granulite basement and forms synclines separated by granite-gneiss domes. It is mainly made up of greenstone volcanics and subordinate metasedimentary rocks: phyllites, greywackes, quartzites, and jaspilites. On the eastern shore of Lake Victoria – the stratotypical area of the group – it is divided into four successively occurring groups of strata (up-section): (1) basic volcanics (mostly metabasalts, of pillow type in places), tuffs, agglomerates, and also the associated epidiorites; (2) acid volcanics, tuffs, agglomerates, and associated granite-porphyries (or intrusive quartz porphyries); (3) jaspilites in association with acid volcanics and their tuffs and also with various slates and greywackes; (4) dacites and quartz porphyries (the Kuria volcanics.) The thickness of the group is not certain, but it reaches several thousands of meters.

The Kawirondo Group is locally developed and overlies the Nyanza Group with a stratigraphic and locally with slight angular unconformities; sometimes it rests directly on the Katarchean gneisses. It is principally built up of clastic rocks: conglomerates, gritstones, greywackes, phyllites (or micaceous schists), and quartzites with

---

13  According to Wilson et al. (1978) and to Orpen and Wilson (1981) the Sebakwian Group in the area of Fort-Victoria is cut by the Mushendyke tonalites (3500 ± 260 m.y.) that are unconformably overlain by the Bulawayan Group rocks. These tonalites reflect either the latest intrusive phase of the Saamian cycle or the earliest manifestations of the Paleoprotozoic granitoid magmatism, that belongs to a separate tectonic episode, it is possibly to be called the Victorian; this latter version is more realistic. At the same time it is nor excluded that the Sebakwian Group is cut by the rheomorphic mobilisates of the tonalite gneisses of the basement

horizons of acid lavas and their tuffs in places among them. Gneisses, granitoids, volcanics, and jaspilites are known in the clastic material of conglomerates.

Both group are cut by adamellites and granites with Rb-Sr age of 2600–2700 m.y., but in Uganda at the border with Kenya granodiorites (Masaba) are observed that cut the Nyanza Group, but are older than the Kawirondo (Bulugwe), their age by the same method being estimated to be 3110 m.y. (Old and Rex 1971). The Mesoprotozoic Buganda-Toro Group unconformably overlies the post-Kawirondo granites.

From the above it can be seen that the correlation of the Paleoprotozoic strata of Equatorial and Southern Africa is more or less reliable: the Nyanza Group corresponds to the Onverwacht Group in the stratotype are of the Geluk Subgroup; the Kawirondo Group to the Fig-Tree Group; the Masaba granodiorites correspond to the analogous rocks of the first group (3150–3200 m.y.) of the Barberton area; these latter seem to have preceded the deposition of the Fig Tree Group there.

**Paleoprotozoic of the Canadian Shield.** The Paleoprotozoic formations are known from many regions of North America, but they are widely distributed, especially in the Superior tectonic province and the Slave tectonic province of the Canadian or Canadian-Greenland Shields.

The sedimentary-volcanogenic strata of the Superior Province, divided as "Ontarian" by Lawson, can be regarded as a parastratotype of the Paleoprotozoic Erathem. In this province, as well as in other parts of the world, these strata form the greenstone belts: elongated isolated portions of synclinorium structure where the linear commonly isoclinal folds alternate with dome-shaped structures. The greenstone belts are separated by fields of granitoids, granite-gneisses, and gneisses partly belonging to the mobilized basement, partly to the post Ontarian or Kenoran instrusions (see Fig. 15). The local stratigraphic schemes exist for any stratigraphic sequence of almost every large area ("belt"), but commonly the greenstone strata exhibit similar threefold sequence structure: in the upper and lower parts they are usually composed of clastic rocks with subordinate volcanics, in the middle part the volcanics are dominant; we do not find, however, such a simple structure in every case.

The lower terrigene strata usually bear different names in different regions (the Couchiching, Manigotagan, Pontiac etc.), the middle volcanogenic part of the group sequence (commonly the thickest) is sometimes given a general name, Keewatin, though this is also applied to a combination of the lower and middle parts of the sequence or even to the whole Ontarian. The Keewatin type of deposits is also widely found. The upper, mostly terrigene, strata usually have a common name, the best-known being: Timiskaming, Dickinson, and Knife Lake. These strata lie on the volcanics with unconformity and are regarded as a separate group.

The Ontarian supracrustals of the largest Abitibi greenstone belt situated in the east of the Superior tectonic province are known in detail (Fig. 21, see also Fig. 15). Principles of stratigraphy of the older strata of this belt were first studied by Gunning and Ambrose as early as 1939–1940. Many sophisticated studies have been done since in this area and many more details of geology of the area have become known, but unfortunately some problems of stratigraphy of the area still exist due to poor cropping, complex tectonics, and common facial changes of the volcanogenic strata.

The lowermost part of the greenstone complex is probably built up of altered terrigene rocks with rare thin horizons of metavolcanics. In the south-east of the belt

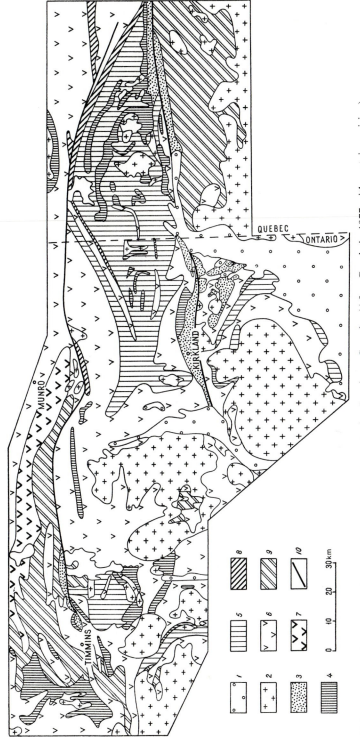

**Fig. 21.** Scheme of geologic structure of the Abitibi belt, Superior Province, Canada. (After Goodwin 1977 with certain revision). *1* Mesoprotozoic strata of the Huronian Supergroup, *2–9* Paleoprotozoic: *2* granodiorite and granite, *3* the Timiskaming Group (conglomerate greywacke, siltstone trachyte); *4–8* the Abitibi Supergroup (the Malartic, Kawagama and Blake River Groups): *4* dacite and rhyolite, *5* andesite, basalt in part, *6* tholeiite basalt, *7* peridotite and basalt komatiite, *8* metasedimentary rocks of the Kawagama Group; *9* the Pontiac Group in the southeast and the Hoyle Group in the north-west (greywacke, sandstone, conglomerate, slate); *10* fractures

they are divided as the Pontiac Group, in the north-west the Hoyle strata possibly correspond to it; they crop out in the core of a large anticline from under the volcanogenic strata. Notable is the fact that to the south the rocks of the group are subjected to high-grade metamorphism down to gneiss formation. It is possible, however, that the underlying rocks of the basement are also erroneously included in the group there. Relations of the Pontiac Group and the overlying strata are either known. In the opinion of some authors (Latulippe 1966) the terrigene rocks of the group laterally substitute volcanics of the Malartic Group, but it cannot be taken for granted.

In the opinion of Holubec (1972), the Pontiac Group is very likely to correspond to the terrigene Kawagama Group resting on the Malartic Group and the terrigene Belcomb Group recognized by this author builds up the lowermost part of the greenstone complex in the Abitibi belt. In addition to greywackes and argillites, this group contains conglomerates with pebbles of granitioids of the basement cropping out near Kinojewis Lake. Judging from geologic mapping, the terrigenous strata called the Pontiac Group are separated by a large fracture and a zone of the Timiskaming terrigenous group from other formations of the Ontarian Supergroup. Its thickness, as well as the thickness of the Belcomb Group, is not certain.

The volcanogenic Malartic Group forms a higher portion of the sequence of the group in the east of the Abitibi belt. Its lower part is composed of basic lavas-tholeiite basalts, partly andesite-basalts and contains also sills of basic and ultrabasic rocks. The upper part of the group is made up of acid lavas and pyroclastic rocks alternate with subordinate basalts and andesites. The thickness of the group is not known, but probably is amounts to several thousands of meters.

In the same locality, in the east of the belt, the Kawagama greywacke-slate strata (Group) conformably overlies the Malartic Group and is partially or fully replaced by volcanics. Its thickness varies, but in some places it can reach thousand meters. Still upward the thick Blake River Group rests conformably; it is principally comprised of tholeiite basalts, and of subordinate andesites, dacites, and rhyolites. Among volcanics interbeds and bands of sedimentary rocks, including the cherts and jaspilites of the Algoma type are known to occur rarely. Goodwin (1977) distinguishes four subgroups in the group with a total thickness up to 28 km. It is certainly strongly overestimated thickness due to isoclinal folding and problems in determination of bed position of volcanic occurrence. It is very probable that the Blake River and Malartic Groups make up a single sequence of volcanogenic rocks separated by the Kawagama sedimentary strata in the east of the Abitibi belt. If this concept is accepted, then the thick strata (up to 1000 m) of basic-ultrabasic komatiite lavas distributed in the north of the Munro district (Arndt et al. 1977) belong to the lower part of this sequence.

The Timiskaming upper terrigenous Group (and the Cadillac Group, its analog) unconformably overlies different rocks of the Blake River and Pontiac Groups. This group is made up of greywacke sandstones, siltstones, and polymictic conglomerates; the volcanics are of subordinate abundance, of trachyte composition in particular. In the sedimentary rocks cross-bedding of stream or deltaic type is common. In the conglomerate pebbles, in company with the volcanics and tuffs underlying the group, granitoids are known to occur that are very similar to those observed as small bodies among the volcanics. Granites and granodiorites of hypabyssal or mesoabyssal type are known among them. Many of these rocks seem to be comagmatic to volcanics.

It is notable that the conglomerate pebbles are characterized by the presence of granitoids with kalispar, and that grains of albitized kalispar, sometimes of fresh microcline, are known in greywackes; the clasts of granite-gneiss and gneiss probably belonging to the basement rocks are much rarer (data of the author for the Timiskaming Lake area). The thickness of the Timiskaming Group in the Abitibi belt amounts to 3000–4000 m.

The Paleoprotozoic strata of the Abitibi belt and of other greenstone belts in the Superior province are cut by different large massifs of biotite and amphibole granites and granodiorites with a reliable age of 2600–2800 m.y. obtained by different methods. They belong to the instrusive formations of the Kenoran orogenic cycle. However, in the souther part of the province, close to the Canadian–U.S.A. border in the Saganaga Lake area, quartz diorites (the Saganaga tonalites) and granites (the Northern Light "gneisses") are encountered, which cut the volcanogenic strata of the Keewatin type (the Ely Group), but are unconformably overlain by the terrigenous strata of the Timiskaming type (the Knife Lake Group). This latter is cut by the Icarus and Gold Island granites of the Kenoran cycle. The Rb-Sr radiometric datings reveal that the Saganaga tonalites and the Kenoran granites are similar in the age values obtained: the first ones are 2710 ± 560 m.y. old (and 2750 m.y. by the U-Th-Pb method), the second ones are 2690 ±480 m.y. old (Hanson et al. 1971). The isotope age of tonalites is almost certainly rejuvenated under the influence of later granites.

It was earlier suggested that the Saganaga tonalites revealed a large diastrophism, the so-called Laurentian orogeny, but later the Canadian and American workers rejected this concept as wrong and thought that the granodiorites and tonalites of the Saganaga type had formed during the entire period of accumulation of the Keewatin lavas and that the break between the Keewatin and the Timiskaming is of no significance. The present author also shared this opinion (Salop 1970, 1973a). However, in the light of new data it seems more reasonable to recognize a small diastrophic episode (of the second-third order) inside the Paleoprotozoic Era. It is very probable that the Saganaga tonalites were intruded approximately together with the analogous granitoids of the first group of the Southern and Equatorial Africa, that is they are of the age of 3150–3200 m.y.

Until recently the problem of basement of the greenstone strata in the Superior province was debatable. Many workers thought the greenstone rocks of the Keewatin type to be the oldest in Canada. The present author, however, came to the conclusion that in the Canadian shield the still older Katarchean rocks that serve as the basement of the greenstone strata are widespread, a conclusion based on critical analysis of data published in different papers (Salop 1970, 1973a, 1977a). Now the Canadian workers have collected many data that confirm this conclusion (see Baragar and McGlynn 1977). The Katarchean supracrustal gneisses and granulites and also the tonalite gneisses cutting them serve as the basement of the Ontarian Supergroup. The gneiss-granulite complex is extensive in the Labrador Peninsula, where it forms the Ungava oldest platform block (protoplatform). A smaller block of the type, the Pikwitonei block, is situated in the north-west of the Superior tectonic province (Fig. 22). Some exposures of supracrustal gneisses and granulites (the Montevideo or Morton complex) are also known in the Minnesota River valley in the south-western portion of the Superior province (U.S.A.). In the Labrador Peninsula and in the state of Minnesota the age of the basement gneisses is more than 3500 m.y. (see Table 3, N 51–53,

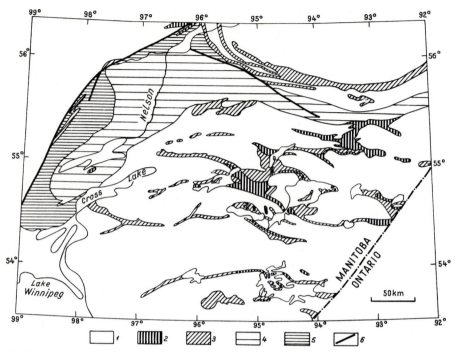

**Fig. 22.** Scheme of geologic structure of the north-western part of Superior Province. (After Stockwell et al. 1970).
*1* granodiorite and tonalite gneiss (Paleoprotozoic and partially remobilized Katarchean), *2–3* Paleoprotozoic, the Heis River Group: *2* metasedimentary rocks, *3* metavolcanics, *4* Katarchean, the Pikwitonei gneiss-granulite complex, *5* the same, diaphthorite after gneiss, *6* pre-Huronian fractures

59–61). Most of the tonalite gneisses separating the outcrops of greenstone rocks in the Superior province probably represent the Katarchean formations remobilized during the Paleoprotozoic, particularly in the areas of Manitoba and borderland areas of Ontario (Smith and Williams 1980). In any case, these rocks are abundant in the boulders and pebbles of basal and intraformational conglomerates of the Ontarian Supergroup.

All major units of a general stratotype of the Paleoprotozoic are recognized in the type sequence of the Ontarian Supergroup in the Abitibi belt, but certain differences are also observed. The Malartic-Blake River volcanogenic complex is very likely to correspond to the Onverwacht Group of Swaziland. The Munro basic-ultrabasic komatiites occur in its lower (?) part, permitting correlation of this portion of the complex and the Tjakasted Subgroup; the upper part of the complex corresponds fairly well to the Geluk Subgroup. The difference lies in the fact that in the Abitibi belt the volcanogenic complex is underlain and separated by terrigenous strata that are not known in Southern Africa. The Timiskaming Group is probably to be compared with the Fig Tree Group

It is to be noted that in other portions of the Canadian shield the normal sequence of the Ontarian Supergroup also varies conspicuously. In some regions the lower terrigenous strata are lacking, locally the terrigenous rocks separating the volcanogenic groups are poorly represented or absolutely lacking. The thick komatiite horizons are not known elsewhere. The twofold structure of the sequence is the only constant feature: in the lower part are situated thick volcanogenic strata of the Keewatin type (one or two cycles, every cycle comprises lavas starting from basic composition in the lower part of the cycle up to acid lavas in the upper part) and terrigenous strata of the Timiskaming type resting on the former with a break.

It is possible that in major locations in Canada the Paleoprotozoic sequence starts only as a level corresponding to the lowermost portion of the Geluk Subgroup of the stratotype. It concerns in particular the Steep Rock area (south-western Ontario) of the Superior tectonic province and also the entire territory of the Slave tectonic province. In the first of the two regions mentioned above, the Steep Rock strata (300–400 m) occur in the lower part of the Keewatin volcanogenic group with a noncontinuous horizon of conglomerates at their base, the latter rest on granites of the basement, up-section a band of dolomites occurs with rare stromatolites, then a horizon of ferruginous-manganese rocks, and finally a band of globular lavas and ash tuffs. The tuffs are conformably overlain by an extremely thick sequence of various lavas; basalts mostly occur in the lower part, basalts and andesites alternating with acid volcanics rest in the upper part (Fig. 23). The whole sequence finishes with an unconformably lying Savant Group – an analog of the Timiskaming Group. In the Slave province, in the west of the shield, the Ontarian Supergroup is represented by the Yellowknife Group. Its lower part, made up of basic and partially of acid volcanics, comprises interbeds of tuffs and agglomerates and a horizon of carbonate rocks with stromatolites, and local thin basal conglomerates resting on granitoids. Terri-

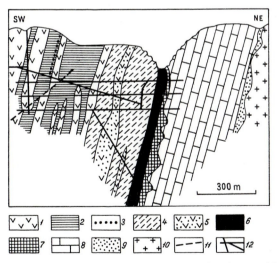

**Fig. 23.** Geologic section in the Steep Rock Lake area (Ontario, Canada). (After Jolliffe 1966). *1–3* the lower part of the Keewatin volcanogenic complex: *1* basic volcanics, *2* tuff, *3* intraformational conglomerate; *4–9* the Steep Rock strata: *4* ash tuff, *5* pillow lava, *6* goethite zone of an ore horizon, *7* manganese zone of an ore horizon, *8* dolomite, *9* basal conglomerate; *10* granite, *11* faults, *12* mine workings

genous strata are observed in the upper portion of the group as in other places. The presence of carbonate rocks with stromatolites in the lower part of the greenstone piles of both regions makes them very similar to the corresponding formations of Zimbabwe and Natal, where the carbonate horizons are compared with the "Middle Marker" at the base of the Geluk Subgroup of Swaziland.

Certain regions of Canada seem to be characterized by equivalents of the highest units of the stratotypical sequence of Southern Africa. Thus, in the south-eastern part of Manitoba (the Rice Lake area) the cross-bedded feldspathic quartzites with conglomerate interbeds belonging to the San Antonio Formation unconformably rest on the Wanipigou terrigenous strata corresponding to the Timiskaming Group. The quartzites are cut by the quartz veins with mineralization of 2720 m.y. old. The San Antonio Formation in its stratigraphic situation corresponds to the Moodies Group (or the Mozaan Group) of the stratotype.

**Paleoprotozoic of the Baltic Shield.** In Europe the rocks of the erathem in question are best known in the eastern part of the Baltic shield in Karelia, Kola Peninsula, and partly in the east of Finland. There they are represented by locally distributed greenstone strata; their composition and structure varies from place to place, but generally they are quite similar to each other and to the co-eval formations of other regions that where examined earlier. The Gimola Group is usually regarded as a regional stratotype; it is developed in some localities of the Soviet Karelia close to the border to Finland, where it is continuously traced and divided as the Ilomanti Group. Three formations are distinguished within the group in the area of the Kostamuksha iron-ore deposit (up-section): Kontoka, Kostamuksha, and Kadiozero (Gor'kovets et al. 1976). The Kontoka Formation (about 4000 m) is made up of basic metavolcanics mainly represented by amphibolites and amphibole schists with relict structure of pillow lavas and also by the lava-breccias; approximately in its middle part the strata are of acid leptite-like volcanics, tuffs, and siliceous schists with jaspilite interbeds. The Kostamuksha Formation (up to 420 m) rests on the Kontoka Formation with an erosion and comprises conglomerates at its base, being built up of quartz-biotite and graphitic schists with jaspilite interbeds. The Kadiozero Formation (more than 700 m) that completes the sequence is characterized by alternation of rhythmically bedded quartz-biotite and graphitic schists (with local garnet, staurolite, and andalusite) und jaspilites; the lower part of the formation is commonly known to comprise a horizon of acid metavolcanics. Thus, the Gimola Group is generally typified by the twofold structure of the sequence: the lower part is mostly volcanogenic of basic composition and the upper largely made up of sedimentary rocks with volcanics of acid composition.

According to data of some workers (Bogdanov etc.) locally the strata (up to 450 m) of polymictic conglomerates with sandstone interbeds occur unconformably on the underlying rocks in the upper part of the Gimola Group, sometimes these strata are divided as the Okunevo Formation. Some think, however, that it may be younger than the group in question.

A similar structure of the sequence is generally typical of the other greenstone groups of the region, in particular of the well-studied Khautavara Group of Southern Karelia. According to data of Robonen et al. (1978) the lower part of this group (the Vietukkalampi and Loukhivaara Formations with total thickness up to 4000 m) is composed of basic metavolcanies with subordinate horizons of intermediate and acid metavolcanics and their tuffs in the upper part of every formation; the lower part of the formation contains rare horizons of altered ultrabasic lavas. The upper part of the group (the Kalajarvi and Kul'jun Formations, up to 3000 m) is made up of grey-wackes, metasandstones and schists alternating with tuffs of acid and intermediate volcanics. Both portions of the group are separated by a stratigraphic break locally fixed by polymictic conglomerates.

In Eastern Finland the Kuhmo or Suomussalmi Group is an equivalent of the units described above (Blais et al. 1977). This group is divided into two parts (sub-groups): the lower one is mostly volcanogenic and the upper one mostly sedimentary with intermediate and acid volcanic strata approximately in the middle portion of it, but unlike the Paleoprotozoic groups of Karelia, many flows of ultrabasic-basic lavas of the komatiite type are observed in the lower portion of the volcanogenic subgroup. It is, moreover, not ruled out that in Karelia certain sheet bodies of ultrabasic rocks, usually regarded as intrusive formations, may appear to be volcanics.

On the Kola Peninsula three Paleoprotozoic groups are recognized: Tundra, Lebyazh'ya, and Kolmozero-Voron'ya, they occur in different structural-facial zones ("greenstone belts"). They are typified by the three-fold structure of the sequence: terrigenous rocks occur in the lower part of the groups, they are altered down to garnet-biotite schists and even gneisses; the middle part of the groups is comprised of thick strata of basic up to acid metavolcanics (amigdaloidal metabasalts, amphibolites, and porphyroids), and in the upper parts of the groups the strata of metasedimentary (terrigenous) rocks alternating with acid metavolcanics and their tuffs in particular are reported to occur. The Kolmozero-Voron'ya Group is also typified by the occurrence of the fourth member of the sequence, it is the Keyvy s. str. (Chervurt) Formation of quartzites, conglomerates, bimicaceous and high-alumina schists (with sillimanite or disthen and staurolite) and amphibolites with a thickness up to 1300 m. This formation rests on different older rocks with a break and a local angular unconformity, but generally its tectonic structure is conformable with the structure of other units of the group. The age of the formation is debatable, earlier I (Salop 1973a) thought that it belonged to the lowermost Mesoprotozoic (the Keyvy horizon), but now I acknowledge that my supposition was erroneous and after analysis of new data and the facts earlier available I share the opinion of the workers who regard it as the upper unit of the Kolmozero-Voron'ya Group (see Fig. 27, column VIII).

All the Paleoprotozoic groups of the Baltic shield discussed above rest unconformably, with local conglomerates at the base, on the Katarchean basement rocks, mainly on the grey tonalite gneisses, and are overlain by Mesoprotozoic rocks with a great angular unconformity. The sedimentary-volcanogenic rocks forming the groups are cut by large bodies of granodiorites, quartz diorites and microcline-plagioclase granites dated in the range of 2600–2800 m.y. by different isotope methods. The same age is obtained for the basement gneisses, indicating a strong effect of thermal-tectonic processes of the Kenoran orogenic cycle. It is to be noted here that the Paleoprotozoic groups of the shield are typified by high-grade metamorphism corresponding to epidote-amphibolite and amphibolite facies. The alteration is mostly intensive in the base of the groups close to the outcrops of granite-gneiss cores and mantled domes where the rocks are commonly transformed into crystalline schists and gneiss-migmatites.

No problem arises in correlation of the Paleoprotozoic groups of the Baltic shield and the stratotype of the erathem. The lower volcanogenic strata are certainly mainly the correlatives of the Onverwacht Group of the stratotype, with its upper Geluk Subgroup. It is only the lowermost part of the Kuhmo Group of Finland with komatiites that possibly corresponds to the lower part ot this subgroup or the Tjakasted Subgroup, this latter possibility being less probable. The upper volcanogenic-terrigenous strata correspond fairly well to the Fig Tree Group of the stratotype. Finally, the Okunevo Formation of the Gimola Group and the Keyvy (Chervurt) Formation of the Kolmozero-Voron'ya Group are the correlatives of the Moodies Group in many aspects. The thickness of the greenstone strata of the Baltic shield (4000–8000 m) is relatively less than that of the corresponding strata of Africa and North America.

**Paleoprotozoic of the Ukrainian Shield.** The rocks of this erathem are best represented in the basin of the Dnieper middle course, where they are recognized as the Konka-Verkhovtsevo (or Verkhovtsevo) Group. According to new data of the Ukrainian geologists (Berzenin etc.), the group is divided into two subgroups: the Konka Subgroup the lower one, and the Belozerka Subgroup the upper one.

The Konka Subgroup lies unconformably on the Katarchean gneisses (the Aulian complex) with local high-alumina or pure quartzites at their base. Generally the subgroup is built up of metabasites and subordinate volcanics of intermediate and acid composition and of metasedimentary rocks in part. A certain difference exists in the composition of its lower and upper parts and thus it became possible to recognize two formations within it. The lower one (up to 2000 m) is made up of amphibolites and chlorite-amphibole-plagioclase schists, metaspilites, and metadiabases; at places rare thin horizons of magnetite-amphibole quartzites and porphyroids are registered. The upper one (up to 3000 m) is of more varied composition: it comprises amphibolites, metaspilites, metadiabases, andesitic metaporphyrites, metakeratophyres, magnetite-amphibole quartzites (jaspilites), different paraschists (quartz-sericitic, bimicacecous, staurolitic, cordierite-bearing and others). The basic metavolcanics are, however, dominant, and the metasedimentary rocks occur mainly in the lower part of the formation.

The Belozerka Subgroup rests unconformably on different horizons of the Konka Subgroup and is principally built up of metasedimentary rocks and of subordinate acid and basic metavolcanics. Three formations are established within it. The lower one (1000–2000 m) is composed of metamorphosed sandstones, siltstones, conglomerates, gritstones, quartz-chlorite-sericitic, and quartz-sericitic schists (or slates), porphyroids, keratophyres, and diabases. Conglomerates are reported at different levels and in pebbles metabasites and ferruginous rocks of jaspilite type occur in addition to gneisses and granitoids. The middle formation (250–800 m) comprises banded magnetite, chlorite-magnetite, chlorite-cummingtonite-magnetite and chlorite-carbonate-magnetite quartzites (jaspilites) and also quartz-chlorite schists, porphyries, keratophyres and sandstones. The ferruginous rocks are of economic importance. The upper formation (up to 2500 m) is made up of various paraschists (the quartz-sericitic, quartz-chloritic, and other varieties); they alternate with metamorphosed porphyries, keratophyres, spilites, and green orthoschists; jaspilite interbeds are common.

The Konka-Verkhovtsevo Group forms composite synclines, they are narrow, keeled, curved in plan, situated (squeezed) between big domes of the Katarchean gneiss-granites (tonalites) that were partially remobilized during the Kenoran orogeny (see Fig. 17). The rocks of the group are metamorphosed under amphibolite facies close to the granite-gneissic cores of domes and locally even migmatizied near the contact, but with the distance growing from the contact the grade of alteration decreases down to the middle or lowest greenschist facies.

The age of the Katarchean tonalitic gneisses of the group basement is known to be 3420 m.y. (see Table 3, N 12) in places of insignificant reworking of rocks by the later thermal-plutonic processes.

The age of granitoids cutting the greenstone rocks of the group and the age of metamorphic minerals enclosed falls within the 2600–2800 m.y. range (Rb-Sr and K-Ar methods). Zircon from the dyke of unaltered massive porphyry related genetically to the volcanic porphyry in the middle formation of the Belozersk Subgroup is dated at 3040 m.y. by the U-Th-Pb method. The value is to be interpreted as the time of volcanism. The rocks of the Belozersk Subgroup are unconformably overlain by the boulder conglomerates of the Pereversevo Formation correlatable with the Gleevatka Formation of Mesoprotozoic of the Krivoy Rog area.

In the Ukrainian shield in the Paleoprotozoic sequence examined the Konka Subgroup seems to be a correlative of the Geluk Subgroup and the Belozersk Subgroup of the Fig Tree Group of Swaziland.

**Paleoprotozoic of the Baikal Mountain Land.** There are many exposures of the Paleoprotozoic in Asia. The sedimentary-volcanogenic complexes of this erathem are well known; they are developed in Eastern Siberia (the Muya Group), in Kazakhstan (the Aralbay and Karsakpay Groups), in the Yenisei Ridge (the Yenisei Group), in China (the Wutai and Anshun Groups), in India (the Lower Dharwar Group)[14]. They are described in brief in my work on the Precambrian of the Northern Hemisphere (Salop 1973 a, 1977 a). Here the Muya Group of the Baikal Mountain Land will be described as an example as it is the best known among other groups; it is developed within the arch-like belt of abyssal fractures traced for a distance of more than 1200 km. The fractures controlled lava outflow and at the same time were the boundaries of the zones differing from the sequence of greenstone strata (Salop 1964–1967).

The Muja Group is divided into three subgroups. The lower one, Parama (1000–1200 m), is locally developed, mostly in the margin of the North-Myua block made up of the Katarchean gneisses. Metamorphosed conglomerates occur in the base of the group, and rest on the gneisses with an angular unconformity. Up-section conglomerates pass into metasandstones with horizons of basic metavolcanics (the Samokut Formation). The upper part of the subgroup (the Bulunda Formation) is built up of marbles alternating with tuffs and volcanics in the base and the top of these

---

14 The lower part of the Dharwar "system", here attributed to the Lower Dharwar Group, is sometimes divided as the Nuggihally subgroup (Srinivasan and Sreenivas 1972). The lower portion of this group is built up of ultrabasic volcanics of komatiite series and the upper one of basic volcanics alternating with metamorphosed fine-grained and chemogenic rocks (including jaspilites). Naqui et al. (1978) think that this group belongs to the "true greenstone belts", in contrast to the overlying two Dharwar groups that are regarded here as Lower Mesoprotozoic (see Chap. 4)

strata. The clastic matter of basal conglomerates locally contains basic metavolcanics in addition to the underlying crystalline rocks, which resemble the rocks of the group. It seems that metabasite strata, older than the Muya Group proper, occur in this region. Their isolated outcrops are probably erroneously assigned to the overlying volcanogenic strata distinguished as the Kilyana Subgroup.

The Kilyana (or the Lower Gorbylok) Subgroup forms the major portion of the group: its thickness sometimes reaches 8–10 thousand meters. It is principally built up of basic, intermediate and acid metavolcanics, forming up to three cyclic strata in each, the lower part being composed of basic and intermediate volcanics and the upper one of acid volcanics and their tuffs. The sedimentary rocks are sharply subordinate; greywackes, tuffites, slates, and jaspilites. Basic-ultrabasic lavas of low abundance are suggested to occur in the lower part of the group.

The upper subgroup is locally developed and its composition varies. In the South-Muya Ridge it is divided as the Upper Gorbylok Subgroup (up to 3500 m), where it is largely composed of tuff-sandstones, tuffs, and agglomerate breccias of acid volcanics (porphyries). Conglomerates, siltstones, and sandstones are also widely distributed and dolomite bands are reported to occur at two–three levels. It rests on the Lower Gorbylok (Kilyana) Subgroup with erosion. In other regions this subgroup is made up of different slates, metasandstones, and quartzites with isolated bands of basic and acid lavas.

The Muya Group contains the lens-like bodies of serpentinites and is pierced by different gabbroids and also by later quartz diorite-plagiogranites. The volcanics commonly associate with small subvolcanic bodies of basic and acid rocks (comagmatic instrusions). Although the age of the instrusive rocks mentioned has not been determined by isotope methods, the situation of the Muya Group in the Precambrian sequence is fairly well known: it is younger than the Katarchean gneisses and older than the Mesoprotozoic sedimentary complex cut by the granites dated at 1900 m.y.

There is no problem in correlation of the Muya Group and the stratotype: the Kilyana Subgroup seems to correspond to the upper subgroup of the Onverwacht Group and the Upper Gorbylok Subgroup to the Fig Tree Group and partly to the Moodies Group. In this case the carbonate strata of the Parama Subgroup correlates with the "Middle Marker" and its equivalents: the carbonate strata of Natal and Zimbabwe and Canada (in the Steep Rock Group). The lower volcanogenic pile that is known only from certain fragments in the basal conglomerates of the Samokut Formation is likely to be compared with the lower subgroup (Tjakasted) of the Onverwacht Group.

**Paleoprotozoic of Western Australia.** The Paleoprotozoic greenstone strata that are well known and extremely complete are located in the west of Australia within the limits of two large blocks or shields: the Yilgarn and Pilbara (Fig. 24). In the Yilgarn block these strata are divided as the Kalgoorlie "system", whose structure in different parts of the shield varies slightly. Below is given a summarized and generalized sequence of the Kalgoorlie stratotypical area or of the Eastern gold-bearing field starting from the oldest rocks (Gemuts and Theron 1975).

Strata 1. Tholeiitic basalts and ultrabasic rocks, with local acid and intermediate volcanics in the top.

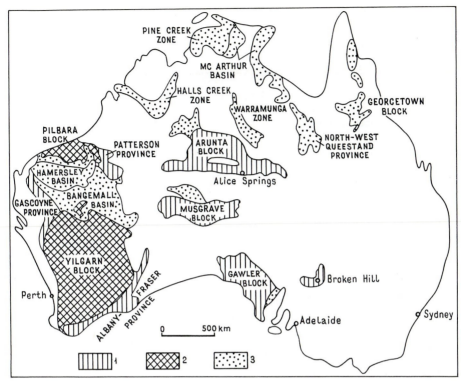

**Fig. 24.** Areas of the Lower Precambrian exposures in Australia.
*1* Katarchean, *2* Paleoprotozoic and partially highly reworked Katarchean, *3* Mesoprotozoic (mainly)

Strata 2. Greywackes, slates, conglomerates, jaspilites, with local acid lavas in the base.

Strata 3. Tholeiitic basalts with several horizons of ultrabasic and basaltic komatiites. The marker of comprised black cherts at the top.

Break

Strata 4. Basic and acid volcanics, tuffs, and greywackes with interbeds of siliceous slates and black cherts

Strata 5. Basal conglomerates; up-section occur arkose greywackes and argillites with subordinate horizon of tholeiite basalts, again conglomerates.

Strata 6. Basaltic and peridotitic komatiites, tholeiite basalts, slates, and cherts.

Strata 7. Tuffs and breccias of acid volcanics, subordinate acid volcanics, and related sub-volcanic rocks.

Break

Strata 8. (Kurrawang). Polymictic conglomerates and greywackes.

The thickness of every group of strata and of all sequences varies greatly and its true value is poorly known. The total thickness, determined by summing up

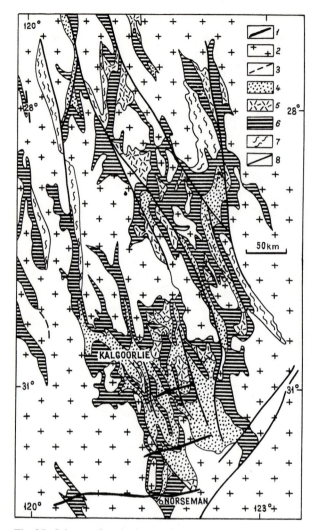

**Fig. 25.** Scheme of geologic structure of the eastern part of the Yilgarn block, Western Australia. [After Gee (see Binns and Marston 1976)].
*1* post-metamorphic dykes of basic rocks, *2* Paleoprotozoic and partially remobilized Katarchean granitoids; *3–6* the Kalgoorlie Supergroup ("system"), *3* jaspilite, *4* terrigenous and volcanogenic-terrigenous strata, *5* acid (gradually turning into basic) metavolcanics, *6* basic and ultrabasic metavolcanics; *7* Katarchean gneiss-migmatite, *8* fractures

thicknesses of different strata from different regions, is estimated at many thousands of meters (probably more than 15–17 thousand meters).

In the Yilgarn block the mode of occurrence of the greenstone strata is generally similar to that of other regions of the world: narrow synclinal folds separated by anticlines and domes, in the cores of which granitoids, gneiss-migmatites with local relict mineral associations of granulite facies crop out (Fig. 25). Many granitoids, mostly of granodiorite composition, are hypogenic formations and form the diapir bodies cutting the greenstone strata, but a certain portion of granitoids is likely to have originated as a result of mobilization of the basement rocks. Archibald and Bettenay (1977) and Archibald et al. (1978) assign gneiss-migmatites to these latter, developed as small inclusions and rather large fields among granitoids in anticlinal structures. The relations of the basement rocks and the greenstone strata are ubiquitously masked because of intensive processes of reactivation of the basement. The Rb-Sr age of the Paleoprotozoic granitoids is within the 2600–2800 m.y. range, that is they belong to the Kenoran cycle of diastrophism. The same method gave values up to 3350 m.y. for the Katarchean rocks of the basement of the Yilgarn block, which are probably slightly rejuvenated (see Table 3, numbers 100 and 101). The greenstone rocks are unconformably overlain by the Mesoprotozic sedimentary piles of sub-platform nature.

Three or four major portions can be distinguished in the sequence of the Kalgoorlie "system". The lower one, comprising the first three groups of strata, is principally represented by basic-ultrabasic volcanics. It is compared with the lower (Tjakasted) subgroup of the Onverwacht Group of the stratotype. The second one, comprising Strata 4, is made up volcanics from basic to acid composition, and is naturally a correlative of the upper (Geluk) subgroup of the same group of the stratotype. The marker, composed of cherts and situated at the boundary between the first and second portions of the sequence, corresponds to the "Middle Marker" of Southern Africa. The third portion of the sequence (Strata 5–7) occurring above the break is mostly built up of terrigene rocks and tuffs (with volcanics in the middle of it) and corresponds to the Fig. 3 Group of the stratotype. The Strata 8 (Kurrawang) overlying the previous portion with a break corresponds approximately to the Moodies Group.

Another remarkable sequence of the Paleoprotozoic sedimentary volcanogenic rocks is known from the eastern part of the Pilbara block. The rock sequence recognized there by some workers (Hickman and Lipple 1975, Ingram 1977) is generally the same and differs mainly by the names of the units and partly by their boundaries. The greenstone strata of the Pilbara Supergroup ("system") are commonly divided into two groups: the lower Warrawoona largely made up of metavolcanics and the metasedimentary rocks overlying them (the Gorge Creek formation), and the upper Mosquito Creek, mostly built up of clastic rocks that unconformably overlie the underlying formations. The lower group embraces two thick sequences of rocks of different composition separated by unconformity, and it seems reasonable to divide it into single groups: the Warrawoona and Gorge Creek.

According to Ingram the Warrawoona Group of that range is divided into three formations (up-section): Chocolate Hill, Sharks, and Miregla. The Chocolate Hill Formation (according to Hickman and Lipple the Talga-Talga Subgroup) is principally made up of nondifferentiated picritic basalts (komatiites) and pyroxenites or peridotites commonly turned into talc-carbonate and tremolite-chlorite schists. Tholeiite basalts alternate with ultrabasic rocks; acid lavas, tufogenic and sedimentary rocks are recorded in extremely low abundance. Sedimentary rocks are represented by quartz-sericitic slates and cherts. Maximal thickness of the formation amounts to 2500 m. The Sharks Formation overlies conformably the Chocolate Hill Formation and is largely built up of tholeiite basalts, in its upper part, however, abundant acid volcanics are registered and in its lower part certain ultrabasic lavas are encountered. The formation is completed with a very thick horizon of cherts with interbeds of acid

lavas. It is easily mapped in many structures and is distinguished as the Marble Bar
Chert marker. Sometimes the cherts enclose interbeds of barite-bearing rocks, which
also contain different microscopic remains of prokaryota and stromatolite structures.
The thickness of the formation reaches 3800 m in some localities. The Miregla Formation rests on the underlying one now conformably, now with a break. It is made up
of pillow basic and intermediate volcanics alternating with acid volcanics and sometimes with sedimentary rocks, mostly cherts. It contains the chert interbeds much
more than the underlying formations. The thickness of the formation varies, its
maximum being 5000 m.

The Gorge Group (Formation) lies unconformably on the underlying rocks; the
unconformity is mainly of stratigraphic nature. The group is made up of clastic rocks:
conglomerate, gritstone, arkose, and greywacke with subordinate interbeds of slate,
tuff of acid volcanics, jaspilite, and chert. Its thickness is more than 2000 m.

The Mosquito Creek Group is locally developed and is likely to rest on different
units of the supergroup. Like the Gorge Creek Group it is mainly made up of clastic
rocks, but they exhibit rhythmical structure and in company with such poorly dif-

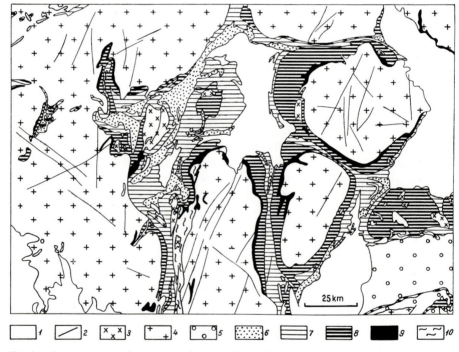

**Fig. 26.** Geologic map of an area in the east of the Pilbara block, Western Australia. (After
Ingram 1977).
*1* Mesoprotozoic strata, *2* dolerite dykes, *3* porphyroid granite, *4* Paleoprotozoic and partially
Katarchean granite-gneiss and gneiss; *5–9* the Pilbara Supergroup, Paleoprotozoic supracrustal strata: *5* the Mosquito Creek Group, terrigenous deposits (turbidite), *6* the Godge Creek
Group, terrigenous deposits and subordinate volcanics, *7–9* the Warrawoona Group: *7* the
Miregla Formation, basic and acid volcanics, *8* the Sharks Formation, basic volcanics, acid
volcanics in the upper part, *9* the Chocolate Hill Formation (the Talga-Talga Subgroup), ultrabasic and basic volcanics; *10* Katarchean (?), gneiss-migmatite

ferentiated rocks as conglomerates, greywackes, and arkoses quartzites and some interbeds of dolomites and rare lava sheets are also registered. The thickness of the group is not known for certain, but probably amounts to several thousands of meters.

The Pilbara block is typified by domal structures (Fig. 26). The domal cores are built up of Paleoprotozoic granitoids with a certain portion of gneiss-migmatites and granites of the Katarchean basement (Fitton et al. 1975). The ultrabasic volcanics of the Chocolate Hill Formation embrace (but only locally) the outcrops of granites and gneisses and in the limbs of domes the sequence is accreted with the overlying units. Usually an age of 2600–2800 m.y. is obtained for the granitoids, but in one dome a value of 3100 m.y. was obtained for these rocks by the Rb-Sr method (De Laeter and Blockley 1972). However, it is not clear if this dating belongs to the early Paleoprotozoic granitoids or to the reactivated granite-gneisses of the basement. The latter (Shaw migmatites) are dated by the U-Th-Pb method on zircon at 3417 m.y. and by the Rb-Sr method on whole rock at about 3000 m.y. (Pidgeon 1978 a). An age of slightly altered red dacites from the Sharks (Duffer) Formation appeared to be 3450 m.y. by U-Th-Pb analysis on zircon (crossing with the concordia; Pidgeon 1978 b); probably this value is to be interpreted as the time of the lava outflow. The Mesoprotozoic strata (the Mount Bruce Supergroup) rest unconformably on the greenstone rocks and the granites cutting them.

Correlation of the Pilbara Supergroup and the stratotype of the erathem and thus of the Paleoprotozoic units of other regions of the world is highly demonstrative; the Warrawoona Group fairly corresponds to the Onverwacht Group and its two lower units: the Chocolate Hill and Sharks Formations correspond to the Tjakasted Subgroup; and the Miregla Formation to the Geluk Group; the Marble Bar marker is a close analog of the "Middle Marker". The Gorge Creek Group is comparable with the Fig Tree Group and the Mosquito Group with the Moodies Group. Correlation of the lower units of the Pilbara block and the Swaziland synclinorium is also evidenced by isotope datings of lavas.

## 3. The Lithostratigraphic Complexes of the Paleoprotozoic

It is clear from the above that Paleoprotozoic strata in different regions of the world are very similar in composition and in sequence structure, as was reported by some workers: Anhaeusser (1971) and Anhaeusser et al. (1968), Glikson (1971, 1976 b, 1978), and many more. At the same time many workers pointed out sufficient differences in the isotope age of the units compared, in Canada and Southern Africa in particular. However, I showed (Salop 1977 b) that these differences are only apparent: they are explained by the fact that Paleoprotozoic rocks are notably altered in many areas and thus in many cases the age values obtained reflect different stages of their transformation. In Southern Africa the slightly altered rocks have generally been dated and commonly these datings give the time of their formation, that is the time of volcanism or sedimentation. The extent of geochronologic data on different areas must therefore be taken into consideration, for it is not of equal validity. Thus for instance, recently some relict datings were obtained for the volcanics of Australia, that evidence that in Paleoprotozoic the lava outflow started on this continent approximately at the same time as it took place in Southern Africa. Previously the greenstone strata of Australia were regarded as younger in age than those of Southern Africa.

Anhaeusser (1973) suggested that the stratigraphic sequence of he supracrustal rocks in the Swaziland synclinorium can be taken as a typical ("model") one for the corresponding formations in other regions of the world. In fact, the three successive groups, ultramafic-mafic, calc-alkaline (basic-to-acid), and sedimentary, which he established are known at present on every continent. The upper sedimentary group, if fully developed, is commonly divided into two parts that represent two single groups of different composition. It permits us to recognize in the Paleoprotozoic four lithostratigraphic complexes of global scale. Their correlation is shown in Fig. 27 and their concise generalized characteristic is given below.

The lower complex I proposed to call the *Komatian* after the typical thick strata of ultrabasic volcanics embracing the Sandspruit, Theespruit, and Komati Forma-tions (the Tjakasted Subgroup) that are distributed in Southern Africa in the Komati River basin (the Tjakasted complex is a synonym). It is widespread but of local distribution: in some places it is reduced by erosion prior to the accumulation of the overlying complex, sometimes it is faintly expressed or was even initially absent. Correspondingly, its thickness varies; nevertheless in many localities, including in the stratotypical area, it amounts to several thousands of meters. Composition and sequence of the Komatian complex in different regions was briefly discussed above. The complex is known to be made up principally of ultramafic and mafic volcanics with peridotite and basalt komatiites dominant. It was suggested earlier that the komatiites were typical only of this stratigraphic level, but now it has been established that these rocks or ones very similar in composition are sometimes registered among the piles of different age. Still, it is well known that the komatiites are of incomparably wider distribution in the lower part of the Paleoprotozoic greenstone strata and thus the complex in question represents to a certain extent an "irreversible" volcanogenic formation. The komatiites were established in the Barberton Mountain Land and are now reported from the lower part of the Paleoprotozoic of other regions of Africa (for instance in Zimbabwe) and also in North America, Australia, India, Finland and in many more areas of the world; in the U.S.S.R. they are known in the Karelian-Kola region and in the Ukraine. Previously many of the rocks now assigned to the koma-tiites had been regarded as intrusive formations. Anhaeusser and Viljoen and Viljoen recognized komatiites as a single group of rocks, and showed that these rocks reveal certain conspicuous aspects of volcanics. In particular, the graphic intergrowths typical of these rocks and determined as a spinifex structure are being formed only under the near-surface environment, it also refers to the polyhedral jointing com-monly observed in komatiites. It is to be noted that the ultrabasic volcanics often associate with the true intrusive bodies of the same composition and with mineraliza-tion peculiar to this group of rocks (platinoids, chrome, nickel, asbest); many of the occurrences are of great economic importance.

The Komatian complex also comprises subordinate rocks like volcanogenic and sedimentary in addition to ultramafic and mafic volcanics. In certain regions, in the Abitibi belt of Canada, and probably in some spots of the Yilgarn block of Australia, the thick strata of clastic rocks are recognized in the lower part of the complex (in Australia it is the Noganyer Group and the Penneshaw Beds), these were formed by destruction of crystalline rocks of the Katarchean basement. In Canada the Cou-chiching, Mackenzie Lake, and pre-Assean Groups probably represent these strata in addition to the Pontiac (Belcomb) Group developed in the Abitibi belt. These groups are made up of metamorphosed greywackes, polymictic conglomerates, and slates with horizons of volcanics and tuffs. Such strata with a local thickness of several thousands of meters can reasonably be recognized as a special *Pontiac complex* (*or subcomplex*), but this is probably premature, as their stratigraphic position is not certain in most places.

The age range of the Komatian complex seems to be between 3500 and 3370 m.y., judging from dating of the rocks from Southern Africa. It should be kept in mind, however, that the value of 3500 m.y. was obtained for the komatiites approximately occurring in the middle (by thickness) part of the complex; the lowest volcanics are naturally older. The upper age boundary is defined according to the rocks from the "Middle Marker" that overlies the complex and is likely to be separated from the underlying or overlying rocks by a masked break (see below). It seems reasonable, therefore, to estimate the age range of the complex in the interval of 3550 (3600?) and 3400 m.y. Thus the Komatiitian complex could have been partially formed as early as at the end of the Saamian orogeny.

The second, *Keewatinian complex* of the Paleoprotozoic is the major unit of the erathem in every region; the Keewatin Group of Canada or the Geluk Subgroup of Southern Africa (including the Hooggenoeg, Kromberg, and Swartkoppie Formations) can serve as its stratotype; its thickness amounts to some thousands of meters and sometimes reaches 7–8 thousand meters. It is of exceptionally wide distribution on every continent and it is not possible or necessary to enumerate here all the units belonging to it; thus, it constitutes a part of all groups and supergroups mentioned at the beginning of this chapter. The Keewatinian complex is largely built up of volcanics alternating with pyroclasts and with subordinate sedimentary rocks: greywackes, slates, cherts, and jaspilites; some single bands of carbonate rocks are also locally registered. Its lower part is typified by basic lavas: tholeiitie basalts and diabases (rare spilites displaying pillow structure). The basalt and peridotite komatiites are evidently subordinate, but in certain areas they form quite thick strata (up to 1000 m) in the base of the complex. Intermediate and acid lavas: dacites, albitophyres, and rhyolites (porphyries), sometimes andesites, are dominant in the upper part of the complex; they constantly associate with abundant pyroclastic rocks. This general background of succession is sometimes interrupted by cyclicity of lower rank when a gradual change of basic lavas into more acid type is registered in every cycle. Volcanics associate with small comagmatic intrusions of plagiogranites, diorites, gabbroids, and ultrabasic rocks.

In many regions in the lowermost part of the Keewatinian complex the strata of sedimentary or volcanogenic-sedimentary rocks are distinguished, they are of exceptional importance in correlation of the greenstone formations. These strata are here proposed to be recognized as a single *Steep Rock subcomplex* after the Steep Rock Group developed in the south-west of the Ontario province (Canada) that is typical of these strata; the subcomplex separates the Keewatin complex from the underlying Komatian complex. Three types of rocks belong to it. One of them, the Steep Rock proper, is characterized by occurrence of rather thick (up to 1500 m) strata of dolomites or limestones (rarely bearing some small stromatolites and oncolites), then sandstones, siltstones, local bands of jaspilites and cherty slates with rare submicroscopic remains of prokaryota. In the top are common horizons of tuffs and lavas, in the base the polymictic conglomerates are usually observed, resting with a

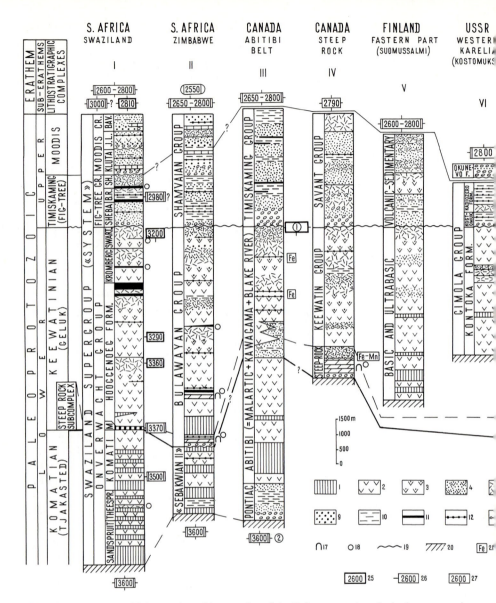

**Fig. 27.** Correlation of lithostratigraphic complex of the Paleoprotozoic principal sequences in different regions of the world.

Stratigraphic columns: *I* after Anhaeusser (1973), *II* after Macgregor (1951); Bliss and Stidolph (1969) and Bond et al. (1973), Bickle et al. (1975) *III* after Gunning and Ambrose (1940), Holubec (1972), Goodwin (1977), Latulippe (1966): *IV* after Jolliffe (1966); *V* after Blais et al. (1977); *VI* after Gor'kovets et al. (1976); *VII* after Robonen et al. (1978); *VIII* after Bogdanov Akhmedov et al. (unpubl. data 1979); *IX* after Berzenin et al. (unpubl. data 1979); *X* after Salop (1964); *XI* after Glikson (1971) and Gemuts and Theron (1975); *XII* after Ingram (1977). Only fundamental works are referred to, additional information (isotope datings etc.) is taken from different sources. Numbers of the units in Column *XI* are given according to Gemuts and Theron (see text).

Symbols: *1* ultrabasic volanics (peridotite komatiite etc.), *2* basic volcanics (basalt, basalt komatiite, tholeiite etc.), *3* intermediate and basic volcanics (andesite, andesite-basalt etc.), *4* tuff of basic and intermediate volcanics, *5* acid volcanics, *6* tuff of acid volcanics, *7* conglomerate, *8* greywacke (partly arkose), *9* quartzite, *10* siltstone and slate, *11* cherty slate, *12* jaspilite, *13* ferruginous-manganese ores (stratiform type), *14* limestone (marble), *15* dolomite, *16* ref-

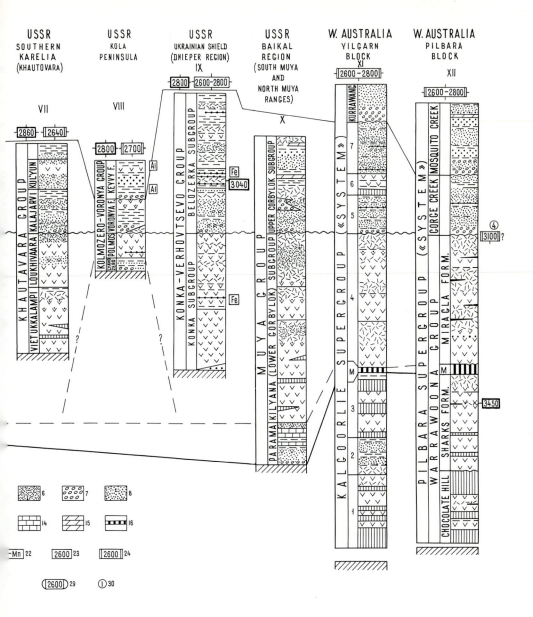

erence horizons of carbonate-cherty rocks (M), *17* stromatolites, *18* microbiota, *19* unconformity, *20* structural unconformity between Paleoprotozoic and Katarchean, *21* some deposits of ferruginous-cherty formation, *22* ferruginous-manganese deposit (Steep Rock); *23–25* methods of isotope datings: *23* K-Ar, *24* Rb-Sr, *25* U-Th-Pb; *26–29* objects of datings (hatchures near the frame of a symbol): *26* volcanics or sedimentary rocks (the time of volcanism or sedimentation), *27* Paleoprotozoic granitoid of Swaziland cycle, *28* granitoid of Kenoran and Saamian cycles, *29* post-tectonic intrusions of basic-ultrabasic rocks ("Great Dyke" of S. Africa); *30* foot-notes.

Foot-notes (figures in circles): *1* in the south of Ontario province the Saganaga granodiorite is situated at this level, cuts the greenstone rocks of the Keewatin Ely Group, and is transgressively overlain by metasediments of the Knife Lake Group (reliable isotope dating is lacking); *2* dating of gneiss of the Katarchean basement from the Montevideo complex in the south of the Canadian Shield (Minnesota, USA); *3* the Okunevo Formation is distributed in the area of the village of Gimola, its Paleoprotozoic age is not certain; *4* dating of granite from a gneiss dome (probably giving the time of thermal events of the Swaziland diastrophic episode) deep erosion now on the crystalline rocks of the Katarchean basement, now on the gneissoid granites of unknown age (see Fig. 23), now on the volcanics of the Komatian complex. Within the sedimentary strata certain breaks and crusts of weathering (in the Steep Rock Group they bear metabauxites) are locally reported. In addition to the stratotypical Steep Rock Group the carbonate strata in the lowermost parts of the Yellowknife, Bulawayan, Insusi (?) and Muya Groups discussed above also belong to this type of rocks. It is probable that the Kamchadal and Arshan (Urik) Formations of the Sayan area (Eastern Siberia) also belong to this type.

Another type of formation is mainly represented by siliceous, locally highly calcareous, rocks that alternate with acid volcanics and their tuffs in some areas. Its thickness varies greatly: from several meters (in the Republic of South Africa) to 3800 m (in NW Australia), depending on the abundance of volcanogenic rocks; however, even if the thickness is quite insignificant, the rocks are easily followed for long distances and serve as markers. Microbiota (prokaryota) and stromatolites are reported to occur in the cherts and calcareous-cherty rocks as well as in these rocks of the Steep Rock type. The siliceous (cherty) strata rest on the Komatian complex with apparent conformity, but in places certain features of breaks are noted, within the complex and in the top of it in particular. In the opinion of Glikson (1978 etc.), the siliceous strata examined (the "Middle Marker" of Swaziland, the rocks at the boundary between the 3 and 4 groups of strata in the Yilgarn block and the Marble Bar Chert in the Pilbara block of Australia) denote a paraconformity between the lower ultramafic and the upper mafic to acid greenstone strata, or in other words between the Komatian and Keewatinian complexes.

Finally, the third type of rocks of the Steep Rock subcomplex not examined earlier is represented by terrigenous rocks rich in alumina. It is known to be distributed on the Kola Peninsula where the Paleoprotozoic rocks are of marked high grade of metamorphism (in the Kolmozero-Voronýa zone the Lyavozero Formation corresponds to this type of rocks).

It will be shown below that the Steep Rock subcomplex marks a break in the subaqueous outflows of volcanics and denotes also an occurrence of oscillating movements of positive sign. Probably it was formed as a result of a small diastrophic episode (Belingwean orogeny of lower order). Glikson suggests that this episode was accompanied by considerable intrusions of granitoids, but this concept does not seem to be adequately grounded up to the moment.

Numerous mineral occurrences and deposits associate with the Keewatinian complex. Among the exogenic type deposits the first to be mentioned are the iron deposits that belong to the banded ferruginous-siliceous formation (BIF) of the Algoma type, characterized by close association of jaspilites and volcanites. Manganese is of less importance in formations of this type. The Algoma-type deposits are rarely large; usually they are of small or middle size, but nevertheless they are of great economic importance, especially if situated in groups. In many regions the hydrothermal gold or gold-polymetallic (Au, Cu, Ag, Zn, Pb, Se, As, Te, Cd, etc.) and arsenic-antimony mineralizations are of exceptionally high interest, they are genetically related to volcanics and their hypabyssal derivatives. Mineralization related to ultrabasic rocks (platinoids, asbest, Co, Cr etc.) is of less importance in this complex than in the Komatian complex. The mineragenetic nature of the Steep Rock subcomplex is characteristic. A large deposit of iron-manganese ores (with high-alumina minerals) is known in the stratotypical Steep Rock Group, the ores having originated in the karsted carbonate rocks. In other regions the magnesite deposits (in Eastern Siberia) and barite deposits (in Australia) associate with dolomites and cherty rocks.

In many regions of the world slight tectonic movements and related small intrusions of granites or granodiorites accompanied termination of accumulation of the Keewatinian lithostratigraphic complex. These tectono-plutonic processes are here attributed to Swazilandian diastrophic episode (orogeny of the second – third order). It was shown above that it occurred 3200 – 3150 m.y. ago. This is very close to the time of formation of the acid lavas making up the upper part of the Keewatinian complex. It seems probable that many of the granitoids belonging to it are comagmatic to these lavas. The Swazilandian diastrophic episode naturally divides the Paleoprotozoic into two sub-eras, evidently different in their character and composition.

The third Paleoprotozoic complex, here called the *Timiskamingian*, belongs to the Upper Paleoprotozoic. The Timiskaming Group of Canada or the Fig Tree Group of Southern Africa can be chosen as its stratotype. Their principal equivalents in other regions of the world were briefly characterized earlier and were shown in Fig. 27. Its rocks usually rest on the Keewatinian complex with a stratigraphic break or with a slight angular unconformity; locally they occur on granitoids that cut the Lower Paleoprotozoic. The complex is mainly made up of clastic rocks: greywackes and slates, tufogenic and arkose sandstones, siliceous slates or cherts (with local microphytofossils), and rare quartzites. Some horizons of polymictic conglomerates are observed at the base of the complex or at its different levels. Among greywackes, in the upper part of the complex in particular, cross-bedding and ripple marks are registered; the rocks are usually poorly sorted. Volcanics are notably subordinate and acid varieties and pyroclasts are dominant among these. In some places, however, the volcanics build up rather thick horizons. Among the sedimentary rocks the interbeds and bands of jaspilites are common, but they do not always associate with volcanics as in the underlying complex. In this aspect they are closer to the Mesoprotozoic jaspilites (of the Krivoy Rog type, see below) that occur in the slate strata. Many of the groups belonging to the complex in question are typified by cyclic structure: in the terrigenous rocks it is seen in diminishing of grain size of clastic matter and increase in sorting grade up the cycle; in the volcanogenic rocks in change of composition from basic or intermediate to acid up the cycle. The thickness of the complex amounts to some thousands of meters.

The mineral deposits of the complex are mainly represented by small to medium (rarely large) sized deposits of iron belonging to ferruginous-siliceous formations (BIF) of the Krivoy Rog and Algoma types. In the U.S.S.R. such deposits are known in the Kadiozero Formation of the Gimola Group in the Kostamuksha area (Karelia).

The age interval of the Timiskaming complex formation is estimated to be in the 3200 (3150) – 3000 (2900) m.y. range. The upper age boundary is not absolutely reliable; the age of the phyllitized shales in the middle part of the Fig Tree Group is 2890 m.y., but this value may equally well reflect the time of rock metamorphism.

*The Moodies complex* is the fourth and uppermost complex of the Paleoprotozoic; it is of local distribution probably because it was destroyed by pre-Mesoprotozoic erosion. It is principally made up of sedimentary rocks and, unlike the underlying complex, is characterized by abundant conglomerates, sub-greywackes, and quartzites in particular, commonly cross-bedded. It also contains calcareous sandstones, dolomites, and limestones with stromatolites in places. The structure of the complex is characterized by high order cyclicity; in the stratotypical Moodies Group of Southern Africa there are three cycles; each starts with conglomerates, upward follow quartzites and carbonate rocks or sandstones and slates with local occurrences of jaspilites and amygdaloidal lavas. The thickness of the complex reaches several thousands of meters. The complex rests on the underlying rocks with a break and locally with a slight angular unconformity. It was suggested that the formation of this unconformity was accompanied by emplacement of granites, however, this is not always well founded. The Mount Kate granodiorites in Western Australia (the Yilgarn block, the Jones Creek area) can be regarded as such intrusions, cutting the strata of the sedimentary-volcanogenic greenstone rocks, but overlain by conglomerates (Durney 1972) that are commonly compared with the conglomerates of the Kurrawang Group (see above and Fig. 27) [15]. However, the age of these granodiorites appeared to be 2690 ± 17 m.y. by the Rb-Sr method (Roddick et al. 1976), and if this value is not rejuvenated, then the intrusion must have taken place during the Kenoran diastrophic cycle after the formation of the Moodies complex, and the Jones Creek conglomerates must be younger in this case and must be attributed to the detractive formations of the Kenoran cycle.

The occurrences of deposits of jaspilites of the Krivoy Rog type (in association with sedimentary rocks) should be mentioned among the minerals related to the Moodies complex.

The lower age boundary of the complex is being debated and was discussed while describing the stratotype. In any case the stratotype of the complex is younger than the Fig Tree Group that here is assigned to the Timiskamingian complex. The upper age boundary of the Moodies complex is reliably drawn according to the manifestation of the Kenoran intensive orogeny that terminated the Paleoprotozoic Era. Granitoids belonging to this diastrophism are always dated in the range of 2800–2600 m.y. as was noted above. Locally some post-tectonic bodies of basic-ultrabasic rocks were emplaced along fractures at the final stage of the diastrophic cycle. An example of these rocks is the "Great Dyke" of Zimbabwe related to large deposits of platinoids and chrome, of nickel and copper in part; its age is 2550 m.y.

# B. Geologic Interpretation of Rock Record

Reconstruction of geologic environment and its change with time during the Paleoprotozoic has some advantages in comparison with the Katarchean Era, as the rocks making up the Paleoprotozoic Erathem are commonly of low alteration and their primary nature is seen clearly almost in every case. The problems, however, lie in the fact that the Paleoprotozoic strata are developed and exposed as isolated portions,

---

15 Durney and other workers thought that the Mount Kate intrusion separates two greenstone piles of the Kalgoorlie Supergroup, but according to new data (Rutland 1978, pers. comm.) the conglomerates do not belong to the "upper" greenstone pile, but are separated from it by a fracture

which limits paleotectonic and paleogeographic reconstructions. The rocks in question are as yet poorly studied in some regions of the world, in South America, Central Asia, and the Antarctic Region.

## 1. Physical and Chemical Environment on the Earth's Surface

The paleotemperature of the planet's surface in the Paleoprotozoic can be judged by knowing the isotope composition of oxygen and the deuterium/hydrogen ratio in silicium hydroxyl of almost nonaltered cherts of the Fig Tree Group (Knaut and Epstein 1976). In Chapter 1 it was stated that these determinations give an everage temperature of 70 °C in the water of the basins where the chemogenic siliceous sediments of this group accumulated. It was also noted there that when the siliceous slates of the Late Archean Isua Group were also subjected to this kind of analysis, the results obtained were not as definite, but the paleotemperature of the sea of that time appeared to be roughly in the 90°–150 °C range. It is also known that the paleotemperature in the Mesoprotozoic was of the order of 60 °C (Knaut and Lowe 1978). Thus in the Paleoprotozoic the temperature seems to varied from 90 °C down to 65 °C at the end of the era. The atmospheric pressure was definitely much lower than in the Katarchean, for great masses of carbon dioxide – the main component of the older atmosphere – were consumed, while thick strata of carbonate accumulated in the formation of the Slyudyankan complex at the end of the Katarchean. It is difficult to give the exact value of the atmospheric pressure; roughly it can be estimated as $10-20$ bar.

The problems of reconstruction of the composition of atmosphere and hydrosphere during the Paleoprotozoic are related to the nature and evolution of these spheres in the Katarchean, for they were inherited from that period of time; the rock records of the Paleoprotozoic are also to be considered. At the end of the Katarchean the concentration of $CO_2$ in the atmosphere was known to decrease sufficiently, but it can nevertheless be suggested that it was quite high at the beginning of the era or in the first half of it. This is evidenced, in pariicular, by the high temperature of the planet surface explained by the hothouse effect, by intense volcanic activity and by the insignificant occurrence of carbonate rocks. In the atmosphere much water vapor must have been present, and it must have lacked free oxygen as was also earlier the case; the concentration of acid smokes in the atmosphere and hydrosphere is likely to have decreased greatly due to interaction with alkali and alkaline earths of rocks (pH of seawater seems to have been near neutral).

The absence of free oxygen is also evidenced by isotope composition of sulfur in the interbeds of unaltered sedimentary barites from the Fig Tree Group. Perry et al. determined (1971) that the sulfate sulfur in these rocks is almost identical in composition with the sulfide sulfur of that time and is very similar to the sulfur of meteorites. It is known, however, that the oxidation processes led to formation of heavier sulfur, that is to enrichment with $^{34}S$ (in relation to $^{32}S$) and that is why the modern sulfate sulfur is heavier than the sulfide sulfur.

It should be noted here that the absence of free oxygen in the atmosphere does not mean its absolute absence in seawater as well: in some portions of the sea bottom "oxygen oases" could exist, with plant microbiota evolving there and sedimentation and other processes occurring there under a different environment. In particular, Dimroth and Lichtblau (1978) described the oxydation processes of crusts of pillow basalt lavas that could have taken place there or oxydation of juvenile sulfur to sulfates (when the sulfate-reducing bacteria are absent) pointed out by Dunlop et al. (1978). The reductive nature of the atmosphere in the Paleoprotozoic is also evidenced by lack of red beds and by very insignificant distribution of photosynthesizing organisms. Probably an extremely low $Fe_2O_3/FeO$ ratio in the Paleoprotozoic volcanics also indicates this. Absence or high deficiency of $O_2$ is likely to have been typical of the Early Precambrian not only in the atmosphere and hydrosphere but also in rather deep horizons of the lithosphere that were connected with them. It will be shown later (Chap. 4) that the stable free oxygen appeared in the atmosphere for the first time only in the Middle Mesoprotozoic (2300–2400 m.y. ago).

Absence or very insignificant concentration of oxygen in the older atmosphere and in the seawater that was in equilibrium with it or, in other words, a very low oxidation reduction potential of the hydrosphere, must have led to a high migrational capacity of protoxide compounds of ferrum and manganese in seawater (Strakhov 1963); for this reason Paleoprotozoic is typified by jaspilite (Fe and Fe-Mn) ores, the deposits are sometimes situated far from the sources of the ore matter, the areas of volcanism.

## 2. Life During the Paleoprotozoic

It was stated that certain probiontes appeared as early as in the Katarchean, but the first remains of prokaryota or any traces of their life activity (phytolites) were registered already in the Early Paleoprotozoic. Naturally they were preserved due to the very low grade of alteration of rocks, but the biogenic factors were of still greater significance, as they provided quick reproduction under an environment that was more favorable than in the Katarchean: the temperature on the Earth's surface was lower, the hydrosphere was less acid and, as a result, less aggressive.

Paleoprotozoic microbiota is quite varied, and many forms of cyanophyta are very similar to those reported from the younger strata of the Precambrian, on one hand proving its high evolution, on the other hand speaking for its slow speed of development and for a certain conservatism. Differences from the younger analogous forms lie mainly in the smaller size of spherical structures (cells). At the same time eukaryotic microorganisms are absent in the Paleoprotozoic, appearing for the first time only in the middle of the Mesoprotozoic. In certain shallow-marine carbonate-cherty rocks of the Paleoprotozoic, the remains of prokaryota are quite abundant, suggestive of the fact that their biomass could have been great in some well-lighted portions of the sea bottom.

## 3. Environment of Supracrustal Rock Accumulation and Evolution of Formations

The Paleoprotozoic lithostratigraphic complexes that replace one another with time really represent the formational complexes or, in other words, the groups of formations genetically related in evolutionary series. Thus it seems reasonable to examine the genesis and environment of the rock formation of every complex separately.

A wide distribution of lavas of the komatiite series is the most typical aspect of the lowest Komatian complex, the lavas being characterized by ultrabasic and basic composition, by high $CaO/Al_2O_3$ ratio (1–2.5), by low Fe/Mg ratio (at given $Al_2O_3$), by low concentration of $TiO_2$ (at given $SiO_2$), and high concentration of MgO, NiO, and $Cr_2O_3$. At the same time, high $CaO/Al_2O_3$, usually regarded as an index of attribution of a rock to komatiite, now appears not to be decisive in distinguishing them from the oceanic tholeiites, as analysis of the latter has revealed that according to this ratio, continuous transitions exist between these two types of rocks (Cawthorn and Strong 1974).

Many workers suggest that komatiites are the products of the "primitive" mantle. An exceptionally high concentration of MgO (up to 40%) in peridotite komatiites indicates their origin from high melting of the mantle matter, as a selctive melting of up to 60%–80% mantle is necessary for the formation of such ultrabasic magma; the basalt komatiites could be formed with 40%–60% of the mantle melted. It was suggested (Cawthorn and Strong 1974) that magma generated at a relatively shallow depth, but this took place due to rather high geothermal gradient.

Metabasalts in the Komatian complex greatly resemble in their composition the modern oceanic tholeiites, but they contain less Al and Ti and more Mn, Ni, Cr, Co, Rb, and total Fe. These features probably reflect some peculiarities of composition of the primitive mantle-crust of the Earth and also of the conditions of magma melting.

The lithostratigraphic complex examined should certainly be recognized as a separate komatiite formation typical of the earliest stage of the Paleoprotozoic Erathem. The tectonic environment of its formation will be discussed below.

The overlying Keewatinian complex is not uniform in formation and can be divided into two formations different in volume and age. The lower one, corresponding to the essentially sedimentary Steep Rock subcomplex, is known to be distributed in different regions and represented by three types of rocks; two of them, the siliceous-carbonate and cherty types, are typomorphic and are of the greatest interest. The third type of rocks is made up of clastic sediments commonly rich in alumina and is widespread in many other Precambrian complexes. All the types mentioned, though exhibiting obvious differences in composition, are characterized by the common environment of formation partially under extreme shallow-water and subaerial conditions. This is evidenced by the occurrence of crusts of weathering and of the older karst with ferruginous, ferruginous-manganese, and high-alumina minerals (meta-bauxites in the Steep Rock Group), by the presence of stratigraphic breaks, of cross-bedding in quartzites, of stromatolites in dolomites, and by many more peculiarities. According to Dunlop et al. (1978) the fine-bedded barite-bearing rocks in the North Pole cherty slates (Australia) were originally evaporites. The shape of the barite crystals shows that this mineral was formed as a result of pseudomorphic replacement of gypsum; in the siliceous country rocks the ripple marks, cross-bedding, intra-formational breccias and rare cherty pseudomorphs after gypsum are common; the same rocks are known to contain the light-requiring microbiota (cyanophytes) and also stromatolites (?), all of which indicates that the sedimentation was taking place in the extremely shallow-water basin highly mineralized.

The siliceous rocks of the Steep Rock subcomplex and of other Paleoprotozoic complexes are likely to represent the chemogenic (colloidal) sediments related to the volcanic events. Their wide distribution is explained by high solubility of silicium (of the dispersion type in particular) in the hot water of the older ocean. Precipitation of the colloidal silica probably occurred as a result of evaporation of saturated solutions and due to pH change (in the mineralized water rich in alkali).

The causes that favored a world-wide distribution of the Steep Rock sedimentary rocks are not yet sufficiently clear. It is possible that simultaneous accumulation of carbonates in the rocks in different regions of the world could have occurred due to a certain decrease in the $CO_2$ concentration in the atmosphere as an effect of lessening of volcanic processes. It seems reasonable to suggest that the uplifting of the Earth's crust was dominant at that time, which partially explains a wide distribution of shallow-water environment and formation of land.

The principal types of rocks of the Steep Rock subcomplex are likely to be recognized as a special carbonate-siliceous ("Steep-Rock") formation.

Greenstone volcanic strata resting on the Steep Rock piles make up the major portion of the Keewatinian complex. Generally the volcanic piles are characterized by homodrome sequence complicated by cyclicity of the second order, rarely of the third order. The lava outflows occurred mainly in the under-water environment, as is evidenced by a wide distribution of pillow structures and by alternation with marine sediments. However, certain moments point to the fact that lavas as a rule had accumulated not very deep in the basins, in particular the intermediate and acid lavas that are usually accompanied by abundant pyroclastic products, including agglomerates. Many of the acid volcanics occurring in the upper part of the complex seem to outflow even under subaerial conditions. It seems that the outflow of basic lavas generally occurred when the basin bottom was submerging and the intermediate and acid lavas had outflowed with the uplifting of the bottom. Thus, cyclicity in the structure of the volcanic piles probably reflects the oscillating movements of the Earth's crust and the general homodrome sequence of lavas in the Keewatinian complex displays its regressive structure. It is to be noted here that the cyclic lava strata in the complex are often completed not only with acid volcanic or tuffs, but also with jaspilites or cherts (Fig. 28) that originated under extreme shallow-water conditions.

Volcanics of the Keewatinian complex belong to differentiates of calc-alkaline magma; their basic and intermediate members are similar to tholeiites of island arcs in many aspects, in the concentration of K, Rb, Sr, Ba, and rare elements in particular (Condie 1975). This similarity becomes still more clear when considering the association of both rocks with acid lavas (and pyroclasts). At the same time, according to Glikson (1976b) and Windley (1977) the volcanics examined differ from tholeiites of island arcs by occurrence of ultrabasic lavas, by bimodal character of distribution of basic and acid lavas (by lower abundance of andesites correspondingly), by dominance of tholeiites with low or middle concentration of $Al_2O_3$ (14%–15%), by high concentration of Ni, Cr, and Co, and also by rare occurrence of alkaline lavas. According to Abramovich and Klushin (1978), all the Paleoprotozoic basaltoids are typified by low concentration of alkalies (of potassium in particular) and aluminum, by low oxidation of ferrum in comparison with the Phanerozoic basaltoids.

All the differences mentioned are certainly quite obvious and reflect the evolution of magmatic processes and partially of the environment (as in the case of oxidation of iron in lavas). It serves as the basis to recognize the Keewatinian complex volcanics as a special bimodal tholeiite-rhyolite greenstone formation.

The Timiskamingian sedimentary or volcanogenic-sedimentary complex reflects a stage of notable extinction of volcanic activity and intensification of oscillating move-

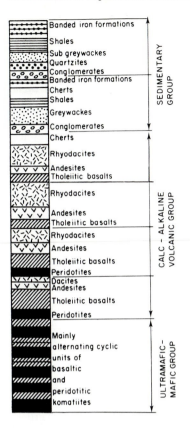

**Fig. 28.** An ideal sequence of rocks in the Paleoproto-zoic greenstone strata based on the model of the Barberton region (S. Africa). (After Anhaeusser 1971 b)

ments that took place during and after the Swazilandian diastrophic episode. The gradients of oscillating movements or of morphostructures are likely to have been great for the coarse-grained, poorly sorted clasts: conglomerates and greywackes in particular accumulated in the quickly subsiding depressions. The intermittence of oscillating movements is registered in the cyclicity of the structure of the complex. The time of deceleration or pause in oscillations is characterized by the accumulation of fine-grained sediments: slates, cherts, jaspilites. Sedimentation took place in shallow-water basins, under partially subaerial conditions. The inner uplifts made up of volcanics and granitoids of the Lower Paleoprotozoic and in part of the Katarchean crystalline rocks were the source areas of clastic material. This tectonic environment prompts a wide distribution of subaerial acid lavas and their tuffs among volcanics.

The second part of the Timiskamingian stage is characterized by a certain decline in tectonic activity, and is reflected in the dominance of fine-grained and better-sorted rocks in the upper portion of the complex. The termination in the formation of the complex is again registered as having witnessed an increase of oscillating movements and probably even a minor tectogenesis (the Barbertonian diastrophic episode).

Geochemical studies showed that the Paleoprotozoic sedimentary rocks are relatively richer in $Al_2O_3$, $Na_2$, O, and MgO than the corresponding rocks of the younger strata of the Protozoic and Phanerozoic.

This complex can be regarded as the oldest greywacke formation, and in different regions of the world the volcanogenic-greywacke, jaspilite-greywacke, and slate-greywacke sub-formations correspond to it.

The Moodies complex that crowns the erathem falls into the time when intensive oscillating movements had resumed. It is mostly typified by the contrast of cyclic sedimentation with the quick change of conspicuously different kinds of rocks: from thick conglomerates at the beginning of every cycle to pure quartzites or carbonate and cherty rocks (including jaspilites) in the middle and at the end of the cycles. Special lithologic studies revealed that sedimentation took place under deltaic conditions and in the littoral-marine tidal zone. During accumulation of sediments in the upper part of every cycle the land was approaching peneplain.

These strata are suggested to be distinguished as the contrast conglomerate-siliceous (cherty) formation that is not common in the younger rocks, least of all in the Phanerozoic. Certain piles conventionally attributed to the Moodies complex and made up almost solely of clastic rocks, conglomerates, and greywackes (like the earlier-mentioned Jones Creek and Kurrawang strata in Australia) form the lower part of this formation or probably represent a younger molasse-like formation: the "early molasse" of the Kenoran diastrophic cycle.

Some workers recognize in the Paleoprotozoic stratotypical sequence the Onverwacht Group as an ophiolite formation, the Fig Tree Group as flysch and the Moodies Group as molasse; however, from the data of Anhaeusser et al. (1969) and from the data examined here, it is possible to state that the composition and environment of formation of the groups mentioned (and of their analogs), and of the well-known Phanerozoic "formational triad", differ greatly, and so this correlation seems highly oversimplified. Naturally, it is necessary to consider the factor of evolution to a great extent. The Paleoprotozoic formations discussed constitute a single vertical range that reflects a peculiar development of the greenstone belts. The lower komatiite formation corresponds to the stage of generation of the mobile zones on the fractured Katarchean basement. The carbonate-siliceous (Steep Rock) formation reflects the origin of tectonic uplifts and a pause in volcanic activity. The bimodal tholeiite-rhyolite formation demonstrates a long stage of geosynclinal submergence and of intensive volcanism with a cyclic change in the products of outflow that gives a clue to the oscillating movements in the greenstone belts. The greywacke formation that replaces the previous one reveals a general inversion of tectonic regime that took place after the Swazilandian orogenic episode. The contrasting conglomerate-siliceous formation fixes a rather strong stabilization of the greenstone belts; it is likely to have been formed in the superimposed near-fracture depressions. The molasse-like formation, that is not reliably distinguished at the moment, completes the vertical formational range. Thus, the formations of the whole tectonic (geologic) megacycle are observed in the Paleoprotozoic.

## 4. Endogenic Processes

The resources of endogenic energy of the Earth must have greatly decreased at the time between Katarchean and Paleoprotozoic as a result of termination of the major stage of the abyssal differentiation of the planet at the end of the Katarchean and due to the "dying out" of many widely distributed radioactive elements (see Fig. 12) that occurred approximately at the same time. The geothermal gradient of the Paleoprotozoic was much less in the Paleoprotozoic than in the Katarchean, as is evidenced by a relatively low-grade metamorphism of many supracrustal strata. However, the value of this gradient was obviously higher than in the Late Precambrian, the more so in the Phanerozoic for comparable ubiquitous and intensive volcanism of every kind has never since been registered in the history of the Earth. It seems that a high geothermal gradient was an obligatory factor in the formation of komatiites, the rocks so typical of the initial stage of Paleoprotozoic volcanism.

Paleoprotozoic plutonic processes were of a lower scale than those of the Katarchean, but were nevertheless intensive. The commonly occurring intrusive bodies of ultrabasites and gabbroids, and also many cutting bodies of granitoids of quartz-diorite and granodiorite composition, are certainly hypogenic rocks, judging from their composition and petrochemical aspects, which originated under melting and differentiation of matter of the upper mantle. Certain diapir massifs of gneissoid granodiorites and migmatites (of the type of the Kaap-Valley or Theespruit massifs in Southern Africa), however, that crop out in the cores of mantled domes and affect the Paleoprotozoic supracrustals thermally and even display intrusive contacts with them, are very likely to have been formed as a result of a partial regeneration of the Katarchean tonalite gneisses of the basement, and the injection of the diapir mass (migma) took place in semi-melted (sub-solidus) state.

Certain small rare intrusive bodies of potassic granites, as a rule confined to the final stages of the Kenoran plutonic cycle, are likely to be of lithogenic nature. The magmatic melts of which these granites were formed are likely to have originated in the anatexis process in which the abyssal fluids rich in potassium were active. The ultrametamorphic (metasomatic) rocks are of subordinate significance among the Paleoprotozoic granitoids.

The explanation of the abyssal environment at the end of the Paleoprotozoic lies in the fact that the post-orogenic intrusions formed after the Kenoran diastrophism are represented by very large dykes and sills of basic-ultrabasic rocks such as: the "Great Dyke" of Zimbabwe (about 500 km long and 5–8 km wide), the Widgiemooltha Dyke Suite in Werstern Australia (one of the dykes is 320 km long), and the Stillwater layered intrusion in Montana (U.S.A.). Thus, at the end of the Paleoprotozoic, after the Kenoran orogeny, the Earth's crust is evidenced to have become so rigid that gigantic, extremely deep fractures could originate in it and serve as leaders for the mantle magma to penetrate. At the same time it is notable that the small post-orogenic intrusive bodies of alkaline granites and syenites, so common in the still later Precambrian and Phanerozoic, are quite rare in the Paleoprotozoic (they are mainly known in the Canadian Shield).

## 5. Tectonic Regime

When elucidating the principal aspects of the tectonic regime in the Paleoprotozoic, it is necessary to examine the origin of the most typical tectonic elements of that time, the so-called greenstone belts and the dome-shaped structures within them.

In the opinion of many workers (Anhaeusser, Viljoen, Viljoen, Glikson, Goodwin etc.), the Paleoprotozoic greenstone strata originated in the narrow troughs between the sialic blocks directly on the oceanic crust of basic-ultrabasic composition or on very thin sialic crust, and the exposures of these rocks we observe now as greenstone belts are the traces of the past troughs. However, on critical examination of the problem, this concept appears to necessitate further specification. Firstly, it is possible to state that as early as in the Katarchean a rather thick sialic crust had been formed and the later supply of granitoid matter during the Protozoic and Phanerozoic tectonoplutonic cycles only increased its thickness. Moreover, geologic observations reveal that the Katarchean gneisses and granite-gneisses ubiquitously underlie the greenstone strata. The basic-ultrabasic volcanics making up the lower part of many greenstone strata also rest on the basement rocks and, only because of this, cannot be regarded as formations of the primary oceanic crust. Finally, the greenstone rocks do not build up narrow zones ("belts") everywhere; in many regions of the world they occupy vast fields, as for instance, in Zimbabwe, in the east of the Superior structural province, in the west of Australia, and in Guiana, where their exposures are separated by granites or older rocks of the basement cropping out in the cores of domes (see Figs. 6, 16, 17, 21, 25, 26) [16]. Even in such places where the greenstone rocks form narrow zones as in Southern Africa, for instance in the Murchison Range area, and in the Barberton Mountain Land, it is evident that these zones are composite synclines separated by large and wide anticlines.

There are no paleogeographical (facial and tectonic) reconstructions sufficiently reliable to prove that the sedimentation and lava outflows during the Paleoprotozoic were localized only within the limits of the so-called greenstone belts. The criteria given by Goodwin for determination of boundaries of paleobasins (belts) according to the oxide facies of ferruginous formations, to psephytic rocks and acid volcanics, are not strict enough and hardly applicable even to the Canadian belts for which they were proposed as Walker showed (1978). The data available rather favor the existence of vast basins in the Paleoprotozoic and the accumulation of sediments and lavas taking place there. This is evidenced by the essentially marine nature of rocks, by the convincing persistence of greenstone strata and by their similarity in sequence; and in particular, by the presence of steady markers within the limits of very large regions (in different "belts").

At the same time in many areas (Zimbabwe, the Baikal Mountain Land, Guiana, etc.) the outcrops of greenstone rocks exhibit very composite and net-like pattern with two or three obvious crossing directions; usually one of these directions is the principal one typical of a group of "belts" that is changing from one place to another either gradually, or more often sharply. It can be suggested that the outcrops of the greenstone rocks are related to a certain regmatic system of fractures that controlled the lava outflow and separated small blocks of the older rocks of the basement – the microcratons. In some regions it is possible to determine that the greenstone belts are confined to abyssal fractures and to related feathered dislocations of lower rank. This aspect is especially striking in the Baikal Mountain Land for instance, where the rocks of the ophiolite series occur in wide and extensive zones (several hundreds of kilometers) of abyssal fractures of north-west and north-east direction. These fractures

---

16 Generally the term "greenstone belt" has no strict definition as applied by Canadian and sometimes by Southern African authors; they give this name to very small narrow outcrops of greenstone rocks and to groups of such outcrops and to vast linear fields of distribution of these rocks (superbelts according to Young 1978)

originated at the very beginning of the Paleoprotozoic and controlled distribution of sedimentary and volcanogenic facies in the greenstone belts during the whole era. Movements along these fractures were recurrently renewed in a later period, up to the Early Cambrian. It is important that many Paleoprotozoic intrusions are situated in the zones of abyssal fractures and higher metamorphism (to amphibolite facies) is observed there (Salop 1964–1967).

It is suggested that the system of fractures and related greenstone belts originated as an effect of a global-scale stretching of the Earth's crust. In the areas of the most intensive processes of shattering and stretching of the lithosphere, a composite net of near-fracture troughs originated, and along their rims the lava flowed out on the surface along the fractures. The greenstone strata were accumulated on the whole territory subjected to destructive processes and subsidence, but most notably in the initial near-fracture troughs that were transformed into synclinorium structures due to folding. The system of troughs and uplifts separating them are to be regarded as the oldest geosynclines or proto-geosynclines and the greenstone belts within their limits as the traces ("gashes" or "roots") of certain near-fracture troughs where the supracrustal strata were preserved and avoided erosion because of their great depth.

It is notable that the greenstone belts and their systems are as a rule observed to be irregularly distributed: in some regions they are abundant, in others few or completely absent. It is difficult to explain in many cases whether this phenomenon is due to irregular primary distribution or to a different level of erosion. However, the first factor seems to be of greater significance, in which case the areas of distribution of the Katarchean gneisses where the greenstone belts are lacking are to be assigned to the oldest relatively stable elements of the Earth (with great thickness of crust?) and are to be defined as protoplatforms. The area of the Aldan shield (including that part buried under the young strata) could have been such a protoplatform (or the "lithoplint" of Dzevanovsky) during the Paleoprotozoic. In the area of this shield are observed only isolated graben-like structures with Paleoprotozoic sedimentary-volcanogenic strata (the Subgan and Olonda Groups), that were formed in the near-fracture troughs that resemble aulacogens of the "younger" platforms. At the same time they differ from the latter by the presence of relatively abundant volcanics, by strong deformation, and by high metamorphism (up to epidote-amphibolite facies) and also by the occurrence of small intrusive bodies of gabbroids and granites. Probably it is reasonable to distinguish these troughs as special structures, protoaulacogens. Another Paleoprotozoic protoplatform is likely to have been situated in Northern Siberia in place of the now-existing Anabar shield.

Two large protoplatforms (the Hudsonian and Minnesotan) separated the wide Ontarian geosynclinal system of sub-latitudinal strike in North America (see Fig. 15). The Hudsonian protoplatform occupied the major portion of northern Canada, embracing almost all of the Churchill tectonic province, the major part of the Labrador Peninsula (the Ungava craton), Baffin Land, and a part of Greenland.

It is possible that a large protoplatform also existed on the site of the Kasai shield in Equatorial Africa, as suggested by the fact that densely situated greenstone belts with smoothly changing general strikes of structures regularly skirt this area (see Fig. 18). A remarkable feature of north-eastern Africa (the area of the Eastern Sahara and the Nile basin) is the absence of greenstone belts. Is it not possible that another protoplatform existed there previously?

The absence of covers on the platforms and of the miogeosynclinal strata in the geosynclinal systems seems at first sight to contradict the conception of large proto-platforms, as their outer zones are usually built up of these piles in the younger systems of the Protozoic and Phanerozoic. However, the real cause seems to lie in the intensive volcanism in all zones of the oldest geosynclinal systems and also in the fact that the Paleoprotozoic rocks are usually subjected to deep erosion and naturally, the thin sedimentary deposits on the platforms and in their framing must have been destroyed by this erosion. The existence of rather large protoplatforms in the Paleo-protozoic is evidenced not only by the tectonic structure, but also by presence of such mature rocks as quartzites and high-alumina schists in the greenstone strata (mostly in their upper part); these rocks were formed of the material supplied from the peneplain. The intra-geosynclinal uplifts and microcratons could not have been covered by the thick, extensive crust of weathering, and could not have provided the necessary supply of weathered material.

Thus in the Paleoprotozoic the geosynclinal systems and platforms are reported as originating for the first time in geologic history and thus the permobile stage of tectonic evolution of the Earth typical of the Katarchean changed to the platform-geosynclinal stage that exists up to the present. This extremely important turning-point in the history of the planet provides the main reason for separating the Paleo-protozoic from the Katarchean and for assigning it to the Protozoic Eon.

Certainly, the tectonic regime and physical environment during the Paleoproto-zoic were to a great extent inherited from Katarchean conditions and were conspic-uously different from the later periods, as will be shown below. The Paleoprotozoic is a transitional era in a sense, though it is quite specific. In the Paleoprotozoic, the Earth's crust was much more stable than it had been in the Katarchean, but at the same time it was more labile than in the following Mesoprotozoic Era. Considerable mobility of the lithosphere is evidenced by exceptionally strong volcanism character-ized by outflow of mainly ultrabasic-basic lava, by common episodes of intrusive events, by very large-scale granitoid diapirism when the hypogenic granodiorites are dominant among its products, which seem to originate as a result of abyssal differen-tiation of basic magma. At the same time, the existence of fracture systems of abyssal origin evidences sclerotization of the crust; both types of the systems are meant here, the initial ones controlling the outflow of lavas and formation of protogeosynclines, and the postorogenic ones controlling large ultrabasic dykes.

Paleoprotozoic geosynclinal systems and protoplatforms do not seem to differ greatly; it is even possible to suggest the existence of transitional types of structures or zones. The geosynclinal systems were wide and within them numerous inner uplifts and small cratonic or "semi-cratonic" blocks resembling the later median masses were situated. Accumulation of lavas and sediments occurred on these massifs, but the piles were much less thick than in troughs. The morphostructures of geosynclinal systems (and of protoplatforms certainly) were of small amplitude and low contrast. The psephytic clastic rocks, conglomerates and greywackes, mainly originated not as a result of erosion of active high uplifts but as a result of washing out of volcanic structures and of margins of inner massifs (blocks) or platforms; they accumulated in troughs or in graben-troughs between relatively stable blocks. This is possibly the cause of the quick change of psephytic rocks into fine-grained and mature sediments.

High arched uplifts could not be formed because of the relatively low strength of the Earth's crust.

It was noted above that the sedimentary-volcanogenic strata of the Paleoprotozoic accumulated principally in marine basins at moderate and shallow depth and the occurrence of clastic rocks among them evidences the existence of extensive source areas. Thus in the Paleoprotozoic, unlike in the Katarchean, the hydrosphere did not cover the entire surface of the planet or the major portion of it; marine basins were, however, extensive at that time, but were separated by large areas of land. The question is posed: where did the water mass of the Katarchean Panthalassa disappear to? The most natural explanation of this seems to be the following: during the Paleoprotozoic the first capacious troughs originated in the sites of the modern oceans, very likely in the place of the Pacific Ocean. The large-scale extension of the Earth's crust, that resulted in a regmatic net of fractures and lava outflow, was possibly the cause of division and submergence of plates of lithosphere and the formation of troughs with oceanic type crust (see Chap. 8).

## 6. Origin of the Mantled Gneiss Domes

It has already been stated that the structures of the mantled dome type are the most typical of Paleoprotozoic tectonics; in the cores of domes the remobilized rocks of the basement and/or the Paleoprotozoic granitoids crop out, in their limbs – the green-stone rocks. Dome swarms are registered not only in the greenstone belts, but also in the anticlinorium zones separating the belts, although the domes in these zones are identified only by the structural elements and by the isometric bodies of gneissoid granites of the Paleoprotozoic.

The origin of these structures was explained by Eskola (1949), who seemed to offer an up-to-date solution, which, however, must now be refined and presented in more detail. Naturally the formation of domes is related to the local rise of mobilized and partially rheomorphosed basement matter or of stocks of hypogenic granite magma against the background of a general migration of the heat front from the interior of the planet. It can be suggested that at the basement-cover boundary an accumulation of heat occurred due to a heat conductivity of crystalline rocks higher than that of sedimentary and volcanogenic rocks; finally this concentration of heat led to a partial melting of rocks and to rheomorphism (palingenesis).

The basement mobilized in the zone of anatexis together with the anatectic granite magma or the hypogenic granite magma uplifted and moved apart the rocks of the supracrustal cover, rising to the surface as diapirs, and thus the structures of the mantled dome type appeared. While moving upward, the rheomorphosed matter can creep laterally because the resistance of the wall-rocks of the cover becomes lower. As a result, the stock of rheomorphosed rocks acquires a mace or mushroom shape, which explains why the cover rocks around the gneiss cores of many domes are registered as exhibiting overturned bedding. As an effect of heat and thermal solutions, the cover rocks near the domes are subjected to metamorphism and different metamorphic zones passing one into the other are created around the core of a dome (Fig. 29). In the Paleoprotozoic domes this metamorphic zonation is poorly seen, due to a strong masking by regional alteration of the rocks, but in younger structures of this type it is strikingly obvious.

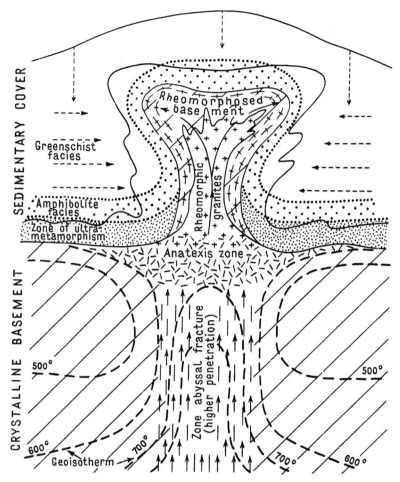

**Fig. 29.** Scheme of formation of a gneiss dome. (After Salop 1971.) Explanation of the figure is in the text

In some cases the mantled (or diapir) domes are situated in the zones of abyssal fractures, and then it is suggested that the fractures served as pathways for the heat or in other words they were the zones of high permeability for magma and thermal solutions. However, it is not a general rule for the domes to be thus related to the fractures, and in most cases the domes are situated irregularly (the dome "swarms"). The causes of this chaotic distribution are not known, but the idea of the fall of a meteorite with the following formation of the Paleoprotozoic (and many younger) mantled domes in these sites seems to be inviting, just as the theory that suggests the falling of cosmic bodies at the early ("Moon") stage of the Earth's evolution and thus explains the origin of gneiss ovals.

At the time examined, however, the Earth's crust was thick enough and rigid, and the meteorite bodies must have become much smaller as the protoplanet nebula had already been greatly reduced. Certain large meteorites or their large clasts must have affected the Earth, but the astroblemes, on appearing quickly, became the victims of erosion and as a rule could not be preserved in the succession of the older strata. Their traces as impactites and their minerals are to be sought in the clastic matter of Paleoprotozoic terrigenous rocks. The effect of the impact naturally resulted in a local crushing of the Earth's crust and in some cases even the magmatic chamber could have been formed by the following outflow of magma along the fractures. In some cases very thick piles of impactite regoliths could have been formed and were very good thermal insulators for the heat to be accumulated under the following warming of the Earth's crust.

Thus the falling of large meteorites had to create a structural and partially thermal heterogeneity of the Earth's crust and it governed the localization of endogenic processes and of the related tectonic patterns. A special effect could be produced by the falling of meteorites during the diastrophic cycles that were characterized by a general rising of the thermal front and by the appearance of magmatic chambers. The impact of large meteorites must have caused an additional emission of heat and even melting of rocks and certainly stimulated the emergence of a granite diapir mass in the crushed portions of the crust.

Certain data favoring the formation of the mantled domes as a result of the impact of meteorites will be discussed in the next chapter when the Mesoprotozoic corresponding structures are examined.

It is therefore suggested here that many (but not all!) mantled domes of the Paleoprotozoic originated as a result of interaction of endogenic and exogenic (cosmic) factors, the former being the most important.


## 7. Principal Stages of Geologic Evolution

Summing up the material discussed in this chapter it is possible to present the succession of geologic events in the Paleoprotozoic in the following scheme (starting from the earliest events):

### Early Paleoprotozoic

1. 3550 (3600)–3400 m.y. Extension of the Earth's crust; formation of regmatic fracture system; separation of protoplatforms – the areas of relatively stable tectonic regime from the mobile zones-protogeosynclines; appearance of oceanic basins (?). Outflow of ultrabasic-basic lavas along the abyssal fractures locally accompanied by accumulation of detrital rocks: formation of the Komatian and Pontiac lithostratigraphic complexes. The organic remains of microfossils (prokaryota) are reported for the first time in cherts.
2. 3400–3370 m.y. The Belingwean diastrophic episode. The oscillating movements of mostly positive character, probably accompanied by subvolcanic intrusions of basic and acid composition.
3. ~ 3370 m.y. Deposition of psephytes, carbonate and siliceous (cherty) sediments, evaporites in part, formation of the Steep Rock lithostratigraphic subcomplex of the Keewatinian complex. First phytolites reported from carbonate rocks.
4. 3370–3200 m.y. Renewal of the movements along fractures and deepening of troughs in protogeosynclines. Intensive polycyclic volcanism: outflows of calc-

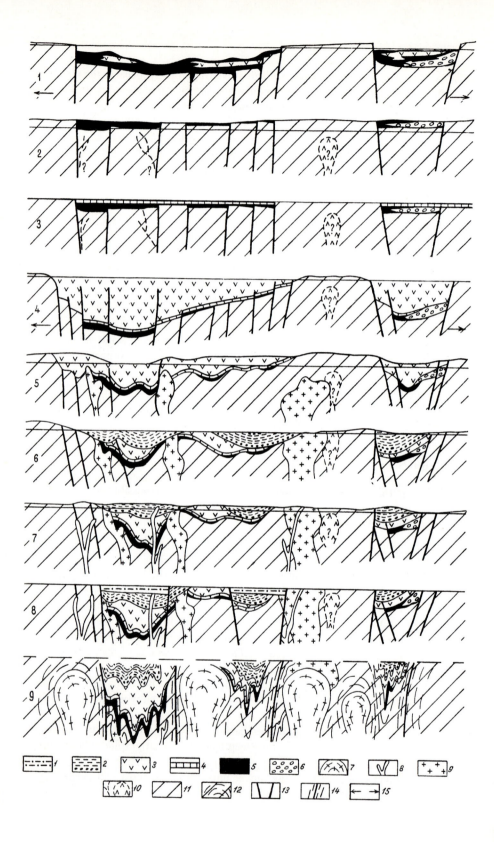

1 ⌐⌐⌐⌐ 2 ⌐∷∷⌐ 3 ⌐ᴠᴠ⌐ 4 ⌐═══⌐ 5 ⌐██⌐ 6 ⌐ᴼᴼᴼ⌐ 7 ⌐∧∧∧⌐ 8 ⌐ᴠ⌐ 9 ⌐+++⌐

10 ⌐∧∧∧⌐ 11 ⌐╱╱⌐ 12 ⌐~~⌐ 13 ⌐╲╱⌐ 14 ⌐∥∥⌐ 15 ⌐←→⌐

alkaline lavas of basic to acid composition (in every cycle); accumulation of terrigenous and chemogenic sediments from time to time, formation of the principal part of the Keewatinian lithostratigraphic complex.

5. 3200–3150 m.y. The Swaziland diastrophism of the second–third order. Strong oscillating movements of mostly positive character accompanied by relatively intensive folding, by partial mobilization of the basement and by intrusions (also protrusions) of granitoids of granodiorite composition.

**Late Paleoprotozoic**

6. ∼ 3150–3000 m.y. Inversion of tectonic regime in protogeosynclines. Deposition of poorly sorted terrigenous sediments in intrageosynclinal and superimposed troughs; local outflow of lavas of mostly acid and intermediate (to sub-alkaline) composition, formation of the Timiskaming lithostratigraphic complex. Subvolcanic intrusions.

7. ∼ 3000 m.y. The Barbertonian diastrophic episode. The oscillating movements of mostly positive character; possible slight folding and small intrusions of granites including of potassic type.

8. 3000–2800 m.y. Deposition of sediments of cyclic and contrasting sequence (conglomerates, quartzites, carbonate, and cherty rocks) in relict and superimposed troughs inside protogeosynclines, formation of the Moodies lithostratigraphic complex.

9. 2800–2600 m.y. The Kenoran diastrophism of the first order. Intensive tectono-plutonic processes: strong folding, formation also of mantled domes, intrusions of granitoids of mostly granodiorite composition (also of the potassic granites at the end of the cycle). Closing of many of the protogeosynclinal systems. Local accumulation of molasse.

10. 2700–2550 m.y. Large-scale splitting of the Earth's crust and injection of ultrabasic (to basic) magma, formation of gigantic dykes and sills of ultrabasites and gabbroids (Fig. 30).

**Fig. 30.** Scheme of evolution of greenstone belts (protogeosynclinal zones) in the Paleoprotozoic. *Numbers* near the profiles correspond to the major stages of tectonic evolution of protogeosynclines. Explanation in the text.
*1–6* rocks of Paleoprotozoic lithostratigraphic complexes: *1* Moodies, *2* Timiskamingian, *3* Keewatinian, *4* Steep Rock (subcomplex), *5* Komatian, *6* Pontiac (subcomplex), *7* diapir intrusions of granitoid of the Kenoran diastrophic cycle, *8* small intrusions of granite of the Barbertonian diastrophic episode (presumed), *9* granitoid (granodiorite) instrusions of the Swazilandian diastrophism, *10* subvolcanic instrusions of the Belingwean diastrophic episode (presumed), *11* Katarchean basement, *12* the same basement rocks, but remobilized, *13* fractures, *14* zones of schistosity and milonitization, *15* stretching of the Earth's crust

# Chapter 4   The Mesoprotozoic

## A. Rock Records

### 1. General Characteristics of the Mesoprotozoic

**Definition.** To the Mesoprotozoic are attributed the supracrustals and plutonic rocks that were formed after the Kenoran diastrophism or at the end of it (the "transitional" units) but previous to termination of the Karelian orogeny that completed the Mesoprotozoic Era 2000–1900 m.y. ago. Thus the duration of the Era is ∼ 800 m.y. Earlier in the U.S.S.R. the strata examined were assigned to the Middle Proterozoic. In 1979 the Stratigraphic Committee of the U.S.S.R. recommended attributing these rocks to the Lower Proterozoic, as is done in many other countries. In Canada it is recognized as the Aphebian, in Australia in most cases as the Nullaginian. Following the world stratotype of the erathem – the Karelian Supergroup – it can also be called the Karelian.

**Concise Characteristics of the Erathem.** The erathem comprises all major formational types of rocks: platform, miogeosynclinal, and eugeosynclinal. The first two types are exceptionally widespread and in this respect the Mesoprotozoic Erathem differs sharply from the previous one, where such rock types were absent. In many regions the platform and miogeosynclinal rocks are closely related and exhibit many common features. Both types comprise several groups separated by breaks and revealing transgressive-regressive succession. The lower groups are usually of local distribution and the overlying groups rest in a tile-like or cover-like manner. Among the rocks of these groups the mature products of sedimentary differentiation are quite abundant. They are, moreover, characterized by very peculiar, sedimentary formations like jaspilites of special types and gold-uranium conglomerates unknown among the strata of different age. In the middle and upper parts of the erathem red beds of low abundance are reported for the first time; evaporites and their traces are also known here. Some groups contain bands (strata) of different slates and carbonate rocks. Volcanics usually occur in many platform and miogeosynclinal groups, as a rule in their upper part. Eugeosynclinal rocks are commonly represented by one thick group and rarely by two groups. It is notable that volcanics are of much lower importance in these strata than in the corresponding strata of the Paleoprotozoic, and komatiites are quite rare or absent among them.

   In many regions of the world tillites are reported from Mesoprotozoic strata, they occur at several stratigraphic levels. The first geologic glacial epochs are fixed by these formations.

The Mesoprotozoic is divided into three sub-erathems by two small-scale (third order) diastrophic cycles recognized as Seletskian and Ladogan; theses three sub-erathems: Lower, Middle, and Upper Mesoprotozoic, differ in composition and inner structure in many respects.

The thickness of the Mesoprotozoic varies greatly. It reaches 15,000 m and even more in the most complete, thickest sequences, but sometimes the values given are evidently overestimated due to errors in calculation and in summing up the thickness of separate isolated units.

The metamorphism of Mesoprotozoic rocks differs in grade according to the tectonic position of the rocks. The platform strata are locally of extremely low alteration which seems close to epigenesis, but are still typified by the lowest grade of greenschist facies. Metamorphic zonation of geosynclinal belts is chracterized by greenschist to amphibolite facies. Certain data on granulite facies alteration of the Mesoprotozoic rocks seem to be erroneous: in such cases the highly metamorphosed Katarchean rocks, subjected to isotope rejuvenation under later thermal events, were attributed to the Mesoprotozoic (Lower Proterozoic).

The plutonic formations of the Karelian cycle are widespread and are represented by pre-tectonic intrusions of gabbroids and ultrabasites, by syntectonic and late-tectonic intrusions of granitoids (including the potassic type) and by the post-tectonic composite (layered) intrusions of basic to acid composition. In some regions intrusive rocks are reported, formed during lower-order diastrophic cycles, but these rocks are of limited extension.

The relations of the Mesoprotozoic rocks and the underlying formations and their geochronologic (radiometric) boundaries are reliably defined in most cases. The lower and upper geologic boundaries of the era are generally drawn along large unconformities. The age of the Karelian cycle granitoids that cut the rocks of the erathem falls within 2000–1900 m.y. (there are abundant datings by different isotope methods). The phenomena of isotope rejuvenation are not typical of the Mesoprotozoic rocks and they are mostly evident in the younger foldbelts.

**Distribution.** Mesoprotozoic strata are developed in all continents and are more widespread there than Paleoprotozoic rocks. The most complete and well-known units among them are the following: in Europe: the Karelian Supergroup of Karelia and Eastern Finland, the Pechenga and Imandra-Varzuga Supergroups of the Kola Peninsula, the Bothnian Group of the Central Finland, the Elveberg Group and its analogs in Sweden, the Krivoy Rog and Frunze Mines Groups of the Ukraine, the Kursk and Oskol Groups of the Voronezh massif (KMA); in Asia: the Udokan Group of Eastern Siberia, the Zama Group of the western Baikal region, the Teya Group of the Yenisei Ridge, the Spassk, Mitrofanovo, and Lysogorsk Formations of the Far East of the U.S.S.R., the Huto Group of Wutai Shan and the Liao Ho Group of the Liangtou Peninsula (China), the Middle-Upper Dharwar Supergroup of the Western Hindustan, the Aravalli and Raialo Groups of Rajasthan; in Africa: the "Witwatersrand Triad", the Transvaal Supergroup and the Matsap Group of the Republic of South Africa, the Marienhof and Abbabis Groups of Namibia, the Ubendian and Konse Groups of Tanzania, the Ruzizi Group of Ruanda and Burundi, the Buganda-Toro Group of Uganda, the Muva Group of Zambia, the Zadinian Group of the Republic of Congo and of Zaïre, the Premayombe Group of

Gabon, the Tannekas Group of Togo and Benin, the Birrimian Group of Western Africa, the Aguelt Nebkha Group of Mauritania, the Tassendjel Group of Algeria; in North America: the Huronian, Animikie, Kaniapiskau, and Great Slave Supergroups, the Penrhyn, Epworth, Goulburn, and Hurwitz Groups developed in different regions of Canada, the Ketilidian Supergroup (the Vallen and Sortis Groups) of Southern Greenland, the Phantom Lake, Deep Lake and Libby Creek, the Vishnu and Big Back Groups in different regions of the U.S.A.; in South America: the Tamandua Group, the Minas Supergroup, and the Itacolomi Group in the Minas Gerais state, the Jacobina Group in Bahia state and the Tocantino Group in Goyas state of Brazil; in Australia: the Mount Bruce Supergroup in the Western part of the continent, the Batchelor, Goodparla, and Finniss River Groups in the north and the Middleback Group in the south.

**Supracrustal Rocks.** Widespread and typical among the platform and miogeosynclinal strata are the monomictic and oligomictic quartzite-sandstones or quartzites with commonly well-preserved psammitic or relict blasto-psammitic structure with local cross-bedding, ripple marks and other aspects of shallow-water environment. In some groups the quartzites associate with quartz conglomerates and also with slates or schists rich in alumina that are altered in pyrophillitic, chloritoid, disthen, and sillimanite types depending on the metamorphic grade. Alongside such mature rocks, arkose, subarkose, and polymictic sandstones are also reported, the latter and the greywackes also being more typical of the miogeosynclinal complexes. The red beds completing the Mesoprotozoic are typified by lithoclastic sandstones and gritstones.

Dolomites are obviously dominant among carbonate rocks. The organogenic dolomites formed of stromatolites are common, in the middle and upper parts of the erathem in particular. Locally the carbon-rich graphitic ("coaly") or shungite slates associate with carbonate or carbonate-slate strata.

The older evaporites are represented by their products of alteration: by the albitized, scapolitized, and zeolitized slates or dolomites with abundant chlorine (up to $5\%-8\%$). In the low-grade rocks the pseudomorphoses after halite and gypsum are observed locally; these minerals are also known as inclusions and fine lens-like laminas in some red beds.

The iron formations of the Krivoy Rog, Animikie (the "Superior Lake") and Urals types, as also the gold-uranium conglomerate formation that are quite typical of the Mesoprotozoic miogeosynclinal strata, will be examined below when describing the lithostratigraphic complexes to which they belong.

Tillites are distinctly distinguished in the Mesoprotozoic Erathem according to their typical features. They were first established in the Huronian Supergroup of Canada (the Gowganda tillites and others) and now are known from many places in the world. Commonly they make up a part of a glacial complex where the glacial-marine or glacial-lacustrine deposits also belong; these are commonly represented by banded slates or sand-clay rocks with isolated ("swimming") boulders with the beds downwarped beneath them (Fig. 31).

Volcanics of the miogeosynclinal and platform strata are mainly represented by the rocks of basic and intermediate composition: diabases, andesites, and porphyrites belonging to the diabase-porphyrite and trappean formation, but in some strata lava of different composition are developed: pillow tholeiite basalts and picrites or plagio-

**Fig. 31.** Gowganda Tillite, Canada, Quirke-Lake area. (Photo by the author)

porphyries and quartz porphyries (rhyolites). Volcanics commonly associate with abundant pyroclastic rocks. Eugeosynclinal groups mostly contain tholeiite basalts, spilites, keratophyres (albitophyres), and plagioporphyries (here the nonaltered rocks are mentioned but in major cases their metamorphosed equivalents are present).

**Organic Remains.** In the Mesoprotozoic the microscopic remains of different pro-karyota and also the products of their life activity, phytolites, are usually known to occur. Some rocks are suggested to contain eukaryota microphytofossils also (see below).

Phytolites are specially abundant in the carbonate rocks of the Middle and Upper sub-erathems of the Mesoprotozoic where they build up large bioherms and even thick, extensive bands or horizons. The cherty rocks there also contain spherical and filamentous microscopic remains.

Attempts at recognizing some specific complex of the Mesoprotozoic have been unsuccessful. In different regions the phytolites are represented by peculiar ("endemic") forms, by forms typical of phytolite complexes of the Upper Precambrian and of the Riphean of the Urals in particular. Thus, in some parts of Siberia the carbonate rocks of the erathem contain stromatolites typical of the Lower Riphean complex with the characteristic *Kussiella kussiensis* form. Kussielids are described from the Mesoprotozoic of Canada (Awramik 1976, Walter and Awramik 1979). The Meso-protozoic strata of other regions contain together with *Kussiella* stromatolites (or without these) different forms of phytolites belonging to the younger phytolite complexes of the Riphean, of the Upper Riphean in particular. Thus, Hofmann (1977), applying the classificational aspects of stromatolites elaborated by Soviet workers, established that the Mesoprotozoic (Aphebian) of Canada is characterized by the following stromatolites: *Gymnosolen, Jacutophyton, Katavia, Minjaria*, and *Tungussia* typical of the Upper and Middle Riphean of the Urals. The stromatolites resembling the Upper Riphean, *Gymnosolen* and *Jurusania*, are also described from the Onega Formation of the Mesoprotozoic of Karelia (Krylov 1966). In Mesoprotozoic strata of the Hamersley basin of Australia, together with local forms, stromatolites close to the *Patomia* (Preiss 1977) typical of the Upper Riphean and Vendian also occur. Many stromatolites recently described from different Mesoprotozoic strata and given new names (see Raaben (ed.) 1978, Semikhatov 1978) exhibit little difference from the ones earlier established in the phytolite complexes of the Upper Precambrian.

**Tectonic Structures.** The structural elements of the Mesoprotozoic are typified by relatively large platforms with angular margins surrounded by broad miogeosynclinal belts and with eugeosynclinal areas situated between the latter. The platform boundaries are usually defined by the marginal abyssal fractures. In some places the narrow near-fracture depressions resembling typical aulacogens penetrate the platforms from the geosynclinal belts. Inside geosynclines, in the zones of abyssal fractures the blocks of the older rocks of the type of median mass are situated (see Figs. 40–44). The tectonic pattern is characterized by the linear folds complicated by local disruptures, including overthrusts. Domes are no longer a necessary element of the foldbelts, as in the Paleoprotozoic, but in some regions they are quite abundant (see Fig. 45).

### 2. The Mesoprotozoic Stratotype and Its Principal Correlatives in Different Parts of the World

**The Karelian Supergroup is the Stratotype of the Mesoprotozoic.** Many comparatively complete and well-studied sequences are known in the Mesoprotozoic Erathem (some of them will be examined in this section), but in my opinion the Karelian Supergroup is the most suitable for a world stratotype, as it answers the requirements, and is widely developed in the eastern part of the Baltic shield in the Soviet Karelia and in the adjacent areas of Finland. In some portions of the Kola Peninsula strata resembling the Karelian are also developed. There they are usually assigned to the Pechenga and Imandra-Varzuga Supergroups, their correlation with the corresponding strata of Karelia, however, is quite evident and many workers attribute them to the Karelian Supergroup and distinguish larger units (groups) within it, giving them the same names. Publications on Karelian formations are quite numerous, with contributions from Soviet and Finnish authors. Among the recently published works the following seem to be the most valuable: Kharitonov (1966 etc.), Krats (1963 etc.), Sokolov (1963 etc.) and Negrutza (1968 etc.). At the moment the stratigraphic sequence of the Mesoprotozoic in Karelia seems to be well-known, but some particulars, as for instance, the problems of nomenclature and volume of the units and the boundaries of these units, are debatable.

The rocks of the Karelian Supergroup in different regions differ slightly in formational aspect, but the nature of the sequences is consistent and in most cases it appears possible to give their larger units (groups) common names. Two types of strata are distinguished: platform (or subplatform) and miogeosynclinal. The first type is distributed within the limits of the so-called Karelian massif ("the Jatulian continent" according to Väyrynen 1954), that probably represented a block of a large Sarmatian platform now buried under the thick cover of the younger strata of the Russian plate.

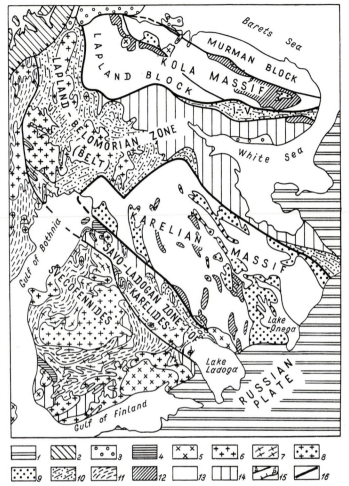

**Fig. 32.** Scheme of geologic structure of the eastern part of the Baltic Shield.
*1* platform cover of the Russian plate, *2* Caledonides of Northern Norway, *3* Upper Precambrian (Hyperborean) strata of Northern Norway and Rybachy Peninsula, *4* Upper Precambrian (Jotnian) strata, *5* rapakivi-granite (Lower Neoprotozoic), *6* Mesoprotozoic late-tectonic granite, *7* Mesoprotozoic syntectonic granitoid, *8* pre-Mesoprotozoic granitoid intensively mobilized in Mesoprotozoic (Central Finland massif), *9* Mesoprotozoic strata of subplatform type, *10* Mesoprotozoic strata of miogeosynclinal type (in zone of Karelides), *11* Mesoprotozoic strata of eugeosynclinal type (in Svecofennian area), *12* Paleoprotozoic greenstone strata, *13* Katarchean gneiss basement of Mesoprotozoic cratons (old platforms), *14* Katarchean gneiss basement of Karelides and Svecofenides, strongly reworked by Mesoprotozoic tectono-thermal processes, *15 a* large fractures, *15 b* frontal thrust of the Caledonides, *16* boundaries of Mesoprotozoic cratons (old platforms) coinciding with abyssal fractures

The second type is developed in the framing of the massif within the limits of the Savo-Ladogan and Lapland-Belomorian geosynclinal zones (Salop 1977a), (Figs. 32 and 43). The platform strata are rather thin, of simple folding (commonly of the near-fracture type), of slight alteration (the greenstone facies), exhibiting a great number of breaks, and wide distribution of mature sedimentary rocks. The miogeosynclinal strata are quite thick, intensely folded and highly altered (the grade goes up to amphibolite facies with the distance growing from the margins of the Karelian massif) and in the upper part of the supergroup rhythmically bedded greywacke-slate strata of flysch type are observed. In the miogeosynclinal zones many massifs of Mesoprotozoic granitoids are registered. On the craton they are lacking. At the same time in both types of strata, abundant volcanics are known, and sills and dykes of basic and ultrabasic rocks are also observed. Later is will be shown that this feature is typical of other Mesoprotozoic strata of platform and miogeosynclinal types and that by this very feature they differ from the corresponding younger formations. At the same time it should be noted that initial volcanogenic formations typical of eugeosynclines are absent in these strata. Volcanics here associate as a rule with comparatively mature sedimentary rocks and usually complete the sedimentary macrorhythm (cycle).

Here six groups are attributed to the Karelian Supergroup (or Karelian complex); starting from the oldest they are the Tunguda-Nadvoitsa, Sariolian, Segozero, Onega, Bessovets (Ladoga), and Vepsian (Kamenoborsk-Shoksha). All these groups were formed during a single geologic megacycle that corresponds to the Mesoprotozoic Era and they generally differ from the underlying and overlying supracrustal complexes. Two lower groups are sometimes divided as the pre-Jatulian formations; the Segozero and Onega Groups are usually combined under a common name "Jatulian" and the remaining two upper formations are regarded as post-Jatulian. This traditional division is reasonable, as will be shown later. Here these three units are attributed to the Lower, Middle, and Upper sub-eras of the Mesoprotozoic respectively.

These groups overlie each other with a stratigraphic break as a rule and sometimes with a slight angular unconformity. The most significant unconformities are observed between the pre-Jatulian, Jatulian, and post-Jatulian. In certain regions (structures) not all the groups are present, as some of them are absolutely missing in the sequence or reduced unter the influence of erosion that occurred prior to deposition of the next group. Local distribution of the groups is typical of pre-Jatulian strata and of the post-Jatulian in part; the Jatulian is widespread, is almost ubiquitous, and the general character of the Karelian formations is known through it, of the Karelian massif in particular. Pre-Jatulian groups are mostly developed in the Lapland-Belomorian zone of Karelides (except the Sariolian Group that is registered as occurring also in some places of the Karelian massif); for the post-Jatulian groups, the Ladoga Group is only developed in the Savo-Ladoga zone of Karelides and the Vepsian Group only in the south of the Karelian massif in the Onega structure (see Figs. 32 and 33).

The Tunguda-Nadvoitsa (Sumian) Group in the Lapland-Belomorian zone and on the adjacent margins of the Karelian massif starts the Mesoprotozoic sequence. It is of binary structure: the lower part is made up mainly of thin terrigenous rocks (80–120 m) represented by basal arkose or polymictic conglomerates and overlying quartzite-sandstones and also of pyritized quartz conglomerates; the upper part is composed of volcanogenic rocks, of basic composition in the lower portion, mainly metadiabases (up to 1800 m), and of acid composition in the upper portion (up to 350 m). Zircon from the quartz porphyry dykes cutting metadiabases is dated at 2440 ± 45 m.y. (Krats et al. 1976). The intrusive bodies (sills and dykes) of basic rocks in the Lapponian Group of the north-eastern Finland have an age of 2425–2450 m.y. (by the Rb-Sr method, determinations made by Kouvo, see Piirainen 1978); the Lapponian Group is an analog of the Tunguda-Nadvoitsa Group of Karelia. It is

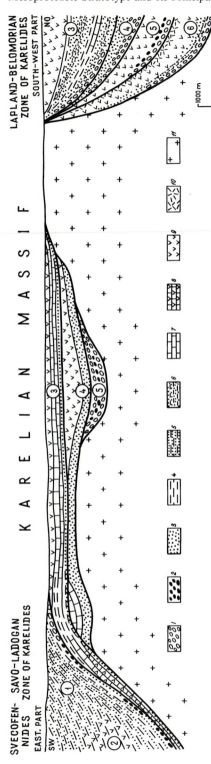

**Fig. 33.** Schematic profile revealing distribution of Mesoprotozoic groups (of the Karelian Supergroup) in the Karelian massif and in the adjacent geosynclinal zones. (After V. Negrutsa, revised).
*1* polymictic conglomerate, *2* tillite and tilloid, *3* quartzite and quartz conglomerate, arkosic and polymictic sandstone, slate and siltstone, slate in part, *4* slate and siltstone, *5* interbedding of slate and greywacke (flysch formation), *6* interbedding arkosic sandstone, quartzite, and slate, *7* carbonate rocks (mostly dolomite), *8* interbedding of carbonate rocks and basic volcanics, *9* basic and intermediate volcanics, *10* acid volcanics, *11* basement rocks (Katarchean and Paleoprotozoic).
Numbers in circles: *1* the Ladogan Group, *2* the Bothnian Group, *3* the Onega Group, *4* the Segozero Group, *5* the Sariolian Group, *6* the Tunguda-Nadvoitsa Group (Paleoprotozoic)

highly probable that these datings give the time of intrusive events younger than the overlying Sariolian Group, and correspond to the Seletskian diastrophic episode. In any case it is certain that the Jatulian quartzites unconformably overlie the intrusive quartz porphyries and basic rocks analyzed; the Sariolian Group is absent in these areas.

The Sariolian Group is locally developed on the Karelian massif, and is of wider distribution in the north and north-east portions of the rocks surrounding the massif. It rests sharply unconformably on the Katarchean and Paleoprotozoic rocks or unconformably on the Tunguda-Nadvoitsa Group. The lower part of the group (300–800 m) is made up of boulder-pebbled conglomerates, gritstones, arkoses, and quartz conglomerates, the upper part (up to 1200 m) is volcanogenic, composed of aphanite, porphyric, or amygdaloidal diabases, variolites and tuffs of basic volcanics with subordinate interbeds of tuffites, sandstones, and conglomerates. The glacial formation occurrences are the most characteristic feature of this group; these are tillites and in most cases lacustrine-glacial (banded schists with dropstones) and fluvio-glacial deposits. According to data of Akhmedov, the tillites are quite conspicuous in the Northern Karelia (the Panajarvi Formation) and in the Pechenga structure of the Kola Peninsula (the Lower Akhmalakhti Formation).

The Segozero Group (the Lower Jatulian) overlies the Katarchean and Paleoprotozoic basement rocks or the lower groups of the Karelian Supergroup, including the different horizons of the Sariolian Group and it is separated from these not only by stratigraphic unconformity, but even by a slight angular unconformity. At the bottom of the group the crust of weathering is commonly observed on the basement rocks. Geologists studying this region call the tectonic movements causing the unconformities the Seletskian. The small intrusions of hypabyssal granitoids and gabbroids were shown to be related to the Seletskian orogeny. The lower-sedimentary part of the group (700–1000 m) starts with quartzites, quartz, and oligomictic conglomerates and arkoses, then up-section follow phyllites (or siltstones) and finally the dolomitic sandstones, slates, and dolomites with locally dispersed stromatolite bioherms. In the north-eastern margin of the Karelian massif, within the limits of the Lapland-Belomorian zone (the Lekhta structure), thin bands of jaspilites are known to occur among the sedimentary rocks. The upper sedimentary-volcanogenic subgroup (up to 750 m) is composed of altered diabases and amygdaloidal cellular and porous lavas and their tuffs and also of quartzites, tuff-sandstones, and slitstones that form a rather thick horizon inside the volcanogenic facies. The sedimentary rocks of the groups are often characterized by extreme shallow-water environment (cross-bedding of flow type, sun cracks, ripple marks). In eastern Finland (the Koli-Kaltimo area) a low uranium mineralization associates with quartzites (Piirainen 1968), probably of primary sedimentary origin, but later affected by the diabase dykes.

The Onega Group (the Upper Jatulian) overlies the Segozero Group with a break within the limits of the Karelian massif and with strong erosion, and even with a slight angular unconformity in the north and north-eastern areas surrounding the massif. In some places in the margin of the massif and throughout the Savo-Ladoga miogeosynclinal zone it lies directly on the basement and starts the sequence of the Karelian Supergroup.

This is divided into the sedimentary and volcanogenic subgroups, as are many of the groups here examined. The lower sedimentary subgroup (800–1500 m) is characterized by a complex structure and is largely composed of metasandstones, siltstones, and argillites alternating with marls and dolomites that dominate in its upper part. Cherty dolomite and hematite-bearing dolomites, shungites, and amygdaloidal diabases occur locally. In Northern Karelia and the Kola Peninsula the synchronous strata contain the banded ferruginous cherty rocks – granular and micro-oolitic jaspilites of the Lake Superier type, ferruginous-manganese and phosphate-bearing carbonate rocks, quartzite with apatite matrix. The rocks, jaspilites in particular, also occur in the Onega Group (the "marine Jatulian") of north-eastern Finland (Sakko and Laajoki 1975). In many regions, in the Karelian massif specially, the rocks are pink-colored; the textural features of shallow water are common. Bioherms are more common in dolomites than in the Segozero Group, they are formed of different columnar and branching stromatolites, some chracteristic forms are also present and even some resembling the Lower and Upper Riphean [Krylov 1966, Raaben (ed.) 1978, Konyushkov 1978, pers. comm.].

The upper subgroup (350–500 m) is widespread in the Onega region, where it is divided as the Suysari Subgroup; it is composed of tuffites, tuffs, and tuff-conglomerates in the lower part, up-section are observed basic and ultrabasic volcanics (diabases, porphyrites, picrites, pillow lavas) alternating with tuff, and the sequence is completed with strata of tuffs and tuff-sandstones with interbeds of cherty-clay slates. In the Pechenga structure of the Kola Peninsula the volcanogenic subgroup (the Upper Kolosjoki Formation) is principally made up of metabasites with thickness up to 1000 m.

The Onega Group in the Savo-Ladogan zone is different from its portion described above and is divided as the Pitkyaranta Group. Near the Karelian massif it differs slightly from the stratotype, but its thickness decreases from 1000 m to 150–200 m (and even to several scores of meters in some places), westward and south-westward, but still it is well traced along the whole extension of the zone and it is suggested that it underlies the eugeosynclinal piles of the Bothnian Group in the area of Svecofennides (Salop 1979 b). It is mainly comprised of metadiabases or amphibolites alternating with carbonate rocks; local quartzites and rare jaspilite interbeds are known from the lower part of the group.

The time of formation of the Onega Group has been determined by radiometric methods with sufficient accuracy. The stromatolitic dolomites yielded an age of $2300 \pm 140$ m.y. by Pb-Pb method (Iskanderova et al. 1978); this value seems to characterize the time of sedimentation. Zircon from the diabase dykes that cut the rocks of the group, but are older than the overlying Ladoga Group strata, gave an age of 2160–2180 m.y. (Sakko 1971; Sakko and Laajoki 1975, Krats et al. 1976). Thus, the Jatulian rocks (the Segozero and Onega Groups) were formed in the time interval of 2450–2180 m.y. Intrusion of the post-Jatulian diabase dykes corresponds to the time of the Ladogan diastrophic episode that separated the Middle and Late Mesoprotozoic Suberas.

As was noted, the Ladoga Group of the Upper Mesoprotozoic is distributed only in the Savo-Ladogan zone of Karelian (in part in the adjacent margin of the Karelian massif in the Northern Ladoga area). Its stratigraphic analogs are, however, known; the platform Bessovets Group developed in the south of the Karelian massif in the Onega area, and specific sedimentary-volcanogenic strata make up the upper part of the Karelian Supergroup in the north of Karelia and in the Pechenga and Imandra-Varzuga zones of the Kola Peninsula.

The Ladoga (or Kalevian) Group is a thick (up to 4000 m) sequence of terrigenous rocks of flyschoid type. Westward the rocks gradually pass into terrigenous-volcanogenic eugeosynclinal strata of the Bothnian zone of Tampere in the Sveco-fennian area. The group lies on the Jatulian rocks (the Pitkyaranta Group), some-times with a masked unconformity, sometimes with a large break that is fixed by the polymictic conglomerates and sometimes even with a slight angular unconformity; locally it rests directly on Katarchean geneisses. The group is typified by rhythmical alternation of dark-grey biotite or andalusite-staurolite-biotite schists and grey-wacke, carbon-bearing and micaceous-quartz schists (or slates), metagreywackes, and metaarkoses; subordinate quartzites are known to increase upward. In the lower part of the Ladoga Group the unsorted boulder-pebbled conglomerates are locally ob-served, in many aspects resembling tillites (the Portanen and Vavasari conglo-merates).

The rocks are strongly dislocated and are marked with zonal metamorphism and the grade increases from the greenschists facies in the north-east (near the Karelian massif) to the amphibolite facies in the south-west. Earlier it was suggested that metamorphism of the group reached the granulite facies, but new studies (Salop 1979 b) revealed that the gneiss-granulite complex developed in the southern part of the Ladoga area is the Katarchean basement and that the Pitkyaranta and Ladoga Groups unconformably overlie it (see Fig. 45).

The Ladoga and Pitkyaranta Groups are cut by the microcline or plagio-micriocline granites and pegmatites with an isotopic age of 1900 m.y.; thus these granites were emplaced during the Karelian diastrophic cycle of the first order. Moreover, both groups mentioned are cut by the younger ($\sim$ 1600 m.y.) rapakivi granites.

The platform analog of the Ladoga Group is the Bessovets Group (up to 1200 m), which rests on the eroded surface of volcanics of the Onega Group (the Suysari Subgroup) and, like the Ladoga Group, is characterized by rhythmical alternation of terrigenous rocks, but unlike this group, does not contain any greywackes; the rocks are represented by quartz and feldspar sandstones, siltstones, shales, and in part by limestones, marls, and tuffites; in the lower part of the group conglomerates and gritstones occur. It is possible to suggest horizons of basalt and andesite porphyrite in the upper part of the group. The rocks are slightly dislocated and hardly altered, due to their ocurrence in the upper part of a large Onega trough (syneclise) on the Karelian massif. The granite intrusions are absent here as in remaining places of this massif.

In the north of the Karelian-Kola region a number of formations (groups) belong to the Ladoga stratigraphic level; in the Pechenga, Imandra-Varzuga, and Pana-Kuolajarvi structures they are recognized under different local names. All of them are characterized by the presence of rhythmically bedded carbonaceous (graphitic) slates, conglomerates, micaceous quartzites, dolomites, or limestones in the lower part and of thick pillow lavas of basic and ultrabasic composition, tuffs and tuff-breccias in its upper part. The thickness of these strata sometimes reaches 2000 m. The lower part is also characterized by distinct lithified moraines (the Vuosnajarvi, Lammas tillites, etc.) and by the related marine-glacial banded (varve-like) slates and puddings (Gilyarova 1964, and pers. comm. of Akhmedov 1980). Thus, in the Ladoga rocks in the south, as well as in the north of the region the second horizon (50–200 m) of glacial formations in the Karelian Supergroup extends for more than 900 km.

The Ladoga Group rocks complete the Karelian Supergroup in every area of their distribution except the Onega region. In the Onega trough the sequence of the group is accreted by the thick (up to 2000 m) Vepsian Group consisting of terrigenous rocks. The lower part of the group (the Kamenoborsk Formation) is principally built up of dark grey and grey feldspar-quartz and arkose sandstones, siltstones, and shales with conglomerate interbeds. The upper part (the Shoksha Formation) comprises red or pink and grey quartzite-sandstones, commonly with cross-bedding of basin and flow type, with interbeds of siltstones and shales. In the uppermost portion the strata of quartzite-sandstones with gritstones, quartz conglomerates, and arkoses are distinguished. The clastic material contains different rocks, the volcanics of the upper (Suysari) subgroup of the Onega Group and shungites that are very typical of the lower subgroup of this group in the Onega region.

The age of the Vepsian sandstones determined by the lead isotope ratios from the mineralized cement is 1990 m.y. (Tugarinov et al. 1963). Diabases cutting the Vepsian Group have been dated at 1830 m.y. by the K-Ar method (Gerling et al. 1966). These data prove that the Vepsian Group belongs to the Karelian Mesoprotozoic and not to the Jotnian (the Middle Neoprotozoic), as was earlier suggested. At the same time the occurrence of detractive red beds in the upper part of the Karelian Supergroup that usually terminates the regressive part of the sedimentary megacycle is an indication of the completeness of the supergroup.

**Mesoprotozoic of the Ukrainian Shield and of the Voronezh Massif.** The Mesoprotozoic strata within these large elements of the East-European platform are mainly known from mining and drilling works (on the Ukrainian Shield, though, there are some exposures), but the largest iron ore deposits occur in these strata, is why they are so well studied.

On the Ukrainian Shield the strata in question are well represented and best known in the Krivoy Rog iron ore basin within the limits of a discontinous submeridional band that extends for about 150 km; from the east it is bounded with plagiogranites with relics of Katarchean gneiss substratum and from the west with stratified rocks of the Katarchean and plagiogranites cutting them. Banded-silicate-magnetite ores occur among the Katarchean gneisses and basic crystalloschists; they are altered under granulite facies, and by appearance they somewhat resemble Mesoprotozoic jaspilites. Some workers suggest that the wall-rocks in this case are high-grade analogs of Mesoprotozoic strata of the Krivoy Rog basin. This correlation, however, cannot be accepted, not only because of the difference of the metamorphic grade (the greenschist to epidote-amphibolite facies for the Mesoprotozoic and amphibolite or granulite facies for the Katarchean), but also because of the differing composition and sequence of both complexes (Kiselev 1977). The complexes are separated by a large abyssal fracture and the related zone of blastomilonites (see Fig. 17). Mesoprotozoic strata are intensely dislocated, near the fractures in particular, and are almost never cut by granitoids; Katarchean rocks are ubiquitously injected by abundant granite matter.

General features of Mesoprotozoic stratigraphy were studied by Svistalsky in the 1930's, but details became known from works of Belevtsev (1957), Belevtsev et al. (1971, etc.) and from many Ukrainian geologists who did surveying and exploration. The Mesoprotozoic strata of the area are usually assigned to the Krivoy Rog Group, but it seems more reasonable to regard it as a supergroup and to recognize two or three groups within it: the Novokrivoyrog, the Krivoy Rog proper, and the Frunze Mines.

The Novokrivoyrog Group embraces the strata (up to 2000 m) of amphibolites or amygdaloidal metadiabases with sharply subordinate interbeds of metasandstones and biotite-amphibole schists; a horizon of quartzites (up to 200 m) rests at the base of these strata, it lies on the so-called Saksagan plagiogranites of the Katarchean. In the stratigraphic scale of Belevtsev this group was previously attributed to the lower ($K_o$) formation of the Krivoy Rog Group. However, the overlying strata of quartzite-sandstones belonging to the Krivoy Rog Group unconformably rest on it and on the plagiogranites. It cannot be excluded that the essentially volcanogenic strata discussed belong to the Paleoprotozoic Konka-Verkhovtsevo metabasite group. This problem has remained debatable up to the present.

The Krivoy Rog Group (s. str.) consists of two formations – the Skelevatka ($K_1$) and Saksagan or Ferruginous ($K_2$). The Skelevatka Formation (50–300 m) is mainly composed of metamorphosed arkose sandstones, quartzite-sandstones (or quartzites), phyllites, carbonate-chloritic, talc-chloritic, and aspide slates (schists). At its base quartz conglomerates or gritstones occur, the clastic matter comprises in addition to quartz some amphibolites and plagiogranites. Clastogenic zircon from the heavy fraction of the conglomerate cement is dated at 2500 and 2900 m.y. (Semenenko et al. 1974). These values seem to closely define the lower age boundary of the group. It is of interest that some rounded grains of pyrite and of possible galenite occur locally in the cement among the small clasts. The Saksagan Formation (750–2000 m) starts with a horizon (10–20 m) of talc schists that represents the altered ultrabasic volcanics and up-section thick strata of alternating jaspilite and slate bands appear; the slates are sericitic, sericite-chloritic, biotite-bearing, cummingtonite, and graphitic. Six or seven horizons of jaspilites are distinguished within the formation as being of economic importance. Ferruginous rocks occur mostly in the lower and upper part of the formation.

The Frunze Mines Rodionovo) Group rests with an erosion and locally with an angular unconformity on the Krivoy Rog Group. Two formations are recognized within it: the Gdantsevo and Gleevatka. The Gdansevo Formation (up to 1600 m) is made up of quartz-biotite-sericite, quartz-carbonate-biotite, and quartz-sericite schists (slates) in the lower part and of quartz-sericite slates rich in carbon and dolomites with local oncolites and catagraphies (Snizhko 1974), and also some problematic remains of the *Corycium* Sederh. type (Belokrys and Mordovets 1968) that were previously described from the Bothnian strata of Finland. Interbeds of massive schistose ferruginous rocks, and also some ferruginous sandstones, partially formed by Krivoy Rog jaspilite erosion, occur in the lower part of the formation. The Gleevatka Formation (up to 2000 m?) lies on the Gdantsevo Formation and probably on the Krivoy Rog Group and on the Saksagan plagiogranite-gneisses with a deep erosion. It is more reasonable to regard it as a separate group. The base of the formation is mainly composed of unsorted boulder-pebbled polymictic conglomerates that locally pass into puddingstone, metasandstones, and metasiltstones comprising dispersed boulders and pebbles. These rocks are especially typical of the strata distributed on the left bank of the Dnieper River, in the area of the Belozersk iron deposits, and the strata are correlatives of the Gleevatka Formation of the Krivoy Rog region. I observed pebbles with downwarped beds of siltstones (dropstones) under them in large drill cores of conglomerates. It is quite possible that the Gleevatka conglomerates are of glacial origin (fluvio-glacial formations and tillites). The upper part of the formation is composed of dark phyllitized or micaceous and carbonaceous (graphite) slate or schists with siltstone interbeds.

The metamorphic minerals (mainly micas) from the rocks from the Krivoy Rog and Frunze Mines Groups are dated in the range of 1700–2000 m.y. by K-Ar analysis (common values are 1800–1900 m.y.), corresponding to the Karelian diastrophic cycle.

A close analog of the Mesoprotozoic groups of the Ukraine is the Kursk and the overlying Oskol Groups of the Voronezh massif or, in other words, of the area of the Kursk Magnetic Anomaly (KMA) [17]. The Kursk Group is divided into two formations, Stoylensk and Korobkovo. The first is made up of quartzites with interbeds of gritstones and conglomerates in the lower part and also phyllitized and micaceous slate (schists) that dominate in the upper part of the formation. Its thickness reaches 1000 m but commonly it is much less, several tens or a hundred of meters. The Stoylensk Formation rests with an angular unconformity on the metavolcanics of the Paleoprotozoic Mikhaylovsk Group that are cut by plagiogranites (2750 m.y.) or on the gneisses of the Katarchean Oboyan Group. The Korobkovo Formation (up to 1000 m) conformably lies on the Stoylensk Formation and is built up mainly of jaspilites alternating with bands of micaceous or phyllitized, sometimes carbonaceous, slates or schists. In the matrix of conglomerates from the basal beds of the Kursk Group syngenetic uranium-bearing pyrite is reported, dated at 2600 m.y. by the Pb-isotope method. This value is not quite exact, but is very close to the time of the beginning of this group rock accumulation.

The Oskol Group lies with a stratigraphic or a slight angular unconformity on the rocks of the Kursk Group and is divided into two formations (or subgroups), their composition varies in different portions (facial zones) and they are given different names. The lower formation (Yakovlevo and its stratigraphic analog, the Rogovo) consists of quartz-sericitic, micaceous and carbonate-micaceous phillite or schists alternating with interbeds and bands of dolomites. In the lower part of the formation ferruginous quartzites, gritstones or polymictic conglomerates with the jaspilite pebbles of the Kursk Group are also known to occur. The carbonate rocks sometimes associate with banded carbonate-silicate-magnetite rocks. The thickness of the formation reaches 1000–1500 m. The upper formation lies on the lower one with an erosion. In some places it is principally made up of quartz-micaceous and carbonaceous slates with interbeds of ferruginous quartzites (the Belgorod Formation), in other places it is built up of quartz porphyries and their tuffs, metasandstones, and conglobreccias (the Kurbakin Formation), or of diabase and andesitic porphyrites, carbonaceous schists, and conglomerates that in some cases resemble the tilloid (?) puddingstone

---

17  The Precambrian stratigraphy of the Kursk Magnetic Anomaly was studied by many geologists and is generalized in works of Golivkin (1967), Polishchuk (1970) and of other authors. here a new scheme of division of the Mesoprotozoic is given, elaborated by Kononov and Petrov (1980, pers. comm.)

(the Voronezh Formation). In the Tim Formation certain black, extremely fine-bedded rocks occur among the carbonaceous slates, the phosphate laminas there alternate with manganese-carbonate or carbonaceous-phosphate interbeds. The thickness of the upper formation (or formations) amounts to hundreds of meters and locally reaches 2000 m.

The rocks of the Kursk and Oskol Groups are intensely folded and cut by gabbroids and also by later granodiorites and diorites with an age of about 1950 m.y.

Correlation of the Mesoprotozoic strata of the Ukrainian Shield and of the Voronezh massif (KMA) and their correlation with the Karelian stratotype does not face any serious problems, though the distance separating them is great. The Novo-krivorog Group of metabasites of the Ukraine corresponds to the Tunguda-Nadvoitsa Group of Karelia that is of similar composition; in the KMA its analogs are not known. The Skelevatka and Saksagan (Ferruginous) Formations of the Krivoy Rog Group are almost identical with the Stoylensk and Korobkovo (Ferruginous) Formations of the Kursk Group. At the same time both these groups generally resemble the Segozero Group of the Karelian Supergroup, especially in the areas where the latter emplaces jaspilites. All these groups are typified by the quartzite and quartz conglomerate occurrences and by the crusts of chemical weathering under them. They mainly differ in lacking volcanics in the upper part of the groups in the Ukraine and KMA, but this seems to be rather a provincial peculiarity depending on the local tectonics. It cannot be excluded that volcanics were eroded prior to deposition of the Frunze Mines and Oskol Groups. Jaspilites are of low abundance in the Karelian-Kola region, which can be explained by the fact that they usually occur in miogeosynclinal strata and the Segozero Group is mostly represented by platform strata; in the miogeosynclinal zone of the Karelides jaspilites are reported and there they belong to the same type as the Krivoy Rog and Kursk jaspilites (see below).

The Frunze Mines Group of the Ukraine and the Oskol Group of KMA are known to consist of two formations separated by breaks (in fact these are two groups or subgroups), which can be compared. The lower formations of both groups containing abundant carbonate rocks, schists, and quartzites are easily compared with the Onega Group. All these formations are also similar in the following respects occurrence of phytolites in the carbonate rocks (the Gdantsevo Formation), the occurrence of ferruginous rocks of the Superior and Urals types (see below), high concentration of iron and manganese in certain carbonates, the presence of shungites (in KMA), and many more characteristic and striking features of these rock compositions. The upper formations of these groups in the Ukraine and KMA, principally made up of dark, commonly carbonaceous schists with conglomerates resembling glacial formations, are naturally compared with the Ladoga Group (and its analogs) of the Baltic Shield. This correlation is also confirmed by many geochemical features, for instance by the occurrence of rocks rich in phosphorus. It cannot be ruled out, however, that some formations of KMA regarded as synchronous may correspond to different levels of the Ladoga stratigraphic horizon; thus the Kurbakin and Tim volcanogenic Formations may correspond to the upper part of the units of this horizon developed on the Kola Peninsula.

Thus, on the Russian platform the analogs of the Sariolian and Vepsian Groups of the Baltic Shield are lacking or unknown, and the formations of the Jatulian level are mainly represented, being also widespread in the Karelian-Kola region.

**Mesoprotozoic of Eastern Siberia.** The best and most complete Mesoprotozoic sequences of this region are situated in the north-east of the Baikal Mountain Land, in the Olekma-Vitim Mountain Land. The Udokan Group is distinguished here; it is developed in the western framing of the Aldan Shield that in the Mesoprotozoic used to be the platform structure (the Aldan platform). In the Udokan Group of a total thickness of 13,000 m, three subgroups (or groups) are recognized: Kodar, Chine, and Kemen; they are separated by breaks (Salop 1964–1967).

The Kodar Subgroup consists of five formations (up-section):

1. The Sygykhta Formation (1600 m): dark-grey metasandstones with abundant interbeds of biotite and andalusite schists and subordinate horizons of metadiabases or amphibolites; in the lower part marble interbeds are reported. The base of the formation is not known.
2. The Orturyakh Formation (1000 m): monotonous dark biotite schists or phyllites with interbeds of dark metasandstones.
3. The Boruryakh Formation (~ 800 m): grey and light-grey quartzites with interbeds of dark biotite graphitic schists.
4. The Ikabiya Formation (1300 m): dark biotite schists or carbonaceous phyllites, metasiltstones; metasandstones, quartzites and a thin marble horizon occur in the lower part.
5. The Ayan Formation (up to 300 m): close alternation of fine-bedded metasandstones and metasiltstones.

All the formations mentioned are known only from the Kodar Range; in the Udokan Range the group sequence starts with the Ikabiya Formation that rests unconformably on the Katarchean gneisses here and there and also on the Paleoprotozoic greenstone rocks (the Olonda Group).

The Chine Subgroup consists of four formations:

6. The Inyr Formation (280–400 m): coarse-bedded massive grey sandstones with rare ripple marks. It overlies the Ayan Formation with a break.
7. The Chetkanda Formation (1000–1500 m): greenish-grey metasandstones in the lower part, alternating with light pink quartzites or quartzite-sandstones; upward and near the top of the formation they become dominant. Local intercalations of fine-bedded magnetite rocks very similar to jaspilites are reported. The ripple marks are common on the bedding surface. In the area of the Middle Vitim the quartzite-sandstone piles correlatable with the Chetkanda Formation (the lowermost part of the Kuzalin Group) rest unconformably on the Paleoprotozoic metavolcanics of the Muya Group.
8. The Aleksandrov Formation (200–460 m): fine intercalation of dolomitic quartzite-sandstones, dolomites, metasiltstones and slates; a horizon of cuprous siltstones is registered. Common ripple marks and sun cracks.
9. The Butun Formation (500–650 m, in places up to 1000 m): grey and lilac-grey massive or poorly bedded siltsones with horizons of stromatolite-bearing dolomites. In the siltsones certain features of past salinity as crystal molds after halite and gypsum, interbeds of albite-scapolite, and zeolite-bearing rocks are common.

The Kemen Subgroup consists of two formations:

10. The Sakukan Formation (3500 m): grey arkose sandstones with common interbeds of altered siltstones and gritstones in the lower part, grey or pinkish-grey oligomictic quartz or arkose sandstones with interbeds of magnetite sandstones (the clastic magnetite formed at the expense of erosion of the magnetite rocks of the Katarchean basement complex). In the upper part of the formation a thick horizon of syngenetic cuprous sandstones and interbeds of calcareous sandstones are observed. Cross-bedding of basin and flow types is quite typical, ripple marks are also known. At the base of the formation a discontinuous horizon of polymictic conglomerates resting on the surface of erosion of the Butun Group is locally reported. Moreover, in the lower part of the formation, in places the unsorted boulder-pebbled conglomerates and puddingstone closely resembling tillites are known to occur (sometimes the downwarped beds are observed under certain boulders).

11. The Namingu Formation (more than 1000 m): grey and dark grey siltstones with subordinate interbeds of slates and sandstones.

The Kodar Subgroup deposits accumulated in a relatively deep marine basin, the piles of the three fromations of the Chine Subgroup formed under extreme shallow-water environment and those of the Butun Formation in a strongly mineralized (salted) lagoon. The sedimentary strata of the Kemen Subgroup accumulated in the terrestrial and subaqueous portions of large river deltas and the supply of clastic material was from the north, from the Chara block of the Aldan platform. It is evident that the Udokan Group comprises all types of rocks of the transgressive-regressive macrorhythm, and thus it is quite representative for the entire sequence of the era-them.

The Udokan Group is folded, its folds surrounding the uplift of the Chara block in the west of the Aldan Shield (Fig. 34). The deformational grade grows southward and south-westward away from the margin of the Aldan platform. The metamorphic grade increases in the same way from the lower and middle degree of the greenschist facies to epidote-amphibolite facies; then the rocks of the lower units are generally of deeper alteration than the younger rocks, but this change is observed to be quite gradual.

The stratigraphic position of the Udokan Group in the Precambrian sequence and its age boundaries have been determined quite well. The group is certainly younger than the Katarchean gneisses (the Aldan Group) and the Paleoprotozoic greenstone rocks (the Olonda Group) that were metamorphosed and cut by granites 2800 m.y. ago. The upper age boundary of the group is determined by the fact that all its units are cut by the syntectonic Kuanda and late-tectonic Chuya-Kodar granites dated at

---

**Fig. 34.** Geologic chart of the Ikabiya-Chetkanda area, Udokan Range. (After Salop 1964–1967, simplified).
1 Quaternary strata, 2 Lower Cambrian strata, 3 granite of the Chuya-Kodar complex, 4–10 the Udokan Group: 4 the Sakukan Formation, 5 the Butun Formation, 6 the Aleksandrov Formation, 7 the Chetkanda Formation, 8 the Inyr Formation, 9 the Ayan Formation, 10 the Ikabiya Formation, 11 Katarchean gneiss-granulite complex of the Chara block (the Aldan Shield), 12 fractures

1900 m.y. by different isotope methods. The Lower Neoprotozoic rocks (the Tep-torgo Group) lie on the erosion surface of these granites.

Stromatolites of the Butun Formation (and of its analog, the Bul'bukhta Formation) are represented by typically Lower Riphean assemblages: *Conophyton garganicus* Korol., *Collenia frequens* Walc., *Kussiella* sp., *Omachtenia* sp., *Baicalia bulbukhtensis* Komar. etc. Thus, in Eastern Siberia the so-called lower Riphean phytolites certainly occur in rocks older than 1900 m.y.

Some problems emerge in correlating certain units of the Udokan Group and the Karelian stratotype, particularly concerning the Kodar Subgroup. It is very likely, however, that its two lower formations (with metabasites in the base) correspond to the the Tunguda-Nadvoitsa Group and the three upper formations, that start with the Boruryakh quartzites, to the Sariolian Group. It is easier to correlate the Chine Subgroup. Certainly it resembles the Jatulian; the quartzite-sandstone strata of the Inyr and Chetkanda Formations correspond fairly well to the Segozero Group (the Lower Jatulian) and the slate-dolomite strata of the Aleksandrov and Butun Formations to the Onega Group (the Upper Jatulian). The units correlated have common structure and many more common details, but they are beyond the scope of this work. The Kemen Subgroup is very close to the Vepsian Group of Karelia in formational aspect, but the tillites in its lower part favor its partial correlation with the Ladoga Group.

**Mesoprotozoic of Hindustan.** Some groups in India are assigned here (Salop 1973a) to the erathem in question, but the most important are the metamorphic strata of the Shimoga-Dharwar fold zone in the south-west of the Hindustan Peninsula (Karnataka state), (Fig. 35). These strata are usually divided as the Middle and Upper Dharwar (Krishnan 1963) and together with the underlying greenstone rocks of the Lower Dharwar are attributed to the Dharwar Supergroup, this latter being regarded as Archean or Upper Archean (Paleoprotozoic) in age. However, I considered (Salop 1966, 1973a) that this combination unwarranted and that the Lower Dharwar only belongs to the Paleoprotozoic and that the Middle and Upper Dharwar is Mesoprotozoic. New data obtained by Indian geologists support this opinion. It was mentioned in the previous chapter that Naqvi et al. (1978) showed that the sedimentary-volcanogenic strata of the Lower Dharwar can only be compared with the "true" greenstone strata of Archean (Paleoprotozoic) and the Middle and Upper Dharwar rocks unconformably overlying them are the Upper Archean-Lower Proterozoic geosynclinal strata that slightly resemble the greenstone belts. Srinivasan and Sreenivas (1972, 1976) assign the entire Dharwar Supergroup to the Paleoprotozoic (Paleoproterozoic according to these authors) but the ultrabasic-basic rocks in the Nuggihalli greenstone belt and in some other belts are suggested to be older – Archean (Naqvi et al. attribute also the rocks of the Holenarasipur and Kolar greenstone belts to the Archean, and Srinivasan and Sreenivas include them in the Paleoprotozoic erathem).

According to Srinivasan and Sreenivas, the metamorphic strata here regarded as Mesoprotozoic belong to four groups (upward): Bababudan, Dodguni, Ranibennur, and Guddadarangavanahalli (abbreviated G. R.).

The Bababudan Group starts with subaerial basalts, then up-section follow pyrite-bearing oligomictic conglomerates with gold-uranium mineralization, then pure quartzites (orthoquartzites) and finally the magnetite quartzites. The group unconformably rests on the Katarchean charnockites and gneisses and also on ultrabasic-basic rocks of the Lower Dharwar (after Naqvi et al.) [18].

The Dodguni Group lies with a break on the Bababudan Group. In the base it contains oligomictic conglomerates overlain by orthoquartzites, phyllitized schists and the sequence finishes with strata of limestones and dolomites.

The Ranibennur Group overlies unconformably the underlying groups or rests directly on the basement. In the base it contains polymictic conglomerates, then follow thick strata of greywackes and schists alternating with covers of pillow basalts ("the strata of grey traps"). Abundant tonalite boulders are observed in conglomerates, their Rb-Sr age being 3250 m.y. (Venkatasubramanian and Narayanaswamy 1974). Tonalites are very typical of the Peninsular Gneisses complex (the Katarchean basement). The time of basalt outflow (?) is obtained by the same method, the value is 2345 ± 60 m.y. (Crawford 1969). The basalts are conformably overlain by slates with jaspilite interbeds.

The fourth and last G.R. Group lies on the older groups with a large break and with an angular unconformity in places. It is composed of ferruginous slates, siltstones, and sandstones, locally colored pink and red. Interbeds of siliceous (cherty) rocks rich in iron and ankerite ores are subordinate to these rocks.

In the opinion of Srinivasan and Sreenivas, the rocks of the two lower groups (Bababudan and Dodguni) correspond to the pregeosynclinal or shelf stage in evolution of the region, the Ranibennur Group to the geosynclinal (flysch) stage and the G.R. Group, regarded as red molasse, to the inversion stage transitional to the platform stage. Possibly it is more reasonable to attribute the two lower groups to the platform formations from the point of view of terminology. Later it will be shown that many Mesoprotozoic geosynclinal strata start with platform-type rocks. It should be noted here that the earlier-mentioned Bothnian eugeosynclinal Group of the Svecofennian area is underlain by thin platform strata of Jatulian.

The above authors think that a large-scale diastrophism preceded deposition of the upper red bed group. The rocks of the three lower groups are intensely folded, metamorphosed under greenschist to amphibolite facies and cut by granites; the upper group is of slighter dislocation and lower grade and is cut only by epidiorites and dolerite or diabase dykes. If this is true, then it seems more reasonable to regard the upper group as a separate unit that may belong to a younger Precambrian erathem. However, this large-scale diastrophism is not certain between the groups; it is very probable that a small-scale tectonic episode separates them.

The lower age boundary of the groups is drawn according to isotope datings of the Lower Dharwar metamorphic rocks underlying them; values in the range of 2700–2900 m.y. are obtained. Their upper age boundary is defined according to datings of metamorphic rocks of the groups and of the granites cutting them. Major values obtained by K-Ar and Rb-Sr methods mostly fall within the 1900–2000 m.y. range, which corresponds to the Karelian diastrophism. The age of 2380 m.y., obtained by Rb-Sr analysis on the isochron, was not reliable for the granites from the southern part of the Closepet massif, which is considered to be younger than the

---

18 The thickness of this group and of other groups is not stated by the authors of the stratigraphic scale. It is known to be some hundreds to several thousands of meters

Ranibennur Group (Crawford 1969), while K-Ar datings of the rocks from other sites of this massif give values of the order of 1900–2000 m.y. It seems that the relations of the Closepet granites and the strata in question, as well as their isotope age, need further refinement.The Closepet massif is built up of various granitoids with abundant pink microcline and pink-grey granites with porphyroblasts of microcline and albite-oligoclase. The massif is of submeridional strike and is followed for more than 300 km; along almost the whole extension it is situated among the Katarchean granite-gneisses, migmatites, and charnockites that are commonly included in the so-called Peninsular Gneisses[19]. The massif adjoins the rocks of the Ranibennur Group only in the northernmost portion of it (see Fig. 35). Unfortunately, little is known about their relations (from the data of the geologic map the group lies on the granites). It is suggested here that the granitoids belonging to the Closepet massif are of different age and that they are partially Paleoprotozoic. However, my observations (Salop 1964) and data of Divakara Rao et al. (1972) suggest a Na and K metasomatosis superimposed on the rocks, as seen from albitization and microclinization and in the formation of poryphyroblasts of feldspars. A zone of abyssal fracture is suggested to have existed in place of the submeridional band where the Closepet massif is situated now: within this zone the emplacements of granites took place in the Paleoprotozoic and later on in the Mesoprotozoic. It is very likely that the supracrustals of the Ranibennur Group are cut by potassium granites, the youngest in the Closepet massif; they belong to the Karelian cycle of diastrophism.

Generally the sedimentary and sedimentary-volcanogenic group of the Middle-Upper Dharwar are very similar to the rocks of the Karelian stratotype, their correlation, though, is problematic, because geochronologic data on the Hindustan rocks are insufficient and detailed lithologic characterization is lacking. It seems realistic to compare the basic volcanic strata from the base of the Bababudan Group and the Tunguda-Nadvoitsa Group; the pyritized oligomictic gold-uranium conglomerates overlying it are likely to be a correlative of the Sariolian Group, while the quartzite piles and the overlying ferruginous piles greatly resemble the Lower Jatulian (the Segozero Group). Gold-uranium conglomerates are the most important level in correlation. It will be shown further that the conglomerate horizons largest and richest in metals occur at the Sariolian (= Witwatersrand) level all over the world, but certain small occurrences of these rocks are known from the lower (Tunguda-Nadvoitsa = Dominion-Reef) as well as from the upper (Lower Jatulian) level. Thus, the Bababudan Group seems to correspond to the three groups of the Karelian Supergroup. The Dodguni Group greatly resembles the Upper Jatulian (the Onega Group) by its composition and structure, as the Ranibennur "flysch" Group does the Ladoga Group in North Karelia and the Kola Peninsula, where it contains abundant basic and ultrabasic pillow lavas. It was noted above that the flysch nature is typical of the Ladoga Group in the entire Karelian-Kola region. The isotope age of the basalts from the Ranibennur Group seems to contradict this correlation, as the value obtained

---

19 Srinivasan and Sreenivas and other Indian geologists regard the "Peninsular Gneiss" and the Champion gneiss-granites as post-Dharwar. However, my observations in South India proved this concept to be wrong; these gneisses appear to be the oldest formations of Hindustan overlain by the greenstone rocks of Paleoprotozoic with conglomerates at the base in the Kolar belt (Salop 1966). This conclusion is shared by many Indian geologists at the moment.

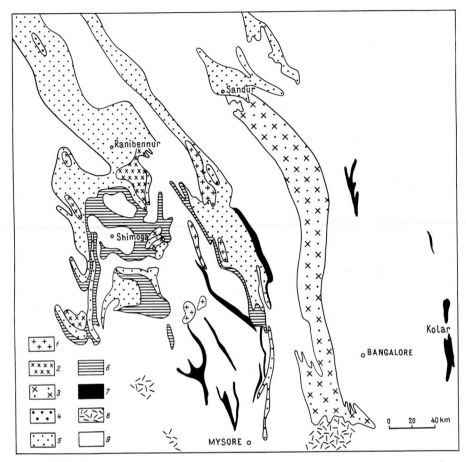

**Fig. 35.** Geologic chart of the Karnataka state, Southern India. [After Srinivasan and Sreenivas (1972), revised, according to data of Naqvi et al. (1978) and Salop (1966)].
Mesoprotozoic: *1* Mesoprotozoic granite, *2* Mesoprotozoic (?) granitoid, *3* Mesoprotozoic-Paleoprotozoic granitoid of the Closepet composite massif; *4–6* Middle and Upper Dharwar: *4* the "G.R." Group, *5* the Ranibennur Group, *6* the Bababudan and Dodguni Groups; Paleoprotozoic: *7* greenstone strata of the Lower Dharwar; Katarchean: *8* charnockite, *9* gneiss, granite-gneiss, migmatite (the "Peninsular Gneiss")

($\sim$ 2345 m.y.) exceeds the lower age limit of the Ladoga Group (2200 m.y.). This dating of basalts is not quite reliable and is not to be regarded as a decisive factor in this correlation. Finally, the G.R. red bed group naturally correlates with the Vepsian red bed group of Karelia, both being assigned to the molasse formation by many workers.

**Mesoprotozoic of Southern Africa.** In Southern Africa the stratotypical sequences of Paleoprotozoic are known as well as the perfect sequences of the Mesoprotozoic subplatform strata that can be chosen as a stratotype of that kind of formation for the whole of Africa and as a world parastratotype of the erathem. These rocks, distributed within the Transvaal craton (Republic of South Africa), belong to five "systems" or groups (which seems to be more correct), and rest successively one on another: the Dominion Reef, Witwatersrand, Ventersdorp, Transvaal (Supergroup), and Matsap. The first three are closely related and sometimes combined under a common name, the Witwatersrand Triad.

All these units form a basin of the synclise type of sublatitudinal strike extending for 650 km with a width of about 350 km, and a huge Bushveld pluton (lopolith) of composite structure and multiphase development is situated in its center. Some African geologists usually distinguish this structure under the name of the Rand basin. As a rule the rocks occur there gently with a common angle of $5°-15°$, but close to fractures and intrusions they are tilted rather steeply. The synclise is complicated by the Freiburg-Johannesburg anticline and the Vredefort mantled dome. In the axial portion of the anticline Katarchean granite-gneisses and Paleoprotozoic greenstone rocks locally group out. The core of the dome is made up of Paleoprotozoic granites and the limbs of Mesoprotozoic strata that are here tilted very steeply and even overturned (Fig. 36).

The Mesoprotozoic stratigraphy of Transvaal is known in detail due to contributions by many African geologists. It is known from many summaries (Du Toit 1954, Haughton 1969, etc. and others) and from numerous papers dealing with problems of composition, structure, and genesis of strata. The normal sequence of Mesoprotozoic units is shown in Table 4. Their details are given below.

The Dominion Reef Group is locally developed. It is probably the result of its initial local distribution, as well as of erosion that preceded deposition of the overlying group. Its thickness varies correspondingly. In the lower part of the group medium-, coarse-grained quartzites and arkoses are dominant, they pass into either quartz-sericite slates, or gritstones and conglomerates. These rocks comprise two continuous beds of gold-uranium conglomerates, though poor in metal concentration. Volcanogenic rocks of the upper and major portion of the group are represented by andesites, commonly amygdaloidal, that up-section pass into porphyries (rhyolites) and their tuffs. The group occurs on the rugged surface of the older granites and contains the products of their physical and chemical disintegration.

The Witwatersrand Group has been studied in detail, as many important gold and uranium deposits occur there. It is divided into two subgroups ("series"), then into formations, bands, horizons, and even beds that have their own names. The simplified sequence of the group is shown in Table 4. The thickness of the group indicated there (5500–7500 m) seems to be overestimated due to certain peculiarities of the structure (cross-bedding etc.).

Quartzites are a dominant component of the group, being commonly crossbedded, and alternate with interbeds and bands of shales, gritstones, and conglomerates; local beds (up to 100 m) of amygdaloidal lavas are known to occur; in the lower part of the group there is a bed of finely banded ferruginous quartzites.

An interesting aspect of this group is the occurrence of puddingstone with faceted boulders in the Government Reef Formation. These rocks are usually interpreted as tillites. In some places two horizons of tillites are observed, separated by other rock types with a thickness up to 180 m. This group is mostly known, however, for the occurrence of horizons ("reefs") of gold-uranium conglomerates or quartzites. These formations are exceptionally important not only from the economic point of view, but also from the point of view of geochemical environment in the Mesoprotozoic and of

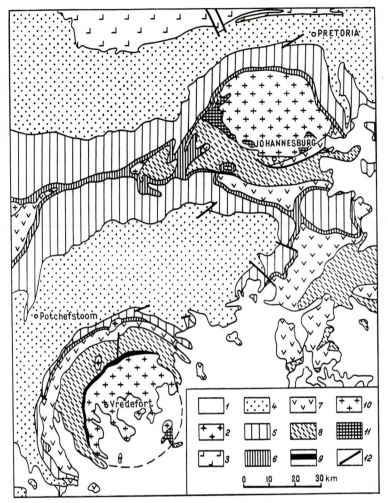

**Fig. 36.** Geologic map of the Johannesburg-Vredefort area. (According to geologic map of Southern Africa 1970, simplified).
*1* Paleozoic strata, *2* Mesoprotozoic alkaline granite, *3* gabbro and norite (the Bushveld massif), *4* the Pretoria Group, *5* the Dolomite Group, *6* the Black Reef Group, *7* the Ventersdorp Group, *8* the Witwatersrand Group, *9* the Dominion Reef Group, *10* Katarchean and Paleoprotozoic granitoid, *11* the Swaziland Group (basic and ultrabasic volcanics), *12* fractures

correlation of the older strata; their genesis will be specially considered later. The terrigenous sediments of the Witwatersrand Group accumulated in the intracratonal basin and the supply of material was more intensive from the northern side of the basin than from the south during deposition of the lower part of the group. River streams carried the material and when discharging into basins they formed the deltaic fans.

The Ventersdorp Group conformably rests on the Witwatersrand Group and virtually constitutes its upper part. It is divided into three groups, or subgroups to be more exact (Winter 1976). The lower Klipriviersberg Subgroup is principally com-

**Table 4.** Stratigraphic Chart of the Mesoprotozoic of Transvaal

| Group or supergroup (thickness in m) | Formation (or subgroup and group) (thickness in m) | Principal rocks | Gold-uranium mineralization |
|---|---|---|---|
| Matsap (up to 4000) | | Red cross-bedded sandstone, quartzite and conglomerate with thick horizon of andesite and its tuff in the middle part of the group | – |
| Transvaal (Supergroup) (4800–8400) | Pretoria Group (4200–7200) | Shale, quartzite, tuff, basic and acid volcanics, tillite | – |
| | Dolomite Group (30–2100) | Dolomite with stromatolites (with hematite ores in the upper part) | – |
| | Black Reef Group (15–150) | Carbonate slate Quartzite, conglomerate | – |
| Ventersdorp (4400) | Pniel Subgroup (1100) | Andesite Conglomerate, quartzite | – |
| | Platberg Subgroup (1800) | Porphyry, andesite, conglomerate, slate | Poor |
| | Klipriviersberg Subgroup (1500) | Andesite and pyroclastic rocks | – |

| Whitwatersrand (5500–7500) | | Formation | | Description | Significance |
|---|---|---|---|---|---|
| Upper Subgroup | Kimberley-Elsburg Formation (2700–4500) | | | Elsburg Reef quartzite and gritstone | Poor |
| | | | | Kimberley Reef conglomerate | Poor |
| | | | | Kimberley slate | – |
| | | Main Bird Formation (up to 3100) | | Bird mandelstone | – |
| | | | | Bird Reef quartzite, conglomerate | Significant |
| | | | | Livingstone Reef conglomerate | Very significant |
| | | | | Main Reef quartzite, conglomerate, slate | Very significant |
| | Jappestown Formation (400–1100) | | | Clay shale, subordinate quartzite, conglomerate lenses, mandelstone horizon | – |
| Lower Subgroup | Government Reef Formation (1200–2000) | | | Quartzite and clay shale with horizon of conglomerate, tillite and gritstone | Little |
| | Hospital Hill Formation (900–1800) | | | Hospital Hill quartzite | – |
| | | | | Water Tower slate and quartzite ("plicated beds"), hematite quartzite | – |
| | | | | Orange Grove quartzite | – |
| Dominion Reef (0–300, up to 1500?) | | | | Basic and acid metavolcanics with subordinate interbeds of metasedimentary rocks | Little |
| | | | | Arkose and conglomerate in the lower part | |

posed of amygdaloidal andesitic lavas and their pyroclastic products. The middle Platberg Subgroup in its lower part is composed of clastic rocks: conglomerates, sandstones, and slates, and in places comprises interbeds of limestones; quartz porphyries dominate in the middle part of the subgroup, in its upper part the amygdaloidal lavas (of andesitic composition mainly), subordinate quartzites, cherty rocks, and fine interbeds of stromatolite-bearing limestones are observed; the *Gruneria* form is determined among the stromatolites, it is also known from the Mesoprotozoic Animikie Supergroup of North America. The upper Pniel Subgroup rests on the underlying rocks with a large break and is made up of two formations: the lower (Bothaville) composed of conglomerates, quartzites, subarkoses, and slates that are locally of calcareous appearance, and the upper (Allanridge) built up of amygdaloidal andesites.

The Transvaal Supergroup unconformably overlies the Ventersdorp Group and still older formations, including the basement rocks. It is usually divided into three groups: Black Reef, Dolomite, and Pretoria. The Black Reef Group is thin and in its lower part comprises quartzites with fine interbeds of gold-bearing conglomerates, its upper part is made up of carbonate slates with quartzite interbeds. The Dolomite Group conformably rests on the underlying one and is divided into three parts (upward): (1) the principal dolomite "stage": bedded grey, blue and pink dolomites, locally siliceous with bands of oolitic (or oncolitic) and stromatolitic dolomites; some varieties contain abundant manganese; (2) the "stage of banded ironstones": banded hematite rocks and jaspilites alternating with cherts and red dolomites; hematite ores and jaspilites are of great economic importance; (3) the upper dolomite stage: dolomites, locally of siliceous aspect. Stromatolites of the Dolomite Group form entire horizons or bands followed for tens of kilometers. They are represented by different forms, including columnar and branching forms; there are also some forms resembling the Lower Riphean (*Kussiella*) and the Upper Riphean.

The Pretoria Group lies unconformably on the Dolomite Group and is divided into four formations ("stages"). The lower Timeball Hill Formation (up to 500 m) consists of basal conglomerates or breccias overlain by shales and siltstone slates with ripple marks and sun cracks, then follow cross-bedded pink quartzites with interbeds of magnetite quartzites and sandy oolitic ironstones. The overlying Daspoort Formation (up to 1600 m) is mainly composed of slates with the Ongeluk tillite horizon with a thickness of 15 to 85 m, with boulders with glacial faceting reported from it. The slates are overlain by andesite lavas, then light cross-bedded quartzites with slate interbeds appear. The Magaliesberg Formation, the third up-section (up to 1300 m) starts with graphitic slate strata that contain amygdaloidal andesites in the lower part and diabase sills all through the sequence; still upward the strata of light cross-bedded quartzites with interbeds of slates and marls are observed. The Smelterskop Formation, the uppermost in the sequence, (up to 350 m) is made up of quartzites in the lower part and of thick amygdaloidal andesitic lavas and their tuffs with quartzite interbeds in the upper part. The Pretoria Group comprises six horizons or bands of carbonate rocks with dome-shaped stromatolites at different levels.

The Matsap Group is the uppermost larger unit of the Mesoprotozoic in Southern Africa. It is distributed in the west of the craton (West Griqualand) and there overlies unconformably the Griquatown Group (the Pretoria Group analog) or the lavas of the Ventersdorp Group. It is formed of thick (up to 4000 m) strata of red or lilac sandstones, quartzites, gritstones, and conglomerates with amygdaloidal andesitic lavas, lava breccias and tuffs with quartzite interbeds in the middle part. The sandstones are typified by cross-bedding that helped to establish the north-east and north direction of supply of the material during their accumulation.

Numerous and reliable isotope datings are available for the Mesoprotozoic rocks of Transvaal. The lower age boundary of the whole complex is defined by its occurrence on the granites cutting the Swaziland Supergroup; their age is known to be approximately 2800 m.y. These granites from the Vredefort dome core are dated at 2850 m.y. by the Rb-Sr method (Slawson 1976). The clastic grains of monazite from the basal conglomerate of the Dominion Reef Group gave an age of 3000–3100 m.y., and are likely to have originated from Katarchean or Paleoprotozoic granites. The age of andesitic lavas from the Dominion Reef Group is $2800 \pm 60$ m.y. on zircon and apatite and $2820 \pm 110$ m.y. on sulfides and zircon from quartz porphyry (Van Niekerk and Burger 1969a). An age of $2300 \pm 100$ m.y. was obtained for the acid lavas from the middle part of the Ventersdorp Group on zircon (Van Niekerk and Burger 1964). However, later detailed geochronological studies of the same authors (Van Niekerk and Burger 1978) revealed that a value of $2640 \pm 80$ m.y., obtained at intersection of isochron and concordia, is closer to the true age. Volcanics from the middle part of the Matsap Group have the Rb-Sr age of $2070 \pm 90$ m.y. (Crampton 1974).

The upper age boundary of the Mesoprotozoic complex is determined by the time of emplacement of the Bushveld composite pluton. There are many datings available for the intrusive rocks making up the pluton and a value of 2050 m.y. seems to be the most suitable in this case and many workers accept it. Finally, it is well-known that the sedimentary-volcanogenic Waterberg Group of the Lower Neoprotozoic unconformably overlies the Mesoprotozoic strata of Transvaal and the rocks of the Bushveld pluton.

Correlation of the South African parastratotype of the erathem and the Karelian stratotype is of essential significance, as it elucidates certain important aspects of the units compared. The correlation is easy and reliable, as it is based on the combination of well-checked, geologic and geochronologic data. The Dominion Reef Group corresponds to the Tunguda-Nadvoitsa Group of Karelia by its situation in the sequence and its composition. The Witwatersrand Group is a correlative of the lower subgroup of the Sariolian Group of the same composition. This comparison is also convincing because of the occurrence of glacial formations in both groups. The essentially volcanogenic Ventersdorp group (excluding its upper Pniel Subgroup) probably corresponds to the upper volcanogenic subgroup of the Sariolian Group. The volcanic products are of similar composition in both groups. The Pniel Subgroup (or Group) separated by a significant unconformity from the underlying rocks of the Ventersdorp Group, can be compared with the Segozero Group (the Lower Jatulian) of Karelia in its composition and structure, and the Black Reef Group, overlying it unconformably and composed of quartzites together with the Dolomite Group, corresponds fairly well to the Onega Group (the Upper Jatulian) of the stratotype. As in the Upper Jatulian, the sheet hematite ores and jaspilites of the Superior type (the oolitic and granular kinds) are also present in the Dolomite Group, a higher concentration of manganese being typical of many rocks that belong to this level. Finally, the Pretoria Group is a reliable correlative of the Ladoga Group. This comparison is reasonable because of the occurrence of tillites in the second from the bottom formation (the

Ongeluk, Griquatown). Such formations are known to occur in the lower part of the Ladoga level of the Karelian-Kola region. They greatly resemble the equivalents of the Ladoga Group that are distributed in Northern Karelia and the Kola Peninsula. As in Transvaal horizons of carbonate rocks and basic lavas occur there also and carbonaceous (graphitic) slates are widespread. The Matsap red bed group corresponds to the Vepsian Group of Southern Karelia.

**Mesoprotozoic of the Canadian Shield.** In the Canadian or Canadian-Greenland shield Mesoprotozoic strata are widespread and are represented by various formational types. Here the North American stratotypical formations developed in the Great Lakes areas will be discussed in brief. They belong to the Huronian platform Supergroup and to the Animikie geosynclinal Supergroup; other important stratigraphic analogs from other parts of the shield will also be examined.

The Huronian Supergroup is assigned to the Lower Proterozoic or Aphebian by Canadian geologists, approximately corresponding to the Mesoprotozoic. Earlier the rock composition was regarded as the stratotype of the erathem, but later works and correlations revealed that it corresponds only to the lower part locally developed in North America. The stratigraphy of the Huronian Supergroup is well known from the very valuable works of many Canadian geologists, Roscoe (1957, 1969) and Young (1966, 1973 b) in particular. The supergroup is now divided into four groups overlying each other with a break but without any significant angular unconformity. The lower Elliot Lake Group consists of the following formations (upward); (1) Livingstone Creek (15–500 m) subarkoses and conglomerates; (2) Thessalon (65–2300 m) basic volcanics, subarkoses, conglomerates; (3) Copper Cliff (320–1200 m) acid volcanics; (4) Matinenda (35–320 m) coarse-grained subarkoses, quartz and oligomictic conglomerates; (5) McKim (65–2300 m) oligomictic sandstones and argillites. All the formations rest conformably, except the Matinenda formation, but in many localities any of the formations can lie directly on the crust of weathering of the basement rocks. A very important and characteristic aspect of the Elliot Lake Group is the occurrence of gold-uranium-bearing conglomerates principally enclosed in the Matinenda Formation (of great economic value), partially in the Thessalon and Livingstone Creek Formations.

The overlying Hough Lake Group is of wider distribution. It is composed of the following formations: (1) Ramsay Lake (15–65 up to 200 m) polymictic conglomerates, of glacial origin in part (tillites or tilloids); (2) Pecors (15–260 m) argillites and siltstones; (3) Mississagi (230–650 m, rarely up to 3000 m?) coarse-grained subarkoses. The following formations are recognized in the next Quirke Lake Group: (1) Bruce (25–65 m) polymictic conglomerates, partially of tilloid appearance, and greywackes; (2) Espanola (160–250 m) limestones, dolomites, and siltstones: (3) Serpent (65–300 m) arkoses and subgreywackes.

The Cobalt Group the uppermost in the Huronian Supergroup, is the most extensive; it lies with erosion not only on the underlying rocks, but also on the basement rocks. It consists of the following formations: (1) Gowganda (650–1300 m) polymictic boulder conglomerates with evident features of their glacial origin (tillites) with some layers of banded siltstones containing dropstones plunged in them, some interbeds of arkoses and argillites; (2) Lorrain (up to 2400 m) quartzites that are locally high-aluminous; (3) Gordon Lake (260–1200 m) variegated siltstones with crystal molds after rock salt and gypsum in places; (4) Bar River (630–2000 m) quartzites and red siltstones, interbeds of hematite quartzites.

The thickness of any Huronian unit, except the Cobalt Group, increases southward with the Penokean fold belt (geosyncline) becoming nearer. The Cobalt Group in contrast is mainly developed in the north. The Huronian strata were accumulated under littoral-marine and continental environment and the rocks commonly display different aspects of shallow-water conditions (ripple marks, cross-bedding, sun cracks etc.). Some features of past salinity are observed in the upper part of the group.

The Huronian rocks are generally slightly altered and form large, gently dipping folds, but to the south of the Murray fracture they are of conspicuous grade, of intense folding, and are cut by different granites in addition to the ubiquitous diabase dykes and sills; in the contact zone the rocks are transformed into schists and even gneisses around these granites. During the Mesoprotozoic the zone to the south of the Murray fracture was situated outside the platform (shelf) and surrounded the Penokean geosyncline.

The following data give a clue to the age of the Huronian strata; they rest unconformably on the Paleoprotozoic greenstone rocks and the granitoids cutting them (these latter are dated in the range of 2800–2600 m.y.) and are pierced by the Nipissing diabase dyke and the Sudbury gabbro-norite pluton with Rb-Sr age of 2150 and 2050–2200 m.y. respectively. They are moreover injected by later granites dated by the same method at 1780–2000 m.y. Prior to the emplacement of the Nipissing diabases the rocks of the Huronian Supergroup, or at least of its lower part, were folded, then subjected to a new deformation during the Karelian (Hudsonian) cycle.

There is no problem in correlating the Huronian strata and the Karelian stratotype as well as the South African parastratotype of the erathem. The lower, essentially volcanogenic portion of the Elliot Lake Group (the Livingstone Creek, Thessalon and Copper Cliff Formations) easily correlates with the Tunguda-Nadvoitsa Group of the stratotype and with the Dominion Reef Group of the parastratotype that is also reported to contain gold-uranium-bearing conglomerates. The Matinenda and McKim Formations of the Elliot Lake Group, the Hough Lake and Quirke Lake Groups, and also the Gonwganda Formation of the Cobalt Group are together to be compared to the Sariolian Group of Karelia and to the Witwatersrand and Ventersdorp Group of Southern Africa. All of them are typified by tillite occurrences and the Canadian and Transvaal rocks examined are also known to contain major horizons of gold-uranium-bearing conglomerates occurring in the lower parts of the strata. The upper part of the Cobalt Group (the Lorrain quartzites, Gordon Lake siltstones, and Bar River quartzites) can be compared with the Lower Jatulian and the Pniel Subgroup. Other common features are added to the common composition with pure quartzites dominating, for instance the occurrence of high-alumina minerals in quartzites, the red color of many rock types, and the presence of hematite quartzites. This correlation is also supported by isotope datings of the Nipissing diabase dykes similar in composition and age to the diabase dykes cutting the Jatulian (but these latter are older than the Ladoga Group).

Large iron ore deposits are known in the Animikie geosynclinal Supergroup that is developed in the Penokean belt (to the south of the Murray fracture mentioned earlier), the supergroup has been studied in detail by American and Canadian geologists (Leith et al. 1935, Grout et al. 1951, James 1958, Goldich et al. 1961, Morey 1973, etc.). The most complete sequence of the supergroup is known in the state of Michigan (U.S.A.), where it is divided into many formations and bands that are combined into four groups separated by breaks. In upward sequences these are: (1) Chokolay – composed of tillites, followed by quartzites and then by slates and dolomites with local stromatolites; (2) Menominee – formed of quartzite, slate, jasper, conglomerates, rare stromatolitic limestone, and volcanics; jaspilite bands of the Superior type are the thickest in this supergroup and of great economic importance; (3) Baraga – composed of quartzite, slate, greywacke, local volcanics, and small interbeds of jaspilite; (4) Paint River – predominantly dark (graphitic) slate alternating with greywackes and siderite bands. Total thickness of the supergroup is 12,000 m, of which 6000–9000 is in the Baraga Group.

Stromatolites of the Animikie Supergroup are various, and according to data of Hofmann (1969) they are represented by the forms registered in the three Riphean phytolite complexes. In particular, the cherty rocks of the Gunflint Formation contain *Kussiella* forms (Awramik 1976) typical of the Burzyan Group (the Lower Riphean) of the Urals. These stromatolites are also characterized by peculiar forms (*Gruneria* etc.) that are not known in Riphean phytolite complexes. Various microbiota is reported from jaspilite chert interbeds (Awramik and Barghoorn 1977).

The Animikie Supergroup unconformably overlies Paleoprotozoic greenstone strata and granitoids and is itself cut by granites which by different methods yielded an age of 1900 m.y. (Aldrich et al. 1965, Banks and Cain 1969). Earlier diabase sills and dykes that intrude the Animikie Supergroup are 2000 m.y. old (Hanson and Himmelberg 1967). The upper age limit of the supergroup is drawn according to the sharply unconformable occurrence of the Neoprotozoic Keweenawan Group on it.

The relationships of the Animikie Supergroup and the Huronian Supergroup were established by Church and Young (1970). According to data of these authors, the Animikie Supergroup lacks the strata corresponding to the major lower portion of the Huronian and its lower Chokolay Group is considered a correlative of the Cobalt Group only, the upper unit in the Huronian. Such correlation is rationalized by the presence of tillites in the lowermost part of the Chokolay Group (Fern Creek and others), which are compared with the Gowganda tillites of the Cobalt Group and also by the similarity between the Lorrain quartzites and the Sturgeon quartzites which overlie tillite in the Chokolay Group. This correlation is also supported by the presence of ferruginous quartzite in the uppermost part of the Huronian (the Bar River Formation).

The units of the Animikie Supergroup are likely to correlate with the Mesoprotozoic stratotype in the following way: tillites at the base of the Chokolay Group and the Sariolian Group and the major portion of the Chokolay Group starting with the Sturgeon quartzite and finishing with stromatolite-bearing dolomites and the Lower Jatulian. The Menominee Group, characterized by oolitic and granular jaspilites (with interbeds of stromatolite-bearing carbonate rocks), and the whole Baraga Group seem to correspond to the Upper Jatulian. The Paint River Group is also similar to the Ladoga Group.

The Kaniapiskau Supergroup, a stratigraphical and in part formational equivalent of the Animikie Supergroup, is developed in the north-east of Canada within the limits of the so-called Labrador trough. The inner structure and rock composition of this supergroup have been summarized by Dimroth (1970, 1972, etc.). The stratigraphy of this supergroup is complicated, and it has been studied in detail, but beyond the scope of this book, so that only a normal sequence is presented here (see Fig. 39, column X). It should be pointed out that the lower Knob Lake Group can reliably be compared with the Chokolay Group and the upper Doublet Group with the Menominee and Baraga Groups (except for the upper part of the group, which is likely to be correlative of the lowermost part of the Paint River Group).

A remarkable and well-studied Mesoprotozoic sequence different from the Animikie type is known in the north-western part of the Canadian shield in the Slave tectonic province of the eastern arm of the Great Slave Lake that structurally belongs to the Athapuscow aulacogen (see Fig. 40). According to Hoffman (1968, 1973) and Badham (1978), the following units are recognized there in upward sequence: the Wilson Island and Union Island Groups and the Great Slave Supergroup. The Wilson Island Group (1600 m) starts with alluvial conglomerates that are overlain by coastal-marine quartzite-sandstone strata, then follow dolomites with siltstone interbeds of reddish-pink color in places, and finally basic and acid volcanics. Local jaspilites and graphite slates are also observed in the group. The rocks are altered under greenschist facies and dip steeply. The group has contact with the Katarchean basement gneiss-granulite complex and with the younger Mesoprotozoic rocks along the dislocation zones. Earlier it was erroneously assigned to the Paleoprotozoic (Archean). It is unconformably overlain by the Lower Neoprotozoic Et-Then Group.

The younger Union Island Group (up to 500 m) also rests on the gneiss basement and comprises regoliths at the base. It is made up of stromatolitic dolomites with local interbeds of acid tuffs, red marls, and calcic slates that are overlain by basic volcanics with higher concentration of alkali.

The Great Slave Supergroup is divided into four groups (upward): Sosan, Kahochella, Pethei, and Christie Bay. The Sosan Group overlies the Paleoprotozoic rocks unconformably and includes the following formations: (1) Hornby Channel (250–1500 m), feldspathic quartzite with interbeds of conglomerate, mudstone and rare dolomite; (2) Duhamel (up to 300 m), stromatolitic and oncolitic dolomite; (3) Kluziai (up to 350 m) feldspathic quartzite. The latter lies on the eroded surface of the underlying dolomite and locally directly on the basement. The Kahochella Group (450–1600 m) consists of finely bedded, commonly red, mudstone, argillite, slate and fine-grained sandstone with rare interbeds of oolitic hematite, and stromatolitic dolomite with local cavities after gypsum. In the western part of the aulacogen the sedimentary rocks are underlain by basic volcanics.

The Pethei Group (400–600 m) is largely composed of dolomite (commonly stromatolitic). To the west it passes into rhythmically unterbedded (flysch-like) thin-bedded carbonate argillite, mudstone and greywacke, with local limestone. The topmost Christie Bay Group includes the following formations: (1) Stark (up to 750 m), red mudstone and argillite with lenses of stromatolitic dolomite and carbonate breccias; (2) Tochatvi (600–900 m), red cross-bedded lithic sandstone with interbeds of red clayey shale containing abundant crystal molds after halite at the top; (3) Portage Lake (100–200 m), argillite, overlain or partially replaced by basalts.

The abundant stromatolites of the Great Slave Supergroup are sometimes similar to the Riphean forms, to the Lower- and Middle Riphean in particular, but the specific ("endemic") forms (Hoffman 1973, Semikhatov 1978) dominate.

The stratigraphic position of different units of the Athapuscow aulacogen is mainly defined according to datings of dykes and diatremes of basic alkaline rocks comprised in the lower part of the Sosan Group, these values are 2170–2200 m.y. In the opinion of Badham (1978) these rocks were emplaced during the deposition of this group. They are similar in composition to the sub-alkaline basic rocks from dykes and sills of the underlying Union Island Group. The Seton volcanics from the Kahochella Group have an age of 1870 m.y., but this value is rejuvenated, as the whole group is cut by the 1900 m.y. old granite. Thus, the emplacement of dykes of basic alkaline rocks took place approximately at the time of formation the Nipissing diabase in the Great Lakes area or similar dykes that accompanied the Seletskian diastrophic episode in Karelia. Thus, the entire Great Slave Supergroup belongs to the Upper Mesoprotozoic. Its three lower groups may be correlated with the Ladoga Group of the stratotype, and the upper Christie Bay Group with the Vepsian sandstone Group that is commonly cross-bedded and red (the cross-bedding of both groups is usually of the flow type).

The Wilson Island and Union Island Groups underlying the Great Slave Super-group are to be assigned to the Middle Mesoprotozoic, as they are characterized by red beds and stromatolite-bearing rocks lacking in the Lower Mesoprotozoic. The lower group greatly resembles the Segozero Group (the Lower Jatulian) and the upper one, the Onega Group (the Upper Jatulian). The Lower Mesoprotozoic is absent in the Athapuskow aulacogen [20].

In Southern Greenland we witness one more interesting and instructive example of the Mesoprotozoic of the Canadian (Canadian-Greenland) shield. The Ketilidian complex (or supergroup) is recognized here. According to Higgins (1970), it includes two groups composed of sedimentary and volcanogenic rocks strongly folded and altered under greenschist to amphibolite facies. The lower, Vallen Group comprises the following formations in upward succession: the Zigzagland Formation (up to 140 m) which begins with conglomerate that unconformably overlies the Paleoprotozoic greenstone Tartoq Group and is followed by quartzites and arkose, then finely bedded limey and dolomitic slate; (2) the Blåis Formation (50–70 m) which is composed of siltstone, slate, arkose, and greywacke with interbedded graphitic slate and dolomite; (3) the Graensesø Formation (up to 500 m), which consists of black locally pyritized slate. Several gabbro sills are present. Dolomites are characterized by various stromatolites, and cherty slates contain microphytofossils.

The Sortis Group includes: (1) The Faselv Formation (100–900 m) which is made up of pillow lava with some interbeds of sedimentary rocks and gabbro sills; (2) the Rendesten Formation (1100–2400 m), composed of banded mudstone, slate, chert, and dolomite, and jaspilite of the Superior type intercalated with thick beds of pyroclastic rocks and diabase sills, with local replacement of sedimentary rocks by thick pillow lavas; (3) the Quernertoq Formation (more than 2000 m) consisting of pillow lava with gabbro and diabase sills.

---

20 Earlier the present author (Salop 1977a) suggested that the lowermost part of the Great Slave Supergroup is to be compared with the lower part of the Huronian; this assumption was based on data of Folinsbee (1972) on the occurrence of gold-uranium conglomerate-like formations in the Sosan Group. Now it has been established that the uranium mineralization in the Sosan sandstones is of hydrothermal type and is not appropriate for correlation. Moreover, new datings of diabases and some more data make this correlation invalid

The K-Ar age of granite cutting the Ketilidian is 1815 m.y., but this value is probably somewhat rejuvenated. The Lower Neoprotozoic Quipisarqo Group, which is transected by 1750 m.y. – old rapakivi granite, unconformably overlies the Ketilidian supracrustal complex and granites.

Rounded pyrite grains occur in quartz conglomerate of the Zigzagland Formation. This kind of pyrite is typical of the gold-uranium conglomerates of the Lower Mesoprotozoic, but locally it is reported from the clastic rocks of the lowermost part of the Middle Mesoprotozoic. By composition and sequence structure, the Zigzagland Formation greatly resembles the Lower Jatulian, and the Blåis and Graensesø Formations also seem to belong here. The Sortis Group is very similar to the Menominee and Baraga Groups of the Animikie Supergroup, and also to the corresponding rocks of the Kaniapiskau Supergroup assigned to the Upper Jatulian.

Higgins considered the Zigzagland Formation as having accumulated in small closed basins, and the overlying formations of the Vallen Group as having formed in a large shallow-water marine basin. Geosynclinal conditions prevailed only during accumulation of the thick upper part of the Sortis Group. Thus, it is possible to state the analogous evolution of the geosynclinal area of Karelides and the Dharwar-Shimoga belt of Hindustan.

It seems reasonable to conclude the examination of the principal Mesoprotozoic sequences of North America with the characteristic Mesoprotozoic geosynclinal strata developed away from the shield in the Eastern Rocky Mountains (Medicine Bow, Wyoming, U.S.A.); these are represented by the Phantom Lake, Deep Lake, and Libby Creek Groups separated by breaks (Houston 1968, Karlstrom and Houston 1979).

The Phantom Lake Group lies with a sharp angular unconformity on the Katarchean gneiss and granite-gneiss. Its lower part is built up of strata (up to 80 m) of metabasalt (amphibolite) and partially of altered tuff of acid volcanics, upward the thick strata (up to 2000 m) of mainly quartzites are situated, these latter commonly cross-bedded and containing bands and horizons of quartz and arkosic conglomerate, phyllite, and sheets of amygdaloidal metabasalt. The sedimentary rocks and volcanics of the group accumulated principally under the continental environment, and quartzite and conglomerate are of fluvial origin. Some rare varieties of quartzite and phyllite represent shallow-water formations. Certain quartz conglomerate contains radioactive minerals.

The Deep Lake Group, with a total thickness up to 3300 m, mainly consists of quartzite, quartz-pebbled, and polymictic conglomerate; in the upper part of the group there is a band of marble alternating with phyllite in quartzite. The group is composed of three cyclic strata (subgroups), each starting with conglomerate, and up-section follows quartzite, commonly cross-bedded. The conglomerate (mixtite) at the base of the middle and upper strata is glaciogenic. Deposits of this group were mainly accumulated in river valleys and deltas and partially under shallow-water environment. Detrital radioactive minerals and pyrite are known to occur in quartz-pebbled conglomerate at the base of the lower strata (the Magnolia conglomerate). Uranium concentration in the rocks is 140–916 g/t.

The Libby Creek Group is of composite structure and eight formations are recognized within it; in ascending order they are: (1) Headquarters Formation (100–900 m) made up of intercalated phyllite or micaceous schist and metaconglomerate with rare interbeds of quartzite and dolomite; (2) Heart Formation (up to 1000 m) composed of quartzite intercalated with phyllite and rare metaconglomerate; (3) Medicine Peak Formation (1900 m), made up of quartzite which is commonly cross-bedded; (4) Lookout Formation (400 m) consisting of micaceous schist, quartzite, and amphibolite; (5) Sugarloaf Formation (580 m) made up of ripple-mark quartzite; (6) Nash Fork Formation (1980 m), comprising dolomitic marble with bands of phyllite and carbonaceous slate, with stromatolites locally present in the marble; (7) Towner Formation (300–550 m), metabasite largely composed of amphibolite and amphibole schist; (8) French Formation (up to 500 m) of muscovite-chlorite schist and black pyritized phyllite with rare quartzite interbeds.

Principal rocks of the group were formed under shallow-water marine conditions and only some quartzites were formed under subaerial conditions. Young (1973 a) showed that most conglomerates and puddingstone of the Headquarters Formation are typical marine-glacial deposits.

Metamorphic rocks of the Deep lake Group gave an age of 1840 m.y. by the Rb-Sr method. Granites cutting both groups are 1715 m.y. old by the same method, but this latter value seems slightly rejuvenated.

Mesoprotozoic strata of Wyoming are easily correlated with the stratotype and parastratotype of the erathem. The Phantom Group of essentially volcanogenic composition is naturally compared with the Tunguda-Nadvoitsa (Dominion Reef) level and the Deep Lake Group, together with the Headquarters and Heart Formations characterized by glaciogenic formations and uraniferous conglomerates, with the Sariolian (Witwatersrand) level; notable also is the similarity of these units with the corresponding Huronian units, where three horizons of tillites are known to occur. The Medicine Peak and Lookout Formations are easily correlated with the Lower Jatulian level, and the Sugarloaf, Nash Fork, and Towner Formations with the Upper Jatulian level; the French Formation is similar to the Ladoga Group.

**Mesoprotozoic of the Brazilian Shield.** The most complete and well-studied Mesoprotozoic sequence is situated in the Minas Gerais state of Brazil, within the limits of the so-called Quadrilatero Ferrifero (Fig. 37). The Minas Supergroup and the overlying Itacolomi Group belong to the erathem in question, the stratigraphy is well known and many workers have made contributions to it, Dorr (1969) in particular.

The Minas Supergroup rests with a great angular unconformity on the greenstone rocks of the Paleoprotozoic Rio das Velhas Group and on the granitoids with an age of 2700 m.y. cutting it. It consists of five groups (up-section): Tamandua, Caraça, Itabira, Piracicaba, and Sabara. The Tamandua Group starts with the Cambotas Formation of quartzite (up to 800 m), in the base it consists of conglomerate and very coarse-grained quartzite overlain by fine-grained sericitic quartzite and inequigranular cross-bedded quartzite. Locally the strata of phyllite with conglomerate cemented by phyllite are registered in the lower part of the formation (Simmons and Maxwell 1961); these conglomerates may appear to be tillite. The Cambotas quartzite is overlain by the strata (up to 280 m) of slate or phyllite locally intercalated with dolomitic phyllite and dolomitic jaspilite (itabirite).

The Caraça Group lies with a break on the Tamandua Group, but commonly it rests directly on the Paleoprotozoic formations. The Moeda Formation is distin-

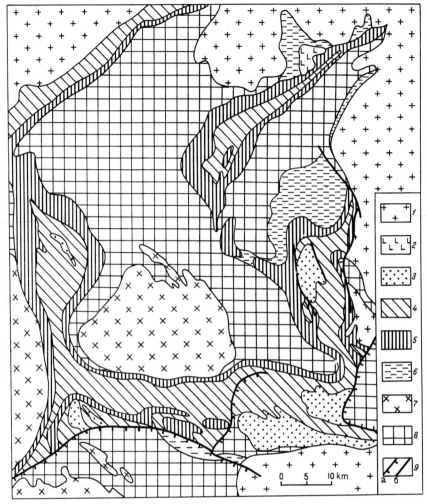

**Fig. 37.** Geologic chart of the Quardrilatero Ferrifero area, Brazil. (After Dorr 1969, simplified).
*1* Mesoprotozoic (post-Minas) and old granite, undivided, *2* Mesoprotozoic gabbroid, *3* the Itacolomi Group; *4–6* the Minas Supergroup: *4* the Piracicaba Group, *5* the Caraça and Itabira Groups (combined), *6* the Tamandua Group, *7* Paleoprotozoic gneiss-granite, *8* the Paleoprotozoic Rio-das-Velhas Group, *9* fractures (*a* thrusts, *b* faults)

guished in its lower part (100–500 m), it comprises coarse-grained quartzite and sandstone, locally cross-bedded, quartz and oligomictic conglomerate and phyllite. In other places the coarse-grained rocks are laterally replaced by thinner and finer-grained deposits. The occurrence of gold-uranium-conglomerate in the coarse-grained deltaic facies is a specification of the Moeda Formation (de Andrade Ramos and Fraenkel 1974). The overlying conformable Batatal Formation (100–200 m) is made up of dark grey carbonaceous phyllite that bears disthene in the zones of high alteration.

The Itabira Group is principally built up of jaspilite that is called itabirite in Brazil[21], and of manganese-bearing banded ores and dolomite. The ferruginous rocks are of immense economic importance. The Caue Itabirite Formation (100–700 m) is recognized in the lowermost part of the group, and consists of itabirite of dolomitic aspect and contains amphibole in its upper portion. The overlying Gandarela Formation (160–900 m) is mainly composed of dolomitic itabirite intercalated with dolomitic phyllite, dolomite, and dolomitic limestone. Abundant manganese is reported in the rocks. Dolomitic itabirite is characterized by a fine intercalation of black laminas of hematite and manganese minerals with the light interbeds of dolomite and quartz. Under tropical weathering the rocks are commonly orange-black or yellowish-brown.

The Piracicaba Group lies with an erosion on the Itabira Group and consists of four formations. The lower Cercadinho Formation (200–900 m) is principally made up of silvery phyllite alternating with quartzite. The basal conglomerate occurs in its base, it contains poorly sorted fragments of quartz, quartzite, and itabirite. Among the overlying phyllites hematite-bearing quartzite is observed as massive, nonbedded, or cross-bedded with ripple marks. The overlying Fêcho do Funil Formation (100–600 m) consists of dolomitic phyllite and siltstone with rare iron-ore rocks; the columnar stromatolites of the *Kussiella* and *Baicalia* groups are reported in dolomitic lenses (Dardenne and Costa 1975). The Taboões light grey quartzite (2–40 m) rests on the rocks and is overlain by the Barreiro Formation (80–300 m), consisting of phyllite, commonly graphite-bearing.

The next thick upward unit is the Sabara Group (up to 3000–3500 m), resting on the underlying rocks with an erosion. It is built up of rhythmically intercalated greywacke, chlorite slate, phyllite, rare quartzite, tuffite, and chert. Many workers state that this sequence greatly resembles the flysch formation. Conglomerate occurs in the base of the formation, upward appears tillite in which large boulders and pebbles of different rocks, of granite mostly, are dispersed in phyllite, chlorite slate, and greywacke.

In the eastern part of the Quadrilatero Ferrifero the Itacolomi Group (1000–2000 m) is spread in the cores of some synclines mainly composed of the rocks of the Piracicaba Group, unconformably overlying these rocks and the older strata of the Minas Supergroup. Two facies are commonly observed in the Itacolomi Group: quartzite and phyllite that, in the opinon of some, replace each other laterally, in the opinon of others, correspond to different levels of the group. Among the rocks mentioned, the conglomerate lenses are known, with pebbles and boulders of different rocks of the Minas Supergroup, including quartzite, phyllite, dolomite, and itabirite. Quartzite of the group is light lilac, pink, to light grey, well-sorted commonly exhibiting cross-bedding of basin type, with ripple marks. Siltstone and phyllite constitute up to 50%–60% of total thickness, they are grey or light lilac, sometimes purple.

Many think that the break between the Minas Supergroup and the Itacolomi Group proves a large tectonic cycle that was accompanied by emplacement of the Itabirito and many other granites. However, the Itacolomi Group is of general conformable occurrence (locally, though, it rests with an angular unconformity); it

---

21 The terms referring to the Precambrian iron-ore formations are poor. Sometimes all high-grade banded ferruginous-siliceous (cherty) rocks are called itabirites. Here these are called jaspilites irrespective of grade or composition of the ore compounds

demonstrates an equal grade of metamorphism and deformation with the underlying rocks (Wallace 1965) together they are cut by granite and pegmatite (Itabiro granite) that are suggested, without sound reason, to be younger than the Itabirito granite. It is very probable that the tectonic events that preceded the accumulation of the Itacolomi Group took place during a small-scale diastrophic episode.

The Itabirito granite is dated at 1340 m.y. by the K-Ar method, but this value seems to be strongly underestimated, as the Quadrilatero Ferrifero is situated in the zone of intensive isotope rejuvenation of the Pan-American (Brazilian) cycle. Tugari-nov and Bibikova (1971) reasonably think that the true age of this granite is 1900 m.y., as obtained by pegmatite cutting granite (the K-Ar method).

The occurrence of gold-uranium-bearing conglomerate in the base of the Caraça Group (in the Moeda Formation) suggests the assignment of this group to the Sariolian (Witwatersrand) or to the Lower Jatulian level, for this conglomerate mainly occurs on these two levels in other parts of the world; in the lowermost part of the second it is, however, less abundant. The Caraça Group matches this second level, judging from lithologic composition and scale of mineralization. The underlying Tamandua Group is then to be attributed to the Sariolian level. This correlation is also supported by the presence of the so-called "slate conglomerates" which are, in my opinion, glacial formations. It is possible that both groups belong to the Sariolian. In the Sierrra de Jacobina mountains (Bahia State), that are situated northward from the Quadrillatero Ferrifero, the Jacobina thick Group corresponds to the Tamandua and Caraça Groups, it contains large horizons of gold-uranium conglomerates (Leo et al. 1964). The Campo Formozo granite cutting the group is dated at 2000 m.y. by the Rb-Sr method (Torquato et al. 1978).

The Itabira Group certainly belongs to the Lower Jatulian stratigraphic level. Its jaspilite (itabirite) is of the Krivoy Rog and Kursk type. The overlying Piracicaba Group corresponds fairly well to the Upper Jatulian in many respects, by the occurrence of iron ores of the Superior type and of stromatolite-bearing dolomite in particular. The Sabara flyschoid group, resting unconformably on the underlying rocks, greatly resembles the Ladoga Group. This correlation is also justified by the presence of tillites in the lower part. Finally the molasse-like Itacolomi Group, consisting of cross-bedded pinkish quartzite, can be compared with the Vepsian Group.

**Mesoprotozoic of Australia.** The Mesoprotozoic strata are distributed in any regions of this continet, but they are best studied in the west of the country within the limits of the so-called Hamersley basin between the older Yilgarn and Pilbara blocks (see Fig. 24). The Mount Bruce Supergroup that extends there can be regarded as a regional stratotype of the erathem. In the north, the rocks of the group formed a slightly deformed cover of the Nullagine plate, in the south, though, they are intensely folded in the Ashburton fold zone, and are thicker and of higher alteration, corresponding to the greenschist facies.

According to Trendall (1975a, b) and many others, this supergroup is divided into three formations: Fortescue, Hamersley, and Wyloo. The Fortescue Group lies transgressively on the greenstone rocks of the Pilbara Group and granite cutting it, this latter being dated at 2600–2800 m.y. (Fig. 38). It consists mainly of basic lavas, intercalated with sedimentary and pyroclastic rocks. Six strata are distinguished in it: (1) Lower Lavas (0–325 m), locally distributed, consisting of amygdaloidal vesicular

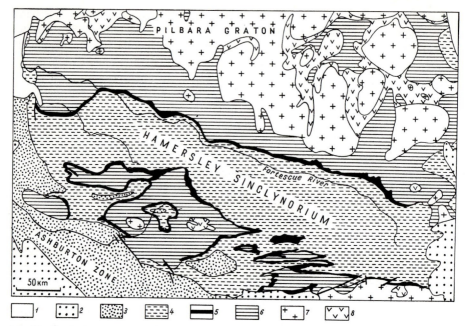

**Fig. 38.** Geologic map of the Hamersley basin. (After Trendall 1975 b).
*1* Phanerozoic sedimentary rocks, *2* Neoprotozoic (the Bresnahan and Bangemall Groups);
*3–6* Mesoprotozoic, the Mount Bruce Supergroup: *3* the Wyloo Group, *4* the Hamersley
Group, *5* the Marra-Mamba ferruginous Formation (the Hamersley Group), *6* the Fortescue
Group: *7, 8* Paleoprotozoic: *7* granite, *8* volcanics

and massive basalts; (2) Lower Sediments (160–1300 m), tufogenic and terrigenic
sandstones and siltstones, local conglomerate and agglomerate; (3) Middle Lavas
(160–600 m), basic and intermediate lavas, sometimes of globular type; (4) Middle
Sediments (130–650 m) made up of bedded tufogenic sandstone and siltstone, cross-
bedded and with ripple marks; local stromatolitic dolomite or limestone; (5) Upper
Lavas (650–800 m), massive basic and intermediate lavas, of globular type here and
there with subordinate tuff-siltstone and chert; (6) Upper Sediments or Jeerinah strata
(115–1000 m) consisting of shale, commonly pyritized, siltstone, subordinate jaspilite,
sandstone, and lava; locally they lie with a break on the underlying strata and are
underlain by basai polymictic boulder conglomerate. The quartz conglomerate with
low gold-uranium mineralization is reported from the middle sedimentary strata
(Robertson 1974). The thickness of the strata increases toward the basin center. The
lower part of the group is characterized by the presence of two large sills, one of
dolerite, the other of granophyre.

The Hamersley Group (up to 2800 m) is typified by the alternation of jaspilite
(47% thickness), shale (14%), dolomite (4.9%), tuff (2%) and chert. Thick strata of
acid lava (30% of the group thickness) are observed in the upper part of the group.
The group also includes many dolerite sills. It is divided in many formations, strata,
and bands. Sometimes it is accepted that its sequence starts with the upper sedimentary strata (the Jeerinah Formation) of the Fortescue Group, which seems quite

reasonable. The Hamersley Group is known by its extremely large iron-ore deposits of the Superior type.

The Wyloo Group lies with a break on the Hamersley Group and is divided into six formations, in upward sequence these are; (1) Turee Creek (300–400 m) consisting of finely bedded greywacke and schist, locally intercalated with quartzite, conglomerate, dolomite, and basalt; conglomerate is sometimes typical tillite with striated boulders (Trendall 1976); (2) Beasley River (up to 180 m) made up of light coarse-grained quartzite-sandstone, locally cross-bedded with ripple marks; (3) Mount McGrat (900–2000 m), alternation of sandstone, greywacke, quartzite, conglomerate, gritstone, shale, and siltstone; it contains a thick horizon of amygdaloidal basalt in its lower part; (4) Duck Creek (up to 900 m), yellow and orange dolomite, dolomitic limestone, locally stromatolite-bearing, interbeds of chert and shale; (5) Ashburton (from 300 m on the Nullagine plate up to 7200 m in the Ashburton zone), fine-bedded and monotonous intercalation of shale and fine-grained greywacke sandstone; rare interbeds of jaspilite, dolomite, chert, acid lava and tuff; (6) the Capricorn Formation (up to 1200 m), alternation of bedded greywacke sandstone, siltstone, shale, dolomite, and acid volcanics.

Stromatolites from the Fortescue Group are represented by forms of the *Gruneria*, *Collenia*, and *Alcheringa* groups, the first being typical of the Animikie Group of Canada. In the Wyloo Group the following forms are known: *Collenia*, *Pilbaria*, and *Patomia*?, in the U.S.S.R. this last only from the Upper Riphean and Vendian strata, while the Mount Bruce Supergroup is older than 1900 m.y.

Several isotopic datings are available for the rocks of the supergroup and for the igneous rocks cutting it. A Rb-Sr age of basic dykes related to basalts of the lower lava strata of the Fortescue Group is 2330 ± 90 m.y. and the slate from the overlying sedimentary strata of this group is 2600 m.y. old by the same method. The authors of this radiometric study (Hickman and de Laeter 1977) suggest that the latter value may be older than the true one due to the occurrence of clastic material in the slate. However, later a still older value of 2700 m.y. which greatly exceeds the age of the group accepted by many workers (Richards 1978), was obtained for galenite from the Fortescue Group by the Pb-model method. The porphyric and granophyric sills in the lower sedimentary strata of the Fortescue Group are dated at 2124 ± 195 m.y. and 2196 ± 26 m.y. and the dolerite sills from the Hamersley Group in the 1690–2200 m.y. range (by Rb-Sr analysis). The Woongarra acid volcanics teterminating with the Hamersley Group yielded an age of 2100 m.y. by Rb-Sr analysis and tufogenic siltstone and acid lava from the Wyloo Group are 1850 and 2020 ± 165 m.y. old respectively (Leggo et al. 1965). The posttectonic Boolaloo granite cutting the supergroup and most likely Lower Neoprotozoic is 1720 m.y. old (Trendall 1975a, b).

The Neoprotozoic Bresnahan and Bangemall Groups unconformably overlie the erosional surface of the supergroup and the granite cutting it.

The Fortescue Group containing gold-uranium conglomerate probably partially corresponds to the upper volcanogenic portion of the Sariolian level (in Southern Africa its analog is the Ventersdorp Group) and in part it is likely to be correlated with the Lower Jatulian (starting probably with the upper sedimentary strata). The Hamersley Group is very likely to be compared with the Upper Jatulian; the basis for such correlation is the occurrence of jaspilite of the Superior type and abundant dolomite with stromatolites (a close analog is the Dolomite Group of Transvaal). The

Wyloo Group, with tillite in the lowermost part and made up of flyschoid strata in its major portion (the Ashburton and Capricorn Formations), is the stratigraphic and formational equivalent of the Ladoga Group. It is suggested that dolerite intrusion with an age of about 2200 m.y. preceded its deposition.

One more Mesoprotozoic sequence that is quite complete and composite extends in the north of Australia in the western portion of the Arnhemland Peninsula and constitutes part of the Pine Creek fold (geosynclinal) zone (see Fig. 24). Three groups belong to the Mesoprotozoic here (up-section): Batchelor, Goodparla, and Finniss River (Walpole et al. 1968, Sweet et al. 1974). The Batchelor Group rests unconformably on the metamorphic Paleoprotozoic Rum Jungle complex and on granite cutting it with an age of 2600 m.y. It is distributed only in the central part of the Pine Creek zone. It comprises four formations, in upward sequence these are: (1) Beestons (300 m) consisting of arkose and greywacke sandstone, breccia-like conglomerate, quartzite-sandstone and mudstone; (2) Celia (300 m) of dolomite-containing *Collenia* stromatolites, commonly silicified, locally marbled; silicified dolomitic breccias; (3) Crater (600 m) made up of oligomictic quartz sandstone, greywacke, arkose, fine-pebbled quartz conglomerate, locally radioactive (with clastic thorite and monazite); (4) Coomalie (300 m) including silicified stromatolitic dolomite, locally marbled, tremolitic slate, and siltstone.

The Goodparla Group is more extensive than the previous one; it unconformably rests not only on the Batchelor Group but also directly on the basement rocks. The group comprises various facies and the formations it includes are distinguished under different names in different parts of the fold zone. It begins with the thick Mount Partridge Formation (more than 3000 m) consisting of arkose, arkose gritstone, conglomerate (with local clastic thorite and monazite), quartzite-sandstone, commonly cross-bedded and with ripple marks, and subordinate ferruginous quartzite and dolomite. This formation is overlain and is partially replaced by the Masson Formation (1200–3600 m) composed of oligomictic sandstone and siltstone exhibiting graded bedding. This latter is also overlain and partially replaced by the Golden Dyke Formation (up to 2700 m) of dolomitic siltstone, dolomite, quartz siltstone, chert and marl.

The Finniss River Group overlies the Goodparla Group conformably or with a break. The flyschoid Noltenius Formation (up to 4200 m) is situated in the lower part of the group. It is built up of siltstone, greywacke, and oligomictic sandstone, in part of quartz-pebble and boulder polymictic conglomerate; in places the rocks are altered and then the greywackes and siltstones are turned into micaceous schist. Many rocks are typified by gradational bedding. The description and illustration of certain conglomerates of the formation is suggestive of their similarity with tillite or marine-glacial formations. It is very likely that the so-called Henschke Breccia also belongs to the Noltenius Formation, as it also resembles tillite. The Noltenius Formation is supposed to be partially replaced by the Berinka Formation (840 m) consisting of acid and intermediate volcanics and clastic rocks. The Noltenius Formation is overlain by the Chilling Formation made up of white medium-grained quartzite-sandstone or quartzite, commonly cross-bedded and with ripple marks.

Three groups are strongly folded and cut by granites and granodiorites; their age is estimated to be 1830 m.y. The geosynclinal complex is unconformably overlain by the Lower Neoprotozoic volcanogenic Edith River Group that is cut by granite-porphyry with an age of 1750 m.y. (Rb-Sr method).

The Batchelor and Goodparla Groups are probably to be compared with the Jatulian, judging from their formational type and composition. It is to be noted that the radioactive psephite of the groups does not belong to the gold-uranium conglomerate formation, as the source radioactivity in this case is the clastic thorium minerals of heavy fraction (and gold is absent). Conventionally it is possible to compare the lower group with the Lower Jatulian, and the upper one, containing abundant dolomite, with the Upper Jatulian. The thick flyschoid Noltenius Formation of the Finniss River Group containing tilloid (?) is a close equivalent of the Ladoga Group, and the overlying Chilling Formation of cross-bedded quartz-sandstone of the Vepsian Group.

## 3. The Lithostratigraphic Complexes of the Mesoprotozoic

The principal Mesoprotozoic sequences of different continents have been examined on the foregoing pages, their correlation with the stratotype and their inter-correlation has also been propsed. Continuing the empirical generalizations, it is possible to recognize global-scale lithostratigraphic complexes. It was shown that six correlative units are distinguished in this erathem all over the world, exhibiting typical features, and thus can be reliably traced in almost any region (Fig. 39). To be sure not a complete lithilogic similarity but only the specific formations of these units is meant. Obviously, every unit ("horizon") distinguished represents a correlative lithostratigraphic complex with typical formations.

All six complexes are observed in the Mesoprotozoic stratotypical sequence of Karelia and the Kola Peninsula, but not all of them are sufficiently demonstrable there. Some are better demonstrated in other parts of the world, in the South African parastratotype of the erathem in particular. Thus the complexes are named according to the typical units (groups) of the stratotype as well as of the parastratotype.

The lowermost *Dominion Reef* (*Tunguda-Nadvoitsa*) *complex* is locally distributed. The Dominion Reef Group of Transvaal can be regarded as its stratotype and its analogs are the Tunguda Nadvoitsa Group of Karelia, the Lapponian of Northern Finland, the Novokrivorog Group (?) of the Ukraine, the lower part of the Udokan Group of Eastern Siberia, the lower (volcanogenic) strata of the Bababudan Group of India, the lower part of the Elliot Lake Group of Canada, the Phantom Lake Group of the U.S.A., and others. In its lowermost part this complex consists of terrigenous rocks, mainly of quartzite, in part of arkose and greywacke and also of conglomerate, but its upper major portion is made up of basic and intermediate volcanics, commonly amygdaloidal; in the topmost part acid lavas are locally reported. In Transvaal and Canada the gold-uranium conglomerate, though of little economic value, occurs in clastic rocks of the lower part of the complex. The complex in question was formed at the very beginning of the Mesoprotozoic: lava from the Dominion Reef Group is dated at 2800 m.y. This complex is likely to be regarded as a transitional unit. However, it cannot be assigned to the Paleoprotozoic ("Archean") as some suggest, as it lies with a sharp unconformity on greenstone strata and on granites cutting them. In the Republic of South Africa the Dominion Reef Group overlies the granite with an age of 2850 m.y. (the Vredefort dome).

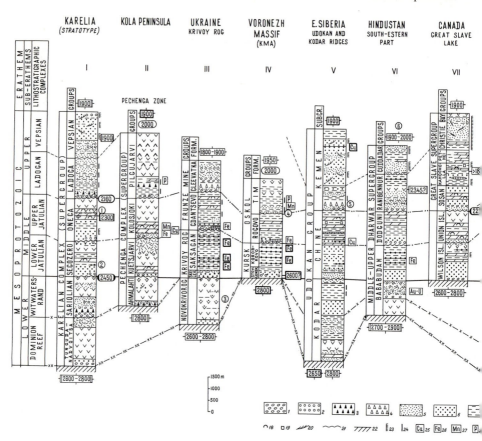

**Fig. 39** Correlation of lithostratigraphic complexes of the Mesoprotozoic principal sequences in different parts of the world.

Stratigraphic columns: *I* after Bogdanov et al. (1971), Salop (1973a) etc.; *II* after Zagorodny et al. (1964), Akhmedov (1979, unpubl. data) etc.; *III* after the Precambrian correlation scheme of the Ukrainian Shield (the Ukrainian Interdepartamental Stratigraphic Committee, 1978); *IV* after the working stratigraphic scheme of the Kursk Magnetic Anomaly, ed. by B. Petrov and N. Kononov (1979, unpubl. data); *V* after Salop (1964–1967); *VI* after Srinivasan and Sreenivas (1972); *VII* after Hoffman (1968, 1973), and Badham (1978); *VIII* after Roscoe (1969), and Young (1973b); *IX* after Goldich et al. (1961) and Morey (1973); *X* after Dimroth (1972); *XI* after Higgins (1970); *XII* after Houston (1968) and Karlstrom and Houston (1979); *XIII* after Dorr (1969); *XIV* after Du Toit (1954), Haughton (1969) and Winter (1976); *XV* after Trendall (1975a, b) etc.; *XVI* after Walpole et al. (1968), Sweet et al. (1974). Only fundamental works are referred to, additional information (isotope datings etc.) is taken from different sources.

Symbols: *1* polymictic conglomerate, *2* quartz conglomerate, *3* tillite, *4* tilloid (mixtite), *5* arkose and polymictic sandstone, *6* quartzite (quartzite-sandstone), *7* siltstone and slate (phyllite), *8* high-alumina slate and quartzite, *9* dolomite, *10* jaspilite of the Krivoy Rog type, *11* jaspilite of the Lake Superior type, *12* sheet hematite and siderite ore, *13* cuprous sandstone, *14* ultrabasic volcanics, *15* basic volcanics, *16* intermediate volcanics (andesite), *17* acid volcanics,

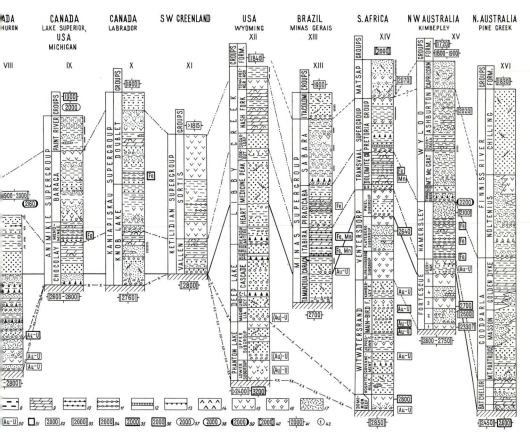

18 stromatolites, 19 evaporite (crystal molds after rock salt and gypsum), 20 cross-bedding, 21 stratigraphic (local slight angular) unconformity, 22 angular unconformity with pre-Mesoprotozoic formations, 23 red-beds, 24 "black-slate formation", 25–31 stratiform deposits (25 copper, 26 iron, 27 manganese, 28 phosphorus, 29 high-alumina raw materials, 30 gold-uranium-bearing conglomerates, 31 large and very large deposits), 32–36 isotope datings by different methods of sedimentary and volcanogenic rocks revealing the time of sedimentation or volcanism (methods: 32 K-Ar, 33 Rb-Sr, 34 Pb-Pb, 35 Pb-model, 36 U-Th-Pb), 37–39 isotope datings of diabase and gabbro-diabase dykes (methods: 37 K-Ar, 38 U-Th-Pb, 39 Rb-Sr), 40 Rb-Sr datings of the Bushveld massif, 41 isotope datings of granitoids and metamorphic rocks (for methods see 32–36), 42 number of a foot-note (see below).

Foot-notes (figures in circles): 1 jaspilite in the Onega Formation (the "Marine Jatulian") in Northern Finland (Puolanka); 2 jaspilite in the Segozero Group of the Lekhta synclinorium of Karelia; 3 the Novokrivorog (metabasite) Group of the Krivoy Rog, of presumed Paleoproto-zoic age; 4 the Voronezh Formation of the KMA – a stratigraphic analog of the Tim Formation, certain conglomerates are mixtites (tilloids?); 5 tilloid in the Upper Khilgando River, the Kodar Range and in the Chetkanda River basin, the Udokan Range; 6 thickness of the groups is approximate in this column; 7 basal conglomerate of the Vallen Group (the Zigzagland Forma-tion) contains rounded clastic pyrite

The overlying second, *Witwatersrand complex* is of much wider distribution than the previous one. Ubiquitously it is built up of clastic rocks, mostly quartzite, oligomictic sandstone, arkose, gritstone, and conglomerate intercalated with subordinate slate and siltstone. Rare interbeds and bands of jaspilite and carbonate rocks are also reported. In the upper part of the complex the strata of basic to acid volcanics alternating with tuff and sedimentary rocks are locally known. The basic lava is usually typified by an amygdaloidal structure. The Witwatersrand Group, together with the major lower portion of the Venterdorp Group, is the stratotype of the complex.

One of the most spectacular features of the complex is the occurrence of one of the three glacial (tillite) horizons separated by rather thick strata of terrigenous rocks. At present tillite is known from many regions of Siberia, of North and South America, of Africa and Asia (in the Bijawar Group, Northern Hindustan). They are most complete in the Huronian formations of Canada and in the corresponding formations of the Rocky Mountains of the U.S.A.; there three tillite horizons occur, the upper one being the most conspicuous.

The gold-uranium conglomerate is a notable and specific characteristic of the complex, being in some regions of great economic value. The largest and really unique deposits of this kind are known in the Republic of South Africa where gold has been mined since 1886. At the end of 1972 150 gold mines were known to operate there and 2.8 milliard tons of the ore or 28,722 tons of the metal had been mined, which constitutes a major portion of gold mining all over the world (in 1932 more than 50% of the world production was mined here). Extraction of uranium started after the second world war and in 1972 76,012 tons of it was produced (Pretorius 1975). Reserves of $U_3O_8$ in the Witwatersrand conglomerates are estimated to amount to 350–370 thousand tons. Uranium is also abundant in the conglomerate deposits of Canada. In 1964 these deposits were estimated to contain more than 200 thousand tons of $U_3O_8$ in the ores, with conditional concentration of metal; approximately the same figure is suggested for low-grade ore.

The gold-uranium conglomerate could have originated under very specific conditions that existed only in the Early Mesoprotozoic, mostly during deposition of the complex in question (see below). For this reason, these formations are a remarkable example of "irreversible" sedimentary formations.

The Witwatersrand complex is distributed in many regions of the world: in Europe it is represented by the Sariolian Group of Karelia and by its correlatives on the Kola Peninsula; in Asia by the lower part of the Udokan Group of Eastern Siberia, by the middle strata of the Bababudan Group of Western Hindustan and by the Bijawar Group of Northern Hindustan; in North America by the middle part of the Huronian Supergroup, by the lowermost part of the Animikie Supergroup, by the Papaskwasati Formation, Chibougamau Group, the lowermost portion of the Hurwitz Group developed in different regions of Canada, by the Deep Lake Group, together with the two lower formations of the Libby Creak Group in the U.S.A.; in South America by the Tamandua Group and Jacobina Group of Brazil; in Africa by the stratotype and probably by the lower horizons of the Muva Group of Zambia and Bugunda Toro Group of Uganda; in Australia by the major portion of the Fortescue Group.

The age of the Witwatersrand complex seems to fall within the range of 2750–2400 m.y. After its formation certain tectonothermal events occurred related to slight deformation of the rocks and to small intrusions of gabbroids and granites. These events belong to the Seletskian diastrophic episode that is registered as having taken place 2450–2400 m.y. ago.

The third, *Lower Jatulian complex* is of a wider distribution than the two previous. Almost everywhere it rests with a stratigraphic and local slight angular unconformity on the underlying Mesoprotozoic complexes or lies with a sharp unconformity on pre-Mesoprotozoic rocks. The stratotype of the complex is the Segozero Group (the Lower Jatulian) of Karelia. Usually it is mostly built up of quartzite (with subordinate quartz conglomerate or gritstone), then of shale and siltstone (transformed into micaceous schist and other kinds of schist in the zones of high alteration of the rocks), and also of dolomite. Local basic volcanics and their tuffs are reported. Clastic rocks make up the lower part of the complex, up-section appear slate and carbonate rocks; if volcanics occur they are situated in the uppermost part of the complex. The sedimentary rocks exhibit shallow-water structures.

The carbonate rocks in the Lower Jatulian complex are usually subordinate. Phytolites are common among them, sometimes making up bioherms and stromatolite horizons, the first in the Precambrian sequence (phytolites are quite rare and are never rock-forming in the Paleprotozoic).

The essentially monomictic clastic rocks are commonly products of chemical weathering and in places they rest on crusts of weathering. In the lowermost quartzite horizons the quartz gold-uranium conglomerate sometimes occurs, the content of ore components in it being low. Quartzite-sandstone with clastic pyrite is also registered. However, this kind of formation is never encountered above the first horizons of carbonate rocks with stromatolites; but there the rare red beds appear that evidence an increase of oxygen concentration in the atmosphere.

The slate or slate-carbonate strata sometimes include sheet bodies of hematite and siderite ores or jaspilite. These latter are quite abundant in some regions, and form large deposits or even iron-ore basins. Among them are the deposits of the Ukraine (the Krivoy Rog basin), of the Voronezh massif (the KMA basin), and of Brazil (the Quadrilatero Ferrifero basin).

Jaspilite of the Lower Jatulian complex does not associate with volcanics, unlike the Paleoprotozoic jaspilite of the Algoma type. This same feature also typifies Upper Jatulian complex jaspilite, but it differs in structure and mineralogical and chemical ore composition (see below). Lower Jatulian jaspilite is characterized by association with sedimentary strata and by common oxide facies; sulfide facies are lacking or quite rare. Ore minerals are mainly represented by magnetite and hematite, rare ferruginous carbonate, and are always crystalline. In some cases the manganese minerals occur together with magnetite and hematite. The phosphorus content in jaspilite is very low; organic remains are absent. Probably this jaspilite is to be distinguished as a special Krivoy Rog type of ferruginous-siliceous formation.

The analogs of the Lower Jatulian of Karelia are: in Europe: the Skelevatka and Saksagan Formations of the Krivoy Rog, Stoylensk and Korobkovo Formations of the KMA; in Asia: the Inyr and Chetkanda Formations of Eastern Siberia (the Udokan Range), the Onguren Formation of the Western Baikal region, the upper strata of the Bababudan Group of Western Hindustan and the Aravalli Group of North-Western Hindustan; in North America: the upper parts of the Cobalt Group of the Huronian Supergroup, the Chokolay Group of the Animikie Supergroup and the Hurwitz Group, the middle part of the Libby Creek Group of the U.S.A.; in South America: the Caraça and Itabira Groups of the Minas Supergroup; in Africa: the Pniel Subgroup of the Ventersdorp Group of Transvaal, the Konse Group of Tanzania, the upper part of the Muva Group of Zambia and the Buganda Toro Group of Uganda, the Misuku Group of Malawi, the Tannekas Group of Togo, the Kusheriki-Kushaka Group of Nigeria and many more; in Australia: the upper part of the Fortescue Group of the Mount Bruce Supergroup in the Hamersley basin and the Batchelor Group in the north of the continent.

The time of formation of the Lower Jatulian complex seems to be in the range of 2450 (2400?)–2350 (2300?) m.y. from the isotope datings available.

The fourth, *Upper Jatulian complex* is typified by abundant slate-carbonate or essentially carbonate rocks with common occurrence of ferruginous-cherty rocks of the Superior type, jaspilite among them. If the ferruginous rocks are of low grade, the ore matter is of granular or micro-oolitic structure and is represented by water silicate of iron (greenalite, minnesotaite, stilpnomelane) and also by hematite, magnetite, geothite, and siderite. In certain deposits the ore interlayers in jaspilite occurring among the carbonate rocks contain a higher concentration of $P_2O_5$ (up to 5%). The chert interlayers in jaspilites contain abundant microscopic remains of algae. It is evident that the Superior type jaspilite is different from the Krivoy Rog type. It is possible that some of these pecularities are due to unequal metamorphic grade, although in some cases ferruginous rocks of both types occur at different levels of the same group and are similarly altered, their difference, however, remaining conspicuous.

The iron deposits of the Superior type are characterized by huge reserves and are of extremely great economic importance in particular the epigenitically enriched ores related to jaspilite. Among these are the gigantic deposits of the Lake Superior in the U.S.A., of the Labrador trough in Canada, the Hamersley iron-ore basin in Australia.

The carbonate rocks are mainly composed of dolomite and dolomitic limestone. Stromatolites and oncolites are more common here than in the underlying complex. In places stromatolites build up rather thick horizons followed for tens of kilometers. The carbonate strata sometimes contain small sheet deposits of hematite and siderite ores and also the stratiform lens-like bodies chracterized by higher concentration of copper or manganese, vanadium, barium, and phosphorus.

At the base of the complex quartzite, arkose sandstone or conglomerate commonly occur; the latter usually comprise clasts of jaspilite from the Lower Jatulian complex. Quartzite is also at places enriched in clastic ore matter. The shallow water environment of sedimentation is registered here as well as in the underlying complex, in every type of rock. Many rocks are pink-colored. In the upper part of the complex strata of basic or acid volcanics are locally known.

The Upper Jatulian (Onega Group) of Karelia is the stratotype of the complex, but many more units in different regions of the world evidently show similarity with the stratotype. Stratigraphic analogs of the Upper Jatulian in Europe are as follows: the Gdantsevo Formation of the Frunze Mines Group of the Krivoy Rog, the Rogovo (Yakovlevo) Formation of the Oskol Group of the KMA; in Asia: the Aleksandrov and Butun Formations of the Udokan Group, Eastern Siberia, the Dodguni Group of Hindustan; in North America: the Menominee Group of the Animikie

Supergroup of the U.S.A., the upper part of the Libby Creek Group of the U.S.A., the lower part of the Doublet Group (the Sokoman Formation) of the Labrador trough, the middle part of the Belcher Group and the upper part of the Manitounuk Group of Canada, the Sortis Group of Greenland; in South America: the Piracicaba Group of the Minas Supergroup of Brazil; in Africa: the Black Reef and Dolomite Groups of Transvaal; in Australia: the Hamersley Group of the Mount Bruce Supergroup in the west of the continent and the Goodparla Group in the north.

Tectonic events are registerd ubiquitously after the formation of this complex, and thus after the formation of the entire Middle Mesoprotozoic Suberathem; they were mostly of oscillating nature, but locally accompanied by slight folding. These movements are here attributed to the Ladogan diastrophic episode; in many regions of the world it was completed with formation of diabase and granophyre dykes and sills. These intrusives are dated at 2200–2160 m.y. Thus the Upper Jatulian complex originated in the time interval of 2350–2200 m.y.

The fifth, *Ladogan (Transvaal) complex* of the Mesoprotozoic rests unconformably on the underlying rocks of the erathem or directly on the crystalline basement in almost every case. It is mainly built up of sandstone, siltstone, and slate, but in some regions dolomite and volcanics of basic, ultrabasic, and rarely of intermediate and acid composition also constitute an important component. It differs from the Jatulian complexes by a notable facial variability and by much more composite inner structure. In many cases, however, the complexes reveal common features. One of these is the occurrence of tillite in its lower part, representing the uppermost Mesoprotozoic level of glacial formations. This is now known in many regions of the world: in Europe it is the tillite of the Ladoga Group of the Ladoga region, of the Vuosnajarvi Formation of Northern Karelia, of the Pilgujarvi Formation of the Pechenga Supergroup of the Kola Peninsula; in Asia, of the Zama Group of the Western Baikal area, of the Kemen Subgroup of the Udokan Range; in South America, of the Sabara Group of the Minas Supergroup and of the Macaubas Group; in Africa, the Griquatown tillite of the Pretoria Group of Transvaal and of the upper part of the Ruzizi Supergroup of Ruanda; in Australia the Turee Creek tillite of the Wyloo Group of the Mount Bruce Supergroup. Moreover, a glacial origin is suggested for certain puddingstone (mixtites) and boulder conglomerates occurring in the lowermost parts of the complex in many areas, for instance in the Gleyevatka Formation in the Krivoy Rog, in the Voronezh Formation of the KMA, in the Kirey Formation of the Sayan area and in the Noltenius Formation of Northern Australia.

The flysch formation is one more typical feature of the complex. It makes up almost the entire complex or some major portions of it. It is characterized by fine rhythmical interbedding of fine-grained greywacke, siltstone, and slate, rare oligomictic sandstone and carbonate rocks. It commonly comprises dark, almost black carbonaceous slate and siltstone, usually enriched in sulfides. In places where they make up single strata they are recognized as a special "black slate" formation.

The sandstone-shale strata with abundant carbonate rocks belong to a special formational type, and are of subplatform or aulacogen nature; the Great Slave Supergroup is an example of this type of strata. Quartzite and dolomite with stromatolites comprise an important portion in their composition, locally red beds and rocks with certain features of past salinity are also known. The occurrence of basic alkaline volcanics and their sub-volcanic equivalents is also registered.

Abundant basic and ultrabasic lavas characterize the upper part of one more type of complex that associates laterally with flysch formation. To this belong the strata distributed on the Kola Peninsula and in the Shimoga-Dharwar zone of Hindustan.

The Ladogan complex rocks sometimes contain horizons of Superior type jaspilites and sheet deposits of hematite and siderite; both are of economic value in some regions, but large deposits are not known anywhere. The intrusive and extrusive ultrabasic bodies containing copper-nickel (or mainly copper) mineralization occur in the flyschoid and black slate strata on the Kola Peninsula and in Eastern Finland; the sedimentary rocks emplacing these bodies are commonly enriched in Ni, V, Cr and P.

The Ladoga Group of south-western Karelia is the stratotype of the complex, its parastratotype the Pretoria Group of Transvaal. They correlate with: in Europe, the Kalevian Group of Finland, the Vuosnajarvi and Kuolajarvi Formations of Northern Karelia, the Pilgujarvi Formation of the Pechenga Supergroup of the Kola Peninsula, the Gleevatka Formation of the Ukraine, the upper part of the Oskol Group of the KMA; in Asia, the lower part of the Kemen Subgroup of the Udokan Group of Eastern Siberia, the Kirey (Belorechka) and Shablyk Formations of the Sayan area, the Ranibennur Group of Hindustan; in North America, the Great Slave Group of Canada and the Paint River Group of the Animikie Supergroup of the U.S.A.; in South America, the Sabara Group of Brazil; in Africa (in addition to the Pretoria Group), the Lomagundi and Umkondo Groups of Zimbabwe and Zambia, the upper subgroup of the Ruzizi Group of Ruanda; in Australia, the major portion of the Wyloo Group of the Mount Bruce Supergroup and the Noltenius Formation of the Finniss River Group.

The formation of the complex is suggested as having occurred in the range of 2200–2000 m.y. judging from the isotope datings that refer to the time of sedimentation and volcansim and give also the age of the lower boundary of the complex.

The sixth, *Vepsian complex* of the Mesoprotozoic is recorded only in areas where the upper units of the erathem were preserved from erosion. Its stratotype is the Vepsian Group of Southern Karelia, and its equivalents are suggested to be the upper part of the Kemen Subgroup of the Udokan Group of Eastern Siberia, the upper part of the Cristie Bay Group of the Great Slave Supergroup in Canada, the Itacolomi Group in Brazil, the Matsap Group in the Republic of South Afrika, the upper parts of the Wyloo Group and the Finniss River Group in Australia. In some of these regions the Vepsian complex rests on the underlying rocks with a stratigraphic unconformity and locally with an angular one, in some places the unconformity is absent or is uncertain.

The complex is characterized by quartz and arkose sandstones that here and there pass into gritstones and conglomerates. The rocks are typified by red color, cross-bedding, ripple marks, and other aspects of shallow-water and subaerial sedimentation environment. In Eastern Siberia the copper deposits of cuprous sandstone type occur in the red beds. It is highly probable that the Vepsian complex represents an early molasse of the Karelian cycle; it was formed approximately 2000 (2050–1950) m.y. ago judging from its situation in the sequence and from certain isotope datings.

# B. Geologic Interpretation of Rock Record

Mesoprotozoic sequences and the rocks forming them provide conclusive and comprehensive information on the geologic evolution of the planet, the evidence at our disposal in this case being even more reliable than in the case of Paleoprotozoic rocks. The environment of supracrustal formation of the erathem is imprinted as a rule in the composition of the rocks and their structural-textural aspects. There is a possibility of successful application of method of actualism in studying almost any type of rock or formation of the erathem, always with due attention to the evolutionary factors. The facial zonation is almost ubiquitously observed and permits a reliable reconstruction of the paleogeographic or paleotectonic environment in the period of rock accumulation; a more detailed stratigraphic division of the Mesoprotozoic in comparison with the older eras reveals the possibility of clarifying the succession of geologic events with more accuracy and detail.

## 1. Physical and Chemical Environment of the Earth's Surface

The average annual temperature of a shallow-water sea in the middle of the Mesoprotozoic Era is suggested to be in the order of 60 °C judging from isotope ratio of oxygen ($^{18}O/^{16}O$) in cherts from the banded iron formation of the Hamersley Group in Australia (Becker and Clayton 1976, Knauth and Lowe 1978). The abundance of carbonate rocks in the Mesoprotozoic shows that the concentration of carbon dioxide in atmosphere and hydrosphere had decreased greatly in comparison with the situation in the Paleoprotozoic and thus could no longer prevent precipitation of carbonates. Moreover, a further decrease in atmospheric pressure must have occurred because, with capture of $CO_2$ by carbonate deposits without due compensation during the volcanic eruptions, the hothouse effect must have lessened as well. The major portion of carbonate rocks, however, is known to have occurred in the Middle and Upper Mesoprotozoic Sub-erathems and thus it is possible to suggest that the temperature of the Earth's surface must have lowered slightly at the end of the era in comparison with that established on the cherts of Australia.

Four global glaciations are registered in the Mesoprotozoic; one at the end of the early subera and an other at the beginning of the late subera were extremely strong. Glacial formations are known to occur among the strata accumulated during the hot climate period and so an exceptionally sharp and short-term temperature reduction must have happened for these deposits to originate, a reduction which could not have been caused by any terrestrial factors. In a special paper (Salop 1977c) I have shown that only cosmic agents could have caused these glaciations. During the Precambrian the glaciations are known not only in the Mesoprotozoic, so it is reasonable to continue the discussion of their causes on examining all these extremely important climatic events.

The gold-uranium conglomerates occur in the Lower Mesoprotozoic Sub-era, and could have originated only under atmosphere lacking or deficient in oxygen (Ramdohr 1961, Roscoe 1969). However, in the Middle Sub-erathem the carbonate strata with stromatolite bioherms are overlain by red beds, a fact registered for the first time in geologic sequence, their abundance increasing up-section in the erathem. The oxidized ferruginous rocks also become widely distributed there; in the cherty interbeds of these, abundant microscopic remains of blue-green algae have been known since the Upper Jatulian complex. Naturally, with a sharp increase of abundance of plant biomass, the biogenic photosynthesis must have intensified proportionally, and as a consequence the atmosphere (and hydrosphere) must have been greatly enriched in oxygen. Its concentration had become sufficient not only for oxidation of iron in deposits but also for its presence in free (molecular) state for long periods of time. Thus, in the middle of the Mesoprotozoic, during formation of stromatolitic carbonate strata of the Lower Jatualian complex ($\sim$ 2400 m.y. ago), the atmosphere suffered an important change, becoming oxygen-bearing after a period of deficiency, but still not very rich in free oxygen. The quantity of it seems to have been a few percent against its concentration in the modern atmosphere.

## 2. Life During the Mesoprotozoic

It was shown that a great increase of plant (mainly algal) biomass occurred in the middle of Mesoprotozoic, resulting in a change of composition of atmosphere and greatly affecting many geochemical processes on the planet surface. Had the outburst of life been accompanied by an equally important change in its organization?

Many workers studying microscopic organic remains in the rocks of the Middle and Upper Mesoprotozoic of North America (of the Animikie and Belcher Groups; Darby 1974) and of Southern Africa (of the Transvaal Supergroup; Nagy 1974) come to the conclusion that certain cellular structures with inner nucleus and mitosis features among microbiota were present in addition to abundant prokaryota. Among these structures are also known to occur these that can be regarded as reproductive organs of the eukaryota algae, of fungi and sponges (Edhorn 1973). Kaźmierczak (1979) suggests comparing the *Eosphaera* from widely distributed in the Animikie jaspilites (and in the synchronous ferruginous-siliceous rocks of different parts of the world) with the *Eovolvox* eukaryota form from the Devonian strata.

It should be kept in mind, however, that the dark opaque grains in the fossilized cellular structures of algae commonly regarded as organellas (nuclei) may appear to be the remains of degraded inner shells of prokaryotic cells (Golubik and Hofmann 1976, Knoll et al. 1978). The problem of recognition of eukaryotic organisms thus becomes difficult. Still, many data indicate an exceptionally important event in life evolution during the Middle Mesoprotozoic: the appearance of the first eukaryotic life forms.

## 3. Environment of Supracrustal Rock Accumulation and Their Evolution

The major portion of supracrustal groups (complexes) of the Mesoprotozoic is made up of sedimentary rocks as well as of volcanics. The formation environment of the first type of rocks is quite certain; this is not so, however, with the second type, for many data are needed to elucidate the genesis of magmatic rocks; in addition to their petrographic composition and mode of occurrence, comprehensive and classified chemical-analytical data are necessary. Such information is identically available only for some objects and it is not sufficiently general to establish any regularities.

The oldest Mesoprotozoic supracrustals are made up of two formations, represented by the locally developed Dominion Reef complex, the lower clastic formation, and the upper volcanogenic one. The first is characterized by the presence of extremely mature rocks like quartzite on the one hand and by the presence of greywacke and polymictic conglomerate on the other. This association suggests the existence of a contrast relief during sedimentation, a certain combination of plain and rugged areas. This environment seems realistic, for it is known that the formation of the Dominion Reef complex took place soon after the early events of the Kenoran orogeny. The accumulation of this contrasting greywacke-quartzite formation occurred in a littoral-marine and alluvial environment and the clastic matter was supplied from different sources. The uppermost thick portion of the complex is represented by andesite-basalt or andesite-basalt-rhyolite formation and the succession of different types of lava outflow was of homodrome nature. The basic lava outflowed mainly under subaqueous (shallow-marine) conditions, the acid lava under subaerial environment, accompanied by explosive phenomena.

The Witwatersrand quartzite-arkose-conglomerate complex and the conditions of its formation are of great interest because of their gold-uranium deposits. Generally this complex is mostly typified by alluvial, proluvial, and deltaic and littoral facies in particular, though some glacial formations are also observed with typical terrestrial moraines (tillites), fluvioglacial, lacustrine-glacial, and probably marine-glacial deposits. The Witwatersrand group of Southern Africa, stratotypical of the complex, accumulated in the Rand basin, a large isometric intracratonnal basin. During deposition of the lower part of the group the supply of clastic material was more intensive from the northern side of the basin than from the southern one. The coarse-grained deposits passed into fine-grained deposits, and the concentration of heavy minerals became lower toward the center of the basin. The river streams carried the material, and on discharging into the basin, they formed the deltaic fans. Accumulation of clastic strata of the complex in the Huronian depression of Canada was similar, with the only difference that the terrigenous material was carried mostly from the Superior platform situated to the north of the depression.

For a long time the problem of genesis of ore mineralization in the gold-uranium conglomerates was debatable. Now, however, it is possible to state that primary mineralization in conglomerates and quartzites is due to the accumulation of clastic gold grains and uranium-bearing minerals in the older placers. At the same time, certain ore minerals and their concentrations originated as a result of epigenetic processes related to the circulation of hydrothermal solutions, including the metamorphogenic type.

The following important facts favor the primary sedimentary genesis of mineralization: (1) the gold-uranium conglomerates are reported only from the Lower Mesoprotozoic and occur mainly at a definite stratigraphic level (stratigraphic control); (2) the gold-uranium mineralization occurs in definite sedimentary rocks: quartz or oligomictic conglomerates and the related quartzites (lithologic control); (3) the metal-bearing conglomerates commonly occur in the older river beds and deltas; (4) the ore components are commonly layered; (5) the unrecrystallized grains of gold, uraninite, and commonly related pyrite are of rounded shape; (6) gold, uraninite, and pyrite are associated with other heavy fraction minerals: monazite, zircon, cassiterite, chromite, spinel, etc.; (7) regular distribution of ore minerals in the rocks and their generally low concentration (in the "reefs" of Witwatersrand an average concentration of gold is 10 g/t and of uranium 280 g/t).

Uraninite and pyrite are known to decompose easily while being carrid as clasts in atmosphere containing free oxygen and thus at present they are never accumulated as placers. The gold-uranium conglomerates could have originated under a quite specific combination of geochemical environment (lack of free oxygen in the atmosphere) and tectonic regime (of relatively stable platform or subplatform type), conditions which existed only during the Early Mesoprotozoic. Certain additional local factors were necessary for this accumulation to happen, for instance, vast areas of peneplained land limited with shallow-water basins, and also the presence of some original source of gold and uranium (it could have been the greenstone complexes of the Paleoprotozoic for the gold and possibly also for the uranium). The upper limit of distribution of the gold-uranium-bearing conglomerates is defined by a great evolution of blue-green algae possessing a photosynthesizing function in the Middle and Late Mesoprotozoic.

Details of genesis of the gold-uranium-bearing conglomerates are discussed in my paper (Salop 1972a) and also in the works of Roscoe (1969), Robertson (1974) and Grandstaff (1980). This latter author came to the conclusion, based on calculation from the dissolution kinetics of uraninite, that this mineral could be present as detrital grains in the older conglomerates not only in the oxygen-deficient atmosphere but also at a concentration slightly higher than 0.01 of its present atmospheric level (PAL). However, Grandstaff used some confusing parameters for his calculations. Thus, he proceeded from the assumption that the uraniferous conglomerates accumulated under subglacial conditions and the temperature of the environment was accepted to have been 15 °C. But it is well known that these conglomerates accumulated during the entire Early Mesoprotozoic and mostly during nonglacial epochs when the climate was exceptionally hot and the annual temperature of the planet surface was of the order of 60°–65 °C. Meanwhile, temperature conditions are an important factor in dissolution kinetics. Additionally, the author did not attach due importance to the constant occurrence of rounded grains of pyrite in the uraniferous conglomerates (and quartzites), whose stability is much lower in an oxidizing environment than that of uraninite. For these reasons, Grandstaff's conclusions do not seem quite convincing. In my opinion, the oxygen concentration equal to the 0.01 modern level (the Pasteur level) or slightly higher was realized only in the second half of Mesoprotozoic when the first eukaryota appeared.

In some regions the formation of the Witwatersrand complex was completed with intensive volcanism with basaltic and andesitic lavas dominant, but locally also the acid lavas outflowed. The eruptions occurred mostly under subaquatic conditions and were commonly interrupted by accumulation of littoral shallow-marine and rare continental sediments. The sedimentary-volcanogenic strata examined make up the transgressive portion of the complex sequence; their accumulation was accompanied by subsidence of the sedimentation basins, by extension of the Earth's crust and simultaneous implacement of diabases and granophyres along the fractures (dykes and sills).

Sediments of the Lower Jatulian complex accumulated after the Seletskian dias-trophic episode, which signaled the start of a wide transgression of the sea toward the peneplained land of the older platforms. Its maximum was registered in the second half of the Upper Jatulian cycle when abundant carbonate strata became common almost everywhere. The monomictic and oligomictic quartz sandstones (quartzites) and quartz conglomerates, that are mostly typical of both Jatulian complexes, repre-sent the mature lateral-marine fomations that originated as a result of a long multifold rewashing, redeposition, and sorting of products of crusts of physical and chemical weathering. In the lower part of the Lower Jatulian complex, gold-uranium quartzites and quartz conglomerates occur with rounded clastic pyrite, evidence of sedimenta-tion in an atmosphere lacking oxygen at the very beginning of the Middle Mesoproto-zoic. Free oxygen appeared in the atmosphere in the second half of the Lower Jatulian stage, due to a rapid evolution of photosynthesizing algae. After that time the Meso-protozoic strata are characterized by red beds, with increasing abundance tn the younger sequences.

An obvious dominance of dolomites among the carbonate rocks is very likely to be related to the higher concentration of $CO_2$ in the atmosphere that led to a high alkaline reserve in the seawater. Under these conditions the dolomite matter, being lighter than calcite, saturated the water and precipitated quickly as the climate became more arid (Strakhov 1963, Vinogradov 1967). The carbonate sediments commonly accumulated in the shallow-water basins, in the littoral zone as a rule, as is evidenced by the occurrence of stromatolites and microphytolites. In certain slate-carbonate strata (for instance in the Udokan Goup of the Eastern Siberia), the occurrence of evaporites (halite and gypsum) is known, showing that some of the older basins contained highly mineralized water.

The genesis of a banded iron formation in the Jatulian complexes, as well as in other Precambrian complexes, is not clear in some respects as yet. The only certain fact is that matter had precipitated chemically and that the microbands in jaspilites are of primary sedimentary origin, only becoming more obvious during the diagenesis of sediments (Trendall and Blockley 1970). However, the sources of silica and ore matter, and the causes of their cyclic precipitation, resulting in fine-bedded rhythmical alternation of interlayers of different composition, still remain debatable.

Two rival concepts exist as to the source of silica and iron: the first regards them as a result of crust weathering on the platforms, the other as the result of subaqueous volcanic eruptions. It seems that both sources are acceptable, though the Paleoproto-zoic jaspilites commonly occurring together with volcanics seem to be more closely related to the second. The difference between the Paleoprotozoic and Mesoprotozoic jaspilites is suggested to be mainly due not to the sources of the matter but to the change in the environment. Iron is known to migrate widely only under defficiency of free oxygen, as only two-valent combinations of iron can transport in water without difficulty. The appearance of a low concentration of free oxygen in the atomosphere of the Middle – Late Mesoprotozoic must have caused a certain intensification of the oxidation-reduction potential (Eh) of the hydrosphere and as a result, a certain lowering of geochemical mobility of iron in the seawater must have occurred. This is probably the reason for a large-scale precipitation of iron in the Mesoprotozoic and the almost constant occurrence of jaspilite in the miogeosynclinal strata at that time. The accumulation of iron in the littoral sea zones was probably due to a wide distribution of blue-green algae producing oxygen in shallow-water areas. The oxidation of iron combinations occurred there, as also their transition into colloidal oxides that precipitated together with colloidal silica.

It was noted above that Upper Jatulian jaspilites are commonly typified by the oolitic structure of the ore matter. This specific feature makes them different from the older banded ferruginous-siliceous rocks and is suggested to be related to the area of their accumulation – the littoral zone densely populated with algae. The precipitating ore particles were covered with the algal mucus, their specific weight and cohesion becoming lower, causing then to roll, and their precipitation to be of a longer duration. The algal mucus was at the same time the medium where the colloidal combinations of iron precipitated. It cannot be excluded that certain blue-green algae and bacteria also played an active part in the precipitation of iron.

Melńik (1973) reasonably thinks that a long period of accumulation of silica and iron in particular must have preceded their precipitation in basin water. The precipitation of iron and silica took place slowly and simultaneously, iron precipitating slightly more quickly. The cyclic precipitation of ore matter was due to the periodical (seasonal) outburst in reproduction ("blossoming") of algae in the opinion of Melńik, and also in the opinion of Perry et al. (1973). Quick precipitation of dense particles of hydroxide of iron and slow precipitation of amorphous clots of silica resulted in the alternation of their interlayers. This attractive hypothesis wins, for it is simple and logic, but still does not explain the microbedding observed in Paleoprotozoic and Katarchean jaspilites, formed at a time when the algal biomass was quite insignificant.

The deposition of sedimentary strata in both Jatulian complexes terminated in some regions with the subaqueous outflow of basic and rare acid lavas. According to data of Svetov and Sokolov (1976), the Jatulian volcanogenic formations of the Karelian massif mainly belong to the "prototrappean" tholeiite-basalt formation. In geosynclinal belts volcanic activity was more intense and synchronous to the sedimentation.

After the Ladogan diastrophic episode, the character of sedimentation changed conspicuously in many parts of the world. The typically Jatulian complexes with essentially epicontinental deposits were replaced by thick greywacke-slate strata, in some areas of flyschoid nature. Sediments were accumulated in quickly subsiding and relatively deep basins, with quick low-amplitude oscillations of the Earth's crust. Certain structural aspects of the rocks favor the existence of turbidity flows during their accumulation. Locally sedimentation was accompanied by subvolcanic intrusions of ultrabasic magma and (or) subaqueous outflow of lavas, mainly of basic and ultrabasic composition, but sometimes also of alkaline. In some regions acid lavas are also recorded.

The character of sedimentation and volcanism described is mostly typical of geosynclinal belts, but it is of interest that certain portions of the Ladogan sequence reveal some flysch and flyschoid features when they are of subplatform or even platform type. At the same time, in the platform and aulacogen successions strata made up of epicontinental sediments rest together with such formations. Carbonate and sandstone-carbonate rocks occurring among them are commonly characterized by a higher concentration of phosphorus, indicating important biogenic processes in their formation. Many stromatolitic bioherms are specially rich in phosphates.

The dark carbonaceous slates with sulfide impregnations (the "black slate formation") occurring in geosynclinal and in part in platform strata are very likely to have been deposited in the deep parts of basins under a sulfuretted hydrogen environment. It is striking that these rocks associate with nickel-bearing ultrabasic intrusions, which seems to be reasonably explained by the fact that the depressions occur in abyssal fractures.

Many conglomerates in the lower part of the Ladogan complex were noted to be of glacial origin, but only some of them are attributable to true tillites, their major portion representing marine-glacial and fluvio-glacial formations.

The Vepsian complex strata – the youngest in the Mesoprotozoic Erathem – are early molasse that originated at the beginning of the Karelian orogenic cycle during a general inversion in geosynclines and under an appreciable uplifting of the Earth's crust in the platform areas. The rocks of this complex are clastic, commonly red, locally cuprous, mainly formed under subaerial environment; alluvial, deltaic in particular, strata are widely distributed and lateral-marine and lagoonal rocks are less common. At the same time, the strata examined differ notably from the typical Phanerozoic molasse which is much less thick, with subordinate psephytic rocks, and common occurrence of monomictic and oligomictic quartz sandstones. All these differences seem to be explained by the lesser amplitude and contrast of tectonic morphostructures than in the Phanerozoic.

## 4. Endogenic Processes

Usually Mesoprotozoic supracrustals of the platform (or subplatform) and miogeosynclinal types were shown to be of low alteration, the grade corresponding to the lowest and middle stages of the greenschist facies. Rocks of the highest grade of this facies and of amphibolite facies are distributed in relatively narrow zones of eugeosynclinal areas or in zones of abyssal fractures and in aureoles of plutonic ("contact-regional") metamorphism in the near-roof portion of batholith intrusions of granites.

Comparison in scale and intensity of metamorphism of Mesoprotozoic and older rocks clearly shows that the thermal regime of the interior of the Earth was continuously lessening with time, probably due to diminishing supply of gravitational energy in the total heat balance of the Earth and to the continued "extinction" of certain radioactive elements.

During the Mesoprotozoic Era the rise of the thermal front occurred three times with intervals of approximately of 200 m.y. and was accompanied by intensification of tectono-plutonic activity. Two early episodes of such activity (the Seletskian and Ladogan) were not strong and the alteration of the rocks then was also not strong, the magmatic processes resulting in emplacement of rare small bodies of meso-hypabyssal granites and in formation of sills and dykes mainly of basic rocks. During the terminal Karelian orogenic cycle, however, the endogenic processes were exceptionally intensive. This orogeny caused pretectonic and early tectonic intrusions of gabbroids and ultrabasites, syntectonic and late tectonic intrusions of large masses of various granitoids, metamorphic and ultrametamorphic alterations of supracrustals, different metasomatic and hydrothermal processes. Granites rich in potassium are mostly widespread among the granitoids: plagiomicrocline and essentially microcline granites and related pegmatites, the mica-bearing, including rare earths and rare metal types. Mesoprotozoic granitoids differ greatly from the Paleoprotozoic ones, which are characterized by occurrence of dominantly plagiogranites-trondhjemites and granodiorites (tonalites) poor in pegmatites.

The phenomena of migmatization and granitization are quite common, especially in the abyssal sections of geosynclinal belts and in the basement rocks underlying Mesoprotozoic strata. Many granites are very likely to have been formed as a result of rheomorphism (palingenesis) of the older granite-gneisses (the abyssal solutions rich in alkali playing a role). At the expense of the older plagiogranites-granodiorites, as a result of potassic metasomatosis, porphyroblastic granites originated with large secretions of microcline enclosed in the quartz-plagioclase matrix.

The thermal processes of the Karelian cycle are registered not only in the geosynclinal belts and adjacent relatively labile portions of the older platforms, but also on the stable platforms with Mesoprotozoic sedimentary cover. These processes led to tectonic activations and low alteration of the cover rocks and, in the basement rocks they also resulted in migration of certain daughter radioactive isotopes, of $^{40}Ar$ in particular. Thus, the K-Ar datings of the basement rocks, irrespective of their true age, commonly give rejuvenated values that correspond to the Karelian orogeny (2000–1900 m.y.). Long, but low-temperature heating (about 300 °C) of the basement rocks did not evoke obvious recurrent structural-mineralogical transformations, but resulted in isotope cryptometamorphism. It is very probable that the sedimentary rocks of the platform cover, being of low heat conductivity, were the cause of a higher temperature beneath them. On certain older shields, for instance in the west of the Aldan shield (of the Mesoprotozoic Aldan platform), the more intensive Karelian thermal activation led to migration of isotopes of uranium, thorium, and lead, and affected the U-Th-Pb and Pb-Pb datings of the Katarchean and Paleoprotozoic rocks. Generally, the isotope rejuvenation of the older rocks related to the Karelian activation is fairly well known.

The plutonic processes on the older platforms were of special nature. In addition to the dykes and sills of diabases that originated during the Early Mesoprotozoic diastrophic episodes (approximately 2400 and 2200 m.y. ago), at the end of the era, in the interval of 2200–1900 m.y., the differentiated metalliferous (Ni, Co, Cr, Ti, Pt etc.) massifs of gabbro-norite-granophire composition were formed. The largest and best-known among them are the Sudbury massif on the Canadian shield, partially occurring among the Huronian rock cover, and the Bushveld huge massif in Southern Africa situated in the platform strata of the Transvaal craton.

Both massifs are lopolith-like in shape and represent layered multi-phase intrusions, typical of platforms of different age. The basic rocks of the massifs (and the ultrabasic of the Bushveld massif also) that were emplaced during the early phases of the massif formation are suggested to be products of melting and differentiation of the basaltic layer or even of the upper mantle of the Earth; certain later granophyres and granites seem to have been formed in the selective melting of the granite layer (lithogenic granites) that is evidenced in particular by the high $^{87}Sr/^{86}Sr$ ratio (up to 0.761) in these rocks. Some of the granite dykes cutting gabbro-norites in the Sudbury massif are likely to be the products of melting of the enclosing older granites (the Murray granites) under the influence of heating produced by basic magma.

## 5. Tectonic Regime

It was noted above that Mesoprotozoic strata are of wider distribution in most regions of the world than Paleoprotozoic rocks, as well as being more complete and better studied. Moreover, different formational types of rocks are reliably distinguished in Mesoprotozoic sequences: miogeosynclinal, eugeosynclinal, and platform, and also the formations of different paleogeographic environments: marine, deltaic, continental, glacial, etc., permitting good paleotectonic reconstructions. They are based on the following criteria:

- distribution of the rocks of the corresponding formational type;
- isotope datings of rocks, including basement rocks; they give the time and the grade of reworking of the basement;
- fold strikes, their curving in plan, virgation, and vergence;
- zonation in distribution of rocks metamorphosed and deformed at different grades;
- situation of marginal abyssal fractures.

Applying these and other criteria, it is possible to reconstruct a general picture of distribution of major tectonic elements on different continents during the Mesoprotozoic. Figures 40–44 (see also Fig. 32) show paleotectonic schemes in more or less detail of North America, Africa, Australia, and of large regions of Europe and Asia. Examination of these schemes is beyond the scope of this more general work, but is found partially in some earlier publications (Salop 1964–1967, 1977a, b etc.), based on data of different regional studies treated according to the above criteria.

In these schemes the areas of distribution of the pre-Mesoprotozoic (Katarchean and Paleoprotozoic) rocks are assigned to cratons, in the Mesoprotozoic they suffered no important reworking. The supracrustal cover of Mesoprotozoic locally overlying these rocks is of platform type and as a rule, slightly deformed. Strong dislocations are registered only in some places and are commonly related to different fractures. The Mesoprotozoic instrusive formations are represented by atectonic layered massifs, dykes, and sills of basic rocks and related acid-derived rocks. The basement of cratons is usually affected by low-temperature activation of the Karelian cycle, it resulted in low-grade alteration of the cover rocks and in the isotope rejuvenation revealed by K-Ar datings.

There are some areas that are from many aspects to be attributed to the craton type, but the pre-Mesoprotozoic rocks composing them were subjected to intensive thermal effect during the Karelian orogeny, as is revealed by their isotope rejuvenation obtained by almost any radiometric method. Isotope datings giving the time of the original metamorphism of the older rocks of the basement are very rare and "relict". In cases where the Mesoprotozoic supracrustal cover is preserved on such cratons, it is sufficiently altered (the middle grade of the greenschist facies) in spite of being of platform nature, and it is deformed (within the aulacogens in particular); locally it is separated by small intrusions of gabbroids and granites. These labile cratons can be called semi-cratons. To these belong certain structural elements on the Canadian shield (the Baker craton in the Churchill tectonic province), in Australia (the Arunta and Musgrave cratons, north-western and south-eastern margins of the Yilgarn craton). Many cratons and median masses of Eastern Europe (the Kola,

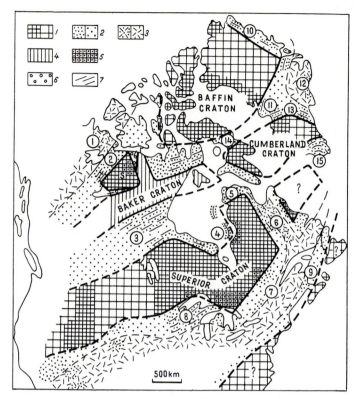

**Fig. 40.** Principal structural elements of North America in the Mesoprotozoic.
*1* cratons (old platforms and median masses), *2* miogeosynclinal belts, *3* eugeosynclinal belts
(*1a–3a* presumed structural elements or their parts), *4* cratons with basement partially reworked by the Karelian (Hudsonian) thermal-tectonic processes – semicratons, *5* cratons with
platform cover, *6* aulacogens, *7* fold strikes.
Abbreviations: *S* Slave cratonal block, *A* Athapuscow aulacogen, *B* Bathurst aulacogen.
Figures in circles – geosynclinal belts: *1* Bear Lake, *2* Coronation, *3* Churchill, *4* Belcher, *5* Cape
Smith, *6* Labrador, *7* Early Grenville, *8* Penokean, *9* Saint John, *10* Carolinian, *11* Rankian,
*12* Scoresby, *13* Nagssugtoqidian, *14* Fox, *15* Ketilidian

Novgorod, Baltic etc.) and partially the Chara, Aldan and Okhotsk cratons in Northern Asia are examples.

Many (if not all) Mesoprotozoic cratons exhibited angular contours due to intersection of marginal abyssal fractures. These boundaries are reliably established for
many well-known cratons in different parts of the world. The shapes of other cratonal
blocks shown in the figures (Figs. 40–45) are to a certain extent a "stylistic generalization", based on well-known examples. The contours of cratons and of the geosynclinal belts separating them are suggested to be governed by the regmatic system of
fractures that appeared soon after the Kenoran orogeny.

The basement of cratons had its own autonomous structure, independent of the
structures of the surrounding folded rocks. Thus, the Katarchean formations display
a specific tectonic pattern of large "gneiss ovals" and the greenstone belts are characterized by "swarms" of mantled domes. The strikes of the basement structure are

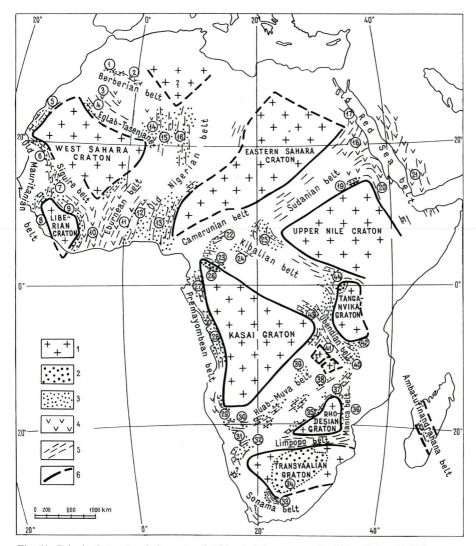

**Fig. 41.** Principal structural elements of Africa in Mesoprotozoic. (After Salop 1977b).

*1* cratons (old platforms); *2–4* formational types of Mesoprotozoic strata: *2* platform, *3* mio-geosynclinal, *4* eugeosynclinal, *5* fold strikes, *6* boundary of cratons.

Figures in circles – name of the Mezoprotozoic groups (in some cases the name of fold zones where several groups are distributed): *1* Zenaga, *2* El-Graara, *3* Yetti, *4* Aguelt Nebkha, *5* Amgala, *6* Akjoujt-Bakel, *7* Diale, *8* Marampa, *9* Siguire, *10, 11* Birrimian, *12* Tannekas, *13* Kusheriki-Kushaka, *14* Tassendjel, *15* Edjere-Anefsa, *16* Edoukel, *17* Abu Ziran, *18* Nafir-deib, *19* Butana, *20* Tsaliet, *21* Bahah, Baish, *22* Lom, *23* Ayos, *24* Mpoko, *25* Madongere, *26* Mbalmayo-Bengbis, *27* Ogooue, *28* Premayombe, *29* Epupa, *30* Huab, *31* Abbabis, *32* Marienhof, *33* Kheis, *34* "Witwatersrand triad", Transvaal and Matsap, *35* Lomagundi, *36* Gairezi, Frontier, *37* Zambezi, *38* Tumbide, *39* Muva, *40* Misuku, *41* Ubendian, *42* Konse, *43* Ruzizian, *44* Buganda-Toro.

usually cut by the Mesoprotozoic mobile belts. The tectonic rebuilding of the older basement rocks and the conformity to the structures of fold belts realized to this rebuilding are characteristic only of some parts in the margins of cratons.

The relations of structures of the cratonal blocks and of the fold belts are most clearly seen in certain median masses of Eastern Siberia. Thus, the median masses (blocks) of the Baikal fold (geosynclinal) system are made up of the same Katarchean rocks as the adjacent large cratons and their inner structure is characterized by fragments of the same "gneiss ovals" that are widespread on cratons (Salop 1964–1967). All of these facts show that the Mesoprotozoic geosynclines (as well as the Paleoprotozoic) originated not on the basaltic or mantle substratum as is widely accepted, but on the crushed basement of mostly granite-gneiss composition.

Mesoprotozoic cratons differ from rudimentary Paleoprotozoic cratonal blocks by conspicuously increased rigidity and by the occurrence of platform cover in many cases. At the same time they are less consolidated than the latest "true" platforms, which is shown by varying, commonly great thickness of the cover rocks, by relatively high alteration and deformation of these rocks, by occurrence of abundant volcanics and other aspects. It is thus not possible to identify the Mesoprotozoic cratons completely with the Late Precambrian and Phanerozoic platforms. They are, however, closer to them than to the Paleoprotozoic protoplatforms, and may be called platforms with reservation. The structural elements examined thus constitute an important link in the evolution of cratons and signify a definite stage in the direct process of a general cratonization (sclerotization) of the Earth's crust.

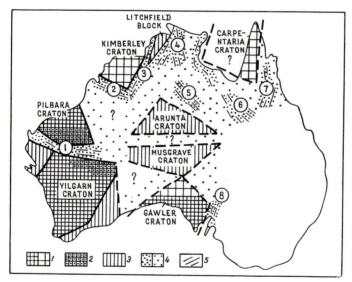

**Fig. 42.** Principal structural elements of Australia in the Mesoprotozoic.
*1* cratons – old platforms (*a* presumed, covered with younger strata), *2* cratons with Mesoprotozoic platform cover, *3* cratons with basement partially reworked by the Karelian tectonothermal processes (semicratons and median masses), *4* geosynclinal belts (*a* presumed, covered with younger strata), *5* fold strikes.
Figures in circles – geosynclinal belts: *1* Ashburton, *2* King Leopold, *3* Halls Creek, *4* Pine Creek, *5* Warramunga, *6* Leichhardt, *7* Georgetown, *8* Middleback

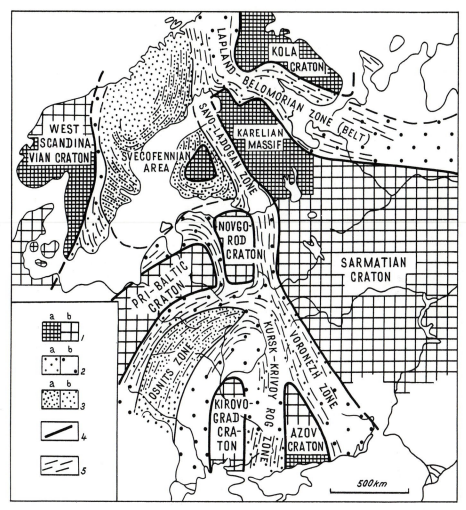

**Fig. 43.** Principal structural elements of Eastern Europe in the Mesoprotozoic.
*1* cratons (old platforms and median masses): *a* exposed, *b* covered with younger strata:
*2* miogeosynclinal zones (belts): *a* exposed or known from drilling, *b* presumed, covered with younger strata; *3* eugeosynclinal zones (areas): *a* exposed, *b* presumed, covered with younger strata; *4* boundary of cratons (commonly coinciding with marginal abyssal fractures); *5* general strike of Mesoprotozoic fold structures according to data of geologic and magnetic surveys

In the Mesoprotozoic, the geosynclinal systems were of higher tectonic differentiation than earlier. In the Paleoprotozoic, the eugeosynclinal regime dominated the whole extension of the mobile areas as in the Mesoprotozoic the miogeosynclinal zones appeared in the periphery of these geosynclinal belts for the first time. These zones also exhibit certain peculiarities: the piles composing them differ from the same type of piles in the younger miogeosynclinal formations, the difference lying in the presence of locally abundant volcanics, in a higher grade of metamorphism, and in the intensive plutonic processes registered in particular close to the eugeosynclinal

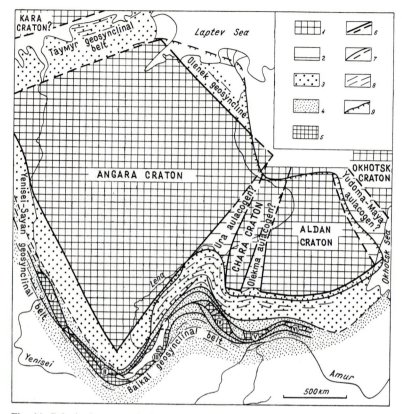

**Fig. 44.** Principal structural elements of Northern Asia in the Mesoprotozoic.
*1* cratons (old platforms), *2* aulacogens, *3* miogeosynclinal zones (belts), *4* eugeosynclinal zones (belts), *5* median masses (blocks), *6* marginal abyssal fractures, *7* intrageosynclinal abyssal fractures, *8* fold strikes, *9* modern boundary of the Siberian platform.
Figures in circles – median masses (blocks): *I* Kansk, *II* Sayan-Khamar-Daban, *III* Baikalian, *IV* North- and South-Muya, *V* Mogocha, *VI* Urkan

zones. These differences, however, are not fundamental enough to recognize the oldest miogeosynclines as a special type of structures.

The eugeosynclinal zones are noted to be of less extension than the miogeosynclinal and the greenstone volcanogenic formations developed within them contain local quartzite interbeds, which seem to be allophylous components (according to Shatsky's concept). The quartzite occurrence shows a close spatial relation between eugeosynclinal depressions and miogeosynclinal zones and median masses, such sedimentary rocks and many other types of mature rocks being widespread in the latter; it indicates moreover a smooth relief on the cratons, pointing out a slight tectonic contrast of morphostructural elements.

The transitional zones characterize the joint of the older platforms and the geosynclinal belts, and are recognized by greater thickness of the strata and by stronger deformation and metamorphism than on the platforms. In most cases, however, the formational type of the rocks and the sequence structure varies only slightly, being close to the platform type as well as to the miogeosynclinal. In some regions (in the southern margin of the Hamersley basin in Western Australia, for instance) the platform-geosyncline boundary can be drawn only arbitrarily. Sharp and more fundamental changes in structural and formational aspects, however, are noted close to certain marginal fractures limiting the platforms. Such a joint can be exemplified by a narrow band near the marginal abyssal fracture separating the Savo-Ladogan miogeosynclinal zone of the Karelides from the Karelian massif (see Fig. 32), typified by the appearance of thick greywacke-slate strata of flysch type, by the strong folding and high alteration of the rocks reaching amphibolite facies at a short distance and by migmatization.

The occurrence of thin platform strata at the base of many geosynclinal complexes is of great significance for knowing the evolution of the Mesoprotozoic mobile belts and of an early stage of this evolution in particular. It was noted above that these formations underlie the geosynclinal complexes of the Savo-Ladogan Karelides zones and of the adjacent Svecofennides zone, of the Kursk-Krivoy-Rog zone, of the Shimoga-Dharwar belt of Hindustan, of the Ketilidian belt of Greenland and of the Penokean belt of Canada. They are likely to be present in many more geosynclinal complexes, and only the Udokan Group of the Kodar-Udokan zone of Eastern Siberia makes an exception among the complexes briefly examined here, for its lower part is of typical geosynclinal nature.

The appearance of Mesoprotozoic geosynclines is suggested as having been preceded by the platform stage in many regions. Sometimes the geosynclinal regime has been established only in the Middle or even Late Mesoprotozoic (as for instance, in the Savo-Ladogan zone of the Karelides and probably also in the Svecofennides area). In many regions of the world the Lower Mesoprotozoic strata are missing in the Precambrian sequence. It must be admitted that the tectonic stable regime dominated on the continents during the Early Sub-era and partly during the Middle Sub-era. The sedimentation of that time occurred mostly under shallow-marine (shelf) and even continental environments, and again the "ill-fated" question arises that has already been discussed when dealing with the paleogeographic situation in the Paleoprotozoic: where did the abundant seawater disappear to? The most realistic suggestion in this case seems to lie in recognizing a sufficiently intense subsidence of large portions of the Earth's crust in places of some of the modern oceans, which explains the facts that are now evident.

## 6. Mantled Gneiss Domes and Astroblems

In an earlier paper (Salop 1971 b) devoted to the concentric structures of the Precambrian, I have shown that the domes of different age (in some cases even up to Mesozoic) reveal an absolute regularity in size and distribution: both diminish gradually with time, particularly the latter. In the Mesoprotozoic the role of these structures in the tectonic pattern is much less than in the Paleoprotozoic: they no longer constitute an obligatory element of the foldbelts as was the case earlier; they are encountered sporadically, though in many places within the geosynclinal (fold) areas as on the older platforms. They are known on the Baltic shield in Europe, in the Baikal fold area, in the Kodar-Udokan zone in Asia; in different regions of Africa (the

Manika, Huab-Muva, Eburnean, Nigerian, and Mauritanian belts, and on the Transvaal craton); in North America (in the west of the U.S.A.); in South America (on the
Guiana and Brazil shields); in some parts of Australia. Usually the domes are isolated,
but sometimes they occur in groups that resemble to a certain extent the so-called
"dome swarms" in the greenstone belts of the Paleoprotozoic.

The mantled gneiss domes of Mesoprotozoic age are evidently recognized on the
older cratons (platforms) and among the miogeosynclinal belts where the crystalline
basement is situated at relatively shallow depth. In cases where the basement is deeply
plunged and thus not exposed in the cores of structures, the domes are not so
conspicuous and are mainly recognized during detailed mapping; in the lithologically
monotonous strata the domes are identified according to the structural elements and
the concentric metamorphic zonation (the "heat domes"). In the Karelian-Kola
region the mantled domes are easily seen on the Karelian massif and in the Savo-
Ladogan miogeosynclinal zone in particular that surrounds this massif from the west.
In the Ladoga region the domes are situated as groups and in their cores the Paleoprotozoic or Katarchean basement mobilized to a different grade crops out, and the limbs
are composed of Upper Jatulian (Pitkyaranta) and Ladoga supracrustals of the
Karelian Supergroup (Fig. 45). The structure of certain domes is complicated by
fractures that are commonly situated along the basement-sedimentary cover contact
and the contact surfaces are usually very steep and in places even overturned.

The occurrence of blastomilonites in many domes of the Ladoga region at the
contact of the basement and metamorphic strata led to a very important concept of
dislocations in dome formation and a peculiar horst nature of the structures. I do not
share this opinion, and the studies done show that deformation at the time the domes
originated was generally of a plastic nature. The occurrence of blastomilonites is
commonly related to incomplete local mobilization of the basement material, in some
cases with subsequent tectonic pressure along the contact. It is to be noted that the
Ladoga domes are the "tectonotype" of mantled gneiss domes, Eskola (1949) being
the first to recognize this type of structure and give them the name now widely
accepted.

Naturally the origin of Mesoprotozoic mantled domes is the same as that of the
Paleoprotozoic analogous structures examined earlier (Chap. 3), that is, they were
formed under the influence of endogenic processes. However, a detailed study of a
typical mantled dome of Mesoprotozoic age in Southern Africa revealed certain
traces of an early history or, to be more exact, pre-history of an origin of these
structures, data which accord well with a suggestion concerning the role of cosmic
impact phenomena in localization of endogenic dome-forming processes. Here I mean
the new data on the genesis of the well-known Vredefort dome on the Transvaal
craton. This dome is absolutely round in shape, with a diameter of 75 km. The
basement granites of the craton are exposed in the core of the dome (the age determined in this case is 2850 m.y.) the limbs are composed of rocks of the Witwatersrand
triad and of the Transvaal Supergroup (see Fig. 36). The beds on the limbs are
overturned away from the center of the structure (Fig. 46). The cover rocks near the
granitic core notably metamorphosed as is usual in mantled domes. The alteration
corresponding to the greenschist facies becomes lower up-section and beside the
granitic core. In the inner part of the dome two small massifs (stocks) of alkaline
granites are situated among the sedimentary rocks.

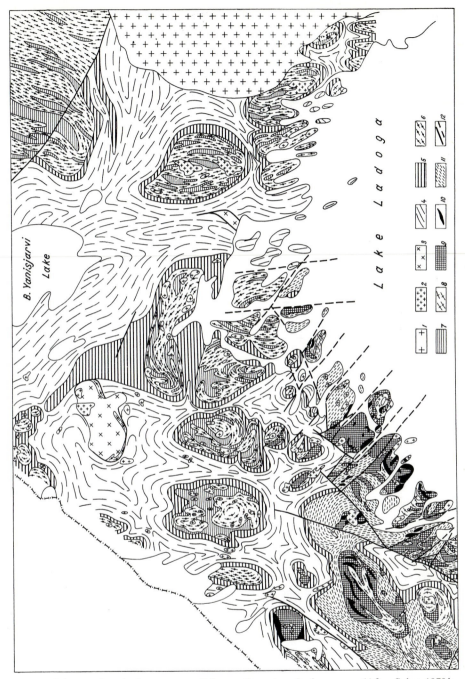

**Fig. 45.** Scheme of geologic structure of the north-western Ladoga area. (After Salop 1979 b, simplified).
*1* Lower Neoprotozoic, rapakivi granites; *2–5* Mesoprotozoic: *2* microclinic granite, *3* gabbro-diorite, *4* the Ladoga Group, *5* the Pitkyaranta (Sortavala) Group; *6, 7* Paleoprotozoic: *6* gneissoid plagiogranite and plagiomigmatite, *7* amphibolite (the Yalonvara Group); *8–11* Katarchean: *8* plagiogranite and granodiorite gneiss, *9* diorite gneiss, enderbite, and charnockite, *10* metagabbro, metanorite, and pyroxenite, *11* supracrustal gneiss and crystalloschist (sillimanitic, garnet-cordieritic, bi-pyroxenic etc.), *12* fractures

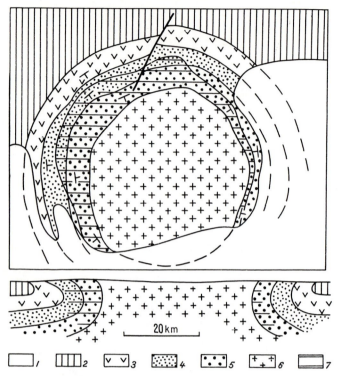

**Fig. 46.** Scheme of geologic structure of the Vredefort dome-astrobleme (map and profile). (After Daly 1947 etc.)
*1* Upper Paleozoic sedimentary cover; *2–5* Mesoprotozoic: *2* the Transvaal Group, *3* the Ventersdorp Group, *4* the Witwatersrand Group (the upper part), *5* the Witwatersrand Group (the lower part) and the Dominion Reef Group; *6* "old granite" (Katarchean or Paleoprotozoic), *7* metamorphic aureole in Mesoprotozoic rocks

The origin of the Vredefort dome has long been debated and opinions vary. According to Bailey (1926), the peculiar mushroom-like structure is due to the centrifugal pressure the Mesoprotozoic surrounding rocks suffered from the quasiplastic older granite. This concept was proposed much earlier than the modern ideas on mantled domes. In the opinion of some other workers, for instance, Boon and Albritton (1937), Daly (1947), and Dietz in particular (1961), the structure originated as a result of meteorite impact. The discussion has continued almost to the present, but new reliable data have now been obtained confirming an impact process in the structure formation, and now many authors share the concept of meteorite origin (Manton 1965, Wilshire 1971, Martini 1978 etc.). Masaytis published a paper in 1981 reviewing recent data on the Vredefort dome as an older astrobleme.

Abundant fine veinlets and dyke-like irregular bodies of pseudotachylites that separate the fragmented blocks (megabreccias) are observed in the core granites of the Vredefort dome and in the surrounding rocks. This fact has been known for a long time, but the pseudotachylites could not be used as evidence of an impact process, as they also originate by tectonic friction. The pseudotachylites of the Vredefort dome, however, are now known to contain minerals such as coesite and stichovite, typical of impactites formed under extremely high pressure conditions; near the Earth's surface they can originate only as a result of meteorite impact. A great number of shatter cones were also found to exist together with pseudotachylites, positive indication of an impact process.

The age of the Vredefort astrobleme is estimated to be 1970 ± 100 m.y. (Grieve and Robertson 1979), indicating that the meteorite fell during the Karelian diastrophic cycle. In my opinion this fact is highly significant in understanding many aspects of the structure and causes of its transformation into a mantled dome. Naturally, during the orogeny a general rise of the heat front from the interior of the Earth to its surface must have taken place. At the same time, the crushed crust at the place of impact must have facilitated the localization of seeping hot fluids and partly also of anatectic melts, which evoked the rise of a mobilized plastic granite basement in the shape of a diapir or protrusion. As a result, an "inversion" of the astrobleme (the impact crater) must have occurred, to transform it into a mantled dome.

Is it possible to compare the formation mechanism of the Vredefort mantled dome and that of other similar structures? Before answering this question, it is necessary to examine briefly the exceptional situation of this dome. The preservation of traces of an impact phenomenon is a result of a concurrence of favorable circumstances. The astrobleme not only originated on the cratonal block with thick sedimentary cover, but the meteorite fall took place during the orogeny when the thermal state of the interior of the Earth differed from usual. If a large meteorite had then fallen on a craton lacking sedimentary cover, no mantled dome could have originated, and if the meteorite had fallen within the geosynclinal area, then in place of the astrobleme the diapir of melted granite magma (i.e., the mantled dome) would have appeared, but there any traces of impact would have been buried under strong contact and regional metamorphism and deformation. If the meteorite fall occurred previously to or after the diastrophic cycle (during the atectonic periods), then the astroblemes formed would have been eroded or buried under the sediments, and the impactites would have been preserved only as fragments in clastic rocks. In such cases the local crushing of the Earth's crust would appear favorable for formation of magmatic diapirs (and mantled domes) in places of meteorite fall, but that would happen only in the period of later orogenic events. It seems necessary to note that impactites or traces of impact of meteorites have little chance of being preserved in the older domal structures even under favorable conditions, as they are quickly eroded.

It is thus concluded that intensive impact phenomenae accompanying the fall of large meteorites and causing a localized crushing of the Earth's crust could have been cause of the mantled dome formation and, during orogenic cycles when abyssal thermal processes are exceptionally strong, this could have been realized more easily. Probably many of the domes originated at sites of previously existing astroblemes. This by no means signifies that all the mantled domes (or the thermal domes generally) were inherited from astroblemes. The localized crushing of the Earth's crust (centers) could be explained by purely tectonic events. These favorable centers could be the intersection of dislocations of different directions, dense packing of dislocations, or some portions of zones of the abyssal fractures. In some geologic publications (see, for instance, Salop, 1974b) certain mantled domes are shown to occur in the abyssal fractures observed in the basement.

Finally, it should be noted that the mantled domes did not originate in the case of a large meteorite falling, even if the conditions were very similar to those of the Vredefort astrobleme formation. A well-known older astrobleme – Sudbury, situated in the margin of the Mesoprotozoic Superior craton (Canada) provides a good example of meteorite fall resulting in an absolutely different geologic structure. It is slightly older that the Vredefort astrobleme; it appeared approximately 2200–2050 m.y. ago (Gibbins et al. 1972), that is during the Ladogan diastrophic episode which preceded the Karelian orogeny.

French (1968) first substantiated the impact origin of the Sudbury structure and later many other workers shared his opinion. In plan the astrobleme is ellipse-shaped with a size of $60 \times 30$ km. This elongated shape is suggested to be of secondary origin and is due to tectonic compression during the Karelian (Hudsonian) orogeny; the original shape seems to have been close to a regular circle. The impact phenomena are observed as abundant breccias with fragments cemented by black glass that appeared during the melting of the breccia material, as numerous shatter cones with the impactites present in particular, of the type of suevite that consists of the lapilli particles of rocks and minerals not fully melted and of clasts of diaplectic quartz, feldspar, and biotite; pseudotachylites with diaplectic quartz are also observed.

The rocks enclosing the astrobleme are Huronian sandstones and slates, granites in part. The meteorite impact was accompanied by strong eruption and caused crushing, partial melting, and rock dusting with formation of thick (up to 1200 m) strata of suevite over the crater, these strata being divided as the Onaping Formation (suevite were regarded as common tuffs previous to recognition of their true nature). The impactites are overlain by slates and sandstones – the products of erosion of the astrobleme accumulated in the lacustrine basin.

Some workers think that formation and emplacement of basic magma is related to the impact effect; it resulted in formation of a large gabbro-norite Sudbury massif and is situated between the surface of the past crater bottom and the Onaping suevite. In my opinion, however, the impact phenomena simply determined the localization of the intrusion, and the formation and emplacement of basic magma occurred due to normal endogenic processes that were mostly characteristic of the Ladogan diastrophic episode. In Canada and other regions of the world the intrusions of mainly basic magma are registered as having occurred during this diastrophism, and abundant dykes and sills of diabases and gabbroids appeared. On the Superior craton the Nipissing diabase dykes then formed (age: 2150 m.y.). During the weak Ladogan diastrophism the thermal front is likely to have been below the level of the granite melts, so that strong mobilization of the craton basement did not occur and consequently the astrobleme did not "invert".

## 7. Principal Stages of Geologic Evolution

Now the succession of the principal geologic events during the Mesoprotozoic will be briefly discussed.

The Early Mesoprotozoic (the Late Paleoprotozoic in part)

1. 2800–2750 m.y. Land and epicontinental shallow-water seas in part are widespread (within the limits of modern continents). Posterior oscillating movements and

formation of the Dominion Reef lithostratigraphic complex in portions where the early Kenoran folding completed. Outflow of lavas of basic, intermediate and partly acid composition. Extension of the Earth's crust accompanied by formation and deepening of the oceanic depressions (in sites of modern oceans).

2. 2750–2450 m.y. Accumulation of thick terrigenous strata formed by destruction of vast areas of the peneplained land in isolated and half-isolated gradually subsiding depressions; formation of the Witwatersrand lithostratigraphic complex. In the supply areas intensive chemical and physical weathering by atmosphere lacking oxygen. Outflow of basic and acid lavas at the end of the stage. Short periods of sharp cooling of climate resulting in glaciations. Formation of certain miogeosynclinal belts.

3. 2450–2400 m.y. The Seletskian diastrophic episode. Oscillating movements of mostly positive sign accompanied by formation of fractures and slight fold deformation. Intrusions of basic magma, local small granite intrusions.

The Middle Mesoprotozoic

4. 2450–2350 m.y. The beginning of extensive marine transgressions. Deposition of quartz sands and dolomites in part under the littoral-marine environments; in the second half of the stage, intensive deposition of chemogenic ferruginous-cherty sediments of the Krivoy Rog type in certain miogeosynclinal belts; formation of the Lower Jatulian lithostratigrahpic complex. Quick reproduction of prokaryota, mainly blue-green algae. Free oxygen, though of low content, appeared in the atmosphere.

5. 2350–2200 m.y. The marine transgression embraces almost all the shelves. Deposition of mature terrigenous sediments (quartz sands etc.) and of carbonate (dolomitic) muds in particular under the littoral-marine environment and of chemogenic ferruginous-cherty sediments of the Superior type in many miogeosynclinal belts; formation of the Upper Jatulian lithostratigraphic complex. Intensive evolution of blue-green algae; first appearance of eukaryota. Increase of free oxygen content in atmosphere; first red beds are reported. The seawater becomes of chloride-sulfate composition.

6. 2200–2160 m.y. The Ladogan diastrophic episode. The oscillating movements of mostly positive sign, locally accompanied by slight folding and emplacement of basic (and acid) magma.

The Late Mesoprotozoic

7. 2200–2000 (2100?) m.y. Formation or deepening of many geosynclinal depressions. Deposition of thick flysch greywacke-slate strata in the miogeosynclinal zones and of greywacke-slate strata with volcanics in the eugeosynclinal zones; of shelf terrigenous-carbonate strata (with local basic volcanics) in the aulacogens and on the platforms; formation of the Ladogan lithostratigraphic complex. World-wide glaciation is registered in the first half of the stage.

8. Approximately 2000 (2050–1950) m.y. Extensive uplifting of the Earth's crust, likely to be accompanied by compensatory deepening of the oceanic depressions. The beginning of the inversion of the tectonic regime in the geosynclinal belts. Deposition of continental and littoral-marine terrigenous commonly red beds in the isolated

intracratonal and intramontane troughs; formation of early molasse of the Karelian cycle: the Vepsian lithostratigraphic complex.

9. 2000–1900 m.y. The Karelian diastrophism of the first order. Intensive tectono-plutonic processes. The geosynclinal areas are characterized by folding and different intrusions: (a) pretectonic intrusions of basic-ultrabasic magma (formation of massifs of ultrabasic rocks and gabbroids), (b) syntectonic intrusions of granitoid composition (formation of microcline-plagioclase granites and granodiorites and related pegmatites), (c) late-tectonic intrusions of granitoid rich in potassium (plagio-microcline and essentially microcline granites or alaskites). Regional metamorphism of greenschist (up to amphibolite) facies, local manifestations of ultrametamorphism as migmatization and granitization in part. In the platform areas the slight alteration of the sedimentary (sedimentary-volcanogenic) cover, emplacement of basic or ultrabasic magma along the fractures; small local intrusions of granite and subalkaline syenite magma. Closing of many Mesoprotozoic geosynclines.

# Chapter 5 The Neoprotozoic

Different supracrustal and plutonic rocks originating after the Karelian diastrophism of the first order (2000–1900 m.y.) and prior to the termination of the Grenville diastrophism of the same order (1000 ± 50 m.y.) belong to the Neoprotozoic Erathem. The erathem is subdivided by diastrophic cycles of the lower order into four sub-erathems: the Lower, Middle, Upper, and Terminal. Ther Lower sub-erathem exhibits a specific composition, being of isolated distribution and as a rule easily recognizable among the remaining three sub-erathems. The latter, on the contrary, have similar composition, commonly occur in the same localities, and sometimes build up common extensive rock sequences (complexes or supergroups). At the same time, they are also characterized by stratigraphic or angular unconformities, locally of significant magnitude; the intrusive magmatism also separates these sub-erathems at places.

All the sub-erathems (sub-eras) characterize a single large stage (megachron) in the evolution of the planet when large platforms originated and started to develop, when new geosynclinal systems appeared and many other events occurred that will be discussed below.

The Riphean complex of the Southern Urals is the best-known large regional Neoprotozoic unit, the age of its lower (Burzyan) group, however, is uncertain and the Lower Neoprotozoic Sub-erathem in the Riphean is moreover lacking or quite poorly represented. The Riphean complex can be accepted only as a stratotype of the three upper sub-erathems. The Neoprotozoic is much more complete in the Baikal Mountain Land, where the Akitkanian of the Lower Neoprotozoic and the overlying Patom Group (or Supergroup) that corresponds to the three upper Neoprotozoic sub-erathems can be regarded as a parastratotype of the Riphean. Thus "Baikalian Erathem" or "Baikalian" is preferable as a regional synonym of the Neoprotozoic to the name "Riphean Erathem" or "Riphean" traditionally used in the U.S.S.R.

As the lower sub-erathem is sufficiently distinct almost everywhere among the remaining units of the erathem, it can be called the Akitkanian according to the stratotype. The recognition of the remaining three sub-erathems is usually difficult, especially in cases where unconformities separating them are faint and the age of the rocks has not been reliably established by isotope methods. The "Riphean" seems to be a suitable name for a combination of all these suberathems. The Akitkanian and Riphean will be discussed separately in the following sections.

# A. Rock Records

## I. The Lower Neoprotozoic (Akitkanian)

### 1. General Characteristics of the Lower Neoprotozoic

**Definition.** The lower sub-erathem of the Neoprotozoic is composed of supracrustal and plutonic rocks that originated after the Karelian diastrophic cycle and previous to termination of the Vyborgian orogeny of the second order, that is in the time interval of 1900–1600 (or 1950–1550) m.y. The sedimentary-volcanogenic strata of this sub-erathem belong to special taphrogenic formations that appeared in the tectonic depressions on the older platforms and on the portions of the older geosynclines stabilized by Karelian folding. Some workers think that formations represent molasse of the Karelian cycle. In particular, this idea was promoted in the working stratigraphic scale of the Precambrian of the U.S.S.R. accepted at the Ufa Conference of 1977 and recommended by the Stratigraphic Commisson of the U.S.S.R., where the rocks of this time interval are attributed to the upper part of the Lower Proterozoic that is mainly built up of Karelian formations. It is not possible, however, to accept this idea as true, as the tectonic elements of the Lower Neoprotozoic are absolutely independent of the structural plan of the Karelides and the rocks of the sub-erathem form a separate large structural stage that accumulated approximately during 300–350 m.y. The Lower Neoprotozoic strata are thus more closely related to the Middle Neoprotozoic strata.

**Brief Characteristics of the Sub-Erathem.** The Lower Neoprotozoic is mostly typified by the continental sedimentary-volcanogenic strata at places laterally replaced or overlain by the continental or littoral-marine sedimentary strata with subordinate volcanics. These strata are similar on every continent and comprise a peculiar stratigraphic level (Salop 1973a 1980, Semikhatov 1974, Moralev et al. 1979). The sedimentary-volcanogenic strata are known for alternation of mainly acid volcanics and tuffs, welded tuffs, and clastic rocks. The thickness of the strata varies, but commonly it is considerable and in places reaches 10,000 m.

**Distribution.** The Lower Neoprotozoic sedimentary-volcanogenic and sedimentary strata are widespread on every continent. In different regions they are recognized under numerous different names, which it is not possible or necessary to mention here, though some of them will be briefly discussed later. Here several well-known units will be given in order to have a more concrete idea of what kind of formations will be examined later on; these are: in Europe the Subjotnian of Sweden, in Asia the Akitkan Group of the Baikal area, in North America the Lower Dubawnt and Letitia Lake Groups, in South America the Uatuma and Cuchivero Groups of Brazil and the Roraima Group of Venezuela, in Africa the Waterberg Group of the Republic of South Africa, the Mayombe Group of Zaïre and the Tarkwa Group of Ghana, in Australia the four successive groups: Whitewater, Speewah, Kimbreley, and Bastion.

**Stratigraphic Situation.** The relations of the rocks in question and the underlying and overlying Precambrian rocks are usually clear: they rest with a sharp unconformity (commonly a structural one) on different older Precambrian rocks and are also overlain by Middle Neoprotozoic rocks unconformably (with a stratigraphic or angular unconformity) and in some places with a masked break or even conformably, these two latter modes of occurrences being typical of the areas of continuous sedimentation where the Vyborgian orogeny terminating the sub-era was of low intensity. It is important that the Lower Neoprotozoic strata usually lie on strongly eroded and folded Mesoprotozoic rocks or on abyssal granites of the Karelian cycle cutting them. The Lower Neoprotozoic rocks, however, are commonly of low alteration and slight dislocation. Everywhere in the world Lower Neoprotozoic strata compirse structures evidently superimposed on the older sequences. In the Baikal Mountain Land they had accumulated in the marginal near-fracture depression that cut the Mesoprotozoic folded structures at an acute angle (see Fig. 47).

The inner structure of the sub-erathem is usually complex. In some regions it is composed of two (rarely three) groups separated by breaks and local angular unconformities that were associated in time with intrusions of subvolcanic and mesoabyssal granites. The lower and the upper units of the erathem usually also differ in composition; the former are made up of volcanogenic rocks as a rule, while the latter are dominantly of sedimentary type.

**Isotope Age.** In major cases volcanogenic rocks of the Lower Neoprotozoic and related granites are reliably dated by different isotope methods. Low-grade metamorphism of many rocks dated facilitates interpretation of the values obtained by Rb-Sr isochron and U-Th-Pb (on zircon) methods; as a rule these values characterize the time of lava outflow or the time of intrusions. The K-Ar datings, however, often give the rejuvenated values.

Principally the Lower Neoprotozoic acid lava and granites are in the age range of 1750–1600 m.y. At the same time, in some localites the volcanics of the lower units of the erathem and also the granites of the early generations reveal certain older age values of the order of 1850–1900 m.y. However, even the oldest volcanics of the erathem rest with a sharp unconformity on the Mesoprotozoic metamorphic and plutonic rocks. The age values less than 1600 m.y. obtained for the Lower Neoprotozoic are usually due to secondary transformations. Values less than 1550 m.y., and even lower (down to 1400 m.y.) are known to have been obtained for granites that do not differ at all from the widely distributed and typical rocks of the sub-erathem. Some of them are likely to belong to the latest phases of the Vyborgian tectono-magmatic cycle while the age of some others needs further checking. On the whole this wide scattering of the true age values of magmatic rocks of the Lower Neoprotozoic seems to be quite natural, for lava outflows and intrusions of comagmatic granites took place during the whole sub-era.

**Supracrustal Rocks.** It was noted above that most groups are typified by alternation of volcanogenic and terrigenous rocks. Acid rocks are dominant among volcanics; quartz and basoquartz porphyries with commonly related dacites, and also syenite and subalkaline varieties of lavas: orthophyries and monzonite-porphyries. The volcanics are sometimes of such low alteration that they are called rhyolites, trachytes, and trachyrhyolites, that is in terms of their cainotypic analogs. However, in major cases the rocks are recrystallized to a certain grade and affected by different processes of alteration of autometamporphic and partly diagenetic character, that become evident in albitization, epidotization, sericitization, chloritization, and silicification. Porphyries associate with welded tuffs, lava-breccias and different kinds of tuffs; the ignimbrite varieties of welded tuffs are specially known. Many volcanogenic rocks are characterized by a red and lilac color that is explained by the microscopic inclusions of iron oxides. Certain varieties of porphyries are dark grey due to many inclusions of magnetite.

Basic volcanics are subordinate, but in some strata they are quite abundant and are mostly typical of the littoral-marine terrigenous rocks that laterally replace the continental sedimentary-volcanogenic strata with porphyries. At the same time they rarely associate with acid lavas. Diabases and basalts, commonly amygdaloidal, are mainly known among these rocks. Lavas of intermediate composition (porphyrites or andesites) are scattered.

The sedimentary deposits are principally built up of psephytic clastic rocks: conglomerates, gritstones, arkoses, or tufogenic, polymictic, oligomictic, and quartz sandstones; the siltstones, argillites, and shales (slates) are also quite abundant. The alternation of rocks of different grade of sorting of the clastic material is characteristic; sometimes in the same sandstone sample there are angular clasts of local rocks and perfectly rounded grains of quartz that were subjected to long transportation. Monotonous strata of well-sorted quartzite-sandstones are usually typical of the marine deposits. Many low-grade terrigenous strata as well as volcanics are red and lilac, rarely variegated. Sandstones commonly display cross-bedding of the stream and basin types; ripple marks and sun cracks are also observed. In the lower part of certain groups interbeds of high-aluminous slates occur among quartzite-sandstones and, depending on the metamorphic grade, are represented by pyrophyllite, chloritoid, disthen, and sillimanite schists. The quartzites emplacing them also contain high-aluminous minerals and local metamorphic concretions made up of diaspore.

The terriginous littoral-marine strata locally contain small sheet ore deposits and lenses of hematite or hematite-magnetite iron ores and also some interbeds of sandstone with highly ferruginous cement. At the same time the ferruginous-siliceous rocks of jaspilite type are lacking. The only exception in the case is the Franceville subplatform littoral-marine group of Gabon (Equatorial Africa), it contains banded oolitic phosphate-ferruginous-cherty rocks as interbeds in pelites. These rocks greatly resemble jaspilites of the Superior type, but they differ in chemism, and by the presence of abundant phosphorus.

**Organic Remains.** The continental essentially volcanogenic and clastic strata making up the major portion of the sub-erathem are lacking any organic remains, though theoretically certain fossil microbiota may occur there in the fine tuffites and cherty rocks. Phytolites (stromatolites and oncolites) are reported from some platform littoral-marine strata with interbeds of carbonate rocks, mostly dolomites; for instance in the above-mentioned Franceville Group.

**Intrusive and Subvolcanic Rocks.** Intrusive magmatism is commonly related to taphrogenic volcanics in space and origin. Different granitoids are widespread: potassic, essentially microcline, granites, granosyenites, granodiorites, and monzonites in part. Certain microgranites and granite-porphyries occur in the lower part of thick acid lava piles and represent the holocrystalline varieties of the latter. In some other cases such granites cut the porphyries and the sedimentary rocks enclosed in them.

At the same time there are granitoid massifs of mesoabyssal type, their relation to volcanics is not so evident, though the kinship of these rocks seems to be realistic on the basis of similarity in mineralogical and chemical composition and also of the occurrence among the exposures of sedimentary-volcanogenic rocks which they cut. Commonly such granitoids are known in the fracture zones that earlier controlled the lava outflow. As a result of recurrent movements along the fractures these rocks, together with the volcanics emplacing them, became highly schistose and even acquired a gneissoid appearance. Tectonic movements are likely to have occurred during the formation of intrusions, which is evidenced by certain features of protoclase commonly observed in such rocks. Many granitoids contain xenoliths of porphyries or granite-porphyries and are characterized themselves by local granophyric microtexture typical of acid lavas and their subvolcanic derivatives. The postmagmatic fluorite mineralization related to acid lavas is also common in granitoids.

The rapakivi granites are typical of this sub-erathem; they locally associate with anorthosites and gabbroids. The formation succession of these rocks is in many cases the same or similar. Firstly the basic magma intruded, and differentiated with formation of stratiform bodies of gabbro-norites and anorthosites, then intrusions of large masses of acid magma occurred that became incipient of the origin of the rapakivi granites or granophyric (and micropegmatitic) granites; then again followed intrusions of basic magma, but these were small-scale and resulted in formation of dykes of diabases, dolerites, and gabbro-norites. In certain composite massifs the still older dykes of diabases are registered as separating two phases of granite magma intrusion (two generations of the rapakivi granites). In some cases porphyroblasts of the microcline are observed in diabases cutting the rapakivi.

In addition to the specific orbicular structure due to development of oligoclase fringes around the potassic feldspar ovoids, all the rapakivi granites are characterized by other common features. These include similarities in chemical composition, iron enrichment of dark-colored minerals, common occurrence of fluorite, in some cases the presence of "exotic" minerals such as "armored" olivine, abundance of xenoliths, and inclusions of basic rocks of the early phase, common occurrence of granophyre or micropegmatite differentiates, and the fact that potassic feldspar is usually represented by orthoclase.

Plutonic bodies of rapakivi are commonly of great extension. The Parguaza rapakivi massif in Venezuela was recently established as occupying more than 30,000 $km^2$, and is the largest in the world pluton of such rocks (Gaudette et al. 1978). The geophysical and partly geologic data show that the rapakivi massifs in Finland and in the Soviet Karelia form asymmetrical laccolith-like or flat, almost horizontal lens-like bodies with their base situated usually at the depth of the Conrad discontinuity (Lauren 1970). The bodies of these rocks in other parts of the world appear to be of the same nature.

The rapakivi massifs and the related anorthosites are established in many regions, on every continent. Almost all of them are situated within the older platforms and are confined to their basement stabilized after the Karelian folding. Usually they associate with taphrogenic strata of volcanics. At least two generations of rapakivi are known to exist. The rapakivi of the early generation cut only the sedimentary-volcanogenic strata and are overlain also by Lower Neoprotozoic strata, mostly of sedimentary type; these latter are cut by the younger bodies of rapakivi. When the age of rapakivi is reliably known according to radiometric methods, it falls as a rule within the interval of 1500–1800 m.y.; for certain single massifs both older values are obtained (up to 1850 and 1900? m.y.) and younger ones (down to 1350–1400 m.y.). The latter are possibly rejuvenated, but it cannot be excluded that single rapakivi massifs belong to the Kibaran tectonic-plutonic cycle of the end of the Middle Neoprotozoic (1400–1300 m.y.). In fact the Berdyaush massif of the rapakivi-like granites is the only one of Middle Neoprotozoic age. In many other cases the existence of rapakivi of this age is not evidenced by geologic data, but only suggested from the isotope datings of rocks from very rare massifs; some of them are even situated among other massifs of the same rocks that reveal the Lower Neoprotozoic age, common to the rapakivi (for instance the Ragunda massif in Sweden). The rapakivi granites are certainly quite peculiar formations of a relatively short time interval and can thus serve as a special stratigraphic bench mark.

Dolerites and gabbro-diabases, dykes and sills of which occur among sedimentary-volcanogenic and terrigenous strata of the sub-erathem and among rocks enclosing basic volcanics in particular also belong to the Lower Neoprotozoic intrusive rocks.

**Metamorphism.** As a rule the supracrustal strata of the Lower Neoprotozoic are of low alteration, usually the lowest and middle grades of the greenschist facies. It was noted that in many cases it is possible to observe only diagenetic and autometamor-phic (for volcanics) changes. In certain acid lavas even the perlitic structure is known; it is quite unstable to any transformations of rocks. However, in the zones of intensive cataclasis and schistosity near the abyssal fractures that commonly associate with granite intrusions, metamorphism increases up to epidote-amphibolite and in some cases up to amphibolite facies. An increase of metamorphic grade is often observed at a relatively short distance across the strike of structures and near the intrusions and zones of differential movements. Sometimes it is possible to observe the transition from gneissoid porphyroids with porphyroblastic biotite to absolutely undifferentiated pink porphyries with perfectly preserved magmatic microtexture at a distance of only 1–3 km and even only several hundreds of meters.

**Tectonic Structures.** The sedimentary-volcanogenic strata and many granitoids of the Lower Neoprotozoic occur in the tectonic depressions that exhibit the nature of grabens, taphrogeosynclines (after of M. Key), rifts, aulacogens, troughs, superimposed depressions, and many other "taphrogenic" structures. Their situation in wide and extensive fracture zones associated with intensive cataclasis of the basement rocks is also quite specific. Some of these zones, with the grabens and depressions within their limits, are followed for several hundreds and even thousands of kilometers. Fractures and taphrogenic structures locally form a system of two directions crossing at right angles. Many depressions are of asymmetrical structure; one of the flanges is surrounded by a large abyssal fracture (or by a system of fractures) and the other joints the basement with no ruptures or with ruptures that are small and step-like. The greatest subsidence of the depression bottom and accordingly the maximal thickness of the sedimentary-volcanogenic strata is confined to the near-fracture portion of the structure. The grade of rock strain increases sharply near large fractures; the folds there are compressed, steep, and isoclinal, dissected as a rule by numerous parallel faults and steep imbricated thrusts. The folds are commonly parallel to the principal fractures or to the margins of depressions and tend constantly to overturing toward the inner part of the latter. Away from the zones of fractures and schistosity the folds are gentle and symmetrical. These features are characteristic of the Lower Neoprotozoic wide-spread tectonic structures that did not suffer later strong re-building. Naturally, in cases of later (superimposed) dislocations, the structures become highly complicated and changed. In particular, the original steep, almost vertical faults are transformed into thrusts and the gentle folds are deformed anew with a local change in their orientation and vergence.

## 2. The Lower Neoprotozoic Stratotype and Some of Its Correlatives in Different Parts of the World

It seems reasonable to give a brief outline of the regional units of the sub-erathem starting with the stratotype – the Akitkan Group of Eastern Siberia – in order to have a better understanding of sequence and structure of the Lower Neoprotozoic.

**Stratotype: the Akitkan Group of the Baikal Area and Its Correlatives Elsewhere in Asia.** This group is developed in the asymmetrical near-fracture trough that is followed as submeridional, slightly curved belt with a width of up to 50 km and a length of more than 500 km; it extends approximately along the limit of the Baikal folded area and Siberian platform. On the flanges the sedimentary-volcanogenic strata making up the group are laterally substituted by essentially sedimentary deposits of the Anay and Teptorgo Groups. This belt of sedimentary-volcangenic rocks is traced for more than 1200 km, embracing these two groups and the Kalbazyk volcanogenic group of the Sayan area. In the Baikal area the trough bounds the older formations (starting from Katarchean up to Mesoprotozoic) along the zone of abyssal fractures. The rocks within the limits of this zone are strongly cataclased, schistose, and cut by subvolcanic and mesoabyssal granitoids of different phases of the Irelian volcano-plutonic complex that formed during almost the whole period of the Lower Neoprotozoic and during the Vyborgian diastrophic cycle in particular. The trough was obliquely oriented in relation to the structures of the older rocks. The fracture zone surrounding it from the east cuts the older structures, but at the same time it is conformal with the structures of the Akitkan Group (Fig. 47). The thickness of the volcanogenic-sedimentary strata and their deformational and metamorphic grade decrease westward, away from the marginal zone of fractures.

In many localities the Akitkan Group adjoins the older formations along the fractures, but there are some portions of undisturbed relations with the greenstone strata of the Muya Group and with the Paleoprotozoic plagiogranites and also with the Mesoprotozoic Zama Group. Ubiquitously the mode of occurrence of the rocks is observed to be with angular unconformities or is transgressive (if the rocks underlying the Akitkan Group are granites). In the west and south of the trough the Akitkan Group and the granites cutting it are overlain by the sedimentary deposits of the Upper Neoprotozoic Baikal Group and of the Eocambrian-Lower Cambrian Mota Formation.

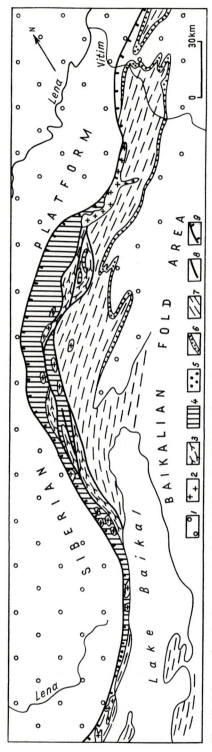

**Fig. 47.** Scheme of geologic structure of the Akitkan near-fracture depression.

*1* Riphean and Paleozoic strata of the Siberian platform and Baikalian fold area; *2–6* Lower Neoprotozoic (Akitkanian): *2* late- and post-tectonic hypabyssal granitoid, *3* protoclastic gneissoid hyp- and meso-abyssal granitoid, *4* the Akitkan Group (mainly acid volcanics), *5* the Anay Group, *6* the Teptorgo Group; *7* pre-Neoprotozoic: Katarchean, Paleoprotozoic, and Mesoprotozoic rocks (orientation of their structures is *hatched*), *8* Early Neoprotozoic abyssal fractures and associated mylonite zones, *9* Riphean-Early Paleozoic marginal fracture (at the Siberian platform-Baikalian fold area boundary)

The age of porphyries from the Akitkan Group is in the 1700–1640 m.y. range according to Rb-Sr analysis. At the same time the K-Ar datings of these rocks, even slightly altered, give obviously rejuvenated values (150–600 m.y.). Granites of the Irelian complex are dated at 1650–1600 m.y. by the Rb-Sr method.

The inner structure of the Akitkan Group is rather complex and varies moreover from place to place. Omitting details it is possible to distinguish three formations or subgroups in it (up-section): the Malaya Kosa, Khibelen and Chaya (Salop 1964–1967). The Malaya Kosa Formation rests with a sharp unconformity on the older strata; it is made up of greenish-grey, lilac and red, polymictic and arkose sandstones, quartzite-sandstones, conglomerates, tuffs, siltstones (or phyllites), with interbeds of quartz porphyries and amygdaloidal diabases in the upper part. The sandstones are characterized by cross-bedding and sun cracks. The formation is known for its strong facial variability, its thickness is from 600 to 1500 m.

The Khibelen Formation (or Subgroup) is built up of grey, dark grey, lilac and red quartz or basoquartz porphyries, or orthophyries and trachyandesites in part, and also of tuffs and welded tuffs (ignimrites). In some thick sheets the porphyries are well crystallized and correspond to granite-porphyries, granodiorite-porphyries and syenite-porphyries in their structure and composition. Sheets of diabase porphyrites are relatively rare. Volcanics locally alternate with interbeds and bands of clastic rocks. In the Akitkan Range area sandstones form quite thick strata (up to 300 m) that, divide the formation into two parts. The formation is mostly thick in the north, in the Akitkan Range, where it reaches 4500 m. Southward (or south-westward) the thickness decreases down to 3000 m in the north of the Baikal Range and 2500 m in the south of it.

The Chaya Formation (or Subgroup) is composed of two sub-formations (or formations). In the North-Baikal Highland the lower sub-formation (2000–2600 m) is made up of lilac and pink-grey arkose, quartz and polymictic sandstones, locally poorly bedded or massive with interbeds and bands of conglomerates, gritstones, siltstones, and tuffs, and also with sheets of porphyries and rarely diabase porphyrites. The upper sub-formation (1000–1300 m) is recognized by some workers as a single Okunevo Formation, and is composed mainly of well-sorted light grey and pink quartzite-sandstones, of quartz gritstones and fine pebbled conglomerates in part, rare fine interbeds of limestones with stromatolites are also registered. The sedimentary rocks enclose numerous sills of diabases and gabbro-diabases.

Generally all the formations of the Akitkan Group rest conformably, though at the boundaries and inside the formations certain breaks are observed. The clastic rocks of the Khibelen Formation lie on the lava sheets with an erosion. The volcanogenic-sedimentary strata of the group were accumulated under continental environments (including the deltaic) judging from many aspects, the only exception is the upper sub-formation of the Chaya Formation that was formed under littoral-marine conditions.

In the south of the Baikal Range the sedimentary-volcanogenic strata of the Akitkan Group are replaced by thick grey and red terrigenous strata of the Anay Group along the strike, these latter also accumulated under sub-aerial and partially under littoral-marine conditions (Salop et al. 1974). The lower part of the Anay Group contains interbeds of high-alumina chloritoid schists and sheets of basic lava are registered in its middle part. The structure of the group is shown on the schematic facial profile of the area situated at the boundary of the Anay and Akitkan zones of the marinal trough (Fig. 48). It is obvious that mainly the Khibelen Formation exhibits strong facial changes; it is gradually replaced by different sedimentary deposits with the lava sheets wedging out. The lowermost Malaya Kosa Formation and the uppermost Chaya Formations (its lower sub-formation to be more exact) are least subjected to any changes. The Anay Group is cut by granites with typical rapakivi, the age of the latter being 1650 m.y. (Manuylova and Sryvtsev 1974).

In the North Baikal and Patom Highlands (the north of the Baikal Mountain Land) the Teptorgo Group (up to 2000 m) is a formational and age analog of the Anay Group; it lies unconformably on Katarchean gneisses and on metamorphic strata of the Mesoprotozoic Udokan Group or on granites cutting it (1900 m.y.); it is overlain by the Patom Group of the younger Neoprotozoic sub-erathems transgressively (Salop 1964–1967). At the base of the Teptorgo Group there are strata of arkose sandstones (the Koganda Formation) or younger strata of quartzites with interbeds of high-alumina chloritoid or disthen schists (the Purpol Formation); upsection appear sandstones and slates with sheets of diabase porphyrites and rare albitized porphyries (the Medvezhevka Formation); the group is completed by quartzites. The relations of the Teptorgo and the Anay Groups cannot be established by direct observations, for nowhere do they adjoin each other. However, the Teptorgo

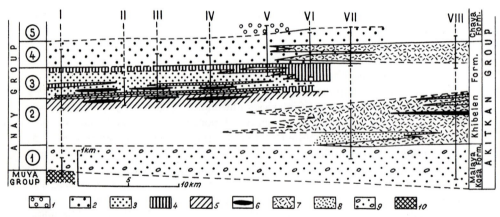

**Fig. 48.** Scheme of change of the Lower Neoprotozoic lithofacies at the boundary of the Anay and Akitkan zones of near-fracture depression. (After Salop et al. 1974).

*1* green-grey stilstone, sandstone, and gritstone with thick beds of dark-grey conglomerate, *2* variegated coarse-bedded (locally cross-bedded) sandstone, gritstone, conglomerate, and siltstone, *3* light-grey quartzite-sandstone with lenses of gritstone, siltstone, and slate, *4* greenish-grey and dark-grey phyllitoid slate, *5* quartzite-sandstone with bands of chloritoid slate, *6* metabasite, *7* porphyry, tuff, and weldet tuff, *8* lilac-grey and grey sandstone, conglomerate, siltstone, and tuff among porphyry, *9* lilac-grey and grey standstone, quartzite-sandstone, siltstone, slate, conglomerate, *10* Paleoprotozoic metamorphic rocks.

Location of sequences: *I* Baikal Range, to the north-west of the village of Malye Kocheriki, *II* the Upper Kheyrem River, *III* the right tributary of the Bolshoy Anay River, *IV* the Upper Bolshoy Anay River, *V* the right tributary of the Bolshaya Lena River, *VI* the left bank of the Shartlay Lena River, *VII* the Upper Solntse Pad' River, *VIII* the Solntse Pad' River.

Figures in circles: strata of the Anay Group: *1* sandstone-siltstone (Malokosa), *2* sandstone-slate (chloritoid), *3* quartzite-sandstone, *4* gritstone-sandstone, *5* siltstone-conglomerate

Group is of the same stratigraphic situation as the Akitkan and the Anay Groups, and it is similar to the latter, the only difference being in a higher abundance of shelf facies. This group extends not only in the near-fracture marginal trough but also in the adjacent areas of the Baikalian fold region. There it rests in the base of a thick miogeosynclinal complex of Neoprotozoic and forms a single structural stage together with the overlying strata of the erathem. At the same time, the Middle Neoprotozoic strata overlie it with a break and locally the group is completely or partially reduced by erosion; the angular unconformity between them, however, is lacking (see Fig. 52).

In Eastern Siberia there are other Lower Neoprotozoic strata similar in composition to those mentioned above. Here only the taphrogenic strata will be briefly discussed; they are developed in the east of the Aldan shield within the limits of the Ulkan near-fracture trough (Gamaleya et al. 1969). These rocks rest on the Katarchean gneiss basement of the shield and underlie a composite sequence of platform strata of the Middle and Upper Neoprotozoic. In the lowermost part of the taphrogenic complex the Toporikan Formation (200–450 m) is situated; it is made up of quartzite-sandstones, up-section lies the Ulkachan Formation (250 m) of sub-areal volcanics of trachyandesite and trachybasalt composition and then the thick Elgetey Formation (up to 4000 m) follows, composed of subareal quartz porphyries and orthophyries with interbeds of tuffs, welded tuffs, and terrigenous rocks. The unaltered porphyries are dated at 1615–1650 m.y. by the K-Ar method, but a single dating of these rocks by U-Th-Pb analysis on zircon revealed an age of 1840 m.y. (by $^{207}Pb/^{206}Pb$ ratio). Granites comagmatic to porphyries gave a range of 1540 to 1800 m.y. by K-Ar analysis (several determinations). The gabbro-granophyric volcano-plutonic complex with rapakivi related to granophyres is 1900 m.y. old on zircon from the granophyres of the Uchur massif, but this value is also obtained only on the Pb isotope ratio and the values obtained on other isotope ratios are sharply discordant. Gamaleya (1968) suggests that the age of the lavas and granites falls within the range of 1900–1800 m.y. If this is so, then the volcanic and intrusive processes involved occurred at the very beginning of the Early Neoprotozoic.

Certain younger strata of the Lower Neoprotozoic in the Ulkan trough are represented by the Birinda Formation (180–600 m) of sub-areal variegated, commonly cross-bedded arkose sandstones with lenses of conglomerates and sheets of amygdaloidal diabases, trachyandesites, and porphyries (rhyolites) in part. The formation lies with a break on the Elgetey porphyries and is unconformably overlain by the Konkulin Formation (up to 400 m) of polymictic conglomerates and cross-bedded red sandstones. Earlier the Birinda and Konkulin Formations were combined into the Uyan Group, now the situation has changed, as many workers think that the unconformity between these formation is significant and that it is more reasonable to separate the Konkulin Formation from the underlying rocks and to join it with overlying Uchur Group that seems to belong to the Middle Neoprotozoic in the light of recent new data.

Southward, in Asia, a number of sedimentary-volcanogenic groups of Kazakhstan and Hindustan belong to the sub-erathem examined. In Kazakhstan the Maytyube Group, of taphrogenic nature, in the Ulu-Tau Mountains is one of these. According to Zaytsev and Filatova (1972) this group is of great thickness (up to 9000 m), of composite structure, and embraces several formations separated by breaks. The major portion of it is composed of porphyries, their tuffs, and welded tuffs, these thick strata being separated by bands of sedimentary rocks, including quartzites, conglomerates and psammitic slates. Sericite-quartz and graphitic slates and also phyllites are situated in its upper part, intercalating rhythmically. Volcanism was mostly active under terrestrial environment and the psephytic strata alternating with lavas accumulated in the shallow-water continental basins and were formed at the expense of erosion of the local rocks, mostly volcanics. Slates of the upper part of the group are possibly of marine origin.

The Maytyube Group rests unconformably on the Paleoprotozoic and Mesoprotozoic ferruginous strata and is unconformably overlain by the Bozdak and Kokchetav Groups of the Middle and Upper Neoprotozoic. Granites cutting the porphyries are determined at 1630 m.y. by the K-Ar method on biotite. The Kuuspek Group (Formation) of the Kokchetav massif is a close stratotypical and formational analog of the Maytyube Group, mainly composed of porphyries with a thickness up to 1300 m.

In Hindustan several groups distributed in different regions of the subcontinent possibly belong to the Lower Neoprotozoic, but their age has not been determined by isotope methods, the only exception being the Delhi and Dhanjory Groups, which are probably of Lower Neoprotozoic age.

The Delhi Group, developed in the Rajasthan state, is commonly divided into two subgroups: the lower – Alwar, composed of rocks of different grade of alteration – arkoses, gritstones, conglomerates and mainly quartzites; and the upper – Ajabhar, made up of phyllites, calcareous slates, and marbles in part. Total thickness of the group reaches sometimes 6000 m, the major portion of it being the lower sub-group. The group lies unconformably on the Mesoprotozoic metamorphic strata of the Aravalli and Raialo Groups cut by the granites with an age about 2000 m.y., but the group itself is cut by the Bairath granites and Kisengarh nepheline syenites that are dated at 1600 and 1490 m.y. respectively (Sarkar 1972).

The Dhanjory Group developed in the Orissa state is principally composed of the amygdaloidal basalts. At the base it there are the strata of quartzite-sandstones and conglomerates that rest transgressively on metamorphic rocks of the Mesoprotozoic upper group of the Orissa Supergroup. Some K-Ar age values are available for the basalts, and fall in the 1600–1700 m.y. range (Sarkar 1972).

**Lower Neoprotozoic of Europe.** Some sedimentary-volcanogenic strata or groups belong to the sub-erathem in this continent, mainly composed of acid volcanics, tuffs, and also clastic rocks. The Sub-Jotnian Group (or Dala Porphyry) of Central Sweden (the Loos Hamra region) is the best-known among these, and can be regarded as a parastratotype of the suberathem as a whole. According to Hjelmqvist (1966), two formations are distinguished in the group; the Lower and the Upper Dala. The Lower Dala is composed of porphyries and porphyrites alternating with red arkose and quartzite-like sandstones, conglomerates, argillites, and shales. At the base of the group sedimentary breccias and conglomerates commonly occur. Sandstones exhibit ripple marks and cross-bedding and argillites are characterized by sun cracks and rain drops. The Upper Dala Formation is mostly composed of red quartz and basoquartz porphyries, ignimbrites, tuffs, and tuffites in part.

According to data of Lundegårdh (see Bogdanov 1967), in the region of the town of Emodalen the Sub-Jotnian sequence starts with the Leksand basal conglomerates, up-section appears a band of porphyrites, then thick strata of light and pink porphyries and ignimbrites, still upward a band of Digerverg conglomerates is observed with interbeds of tuffites, in its turn overlain by a band of porphyrites cut by small bodies of quartz syenites. The sequence is completed by Bredvord porphyries and ignimbrites. The total thickness of the sequence reaches 2000 m.

The Sub-Jotnian rocks form gentle folds and rest with a sharp structural unconformity on the metamorphic and plutonic Mesoprotozoic rocks of the Svecofennian complex, on the Late Svecofennian (Sveco-Karelian) granites with an age of 1900 m.y. including. The Sub-Jotnian porphyries are cut by the comagmatic microcline granite-porphyry and granite, including rapakivi, whose age is shown to be 1620–1650 m.y. The Rb-Sr age of porphyries is 1670 m.y. (Welin and Lundqvist 1970). The Sub-Jotnian rocks and the granites cutting them are unconformably overlain by the Jotnian sandstones (the "Dala sandstones") which belong to the Middle Neoprotozoic judging from datings of diabases cutting them.

In the still more southerly areas of Sweden, the Smoland-porphyry (1695 m.y. by the Rb-Sr method) corresponds to the Sub-Jotnian, in the south of Norway it is the Engerdal porphyry (1628 m.y., Rb-Sr method), and in the eastern Soviet part of the Gulf of Finland (on the Hogland Island) it is the Hogland porphyry. In the Ukrainian shield in the region of the town of Ovruch the sedimentary-volcanogenic strata (with porphyries and diabase porphyrites) possibly belong to the Lower Neoprotozoic, they are recognized as the Pugachev Group (the upper) and the Ovruch Group (the lower). The Pugachev Group is composed of two formations. The lower Belokorovichi Formation (1200 m) is largely built up of grey or pinkish-grey sandstones and quartzites with small sheets of diabases; the overlying Ozeryansk Formation (700 m) principally comprises slates and argillites that enclose a sheet of diabase porphyrites near the base. The Ovruch Group also consists of two formations (up-section): the Zbran'kovo Formation (340 m) composed of red porphyries alternating with amygdaloidal diabases and trachyandesites and sandstones, and the Tolkachev Formation (up to 900 m) with dominant red cross-bedded quartzite-sandstones. Porphyries of the Zbran'kovo Formation fall within the age range of 1464 to 1530 m.y. by the K-Ar method and they are 1400 m.y. old (Rb-Sr analysis), but these values seem to be strongly underestimated, as the diabases from the same formation gave an age of about 1700 m.y. (K-Ar method).

In the Baltic shield and in the adjacent areas of the Russian plate there are many massifs of rapakivi granites associated with local anorthosites and alkaline syenites. In major cases the rapakivi occur together with sedimentary-volcanogenic strata of the Sub-Jotnian type, in other cases they are isolated, but there are nevertheless some reasons suggesting that the supracrustal formations in the area of their distribution was destroyed by erosion (xenoliths of porphyries and tuffs are locally reported in the granites). In the Ukraine the Korsun'-Shevchenko large pluton of anorthosites and rapakivi (1720 ± 70 m.y., K-Ar analysis) contains xenoliths of quartzites of the Pugachev Group, but its erosional surface is overlain by porphyries of the Zbran'kovo Formation with local conglomerates at the base. The porphyries are cut by veins of granophyric (micropegmatitic) granites that are also quite abundant among the rapakivi (Salop 1973a).

**Lower Neoprotozoic of North America.** Many units here belong to the sub-erathem examined. Among them the best-known and best-studied are the Lower Dubawnt, Martin Lake, and Letitia Lake Groups developed in different regions of Canada, the Qipisarqo Group of Greenland, the Red Rock Group, the Hogan porphyry and Royal Gorge porphyry in the U.S.A.

The Lower Dubawnt Group of the northern part of the Churchill tectonic province on the Canadian shield can be regarded as a regional (North American) stratotype of the sub-erathem. According to Donaldson (1967) the following formations are recognized in it (upward): (1) the Southern Channel Formation (1650 m): conglomerates with sandstones interbeds; (2) the Kazan Formation (4000 m): red cross-bedded feldspathic sandstones and siltstones; (3) the Christopher Island Formation (up to 200 m): lavas of andesitic, latitic, liparitic, and trachytic composition alternating with red tufogenic sandstones and agglomerates; (4) the Pitz Formation (80 m): red and brown porphyries. There is a break at the base of the Christopher Island Formation and the sills and laccoliths of the Martel alkaline syenites occur between this formation and the Pitz porphyries, which are very likely to be the comagmatic intrusive derivatives of trachytes. The sedimentary-volcanogenic strata rest subhorizontally on the older folded rocks, including the Mesoprotozoic Hurwitz Group, and they are unconformably overlain by the sedimentary rocks of the Middle Neoprotozoic Upper Dubawnt Group. In addition to the Martel syenites with an age of 1725 m.y. (Rb-Sr method), the Lower Dubawnt Group is cut by the diabase dykes with the K-Ar age of 1560 m.y. Rb-Sr and K-Ar methods gave an age of 1730–1770 m.y. for the unaltered Pitz porphyries; with the geologic data combined it reveals a Lower Neoprotozoic age for the group.

The Letitia Lake Group is developed on the Labrador Peninsula in the Grenville tectonic province. It is made up of quartz porphyries, tuffs, quartzites, argillites, and slates, commonly red-colored (Brummer and Mann 1961). At the base of the group conglomerates rest unconformably on the granites that were formed in the Mesoprotozoic or even earlier, but suffered strong Mesoprotozoic mobilization, judging from the isotope datings ($\sim$ 2000 m.y., K-Ar method). The rocks of the group are cut by alkaline syenites, their age in the adjacent areas is determined by K-Ar analysis to be 1700–1750 m.y.

Abundant anorthosites associated with rapakivi granites are registered in the distribution region of the Letitia Lake Group and the Croteau Lake Group, a close analog of the former. The age of the rapakivi granites has not been dermined, but the anorthosites are older than 1540 m.y. (K-Ar analysis). The age of the lead-zinc mineralization in the rocks of the Croteau Lake Group is 1685 $\pm$ 160 m.y. (Pb-model method).

The sedimentary-volcanogenic Martin Lake Group of the area of the Athabasca Lake, the terrigenous Et-Then Group of the area of the Great Slave Lake, the Fox River porphyry in Wisconsin (U.S.A.) and others also belong to the Lower Neoprotozoic in addition to the groups discussed above; all of them are distributed within the Canadian shield. All of them exhibit the same stratigraphic situation in the sequence of the Precambrian shield. Volcanics from the middle part of the Martin Lake Group are dated at 1630 m.y. by the Rb-Sr method and the diabases cutting them at 1560 m.y. (K-Ar method). The Fox River porphyry (rhyolite) appears to be 1670

(1640) m.y. old by the Rb-Sr method and 1760 m.y. old by the U-Th-Pb method on zircon. (Van Schmus 1978).

The Qipisarqo Group of Southern Greenland, composed of terrigenous and volcanic rocks, lies on the Ketilidian Mesoprotozoic formations with conglomerate at the base. The anorthosites and the rapakivi are younger, the age of the latter is 1750–1760 m.y. (Rb-Sr and U-Th-Pb methods; Van Breemen et al. 1972).

The Lower Neoprotozoic strata are known for certain in the Ozark Plateau (Missouri) and in the Mazatzal Mountains region (Arizona) of the U.S.A. At the former site the Lower Neoprotozoic is represented by the subaerial Hogan and Royal Gorge porphyries (rhyolites), their age is 1530 m.y. on zircon (Bickford and Mose 1975), and at the latter it is represented by the sedimentary-volcanogenic Red Rock (Texas Gulch) Group. This latter is made up of sub-aerial red porphyries, their lava-breccias and tuffs, basalts and terrigenous rocks, that rest unconformably on the Mesoprotozoic metamorphic Big Back (Ash Creek) complex and are overlain by Middle Neoprotozoic strata. The porphyries are dated on zircon at 1715 m.y. and the age of the granites cutting them is 1640–1655 m.y. (Rb-Sr, U-Th-Pb and K-Ar methods, Lanphere 1968, Livingston and Damon 1968, etc.). In both areas rapakivi massifs are known to occur, situated not far away from the exposures of the Lower Neoprotozoic; their age is 1500–1600 m.y. (Bickford and Mose 1972 etc., see also Salop 1973 a). A large pluton of rapakivi in Wisconsin reveals the same age (1510 m.y. on zircon) (Van Schmus et al. 1975).

**The Lower Neoprotozoic of South America.** The rocks of this sub-erathem are widespread on the Guianan and Brazilian shields, in every region they are composed of two complexes: the lower is terrigenous-volcanogenic and the upper essentially terrigenous; they are separated by an unconformity. The lower complex is largely built up of acid and partially intermediate volcanics: porphyries (rhyolites), dacites, rhyodacites, and ignimbrites and tuffs in particular, andesites and their tuffs are subordinate, basic lavas are rare. Volcanics are interbedded with arkose and polymictic sandstones, quartzites, conglomerates, siltstones, slates, and tuffites. In certain localities the sedimentary deposits form rather thick bands among volcanics and in places they underlie the complex. Many volcanogenic and sedimentary rocks are red-colored, the sandstones are characterized by cross-bedding and ripple marks. The thickness of the complex varies greatly, in some regions it reaches 5–6 thousand meters, for instance in Rio-Grande-do-Sul (Brazil).

These rocks occur in the near-fracture depressions or grabens and rest with a sharp unconformity on the older metamorphic and plutonic rocks (starting from Katarchean up to Mesoprotozoic) of the basement of the shields. In many places they are cut by subvolcanic granite-prophyries and granophyric granites and also by meso-abyssal microcline granites, including rapakivi granites.

The rocks forming this complex are known in different regions under different local names: the Cuchivero or Cairaça Groups in Venezuela, the Surumu Group and Iricoume Formation in Northern Brazil, the Ivocrama and Cujuvini Groups in Guaiana, the Dalbana Formation and the Sipalivini porphyry in Surinam, the Iriri and Marioa Groups in Brazil (Beurlen 1970, Mendoza 1974, Priem et al. 1971, Verhofstad 1970, Gaudette et al. 1978, Moralev et al. 1979).

Datings of volcanics by different isotope methods give great scattering of values: from 1200 to 1800–1900 m.y. The K-Ar datings, as usual, give the lowest values; the Rb-Sr isochron datings commonly fall within the 1640–1840 m.y. range. The age of the granites cutting the terrigenous-volcanogenic complex for various massifs varies from 1500 to 1750 m.y. (Rb-Sr method). An age of 1545 m.y. was obtained for the rapakivi forming a huge Parguaza massif in Venezuela by the U-Th-Pb method. (zircon, intersection of isochron and concordia) and a whole-rock age of 1530 m.y. by the Rb-Sr method (Gaudette et al. 1978). In Northern Brazil the age of the Sirocucu rapakivi massif is determined to be 1500 m.y. and the age of the granophyric granites distributed in the same region and related to volcanics is 1680 m.y. (Rb-Sr method). The formation of the complex examined, judging from datings of almost unaltered acid lavas, started soon after the Karelian (Trans-Amazonian) folding and continued for more than 150 m.y.

The upper, essentially terrigenous, complex of South America is represented by the Roraima Group that occupies an extensive territory in Venezuela, Guiana, and Brazil and by the synchronous Uatuma Group that is widely distributed in Northern Brazil. Unlike the lower taphrogenic complex, this complex rests like a mantle on the older formations and forms the platform cover, commonly representing table mountains. The two complexes are usually separated by a stratigraphic or angular unconformity, but in some localities this is not quite evident.

The Roraima Group occupies a square of about 250,000 km² in the central part of the Guiana shield. As a rule it rests subhorizontally, but the faults within it are common and near these certain disturbances are observed (the near-fracture folds and flexures). The group is made up of pink, red and light grey arkose and quartz sandstones, partly of green and red shales and cherty rocks. Conglomerates are encountered at different stratigraphic levels, including at the base of the group; the pebbles are characterized by abundant different rocks of the lower complex, porphyries and subvolcanic granitoids are specially conspicuous. Cross-bedding and ripple marks and local sun cracks and traces of rain drops are observed in the sedimentary rocks. The common jasperoids or phtanites are regarded by many as silicified tuffs of acid volcanics deposited among psephytic sediments as a result of wind activity that supplied ash material from the areas of synchronous volcanism (Kloosterman 1975). In some places typical crystalloclastic tuffs are observed and in the area of the Tafelberg mountain in Surinam, ignimbrites are also registered.

In Guiana the Roraima Group is divided into three units: the lower one mainly composed of conglomerates, the middle largely made up of bedded sandstones with cherty rocks interbeds, and the upper formed of massive sandstones. In the stratotypical sequence, in the area of the Roraima Mountain at the boundary of Venezuela, Brazil, and Guiana, four formations are recognized up-section within the group: (1) the Uairen Formation (850 m): alluvial conglomerates and sandstones, (2) the Kukuenan Formation (up to 100 m): mainly clay shales, (3) the Uaimapué Formation (250 m): phtanites alternating with red arkoses and siltstones of deltaic origin, (4) the Mataui Formation (up to 600 m): light grey and pink cross-bedded quartz sandstones (Reid 1974). Fossil placers of gold and diamonds occur in the Uairen Formation. Total thickness of the group sometimes reaches 1800–2000 m.

Numerous dykes and sills of dolerites and diabases occur among the sedimentary rocks, and are sometimes thick and differentiated. These bodies are usually pierced by dykes and irregularly shaped inclusions of granophyres and microgranites that probably represent the ultimate products of differentiation. The age of the dolerites is estimated to be 1695 ± 66 m.y. (Snelling and McConnell 1969) and 1600 ± 50 m.y. by Rb-Sr and K-Ar methods (Priem et al. 1973). Ignimbrites from the Tafelberg Mountain gave an age of 1600 ± 18 m.y. by the Rb-Sr method on isochron for 14 samples (Priem et al. 1973). Thus the isotope datings of the Roraima rocks differ slightly from the datings of the underlying sedimentary-volcanogenic complex. However, considering the unconformable occurrence of the Roraima Group rocks on the rocks of the complex, it is possible to suggest that many datings of the latter are rejuvenated. At the same time, the occurrence of tuffs and welded tuffs in the Roraima Group indicates that volcanic activity had not ceased completely during its deposition. Thus the group and the underlying complex become more closely related. It should be noted that prior to radiometric studies the Roraima Group was considered to be of Lower Paleozoic age (earlier on the geologic maps of South America its index used to be Cambrian-Ordovician).

The Uatuma Group, distributed on both sides of the Amazonian depression in Brazil, is similar to the Roraima Group in major features. Certain differences lie in the occurrence of more abundant tuffites, tuffs, or ignimbrites and local rhyolite and dacite lavas among red terrigenous rocks. The thickness of the group exceeds 1000 m. It lies transgressively on the older Precambrian and is unconformably overlain by the Silurian strata; its isotope age is not determined. Kloosterman (1975) demonstrated the analogous nature of this group to the Roraima Group.

**Lower Neoprotozoic of Africa.** The erathem of this large continent is principally represented by continental taphrogenic sedimentary-volcanogenic strata mainly composed of acid lavas and their tuffs and also of terrigenous rocks. In many regions they extend, but they are poorly divided because of monotonous composition or due to irregular alternation of lavas and sediments and their lateral substitution. At the same time, in major cases the stratigraphic situation of these strata in a general Precambrian sequence and the radiometric age of the rocks composing them is known for certain. A brief outline of some strata of the sequences best studied will be given below; some will be only mentioned, in certain cases with their datings. All these sequences are described in my book on the Precambrian of Africa (Salop 1977b).

The stratigraphy of the Waterberg Group of Southern Africa is known in detail, it extends in the northern margin of the Transvaal craton. According to De Vries (1969) the group ("system") divides into two subgroups (groups): the Nylstroom Subgroup with abundant volcanics and the overlying Kransberg Subgroup of essentially terrigenous composition: each is divided into three formations. The Nylstroom Subgroup consists of (up-section): (1) the Swaershoek Formation (1000–2300 m): felsitic, rhyolitic (quartz-porphyric), trachytic and amygdaloidal andesite lavas, agglomerates, tuffs, sandstones, conglomerates (of basal type); (2) the Alma Formation (0–2700 m): coarse sandstones and conglomerates upward passing into fine-grained sandstones, commonly cross-bedded, with ripple marks, interbeds of felsites; (3) the Langkloof Formation (875 m): medium- and coarse-grained sandstones and conglomerates. The Kransberg Subgroup is divided upward into: (1) the Sandriviersberg Formation (1250 m): coarse-grained cross-bedded sandstones, with a band of siltstones and local thin horizons of amygdaloidal lavas, (2) the Cleremont Formation

(125 m): coarse-grained white essentially quartz sandstones, (3) the Vaalwater Formation (485 m): shales and sandstones. The major proportion of acid volcanics and
sandstones are red or purple-red. The rocks of the group accumulated under continental environment (under alluvial, partially deltaic and lacustrine conditions); the
rocks of the two upper formations are possibly of littoral-marine origin.

The Waterberg Group lies on the Mesoprotozoic different rocks, including gabbroids and granites of the Bushveld complex and consequently it is younger than
2000 m.y. However, the acid volcanics are cut by the comagmatic granophyric
granites that are dated at 1790 ± 70 m.y. on zircon. This value is regarded as the time
of formation of the lower volcanogenic portion of the group. The entire group is cut
by alkaline rocks and syenites that yielded the Rb-Sr age of 1500 m.y. and an age of
1420 ± 70 m.y. by U-Th-Pb method on sphene and zircon (Burger and Coertze 1973).
Thus by time of formation and composition, the Waterberg Group corresponds fairly
well to the Lower Neoprotozoic of other continents.

In addition to the Waterberg Group the following groups in Southern Africa are
likely to belong to the erathem discused: Nagatis, Dordabis, and Khoabendus distributed in different regions of Namibia. All these groups are represented by terrigenous-volcanogenic commonly red strata with abundant volcanics that belong to the
trachyte-rhyolite formation. They are certainly younger than the Mesoprotozoic
metamorphic rocks and granites (the Fransfontein granite and others) with an age of
1900 m.y., but they are overlain by the Middle–Upper Neoprotozoic strata (the
Kunjas, Sinclair, Stinkfontein,Tsumis, Auborus Groups etc.). Unfortunately, their
isotope age has not been determined.

In Equatorial Africa a number of sedimentary-volcanogenic strata belong to the
Lower Neoprotozoic, these are: the Kate porphyry (1790 m.y.), distributed in the
western part of Tanzania, the Ndembera porphyry (∼ 1900 m.y.) in the eastern of the
same country, the lower sedimentary-volcanogenic group of the Burundi Supergroup
in Ruanda and Burundi, the Marungu or Luapula porphyry (1790 m.y.) in the Northern province of Zambia, the Shamazio Group in the Southern province of the same
country, the Mayombe Group of Congo and Zaïre, and the Oendolongo Group
(s. str.) of Angola. The stratigraphic situation of all these units is well defined: they
rest on the Mesoprotozoic metamorphic strata and granites and are unconformably
overlain by the Middle Neoprotozoic (see Salop 1977 a). Some of the units (for
instance, the Shamazio Group) rest at the base of the Neoprotozoic miogeosynclincal
complexes, but like many others they reveal a turning taphrogenic stage in geologic
evolution of the corresponding regions. They differ from the remaining units by more
complicated structure and composition, in particular by presence of abundant slates
of possible marine origin in their upper parts, at the same time, the widespread acid
volcanics of the lower portions of the groups outflowed in the near-fracture troughs
and exhibit a sub-aerial nature.

In addition to the sedimentary-volcanogenic strata mentioned, two more subplatform groups mainly composed of sedimentary deposits can be assigned to the
Lower Neoprotozoic in Equatorial Africa, these are: the Franceville Group of Gabon
and the Karagwe-Ankolean Group of Uganda; their close analogs in the adjoining
countries are also to be included. The Franceville Group is situated in the south-
western margin of the Nile craton and occurs in the near-fracture trough representing
aulacogen that is opened south-eastward, in the direction of the Oubangui-Burundi

geosynclinal belt (see Fig. 62). The stratigraphy of this group was studied in detail by Weber (1968). Generally the group is made up of cross-bedded arkose sandstones and conglomerates in the lowermost part, then follow argillites and shales with fine interbeds of dolomites and fine-bedded jaspilite-like rocks rich in phosphorus that are laterally substituted by strata of shales with tuffs and basic-alkaline volcanics (trachy-basalts) and also dolerite sills in their upper part. Still further upward appear cherty slates, tuffs and tuffites, then carbonaceous slates with tuff horizons of acid and sub-alkaline (trachyte) composition and finally in the uppermost part sandstones and shales and local conglomerates occur. The uranium deposits of syngenetic type are related to the carbonaceous slates, acid tuffs and cherty rocks enclosed in the middle and upper parts of the group; the mineralization is very likely to be the result of leaching and redeposition of uranium from the acid tuffs during sedimentation. The ores of some deposits are known to contain a very low concentration of $^{235}U$ that is probably due to the process of "burning down" of this isotope during the natural nuclear reaction (the Oklo "natural nuclear reactor"). The thickness of the group varies from 520 m up to 2200 m. According to Weber, the psammites of the lower part of the group accumulated in a continental environment, the overlying strata are of deltaic and littoral-marine origin. Sedimentation took place under unstable tectonic conditions and was accompanied by volcanic activity.

The Franceville Group lies unconformably on different older rocks, including the Mesoprotozoic Ogooué Group, this latter was metamorphosed 2000 m.y. ago. The Franceville rocks rest subhorizontally or form gentle folds cut by faults. The Rb-Sr datings of unaltered pelites and tuffs gave values of about 1800 m.y., which is probably the time of formation of the group. The same age was obtained for the microsyenites cutting the rocks of the group but probably comagmatic to the sub-alkaline volcanics also (by the Rb-Sr method).

The Karagwe-Ankolean Group is distributed in Uganda between the Victoria Lake and the Edward Lake in the southern margin of the Nile great platform, which is a continuation of the shelf where the Franceville Group extends (see Fig. 62). The sub-platform analogs of these groups are situated between them, in Cameroun it is the Sembe-Ouesso Group (the Lower Dja) and in the Central African Republic it is the Liki-Bembi Group.

The Karagwe-Ankolean Group is typified by alternation of thick strata and interlayers of quartzites and phyllites, the quartzites are commonly cross-bedded, locally red-colored, with interbeds of quartz-pebbled conglomerates at places. The thickness of the group is estimated to be from 5500 up to 8500 m. In the opinion of some workers the group accumulated under a deltaic environment that at times changed into the littoral-marine type, under conditions of a vast shelf surrounding the land area situated in north, the source area of clastic material.

The group rests on the eroded surface of the older Precambrian rocks from the Katarchean to the Mesoprotozoic Buganda-Toro Group and on the granites cutting it (these latter are dated at 1930 m.y.). The tectonic pattern of the group is typified by relatively small mantled domes with basement granites and gneisses and newly formed rheomorphic granites cropping out in their cores; both gave a very similar age of 1250–1360 m.y. (Rb-Sr method). The metamorphic grade of sedimentary rocks increases toward the base of the group, close to the cores of the domes in particular; in the distance the rocks are altered just under the lowest grades of the green-schist facies. The K-Ar age of the phyllites from the Karagwe-Ankolean Group is 1340 m.y. All the values obtained characterize the time of remobilization of the basement and metamorphism of the group during the Kibaran orogeny. The Middle Neoprotozoic Sub-era was terminated with this orogeny and thus the Karagwe-Ankolean Group may belong to the Lower as well as to the Middle Neoprotozoic. However, correlation with other Precambrian strata of Equatorial Africa whose age is more certain and the tectonic evolution of the Karagwe-Ankolean Group suggests that the group belongs only to the Lower sub-era (Salop 1977 b).

In Western Africa the Tarkwa Group is assigned to the Lower Neoprotozoic, extending within the limits of the Eburnean fold belt of the Mesoprotozoic in Ghana, the Ivory Coast, and the Upper Volta. The group is composed of clastic rocks: conglomerates, arkoses, shales (or phyllitized slates), and quartzites with rare thin layers of basalts, andesites, porphyries, and their tuffs. The rocks are of relatively low grade and locally are violet or red-colored. The Tarkwa formations commonly form relatively large and simple synclines exhibiting superimposed synclinal folds, but at places they are strongly folded and cut by small bodies of granites and microgranites.

In the area of the typical section in Ghana, in the basin of the Tarkwa River, the group extends for about 250 km with the north-eastern strike, it is generally of synclinal structure, locally complicated with marginal fractures. The lower part of the group or the Kawere Formation (200 m) is made up of conglomerates with interbeds of arkoses and quartzites that rest unconformably on the Mesoprotozoic Birrimian Group and on the pre-Birrimian rocks. Upward the Banket Formation (500 m) lies conformably, composed of feldspathic sandstones with horizons of gold-bearing conglomerates with rare detritic diamond inclusions. Further up-section the Tarkwa phyllites (300 m) appear, then follow the Huni fine-grained quartzites with the peculiar Dompim quartzite in their upper part. In this section the total thickness of the group reaches 200 m.

The Tarkwaian strata in Ghana are cut by dolerite and gabbro-diabase bodies and by rare veins of pegmatite granite or micropegmatite that are likely to be related to basic rocks. The K-Ar and Rb-Sr age of feldspar and muscovite from these veins is 1600–1700 m.y.

In North-Western Africa the Lower Neoprotozoic formations are represented mainly by the sedimentary-volcanogenic strata with abundant porphyries and granites and granite-porphyries related to volcanics. They are mostly complete in the Reguibat massif. In Mauritania in the central part of this massif the Aiuon-Abd el Malek (or Aioun-Malek) Group belongs to these strata, in the lower part it is built up of pink and violet arkoses that locally pass into conglomerates, upward these are replaced by porphyries, dacites, ignimbrites and tuffs of the same color (Sougy 1972). The rocks are only locally altered, usually they are almost unaltered, but ubiquitously are rather strongly folded, however. The thickness of the group is estimated to be of several hundreds of meters, but probably it may exceed 1000 m.

The detritic rocks underlying the group lie unconformably on the Mesoprotozoic Aguelt Nebkha Group and on the granites cutting it. At the same time the porphyries associate with microgranites that represent the holocrystalline varieties of lavas or the subvolcanic intrusions. The Aioun Malek Group is then cut by pink and red leucocratic granites with alkaline varieties and rocks of the rapakivi type. Rapakivi from the Bir-Morgrein massif are dated at 1813 m.y. by the Rb-Sr method.

In Algeria, in the eastern part of the Regibat massif, the Lower Neoprotozoic is represented by the Eglab Group and by the Guelb el Hadid Group overlying the former with a break. The Eglab Group is principally composed of acid volcanics and tuffs and is an analog of the Aioun Malek Group of Mauritania. It lies transgressively on the granites of the Karelian (Eburnean) cycle. The Guelb el Hadid Group is made up of strata (150–300 m) of conglomerates and arkoses with thin covers of porphyries and tuffs, these strata being overlain with an erosion by clastic rocks with a thickness of about only 10–20 m. Both groups are cut by granites and granite-porphyries and also by the earlier gabbroids. The platform Hank Group of the Upper (?) Neoprotozoic overlies all these rocks with a large break.

The sedimentary-volcanogenic Panampou Group is a stratigraphic and formational analog of the Reguibat massif groups mentioned, it is developed in Senegal within the Mauritanian fold belt, where it rests unconvormably on the Mesoprotozoic metamorphic strata and is also overlain unconformably by the platform Youkounkoun Group of the Upper or Terminal Neoprotozoic.

**The Lower Neoprotozoic of Australia.** In this continent the sub-erathem is distributed in many regions and is represented by sedimentary-volcanogenic and sedimentary strata of taphrogenic type, as in other parts of the world.

The most complete and thickest sequence of the sub-erathem is known from drilling in the area of the Kimberley plate (craton), situated in the north-west of Australia (see Fig. 42). Four successively lying groups belong there to the Lower Neoprotozoic, as follows: the Whitewater, Speewah, Kimberley, Bastion Groups (Dow and Gemuts 1969, Plum and Gemuts 1975).

The Whitewater Group extends mainly in the Halls Creek fold zone that surrounds the Kimberley plate from the south. It is built up of slightly altered and dislocated red porphyries (rhyolites) and dacites with tuffs, welded tuff, tuff-conglomerates and rare sandstones. Locally its thickness reaches 3500 m. It rests with a sharp unconformity on the metamorphic rocks of the Halls Creek Group and on the granites cutting it, the age of the latter being 1900–1960 m.y. The porphyries and the related subvolcanic granites revealed an age of 1800–1820 m.y. by the Rb-Sr method.

Three overlying groups are mainly situated on the Kimberley plate and in the south-west of the Halls Creek zone in part. The Speewah Group (up to 1200 m) rests on the Whitewater Group with a break and with a local angular unconformity. Five strata (formations) are recognized in it; they are composed of feldspathic-quartz sandstones, siltstones and argillites, commonly cross-bedded and with ripple marks. The supply of clastic material is suggested as occurring mainly from the north-eastern part of the plate and of the Litchfield block. In the middle part of the group the thin strata (5–80 m) of acid tuff and welded tuff are recognized, they are regarded as a local marking horizon. Argillites are dated at 1730 and 1760 m.y. by the Rb-Sr method.

The Kimberley Group (up to 3500 m) is separated from the Speenwah Group by a break, it is characterized by a persistent composition and sequence and comprises five strata (upward): (1) King Leopold: red quartz sandstones, locally glauconite-bearing; conglomerates at the base; (2) Carson: tholeiite basalts with interbeds of cross-bedded arkoses; (3) Warton: light quartz and feldspathic sandstones with interbeds of shales; (4) Elgee: massive red siltstones, sandstones, and shales with dolomite interbeds; (5) Pentecost: quartz glauconite sandstones and siltstones, commonly cross-bedded. Stromatolites are reported from the Elgee dolomites and the *Kussiela* sp. form is determined among them, usually known from the Lower Riphean (Semikhatov 1974). The Carson lavas and tuffs are dated by the Rb-Sr method, the values obtained are 1810 and 1700 m.y. Transportation of terrigenous material mainly occurred from the north-west during the deposition of clastic rocks of the group.

The uppermost Bastion Group (up to 1000 m) is made up of variegated shales and siltstones with interbeds of dolomites and calcareous sandstones. Shales from this group are 1710 m.y. old (Rb-Sr method). All the groups, and the Speewah Group in particular, are pierced by thick (up to 1700 m) sills of the Hart dolerite with an age of about 1800 m.y. Upward appear the Colombo red sandstones; they lie with a break, and then the platform Glidden Group of the Terminal Neoprotozoic rests unconformably.

This extremely thick sedimentary-volcanogenic complex (up to 9000 m) easily divides into two sub-complexes: the lower made up of subaerial acid volcanics and the comagmatic granites (the Whitewater Group) and the upper essentially sedimentary with interbeds of acid lavas in the lower part and with Carson basalts (up to 900 m) in its middle part. This complex comprising the Speewah, Kimberley, and Bastion Groups accumulated under the shallow-water marine enviroment and only partially under continental conditions (mainly of deltaic type).

The Red Rock thick strata (up to 1800 m) are an analog of the Speewah Group, and are mostly built up of quartz sandstones and siltstones, commonly red-colored with interbeds of conglomerates (Dow and Gemuts 1969). The strata are situated eastward from the Kimberley craton and the Halls Creek zone, in the part of Australia that is sometimes recognized as the Sturt plate. These strata lie with a sharp unconformity on the metamorphic rocks and are overlain with an erosion by the Mount Parker sandstones and then by the Bungle-Bungle dolomites of the Upper Neoprotozoic.

In the central part of the continent, near the boundary of the Arunta craton and the Mesoprotozoic Warramunga fold zone, thick strata (up to 6000 m) of almost unaltered light grey and pink quartz and feldspathic sandstones and siltstones are encountered, enclosing horizons of volcanics of intermediate and acid composition in the upper part. These strata are recognized as the Hatches Group, which rests with a great angular unconformity on the Mesoprotozoic metamorphic rocks (the Warramunga Group) and on the Katarchean gneissic Arunta complex and is conformably overlain by the Upper (?) Neoprotozoic. It is cut by gabbroids and granites, including granophyric granites. The Rb-Sr age of the granites is 1690 m.y. (Compston and Arriens 1968).

The Edith River Group also belongs to the Lower Neoprotozoic; it extends in the Pine Creek fold zone in the north of the continent. It is mainly composed of rhyolites, ignimbrites, dacites, and rare andesites, sandstones, conglomerates, and shales. Many volcanics as well as sedimentary rocks are red-colored. This acid volcanics associate with subvolcanic intrusions of granites and granite-porphyries. The granites are 1760 m.y. old according to Rb-Sr analysis (Walpole et al. 1968). The Edith River volcanics are overlain with a break by the thick Catherine River sedimentary-volcanogenic Group (up to 5000 m), that seems to represent in this region the upper part of the Lower Neoprotozoic. It is unconformably overlain by Lower Riphean strata. The sequence of this group is shown in Fig. 56 (Column XXIII).

The Lower Neoprotozoic sedimentary-volcanogenic strata are of exceptionally wide distribution in the north-east of Australia. There in the Leichhardt fold belt (the north-western part of Queensland) the Mesoprotozoic Leichhardt metavolcanics and the Kalkadoon granites (1900 m.y. old) cutting them and the Magna Lynn metabasalts are unconformably overlain by the Argilla Group (2000 m) composed of red or pink-grey rhyolites, ignimbrites, rhyolite-dacites, and dacites (rare andesites) that alternate mainly in the upper part of the group with feldspathic sandstones and siltstones, commonly cross-bedded and with ripple marks. Abundant large dykes and sills of microgranites and granite-porphyries are reported among volcanics, and are genetically related to lavas. The rocks are deformed and altered only in the zones of fractures and at the contact with the Wonga hypo- and mesoabyssal granites cutting them. The Rb-Sr age of the latter is 1700–1780 m.y.

The Argilla Group is unconformably overlain by a very thick sequence of sedimentary and volcanogenic rocks; in the west of the region it is called the Haslingden Group, in the east the Soldier Kap Group. The normal sequence of these differs to a certain extent, but generally they are made up of thick strata of quartzites, of basic and acid volcanics in part. The groups are cut by the Weberra and Sybella granites with an age of 1650–1660 m.y. by the Rb-Sr method (Wilson 1978).

The Haslinden and Soldier Kap Groups and the granites cutting them are overlain unconformably by the McNamara Group and the synchronous Mt. Isa Group that belong to the Middle Neoprotozoic. Thus, in the north-east of the continent, the Lower Neoprotozoic is represented by two groups of different age, separated by an unconformity and granite intrusions that are likely to be comagmatic to volcanics. The lower group is essentially volcanogenic, principally made up of sub-aerial acid lavas; the upper one is of quartzite-sandstone composition with abundant basic lavas, accumulated mainly under the littoral-marine environment. In the opinion of Wilson (1978) the sedimentary-volcanogenic groups of Queensland continue under the cover of the younger strata far southward, marking the eastern boundary of the older (Neoprotozoic) continental massif (platform) of Australia.

In the Southern extremity of the supposed volcanic belt are situated the Eyre Peninsula and the Moonabie volcanogenic Groups. The group lies unconformably on the Katarchean gneisses of the Gowler shield that suffered Mesoprotozoic activation and is cut by granites of the Karelian cycle (∼ 1900 m.y. old). The group is principally built up of acid lavas (rhyolites) alternating with quartzites, feldspathic gritstones, and sandstones. The group is overlain by the Corruna strata (150 m) comprising conglomerates that alternate with sandstones, argillites and clay shales, commonly red-colored; the conglomerate pebbles contain jaspilites of the Mesoprotozoic Middleback Group that also extends in the Eyre Peninsula. The conglomerate strata are overlain by the Gawler Range rhyolites that closely associate with subvolcanic granite-porphyries.

Some of the rocks of the Eyre Peninsula are dated by the Rb-Sr method (Compston and Arriens 1968): the Moonabie rhyolites are 1630 m.y. old, argillites from the Corruna strata are 1530 m.y. old, the Gawler Range rhyolites also 1530 m.y. The Charleston and Berkitte granites cutting the Corruna strata and probably related to the Gawler Range rhyolites gave an age of 1550 and 1590 m.y. respectively. These datings indicate that all of these formations belong to the Lower Neoprotozoic.

### 3. The Lithostratigraphic Complexes of the Lower Neoprotozoic

The outline above of the principal units of the Lower Neoprotozoic in different parts of the world had the aim of justifying supposition expressed earlier on the exceptional character of the erathem in question and of demonstrating also that it forms a separate large structural stage in the Precambrian. At the same time it is evident that two sub-stages are observed in its most complete sequences, usually separated by a break, and the lower sub-stage is largely built up of acid volcanics, laterally substituted by terrigenous rocks at places and the upper one – essentially sedimentary in nature (Fig. 49). Both sub-stages are here regarded as lithostratigraphic complexes. They are called the Khibelenian and Chayan complexes according to the corresponding major units of the stratotypical Akitkan Group.

The age boundary of these complexes does not seem to correspond in every region, though this is problematic due to a great scattering of radiometric datings because of the different grade of alteration of the rocks analyzed and also due to laboratory errors. In certain cases it is difficult to interpret the age obtained on the hypabyssal granites, as it is not clear whether these rocks are synchronous to the lava outflows (rhyolites) or whether they belong to the intrusion of different formation time. Sometimes it is not evident if the granites cut both complexes or only the lower one. Finally, there are discrepancies between isotope datings and the geologic position of the rocks analyzed. Thus, for instance, the rapakivi granites of the Parguaza massif with the Rb-Sr age of 1545 m.y. are unanimously considered by students of South America to be older than the platform Roraima Group, whose age is estimated to be 1600 m.y.

**Fig. 49.** Correlation of the Lower Neoprotozoic (Akitkanian) units in some regions of the world.
*I* Baikal Mountain Land, Akitkan Range (the Akitkan Group), *II* Baikal Mountain Land, Baikal Range (the Anay Group), *III* the Aldan shield, Ulkan depression (the Elgetey Formation and some others), *IV* Central Sweden (Subjotnian), *V* the Ukraine, Ovruch Ridge (the Pugachev and Ovruch Groups), *VI* Canada, Churchill province (the Lower Dubawnt Group), *VII* Canada, Labrador Peninsula (the Letitia Lake Group), *VIII* Southern Greenland (the Qipisarqo Group), *IX* USA, Arizona (the Red Rock Group), *X* South America, the Guiana shield (the Cuchivero and Roraima Groups), *XI* Southern Africa (the Waterberg Group), *XII* Equatorial Africa, Tanzania (the Kate strata), *XIII* North-Western Africa, Reguibat (the Aioun Malek Group), *XIV* Northern Australia, the Kimberley plate (the Whitewater Group and others), *XV* North-Eastern Australia, the Leichhardt belt (the Argilla Group and others), *XVI* Southern Australia, the Goler plate (the Moonabie Group and others).
Symbols: *1* lower continental terrigenous strata, *2* acid volcanics, *3* basic volcanics, *4* upper continental terrigenous strata, *5* littoral-marine (shelf) essentially terrigenous strata, *6* stratigraphic unconformity, *7* angular unconformity (with the underlying pre-Neoprotozoic rocks), *8* rapakivi-granite, *9* granite, *10* syenite; *11* diabase (dykes and sills), *12* anorthosite, *13–15* figures give isotope datings in m.y.: *13* Rb-Sr method, *14* U-Th-Pb method on zircon, *15* K-Ar method.
Indexes: *KA* Katarchean, *MP* Mesoprotozoic, *NP₂* Middle Neoprotozoic, *NP₃* Upper Neoprotozoic. *Correlation solid line* boundary between the Khibelenian Chayan lithostratigraphic complexes: *hatched line* between the lower und upper sub-complex of the Chayan complex. Thickness of strata is approximate

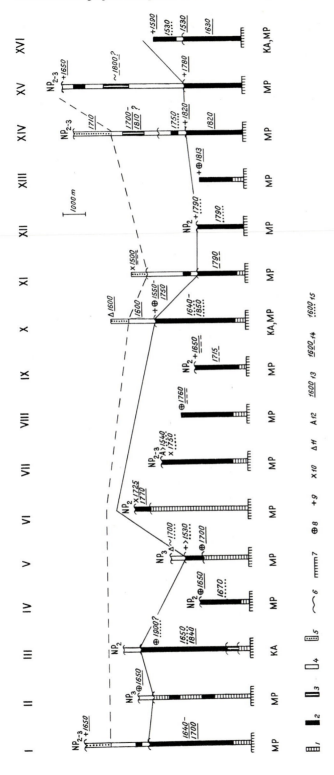

In fact the Roraima Group is never cut by the granites and lies locally on the Lower Neoprotozoic granites, including the rapakivi granites, and their isotope age is evaluated at 1750 m.y. It is highly probable that the Parguaza rapakivi are older than is revealed by dating.

The difficulties in determining the age boundaries certainly exist, but still certain facts (see for instance, Fig. 49) suggest that asynchroneity between the boundaries of the complexes does not much exceed a range of 1820–1700 m.y.

Characteristics of the formations making up both complexes were given above and here the data available will be generalized briefly.

The lower-*Khibelen complex* is made up of continental strata that belong to two lithogenic types. The principal type includes the subaerial essentially volcanogenic rocks consisting mainly of acid volcanics and their tuffs with subordinate different clastic rocks. The latter build up interbeds and bands at different stratigraphic levels, but their thickest strata commonly occur at the base of the complex. The other type is much less extended and is made up mainly of terrigenous red beds that accumulated under the continental environment; acid volcanics are obviously of subordinate significance; the basic lava sheets are locally registered. In some regions volcano-terrigenous strata of transitional type are developed, binding the two types of strata (for instance, the Anay Group in the north of the Baikal Range), volcanics are quite important among these rocks. Within the Khibelen complex local breaks are quite common. The thickness of the complex is great, reaching 5–6 thousand meters.

The acid volcanics are genetically related to different subvolcanic granites and possibly to certain mesoabyssal intrusive bodies of these rocks that cut the volcanogenic strata. Both types of rocks of the complex are cut by rapakivi and by earlier gabbroids and anorthosites that also mark a part of the "rapakivi formation".

The Khibelen complex comprises partially or fully many of the units discussed above. Here only some of them will be enumerated: in Asia: the Malaya Kosa and Khibelen Formations of the Akitkan Group, the Anay Group (excluding the upper strata) and the Teptorgo Group, distributed in different regions of the Baikal Mountain Land, the Toporikan, Ulkachan and Elgetey Formations of the Aldan shield, the major portion of the Maytyube Group of Kazakhstan; in Europe: the Sub-Jotnian of Central Sweden and its equivalents in other regions of the Baltic shield; in North America: the Lower Dubawnt and Letitia Lake Groups (Canada), the Qipisarqo Group (Greenland); in South America: the Cuchivero, Caiçara, Surumi Groups and others; in Africa: the Nylstroom Subgroup of the Waterberg Group (the Republic of South Africa), the Nagatis and Dordabis Groups (Namibia), numerous essentially volcanogenic strata of Equatorial Africa: the Kate, Ndembera, Marungu porphyries and others, the Franceville sedimentary Group (Gabon), the Ajoun Malek Group (Mauritania), the Eglab Group (Algeria); in Australia: the Whitewater Group (the Kimberley area), the Edith River Group (the Pine Creek belt) and the Argilla Group (the Leichhardt belt or Mt. Isa).

Mineral deposits in the Khibelen complex are not numerous. Certain hydrothermal uranium deposits (in Gabon and other places) are probably genetically related to acid lavas. Banded ferruginous-manganese ores rich in phosphorus (in Gabon), small sheet deposits of massive hematite or hematite-magnetite ores (in many regions) and high-alumina slates in the Baikal Mountain Land are situated in sedimentary strata.

The upper-*Chaya complex* mostly rests on the Khibelen complex unconformably (with a break or with an angular unconformity) and on the granites cutting it and locally directly on the older formations (starting from the Katarchean to the Mesoprotozoic). The unconformities are due to orogenic processes, though of relatively low intensity, and were the result of uplift of the Earth's crust, gentle folding and intrusion of granitoid (partially basic) magma. These events completed the formation of the lower complex (1820–1700 m.y. ago) and belong to the Parguazan diastrophic episode (see Martin 1974).

The Chaya complex, when represented fully, consists of two sub-complexes that are of different stratigraphic situation, the lower sub-complex being made up of continental terrigenous commonly red beds with covers of basic (basaltic) and acid lavas. The upper subcomplex comprises the shallow-marine, partially deltaic terrigenous deposits with local interbeds of carbonate stromatolite-bearing rocks.

In different regions the Chaya complex is represented by the deformed as well as by horizontal typical platform strata; these latter build up the base of a younger, Riphean cover of the large Late Precambrian platforms. The thickness of the complex varies greatly because it is reduced by erosion in many places. On the platform it does not exceed 2000 m, in foldbelts, however, it reaches several thousands of meters, for instance in the north-east of Australia, in the Leichhardt foldbelt the thickness only of the lower sub-complex exceeds 6000 m.

The following units belong to the complex discussed: in Asia: the Chaya Formation of the Akitkan Group in the Baikal area, the Birinda Formation of the Aldan shield, the upper part of the Maytyube Group of Kazakhstan; in Europe: the Tolkachev Formation of the Ovruch Group (the Ukraine); in South America: the Roraima Group (Venezuela, Guyana), the Uatuma Group (Brazil); in Africa: the Kransberg Subgroup of the Waterberg Group (the Republic of South Africa), the Tarkwa Group (Ghana etc.), the Guelb-el Hadid Group (Algeria); in Australia: the Speewah, Kimberley, and Bastion Groups (the Kimberley plate), the Red Rock Group (the Sturt plate), the Haslingden Group (the Leichhardt belt), the Katherine River Group (the Pine Creek belt).

As to mineral deposits occurring in the Chaya complex, the diamond and gold deposits of the type of fossil placers are worth mention, and are known in South America (Guyana) and Africa (Ghana).

The strata of the Chaya complex are pierced by dolerite and gabbro-diabase dykes and sills in particular. They are much more abundant than in the underlying complex, which is suggestive of the fact that they are comagmatic to the basic volcanics occurring among the sedimentary rocks of the complex. The Chaya complex is then cut by small bodies of granophyric or micropegmatitic granites and granite-porphyries. Large intrusive bodies of mesoabyssal granitoids, including rapakivi granites, are reported only from the deformed strata; they are lacking in the platform strata of the type of the Roraima Group. Isotope ages of dolerites commonly fall within the 1600–1700 m.y. range and of granitoids, including rapakivi, within 1600–1700 m.y. Deformation of the Chaya supracrustals and emplacement of granitoids occurred during the Vyborgian diastrophic cycle of the second order, that orogeny completed the Early Neoprotozoic Subera. Proceeding from this, it is possible to assume the range of formation of the Khibelen complex to be 1900(1950)–1750(1800–1700) m.y. and that of the Chaya complex 1750–1600 m.y.

# II. The Middle, Upper, and Terminal Neoprotozoic (Riphean)

## 1. General Characteristics of the Sub-Eras

**Definitions.** To the Middle Neoprotozoic are assigned the strata that were formed after termination of the Vyborgian orogeny and prior to the beginning of the Kibaran orogeny of the second order that occurred in the range of 1400–1300 m.y.; the plutonic rocks of this cycle also belong to this sub-erathem. The Upper Neoprotozoic Sub-erathem encloses the strata formed between the Kibaran orogeny and the time of termination of the Avzyan diastrophism of the second-third order that occurred approximately 1200 (1250) m.y. ago. Finally, to the Terminal Sub-erathem belong the strata formed after the Avzyan diastrophism and prior to the end of the Grenville orogeny of the first order, 1000 ± 50 m.y. ago, that terminated the whole Neoprotozoic Era.

**Distribution.** The Riphean strata are of exceptionally wide distribution all over the world. There is no necessity to enumerate the Riphean units, as these are well-known, and some typical or complete sequences of the Riphean sub-eras will be discussed later on.

**Brief Characteristics of the Sub-Eras.** It was noted at the beginning of the chapter that the three upper Neoprotozoic sub-eras are similar in many aspects and that it is difficult to separate them without detailed analysis of their stratigraphic situation, lithologic composition, sequence structure, and without isotope age values. This is true in particular of the cases when it is impossible to determine which of the unconformities observed characterise the boundary sought for. At the same time the Riphean Sub-eras are generally conspicuously different from the other Precambrian units and it seems reasonable to give their summarized characteristics.

The platform terrigenous-carbonate strata are much more abundant in these formations than in any older units, in Riphean they build up the cover of large cratons, the miogeosynclinal formations being dominant among the geosynclinal types of rocks. The latter differ from the equivalent Mesoprotozoic formations by higher abundance of psephytic rocks and by more common occurrence of thick carbonate or slate-carbonate strata. The eugeosynclinal strata demonstrate an appreciable dominance of greywacke-slate strata over volcanogenic ones and are typified by the presence of principally acid lavas and their pyroclastic products among the latter. At the same time localities are known with continental sedimentary-volcanogenic taphrogenic formations closely resembling the corresponding Akitkan formations.

The shallow-marine deposits are exceptionally widespread among the platform and miogeosynclinal formations; continental strata are mainly registered among platform formations and red beds are quite characteristic of them. Glauconite-bearing rocks are common in the littoral-marine strata. Evaporites are also much more common here than in any other older sedimentary strata of the Precambrian.

The sedimentary iron ores are known to be a perfect indicator of the geochemical environment, in these strata they are represented exceptionally by hematite (rarely hematite-magnetite) and siderite sheet deposits that are here attributed to the Urals type that appeared for the first time in the middle of the Mesoprotozoic or became distinctly recognized at that time. Dolomites are dominant among the carbonate rocks as before, but the abundance of limestones is slightly higher. The miogeosynclinal formations are usually characterized by rhythmically-bedded strata of flysch type.

All the sub-erathems comprise volcanics, but their abundance is much less than in any of the corresponding formational types of the older Precambrian units. Basalts belonging to the trappean formation (the"older trapps") are mainly present in the platform strata; andesite and rhyolite lavas, together with basaltoids, occur in the geosynclinal and taphrogenic strata; all these rocks combined represent the rhyolite-andesite-basalt formation typical of the so-called "secondary" geosynclines or of the final (orogenic) stages in evolution of a normal geosyncline.

The thickness of the Riphean sub-erathems varies greatly according to the formational (tectonic) type of strata and to the grade of preservation under erosion. The thickness of geosynclinal and taphrogenic formations of any of the sub-erathems may reach sometimes 10–12 thousand meters, but commonly it is much less, about 3–5 thousand meters.

Among the mineral deposits related to the supracrustal rocks of the Riphean sub-erathems, the following are worth mention: the same sheet iron ores of the Urals type already referred to, known in many regions, in the Southern Urals in particular; then the magnesite deposits occurring among the high-magnesial dolomites, formed at the expense of a secondary enrichment of the latter (Southern Urals, Korea etc.), the sheet deposits of phosphorites (Western and Eastern Siberia, Mongolia, India and other regions), the fossil placers of gold and diamonds in terrigenous strata (India, Brazil, Siberia, Africa), manganese ore deposits in carbonate rocks (many regions). However, the stratiform (partially hydrothermal-redeposited) copper and polymetallic (Ag-Pb-Zn) ore deposits in terrigenous and terrigenous-carbonate rocks are of the greatest economic importance (for instance in the Belt Supergroup of the U.S.A., in the Mt. Isa Supergroup of Australia and in other units in other continents). Large deposits of sulfide and native copper are related to the basalt volcanism of the trappean type and are known in some regions, in North America in particular (in the Upper Keewenawan Group of the Lake Superior area, in the Coppermine Group of the area of Coronation Bay, in the Seal Group of the Labrador Peninsula).

**Tectonic Structures.** The Riphean is typified by a high grade of differentiation of tectonic elements. The platforms and geosynclines are usually surrounded by abyssal fractures. The platform deposits make up an extensive cover or they occur in the tectonicallly isolated depressions-grabens, where they commonly represent outliers of this cover preserved from erosion. In some places they occur in the deep aulacogens and there they are of great thickness compared with that of the synchronous strata in the adjacent miogeosynclinal belts. Unlike the Mesoprotozoic strata of the same type, the Riphean platform strata are almost undeformed: they rest horizontally or form very gentle folds with an angle of inclination of the limbs of several grades; near-fracture dislocations make an exception.

The geosynclinal systems and belts are separated into zones and sub-zones by the inner uplifts and abyssal fractures, and have a different tectonic regime. In some geosynclinal systems the deep marginal troughs are quite evident, the structural elements of this type are lacking in the older mobile areas or they are faintly expressed there. Folds in the Riphean strata are of the linear type, and conform to the boundaries of geosynclinal belts and platforms or to the boundaries of the median masses, inner geosynclinal uplifts (intrageanticlines) and abyssal fractures. Dome-shaped structures are quite rare, but in some places, for instance, in Equatorial Africa and in British Columbia, some groups of small mantled domes are reported, and it is very tempting to explain their genesis by the meteorite hypothesis (of the domes or "arenes" developed in the south-west of Uganda in particular).

**Metamorphism.** The Riphean rocks of the platform areas are altered only by diagenesis or epigenesis and rarely by metamorphism corresponding to the lowest grade of greenschist facies. In the geosynclinal areas metamorphism is of a zonal nature and it increases from greenschist to amphibolite facies, locally even granitization and migmatization are reported. However, the rocks under the greenschist facies dominate among the rocks of different grades of alteration.

**Intrusive Rocks.** Plutonic processes are associated with every orogenic cycle during the Riphean. These processes were most intensive during the final cycle, the Grenville orogeny of the first order; in the geosynclinal areas they were accompanied by syntectonic and late-tectonic intrusions of appreciable mass of basic and particularly acid magma. The basic magmatism in different belts is of different significance, but the intrusive bodies of granitoids are ubiquitously widespread. Among the latter the normal calc-alkaline granites are dominant, granodiorites are of less abundance, locally the leucocratic, alaskitoid, and pegmatoid granites are also known. Numerous dykes (rare sills) of diabases and dolerites occur on the platforms, and are the youngest generation of basic dykes in the Riphean (1000–1100 m.y.).

Relatively small or medium-size intrusive bodies of granites, granodiorites, granosyenites, and syenites are related to the Kibaran diastrophism of the second order in geosynclinal areas, and in activation zones of platforms. In areas of exceptionally wide distribution of Kibaran granitoids, in Equatorial and Southern Africa, for instance, the leucocratic muscovite granites and associated muscovite and rare metal including tin-bearing pegmatites are observed among them. The typical "red granites" or alaskites and also granite-porphyries and syenite-prophyries are reported there. Certain single bodies of rapakivi or rapakivi-like granites seem also to belong to the Kibaran cycle, and are known in Sweden (the Ragunda massif?) and in the Southern Urals (the Berdyaush massif), their Rb-Sr age being 1320–1350 m.y. In some localities alkaline or nepheline syenites are encountered. In platform and geosynclinal areas dykes and sills of diabases or dolerites are abundant, characterizing the earliest generation of such dykes in the Riphean (their age is $\sim$ 1400 m.y.).

Avzyan diastrophism of the second-third order is related to weak manifestations of the intrusive magmatism. As a rule, the magmatic formations are represented by dykes and sills of diabases, dolerites, and gabbroids with an age of 1200–1250 m.y. They are distributed on the platforms and in the geosynclinal areas as well, but in the latter small or medium-sized intrusive bodies of granites occur locally.

It should be noted that the intrusive bodies of different age mentioned here provide an exceptionally valuable opportunity for isotope datings, they serve as good geologic and geochronologic bench marks that permit subdivision and recognition of Riphean sub-erathems.

Riphean endogenic metallogeny related to the intrusive complexes of the Grenville and Kibaran orogenies is extremely important and various; its examination, however, can be useful only on analyzing much material on petrology and geochemistry of magmatic rocks and that is beyond the scope of the present work. It should, however, be noted that the major portion of endogenic mineral deposits is in no way specific to the Neoprotozoic, such deposits being widely known in different Precambrian eras and in the Phanerozoic in particular.

**Fossils and Their Stratigraphic Value in the Riphean.** Stromatolites and microphytolites are sometimes quite abundant in carbonate rocks of the three upper Neoprotozoic sub-eras. If some particular cases are ignored, then it is possible to state that they are more abundant than in the Mesoprotozoic and that their content is higher in the Terminal Sub-era than in the Middle and Upper ones.

Krylov (1963) established that stromatolites from different groups of the Riphean complex of the Southern Urals are evidently different and thus recognized three successive associations or complexes of these structures. Later this worker and others (Semikhatov, Komar, Nuzhnov, Raaben) made a comparative study of Urals stromatolites and those in the Upper Precambrian strata of other parts of the U.S.S.R., mostly in Siberia, and found out that the same succession of their associations is kept through the sequence. As a result, certain new groups and forms of stromatolites were added to the Riphean associations that were not known in the Urals, and in the uppermost part of the Precambrian the fourth phytolite complex was recognized and called Vendian. Zhuravleva (1964) and other authors compiled long lists of different forms of microphytolites (oncolites and catagraphia) adding them to the list of typical stromatolites of the four associations (complexes).

Earlier the guiding forms of the phytolite complexes used to be the stromatolites with the same or similar combination of the outer features (so-called groups or formal genera), but later it appeared that many, if not all of them, have a very great range of distribution, and so the forms of the formal "species", the taxons of the lower order, appeared in later lists, exhibiting microstructural differences. The number of guiding forms also diminished appreciably and the trend to decrease further is evident. At the moment the simplified list of guiding forms of stromatolites sufficiently refined is as follows (Semikhatov et al. 1979):

1. the Lower Riphean complex: *Kussiella kussiensis* (Masl.), *Gongylina differenciata* Kom., *Conophyton cylindricum* Masl.;
2. the Middle Riphean complex: *Baicalia* (some forms), *Conophyton metula* Kir., *Jacutophyton* (some forms);
3. the Upper Riphean complex: *Inseria tjomusi* Kryl., *Jurusania cylindrica* Kryl., *Katavia katavica* Kryl., *Minjaria uralica* Kryl., *Gymnosolen ramsayi* Steinm., *Malginella malgica* Kom. et Semikh. and others;
4. the Vendian complex: *Boxonia grumulosa* Kom., *Linella simica* Kryl., *Paniscollenia emergens* Kom., *Colleniella singularis* Kom. and others. (Microphytolites are not included in the list of the guiding forms of phytolite complexes, as there are grave doubts as to the organic nature of many microstructures).

This list is not accepted by all workers. Stromatolites are known to occur mainly in Siberia in the Upper Precambrian of the U.S.S.R. and so the conclusions of Shenfil' (1978) on the vertical distribution of stromatolites in this region are of special interest.

Shenfil' also made a comparison of the Upper Precambrian sequences of Siberia and the typical Riphean sequences of the Urals. According to this author, the following forms are guides to the Lower Riphean complex: *Kussiella kussiensis, Omachtenia omachtensis, Gongylina differenciata*; to the Middle Riphean complex: *Svetliella svetlica, S. totuica* and some forms of the *Baicalia* group; to the Upper Riphean complex: *Inseria tjomusi, Jurusania cylindrica, Gymnosolen ramsayi, Minjaria uralica* and other forms of these groups. The guides are seen to be limited, are not distributed everywhere, and in certain sequences they are absolutely lacking. Thus, the practical value of using stromatolites for division and correlation of the Precambrian becomes limited due to these facts. The appearance of new forms in different regions is moreover not simultaneous mainly because of the absence of favorable conditions for the stromatolite structures to grow; and a transitional zone of different vertical range exists below the level of appearance of these new forms, separating the subsequent complex from the previous one. The definition of boundaries and correlation of the units is to be done inside this transitional zone by nonpaleontological (historical-geologic, radiological) methods (Shenfil' 1978). Shenfil' notes that the possibilities of using phytolites for division and correlation of the Upper Precambrian are not to be overestimated to the detriment of historical-geologic method and of the tasks of stratigraphy of the Late Precambrian as a whole.

I fully support this idea. It must also be kept in mind that the Riphean stromatolites were established in the Mesoprotozoic (older than 1900 m.y.) of Eastern Siberia and Canada; later it will be shown that they also occur in the strata younger than Riphean of Australia and Africa that are assigned to the Epiprotozoic. All these facts make me think that the possibilities of phytolite usage for correlation of the Precambrian strata situated in distant areas and especially on different continents are very doubtful, at least in their present state of knowledge. At the same time, the fact of successive change in stromatolite associations in the Upper Precambrian sequence is not to be refuted, but this vertical zonation is likely to be only of statistical character: the dominance of these or other forms at different stratigraphical levels. The problem of synchroneity of the strata characterized by similar associations of phytolites still remains debatable, as the geochronological (radiometric) data necessary for detailed correlation are lacking. It should be noted that synchroneity of the Neoprotozoic strata of Siberia and the Urals was usually based on K-Ar datings of glauconite, and these values are poor for establishing age and for correlation of the older rocks, and are, moreover, of different grades of alteration.

It is highly probable that the stratigraphic value of the true organic remains that are usually assigned to microfossils and microbiota is much greater than that of phytolites in the Neoprotozoic strata. Neoprotozoic microfossils include various algae, fungi, accumulations of different trichoms and the so-called acritarchs; but the stratigraphic significance of the paleontological material collected has not yet been studied properly. Many new schemes of zonal distribution of acritarchs are proposed by Soviet workers, but they are still only of local significance. A world-scale regularity that can be noted at present is the appearance of acritarchs of the *Kildinella* group and of their large forms recognized as *Chuaria*, particularly in the Upper Neoprotozoic, and their quick evolution in the Terminal Neoprotozoic. Naturally, this fact does not cover the whole stratigraphic value of acritarchs that are quite common and varied in form in the Riphean.

It should be noted that lately in many regions of the world and in Siberia in particular the abundance of local Neoprotozoic microbiota is increasing conspicuously, and that the remains found differ obviously in form, inner structure, and characteristic cell division.

All the true paleontological objects and acritarchs in the first place and then other organic remains of microbiota must certainly be studied in detail and a thorough analysis of their vertical distribution should be undertaken. Only then will it be possible to claim an important role for the paleontological method in Neoprotozoic stratigraphy.

## 2. The Riphean Stratotype and Its Principal Correlatives in Different Parts of the World

**Stratotype: the Riphean Complex in the Southern Urals and Its Correlatives in Europe.**
The Riphean complex, distributed within the limits of the Bashkir anticlinorium, has been regarded for many years as a typical or stratotypical unit of the Upper Precambrian (Upper Proterozoic, Riphean). Pioneer studies in this area were done by Garan' (1946), Olli (1948), and Shatsky (1945), who have summarized the available information. This sequence is in fact one of the best proposed for the stratotype, because it is complete and studied in detail, various phytolites and microphytofossils being observed at different stratigraphic levels. The geochronological data, however, are not sufficient, with the exception of one or two geochronological bench marks, other dating not being reliable or their interpretation being problematic, which is especially true for K-Ar datings of glauconites that principally served as a base for evaluating the age of the Riphean strata. There are moreover some doubts as to the assignation of the lower group of the Urals Riphean to the Neoprotozoic. It somehow reduces the importance of the Urals sequence as a stratotype. However, almost the same situation is observed in any other larger unit of the Upper Precambrian in different regions of the world.

The Riphean complex is a very thick and composite sequence of miogeosynclinal strata divided into four groups (upward): the Burzyan, Mashak, Yurmatau, and Karatau Groups. The first is commonly assigned to the Lower Riphean, the second and the third to the Middle Riphean and the fourth to the Upper Riphean. Their characteristics are briefly given below according to data of many workers, the principal study being done by Garan' (1946, 1963).

The Burzyan Group is divided into three formations: the Ayya, Satka, and Bakal Formations. The Ayya Formation (1700–2000 m) is of composite structure; three subformations are recognized within it, the lower one made up polymictic, commonly tufogenic sandstones, conglomerates, and subordinate alkaline basaltoids; the middle one composed of arkose and polymictic sandstones, quartzite-sandstones and conglomerates with pebbles of granite, jaspilite, basic and acid volcanics, of rare pink quartzite-sandstones; the upper represented by quartz-micaceous and phyllitoid (locally carbonaceous) slates that alternate with dolomites near the top. Certain traces of erosion are observed between the Lower and the Upper Sub-formations.

The Satka Formation (up to 2400 m) is principally composed of dolomites, of limestones in part with slate interbeds; the uppermost Bakal Formation (1300 m) is made up of phyllitoid slates with interbeds of dolomites and limestones. The Bakal

Formation is typified by a cyclic and rhythmic structure with every cycle and rhythm starting with terrigenous rocks and concluding with carbonate rocks. The Satka and Bakal Formations enclose sheet bodies of siderite and hematite ores (the Urals type of ores) and also syn-epigenetic magnesite deposits occurring in high-magnesial dolomites. Carbonate rocks of both formations contain stromatolites *Kussiella kussiensis* Kryl., *Collenia frequens* Walc., *C. undosa* Walc., *Conophyton cylindricus* Masl., and also oncolites and catagraphs, that together constitute the so-called Lower Riphean phytolite complex.

The situation of the Burzyan Group in the Precambrian sequence of the Southern Urals is defined according to the following data: the group or its Ayya Formation rests unconformably on gneisses of the Katarchean Taratash complex and its complete stratigraphic analog in the same Bashkir anticlinorium – the so-called Yamantau Group (complex) – is unconformably overlain by the Mashak Group conventionally assigned to the Middle Riphean.

The radiometric datings that give the age of the Burzyan Group show some discrepancy. The Taratash gneisses reveal a great scattering of isotope datings covering a range of 3420 to 2400 m.y. ($\alpha$ Pb and U-Th-Pb methods on zircon). The youngest values of the order of 1750–1900 m.y. are obtained by K-Ar analysis on micas from small bodies of granites and metasomatic rocks situated among the gneiss field that has no contact with the Burzyan Group. The Satka Formation of the Burzyan Group is cut by a small Berdyaush massif of composite structure largely built up of rapakivi-like granites and partially by rapakivi, gabbroids (the early phase), and nepheline syenites (the late phase). The Rb-Sr age of granites was determined to be 1560 m.y. and the age of the same granites on zircon 1570 m.y. (analysis from the VSEGEI laboratory). It permitted the assignment of the Burzyan Group to the Mesoprotozoic (Salop and Murina 1970). Still more reliable results were obtained recently (Krasnobaev 1980). They are more refined, and by applying a new laboratory technique and new methods of preparation of samples for analysis on zircon, very close values became available of the order of 1350–1400 m.y. by Rb-Sr analysis of different rocks of pluton, by the U-Th-Pb method on zircon from granites (crossing with concordia), and by the K-Ar method on micas. However, the values obtained give only the age of the igneous rocks and of the possible upper boundary of the Burzyan Group, but the group itself could be much older, for nowhere does the Berdyaush massif have contact with any of the younger Riphean groups. Thus, although formerly I shared the opinion as to the Mesoprotozoic age of the Burzyan Group, now in the light of new geochronologic data on the Berdyaush massif this assumption is substantially shaken.

The above-mentioned Mashak Group is known to be poorly studied in its stratigraphic aspect. It is largely built up of subaerial basic and acid volcanics and their tuffs, then of quartzite-sandstones (locally of red color), conglomerates, and phillites, its total thickness being 2300 m. The Rb-Sr ages obtained for volcanics are very low and have no stratigraphic significance, the K-Ar datings of the same rocks reveal very great scattering from 420 to 1335 m.y. and thus are of no value in elucidating the true age of the rocks. The Mashak Group is overlain by the Yurmatau Group; in places a significant break and even a slight angular unconformity is observed and a gradual transition is registered. This latter relationship led to the assumption that the Mashak Group is to be regarded as the lower formation of the Yurmatau Group, but this seems highly improbable.

The Yurmatau Group (up to 4000 m) is divided into three formations (up-section) the Zigalga, Zigazino-Komarov, and Avzyan. The Zigalga Formation (200–1100 m) is made up of grey quartzite-sandstones and quartzites with conglomerate interbeds in its lower part and phillitized siltstones and slates or phillites at different levels. The Zigazino-Komarov Formation (800–1600 m) is mostly comprised of different phillitoid slates with dark carbon-bearing varieties; siltstones, dolomites, and limestones are subordinate, sheet siderites are reported from the upper part. The Avzyan Formation (1100–1650 m) is built up of dolomites, limestones, and subordinate sandstone-slate rocks. Stromatolites are registered in carbonate rocks: *Baicalia baicalica* Kryl., *Conophyton* Masl., *Svetliella totuica* Kom., *Collenia frequens* Walc., *C. columnaria* Fent., and microphytolites and phytofossils are also observed. Among the latter acritarchs of primitive structure are known; *Kildinella*, *Turuchanica*, *Margominuscula* and others. Significant deposits of siderite and hematite ores are observed to be enclosed among the rocks of the formation.

The Pb-Pb age of the phosphate-bearing sandstones from the lower part of the Zigalga Formation is 1430 m.y. (Yershov et al. 1969), but this very important dating needs checking by new methods of laboratory analysis. An age of 1260 m.y. was obtained by the K-Ar method for glauconite from the upper part of the Avzyan Formation; this value probably characterizes the minimal age of the Yurmatau Group. The Yurmatau Group is cut by the diabase dykes with an age of 1200 m.y. (K-Ar analysis), but the upper stratigraphic boundary of these dykes is not known. It is very likely that they are pre-Karatau in age; in any case, dykes of that age are not certain in the Karatau Group.

The Karatau Group rests with a significant unconformity (commonly of angular type) on the older units of the Riphean complex, even on the Burzyan Group and on the Katarchean Taratash gneisses. In upward sequence six formations are distinguished in this group: (1) the Zil'merdak Formation (100–3000 m): arkose and quartz sandstones, commonly red-colored, with lenses of small-pebbled conglomerates and gritstones, subordinate siltstones and argillites; the rocks are typified by shallowwater structures, crystal molds after halite; (2) the Katav Formation (250–600 m): variegated limestones and marls; (3) the Inzer Formation (150–800 m): siltstones, slates, sandstones with glauconite; (4) the Min'yar Formation (400–800 m): dolomites, rare limestones, interbeds and lenses of cherts in the upper part; (5) the Uka Formation (400–600 m): oncolitic, pelitomorphic and clastic limestones, quartz sandstones and siltstones; (6) the Krivaya Luka Formation (250–400 m): siltstones and shales with a horizon of quartzite-sandstones approximately in the middle of the formation.

Carbonate rocks of the Karatau Group are characterized by stromatolites; the Katav Formation contains *Inseria tjomusi* Kryl. and *Jurusania cylindrica* Kryl., the Min'yar Formation *Minjaria uralica* Kryl., *Gymnosolen ramsai* Steinm. and *Katavia karatavica* Kryl., the Uka Formation *Linella ukka* Kryl. and *L. simica* Kryl.; microphytolites are also known from every formation. Acritarchs with costa sculpture: *Protosphaeridium vermium* Tim., *Kildinella veslajanica* Tim. and large form similar to *Chuaria* and trichomes of blue-green algae are known from the Katav Formation. The *Kildinella* associate with accumulations of trichomes – algae: *Leiotrichoides typicus* Herm., *Tortunema sibirica* Herm., *Polytrichoides lineatus* Herm. (Yankauskas 1978).

Glauconite from different formations of the Karatau Group is dated by the K-Ar method in the range of 610–1180 m.y. and sometimes a scattering of up to 100 m.y. is registered for the same stratigraphic level, which evidences an important migration of argon and poor stratigraphic value of the ages mentioned. The oldest datings are as follows: 1180 m.y. for the Zil'merdak Formation, 965 m.y. for the Katav Formation, and 930 m.y. for the Inzer Formation; this latter is suggested to be most true to the time of deposition of the corresponding units, but still seems to be slightly rejuvenated.

All the formations of the Karatau Group generally lie conformably and only local erosion is observed between the Min'yar and Uka Formations. Some workers wrongly attach great importance to this break because of the occurrence of the *Linella* stromatolites and microphytolites of the so-called Vendian or fourth complex in the Uka Formation and thus they exclude the Uka and the overlying Krivaya Luka Formations from the Upper Riphean and regard them as a separate unit under the name Kudash. However, unprejudiced workers think that in the Urals Precambrian the Uka and Krivaya Luka Formations are a part of the Karatau Group (as was considered earlier).

The Karatau Group and the older groups of the Urals Riphean and the diabases cutting the Karatau Group (Chumakov 1978b) are unconformably overlain by the Epiprotozoic Kurgashli Formation enclosing tillites.

In the Bashkir Urals area that is adjacent to the Southern Urals and belongs structurally to the Kama-Belsk aulacogen of the Russian plate, the platform analogs of the Lower and Upper Riphean are recognized, the Middle Riphean ones possibly being absent there (Aksenov et al. 1978, Keller and Shul'ga 1978).

To the Lower Riphean belongs the Kyrpinsk Group, known from a number of deep boreholes. In upward sequence it is divided into four formations: the Tyuryushevo, Arlan, Kaltasinsk, and Nadezhdinsk Formations. The Tyuryushevo Formation is known from drilling only in the western margin of the aulacogen and its relations with the overlying formations are not certain, according to some indirect evidence, including geophysical data, it is assumed to be situated in the lowermost part of the group sequence. The formation is made up of red sandstones, quartzite-sandstones and conglomerates that rest on the Katarchean gneisses of the platform basement. In the lower part of the formation a horizon of almost unaltered basalts is known to occur among the sandstones, the K-Ar age of these basalts being 1650 m.y. The apparent thickness (known from drilling) of this formation is just 240 m, but if it is really situated in the lower part of the Kyrpinsk Group, then a great interval (up to 2000–3000 m), not known from drilling, must exist between it and the overlying Arlan Formation, judging from geophysical evidence. The Arlan Formation (more than 1500 m) is mostly built up of argillites and siltstones with interbeds and bands of dolomites; in its lower part, as is known from drilling, strata of fine-grained pink quartzite-sandstones occur and probably wedge out in the east. The rocks of this formation contain glauconite; the K-Ar datings of this mineral gave values up to 1530 m.y. The Kaltasinsk Formation (up to 1720 m) is made up of light dolomites which yielded the Pb-Pb age of 2100 m.y. (Iskanderova et al. 1978, L. Neimark 1980, pers. comm.). Geologic interpretation of this value obtained for two groups of samples from different holes is problematic, for there are no doubts as to the Riphean age of the Kyrpinsk Group. Microphytolites are known from the Arlan and Kaltasinsk Formations. The Nadezhdinsk Formation (up to 230 m) consists of siltstones, argillites, and marls with glauconite K-Ar age in the range of 1013–1420 m.y.).

The Tyuryushevo and Arlan Formations are correlative with the Ayya Formation, the Kaltasinsk Formation with the Satka Formation and the Nadezhdinsk Formation with the Bakal Formation of the Burzyan Group of the Southern Urals.

The Gozhan Formation overlies the Nadezhdinsk Formation with a great break, and it is made up of red sandstones. Its complete analog is the Chamayevo Formation (more than 440 m), that starts the sequence of the Chishmalin Group, a close equivalent of the Karatau Group of the Upper Riphean of the Southern Urals. The Chamayevo (Gozhan) Formation is successively overlain by the Sedyash Formation (610 m) of siltstones, argillites, and marls; the Kush-Kul Formation (660 m) of quartz sandstones, and the Shtandinsk Formation (450 m) are characterized by alternation of argillites, limestones, and dolomites in the lower part and dolomites and marls in the upper part. All formations of the Chishmalin Group, except the upper (dolomitic) part of the Shtandinsk Formation, can be reliably compared with different subformations (strata) of the Zil'merdak Formation, and the dolomites completing the group with the Katav Formation of the Southern Urals (Aksenov et al. 1978).

The Chishmalik Group is cut by the diabase dykes dated at 1000 m.y. by the K-Ar method on rocks. These values define the upper age boundary of the Upper Riphean boundary of the Urals and of the Near-Urals region. It should be noted that almost the same age (1100 ± 50 m.y.) was obtained for the diabases that cut the Upper Riphean Bystrinsk Formation on Timan (this latter contains *Gymnosolen ramsai* Steinm. etc.): the datings mentioned are numerous, obtained by the K-Ar method on amphiboles in the VSEGEI laboratory.

Great similarity is observed between the Tyuryushevo red volcanogenic-sedimentary formation and the upper part of the Lower Neoprotozoic, in other words with the Chaya lithostratigraphic complex. The basalt age from the lowermost part of the Tyuryuschevo Formation (1650 m.y.) agrees well with the age of the rocks of this complex. At the same time the Tyuryushevo Formation greatly resembles the lower sedimentary-volcanogenic strata of the Ayya Formation that are separated from the overlying rocks by a stratigraphic unconformity. Pebbles of pink quartzite-sandstones are known to occur in conglomerates of the middle strata of the Ayya Formation; this kind of rock greatly resembles the red beds of the Tyuryushevo Formation. In case this correlation is true, the lowermost Riphean of the Southern Urals is to be assigned to the uppermost part of the Lower Neoprotozoic (Akitkanian).

The major portion of the Burzyan Group and of the Kyrpinsk Group is accordingly belongs to the Middle Neoprotozoic. The Berdyaush rapakivi-like granites with isotope age about 1400 m.y. correspond to the Kibaran orogeny. In this case the Mashak and Yurmatau Groups are to be attributed to the Upper Neoprotozoic; the available datings of the rocks of these groups do not contradict this conclusion. Diabases cutting the Yurmatau Group (1200 m.y.) were emplaced during the Avzyan diastrophic episode. Finally, the Karatau Group and the Chishmalin Group naturally belong to the Terminal Neoprotozoic, and the diabases cutting them (~ 1000 m.y.) to the Grenville orogeny that concluded the Neoprotozoic Era.

Thus the Lower Riphean of the Southern Urals (except its lowermost part) corresponds to the Middle Neoprotozoic, the Middle Riphean to the Upper Neoprotozoic, and the Upper Riphean to the Terminal Neoprotozoic of a general stratigraphic scale of the Precambrian.

The Urals Riphean correlates with different local units of the Upper Precambrian distributed on the Russian plate, on the Baltic shield and in Western Europe. Their correlation is to be looked for in some generalized papers (Salop 1973a, Semikhatov 1974, etc.). Aksenov (see Keller and Shul'ga 1978) gave the most detailed correlation of the Riphean strata of the Russian plate and of the Urals stratotype based on recent data. It should be noted here that in the light of new data the Dalsland Group of Sweden belongs to the Terminal Neoprotozoic: it rests unconformably on the Early Gothian (Kibaran) granites of the Ljungbergen type (1400 m.y.) and the Late Gothian (Avzyan) granites of the Hästefjorden type (1240 m.y.), but it is cut by the Sveco-Norwegian (Grenville) granites of the Bohus type (1000–950 m.y.) (Skiöld 1976). The Torridonian and Moine Groups of Scotland of terrigenous composition probably correspond to the Dalsland Group (the Moine Group was metamorphosed ~ 1000 m.y. ago; certain slightly altered Torridonian rocks reveal a very similar age) and also the terrigenous, commonly red-colored strata of the eastern Finmarken, of the Sredny and Rybachy Peninsulas at the shoreline of the Barents Sea are synchronous to the same group. The Jotnian Group of Finland, the Priozersk and Salma Groups of the Ladoga region, the Sherovich Group of the Polessian Supergroup of Byelorussia, the Kaverino Group of the Pachelma aulacogen and other continental terrigenous, essentially red-colored, strata of the southern margin of the Baltic shield and of the Russian plate are older; judging from the age of basalts (1300 m.y.) enclosed in some of these groups they belong to the upper sub-erathem (the Middle Riphean).

**The Riphean of Asia.** The sequences of the Riphean sub-erathems are exceptionally good in Siberia, where they are represented by geosynclinal as well as by platform strata. One of the best sequences is situated on the Patom Highland in the Baikal Mountain Land, and can be regarded as a Siberian parastratotype. There the Patom Group (or Supergroup) with a thickness of up to 12,000 m of miogeosynclinal type belongs to the Upper Precambrian; three subgroups are recognized within it (up-section): the Ballaganakh Subgroup, the Kadalikan Subgroup, and the Bodaybo Subgroup (Salop 1964–1967). The two lower belong to the Neoprotozoic and the upper (Bodaybo) is likely to be Epiprotozoic.

The Patom Group rocks are folded and the folds are arch-shaped with their convex side to the north, to the protruding angle of the Siberian platform. Structure and composition in part of the units of this group vary in relation to its position in one or another structural-facial zone (Figs. 50 and 51). In the near-platform margin of the Patom arch the Pri-Lena zone of the marginal trough is situated, characterized by great thickness, gentle folding (with the faint vergence toward the platform), and very low grade of rock alteration (the lowest grade of the greenschist facies).

The next three zones southward belong to the Bodaybo inner trough that is separated from the marginal trough by the Tonoda geoanticlinal rise. In the Zhuya zone joining the Tonoda rise the thickness of the strata is greatly reduced and the rocks are intensely folded and appreciably metamorphosed (middle grades of

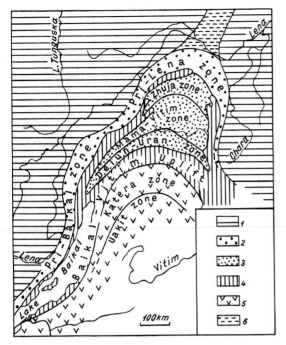

**Fig. 50.** Scheme of principal structural elements of the Baikal Mountain Land in the Neoproto-zoic. (After Salop 1964–1967).
*1* Siberian platform, *2* Baikal-Patom marginal trough, *3* Bodaibo intrageosynclinal trough, *4* intrageosynclinal uplift, *5* eugeosynclinal belt, *6* Ura aulacogen

greenschist facies). There the folds are asymmetrical, even isoclinal, tilted southward to the Mama-Vitim zone. Within the limits of the latter the strata of the two lower (Riphean) subgroups acquire a great thickness again and reveal frequent rhythmical alternation of the flysch-type rocks that are characterized by upright folds, strong metamorphism with local epidote-amphibolite and amphibolite facies and the sequence of the group is accreted with the upper Bodaybo Subgroup; in these same places the rocks are cut by granites. Still southward the Delyun-Uran zone is situated, in some parts of which abundant basic and acid volcanics are reported from the lower formations of the Patom Group. These rocks are transitional to the eugeosynclinal type that is developed in the southern part of the Baikal folded area and is recognized as the Katera Group.

This brief description of the structural-facial zonation in the Patom Group is given to show a sharply increased grade of tectonic differentiation that characterizes the Neoprotozoic geosynclines and differentiates them from the older ones. Only the strata making up the Pri-Lena zone of the marginal trough and representing most completely the two lower subgroups of the Patom group will be briefly discussed here. In Figs. 52 and 53 the sequences of the remaining zones and correlation of the rocks with the Pri-Lena zone strata are shown.

The Ballaganakh Subgroup (Group) comprises two formations: the Ballaganakh Formation proper and the Mariinskaya Formation conformably overlying it. The

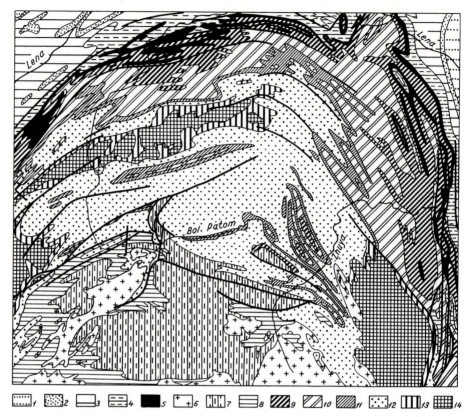

**Fig. 51.** Geological map of the Patom Highland. (After Salop 1964–1967).
*1* Ordovician, *2* Upper Cambrian, *3* Lower Cambrian, the Lena stage, *4* Lower Cambrian, the Aldan stage, *5* Eocambrian, the Zherba Formation, *6* Epiprotozoic, granite, *7* Epiprotozoic, the Bodaybo Subgroup of the Patom Group; *8–18* Neoprotozoic: *8–10* the Kadalikan Subgroup of the Patom Group: *8* the Kadalikan Subgroup undivided, *9* the Chencha and Zhuya Formations combined (the Terminal Neoprotozoic), *10* the Valyukhta, Barakun, and Dzhemkukan Formations combined (the Upper Neoprotozoic); *11,12* the Ballaganakh Subgroup of the Patom Group (the Middle Neoprotozoic): *11* the Mariinskaya Formation, *12* the Ballaganakh (s.str.) Formation; *13* the Teptorgo Group (the Lower Neoprotozoic), *14* Mesoprotozoic and Katarchean formations (undivided)

Ballaganakh Formation is made up of rather monotonous greenish-grey and light-grey inequigranular oligomictic and arkose sandstones, quartzites, gritstones, and polymictic conglomerates with interbeds and bands of dark carbonaceous slates (phyllites) and siltstones. Conglomerates are abundant in the margin of the Tonoda rise and also in both extreme flanks of the zone, occurring there mainly in the base of the formation. According to the dominant rocks and to the nature of their alternation, the formation is divided into the following sub-formations (or formations in other schemes): the Khorlukhtakh, Khayverga, and Bugarykhta. The latter is typified by abundant interbeds of quartzites. Thickness of the subformation is 400 m in the flanks of the zone and varies up to 3700 m in its central part. The Mariinskaya

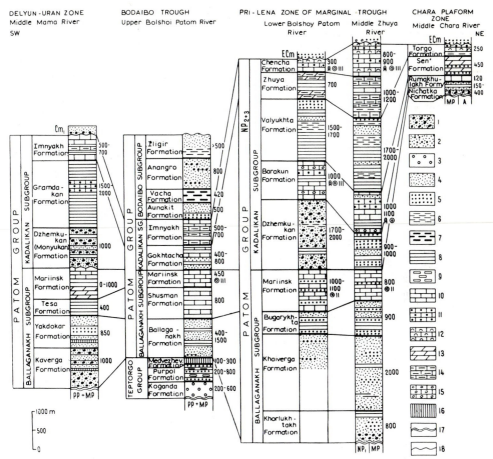

Fig. 52. Correlation of the Neoprotozoic sequences in different zones of the miogeosynclinal belt and near-platform margin of the Baikalian fold area. (After Salop 1973a).
*1* conglomerate, *2* gritstone (among sandstones), *3* arkose, *4* oligomictic and polymictic sandstone, *5* quartzite-sandstone (quartzite), *6* siltstone *7* carbonaceous slate and quartzite, *8* shale (slate, phyllite), partially siltstone, *9* chloritoid slate, *10* limestone, *11* oncolitic limestone, *12* stromatolitic limestone and dolomite, *13* dolomite, *14* marl, calcareous slate, *15* carbonate breccia and conglomerate, *16* metadiabase (amphibolite), *17* angular unconformity, *18* stratigraphic unconformity.
Indexes: *PP* Paleoprotozoic, *MP* Mesoprotozoic, *NP₁* Lower Neoprotozoic, *EP* Epiprotozoic, *ECm* Eocambrian

Formation consists of dark, almost black limestones, locally sandy and cross-bedded with interbeds and bands of calcareous and quartzite-like sandstones and also of phyllitoid slates. The limestones are replaced by the light dolomites in the flanks of the zone. Thickness of the formation varies from 300–400 m from the flanks of the zone and near the Tonoda rise up to 1200 m in the central and axial parts.

The Kadalikan Subgroup (Group) is divided into five formations, in upward sequence: the Dzhemkukan, the Barakun, the Valyukhta, the Zhuya, and the Chen-

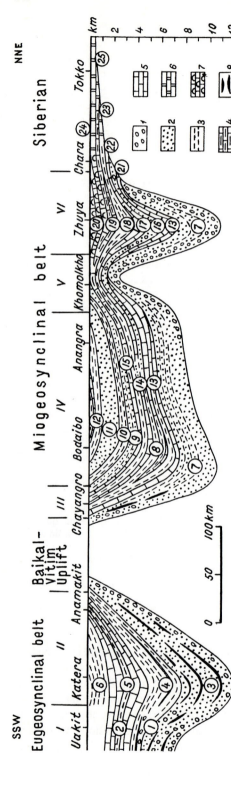

**Fig. 53.** Schematic lithofacial profile of the Neoprotozoic (Riphean) strata across the strike of the Baikal geosynclinal system. (After Salop 1964–1967). *1* conglomerate, *2* sandstone, *3* slate and siltstone, *4* calcareous slate and marl, *5* limestone, *6* dolomite, *7* phytolitic limestone and dolomite, *8* volcanics. Structural-facial zones: *I* Uakit, *II* Katera, *III* Delyun-Uran, *IV* Mama-Vitim, *V* Zhuya, *VI* the Pri-Lena zones of the marginal trough. Formations (figures on the profile are in circles): *1* Mukhtun, *2* Nerunda, *3* Ukolkit, *4* Nyandoni, *5* Barguzin, *6* Yanchuy, *7* Ballaganakh, *8* Kadalikan Subgroup (flysch, undivided), *9* Aunakit, *10* Vacha, *11* Anangro, *12* Iligir, *13* Mariinskaya, *14* Gokhtacha, *15* Imnyakh, *16* Dzhemkukan, *17* Barakun, *18* Valyukhta, *19* Zhuya, *20* Chencha, *21* Nichatka, *22* Kumakhulakh, *23* Sen, *24* Torgo, *25* Lower Tolba

cha. The Dzhemkukan Formation overlies the Mariinskaya Formation conformably without any evident break, it rests with a sharp unconformity on the lower strata of the Patom Group and occurs with an angular unconformity on the older rocks underlying the group. The formation is largely made up of light and dark oligomictic and quartzite-like sandstones with subordinate abundance of phyllitoid slate and siltstone interbeds. In the Lower Bolshoy Patom River these rocks are laterally substituted by thick strata of polymictic conglomerates that originated on the destruction of the older rocks of the Tonoda rise and of the platform basement and partially by erosion of the Mariinskaya Formation rocks and syngenetic rewashing of lithified sediments that were almost simultaneous to the conglomerates. Thickness of the formation varies from 600–700 m to 2000 m (in the Lower Bolshoy Patom River). The Barakun Formation generally lies conformably on the Dzhemkukan Formation and is made up of black limestones, locally oncolitic and breccia-like with interbeds of dark siltstones and phyllites; in places, in the flancks of the zone in particular, the interbeds of quartzite-sandstones and fine-pebbled conglomerates occur and the limestones are partially replaced by dolomites. The thickness of the formation is fairly constant, being about 1000–1100 m, and only in the flanks does it decrease to 650 m. The Valyukhta Formation exhibits gradual transitions with the underlying formation and is built up of monotonous black or dark grey phyllites with rare thin interbeds of fine-grained sandstones and dolomites. The latter locally contain stromatolites of the *Baicalia* group that are most typical of the Middle Riphean phytolite complex. The thickness of the formation varies from 500–600 m to 2000 m.

The Zhuya ("green-red") Formation is made up of greenish-grey and (or) lilac marls, clayey and sandy limestones and calcareous-clay shales and also of sandstones and siltstones with cross-bedding and ripple marks. The carbonate rocks locally contain microphytolites, mostly oncolites. The formation appears to lie conformably on the Valyukhta Formation, but it differs greatly from the latter in composition and color of rocks and also in its shallow-water environment. It should be noted here that the Zhuya Formation is commonly underlain by a horizon of quartzite-sandstones that is conventionally attributed to the Valyukhta Formation. In reality, the Zhuya Formation starts a new large sedimentary cycle that is to be recognized as a separate sub-group (or group). The thickness of the formation varies from 500 m to 1200 m, reaching its maximum in the Zhuya River basin, in the east of the zone. The Zhuya Formation is conformably overlain by the Chencha Formation that makes up the upper part of the sedimentation mesorhythm. This formation comprises the aphanite limestones of grey with lilac hue or brown color and dolomites. The oncolitic varieties of these rocks are typical of the formation, of its upper part in particular. Stromatolites are mostly known from the lower part of the formation. Some forms of the *Inseria* group characteristic of the Upper Riphean of the Urals are established there. Thickness of the formation varies from 300 to 900 m.

The Patom Group rests unconformably on older Precambrian formations, including the Lower Neoprotozoic Akitkan Group. Its relations with the Teptorgo Group of the Lower Neoprotozoic (the stratigraphic analog of the Akitkan Group) are complex: in the Pri-Lena zone of the marginal trough it overlies the Teptorgo Group with a marked break and locally even with a slight angular unconformity, but in the inner parts of the miogeosynclinal belt (in the south-west of the Mama-Vitim zone in particular) the rocks of both groups seem to exhibit a conformable mode of occurrence, although in detailed mapping a certain break is reported between them (Salop 1974 b). In the Pri-Lena the upper boundary of the Patom Group is determined according to its unconformable occurrence (but without any angular unconformity) on the Chencha limestones, the Zherba quartzite-sandstone and conglomerate of Eocambrian, this latter being overlain by the oldest beds of the Lower Cambrian (that are fossiliferous). It was mentioned above that in the inner part of the miogeosynclinal belt the sequence of the Patom Group is accreted by the Bodaybo Subgroup of Epiprotozoic.

Isotope datings are not available for any units of the Patom Group and so geohistorical criteria and sometimes phytolites may be useful in defining their situation in a general stratigraphic scale of the Precambrian. The lower, Ballaganakh Subgroup is very likely to belong to the Middle Neoprotozoic (Lower Riphean) Sub-erathem. The break between the Ballaganakh and the Kadalikan Sub-groups is suggested to be due to the Kibaran orogeny. In this case, the Dzhemkukan, Barakun, and Valyukhta Formations corresponding to the second single transgressive mesorhythm of strata are to be assigned to the Upper Neoprotozoic. The Zhuya and Chencha Formations of the next mesorhythm belong to the Terminal Neoprotozoic, judging from their situation in the sequence and from the stromatolites typical of the Upper Riphean phytolite complex that occur in the Chencha Formation. Certain changes in sedimentation at the Valyukhta-Zhuya boundary are likely to be related to the Avzyan diastrophic episode.

The Neoprotozoic eugeosynclinal strata distributed in the inner part of the Baikal Mountain Land are well exemplified in the Katera structural-facial zone of the eugeosynclinal belt (see Fig. 50). These rocks are recognized there as the Katera Group, the strata are extremely thick (up to 11,500 m) and continuous, made up of volcanogenic-sedimentary rocks (Salop 1964–1967). The units comprising the group are as follows (in upward sequence): the Ukolkit Formation (3000–5000 m) of polymictic metasandstones (greywackes) and metagritstones, locally tuffaceous with interbeds of phyllites and of volcanics of acid to basic composition, with basal conglomerates in the lower part; (2) the Nyandoni Formation (1500–3000 m) of various slates and metasiltstones with interbeds of metasandstones, graphitic slates (phyllites), crystalline limestones and with covers of acid to intermediate volcanics; (3) the Barguzin Formation (500–4000 m) of alternating limestones (or marbles) and slates and phyllites, commonly graphite-bearing; (4) the Yanchuy Formation (more than 1500 m) of phyllites and porphyroblastic ankerite-siderite slates. Certain poorly recognizable stromatolites and oncolites are reported locally from the Barguzin Formation limestones.

The Katera Group lies unconformably on different rocks of the Paleoprotozoic and Mesoprotozoic, and also on acid volcanics of the Padra Group that are conventionally compared with the Akitkan Group volcanics. The rocks of the group are cut by quite large sized granite bodies of the Barguzin intrusive complex that belongs to the Grenville orogenic cycle. Recently an age of $1000 \pm 50$ m.y. was obtained for the granites from the left bank of the Barguzin River (the Rb-Sr isochron, VSEGEI laboratory, pers. comm. of Murina and Shergina 1980). The Ukolkit and Nyandoni Formations are reliably compared with the Ballaganakh Subgroup, and the Barguzin and Yanchuy Formations with the lower part of the Kadalikan Subgroup of the Patom Group (see Fig. 53).

One more complete sequence of the Riphean geosynclinal strata is situated in the Yenisei Ridge, where a part of the Yenisei-Sayan fold belt is exposed, this portion of the belt surrounding the Siberian platform from the south-west (see Fig. 61). Two zones are established within this area: the miogeosynclinal one adjoining the platform and the eugeosynclinal one situated westward. Here only the miogeosynclinal strata will be discussed in brief; the eugeosynclinal rocks are similar to these and differ in occurrence of volcanics. Three groups belong to the Riphean in the miogeosynclinal zone, separated by stratigraphic and local slight angular unconformities; up-section these are: the Sukhoy Pit, the Tungusik, and the Oslyansk Groups (Kirichenko 1967, Semikhatov 1974 etc.).

The Sukhoy Pit Group (7000 m) rests with an angular unconformity on the Teya Group of the Mesoprotozoic and comprises six formations: (1) the Korda Formation (up to 1000–1500 m) of arkose sandstones, quartzites, dark carbonaceous phyllites, siltstones, and dolomites; (2) the Gorbylok Formation (600–1500 m) of quartz-chlorite magnetite-bearing slates and quartzites; (3) the Uderey Formation (1000–2000 m) of phyllitized slates, quartzite-like sandstones, interbeds of calcareous siltstones; (4) the Pogoryuy Formation (1000–1600 m) of siltstones, phyllitized slates, quartzite-like sandstones, interbeds of calcareous siltstones; (5) the Kartochki Formation (300–500 m), of variegated slates, limestones, calcareous breccias; (6) the Aladinsk Formation (250–800 m) of dolomites, breccia-like dolomites, magnesites. Some phosphate-bearing rocks are reported from the Uderey and Pogoryuy Formations. Glauconite from the Pogoryuy Formation is dated in the range of 750–1300 m.y. (nine determinations from different samples); then, there is one more dating of 1630 m.y. (Salop 1973a); this great scattering evidences a significant loss of argon in many samples.

The Tungusik Group (up to 5000 m) comprises the following five units: (1) the Krasnogorsk Formation (120–500 m) of slates with interbeds of quartzite-like sandstones, the slates at the base are high-alumina (with chloritoid); (2) the Dzhur Formation (500–600 m) of dolomites and limestones, locally stromatolitic, sandstones and slates, commonly red-colored; (3) the Shuntar Formation (900–2000 m) of slates locally carbonaceous or chloritoid with interbeds of stromatolitic dolomites and siltstones; (4) the Sery Klyuch Formation (500–650 m) of dolomites, limestones, rare interbeds of quartzite-sandstones and slates; (5) the Dadykta Formation (500–800 m) of slates, siltstones, and sandstones, locally red-colored, with interbeds of dolomites and limestones in the upper part.

The Oslyansk Group (up to 3000 m) comprises the Nizhneangarsk Formation (500–600 m) of slates (shales), argillites, siltstones, sandstones, locally red-colored, hematite gritstones and hematite ores, and the overlying Dashka Formation (1500–2500 m) of clay and pure limestones, dolomites, calcareous slates, marls, interbeds of sandstones and argillites; the carbonate rocks contain stromatolites and microphytolites in particular. The Epiprotozoic postgeosynclinal (orogenic) groups overlie the Oslyansk Group with a great break and angular unconformity.

Two lower Riphean groups of the Yenisei Ridge are cut by granites that are usually assigned to two complexes: the Teyan complex made up of leucocratic and gneissoid granites and pegmatites, these granites being situated among the rocks of the Sukhoy Pit Group, and to the Tatar-Ayakhtan complex composed of granites, adamellites, and diorites located among the Tungusik Group rocks. They are likely to have the same age of 950–1000 m.y. by K-Ar analysis on biotite and U-Th-Pb analysis on zircon and orthite from the Teyan granites (Volobuev et al. 1964). The Oslyansk Group is not cut by the granites, but is separated from the Tungusik Group only by a stratigraphic unconformity and thus it is possible to suggest that it is also older than the granites. The granite intrusions very probably occur during the formation of a large angular unconformity separating the Neoprotozoic (Riphean) strata from the Epiprotozoic molasse piles. Studies done by Volobuev et al. (1976) proved that the Grenville tectono-magmatic cycle was of great significance in the formation of Ripheids of the Yenisei Ridge. The Grenville granites are known to give commonly rejuvenated K-Ar datings of the order of 650–880 m.y., which is suggestive of the thermal events having taken place after deposition of molasse, during the Katangan orogeny.

In a general scale of the Precambrian it is possible to allocate the Neoprotozoic units of the Yenisei Ridge, applying mainly the geohistorical method and considering certain manifestations of diastrophism that are registered as separating the groups and being of a global scale. The Sukhoy Pit Group is very probably of Middle Neoprotozoic judging from its lowest position in the sequence and from the oldest (relict?) ages of glauconites. A general similarity is observed between its sequence and the Ballaganakh Subgroup of the Siberian parastratotype and the Burzyan Group of the world stratotype. Moreover, an occurrence of magnesite typical of the Satka Formation of the Urals is known in the Aladinsk Formation. The Sukhoy Pit Group lacks any stromatolites.

The Tungusik Group greatly resembles the Kadalikan Subgroup of the Patom Highland and its three lower formations in particular; by lithological composition it seems possible to compare the Krasnogorsk and Dzhemkukan Formations, the Dzhur and Barakun Formations, and the Shuntar and Valyukhta Formations, but it is possible that this latter formation corresponds to the Shuntar, Sery Klyuch, and Dadykta all combined. The Dzhur Formation contains the following stromatolites: *Conophyton lituus* Masl., *C. cylindricus* Masl., *C. metula* Kir., *Baicalia ampla* Semikh. and *Tungussia nodosa* Semikh.; in the Sery Klyuch Formation: *Conophyton baculus* Kir., *Baicalia unca* Semikh., *Tungussia nodosa* Semik., *Minjaria uralica* Kryl., *M. nimbifera* Kryl., *Gymnosolen confragosus* Semikh. The stromatolite assemblage of the Dzhur Formation is typical of the Middle Riphean phytolite complex and that of the Sery Klyuch associated with certain forms of the Middle and Upper Riphean. This was the basis for many workers (Semikhatov 1974, Semikhatov et al. 1978) to think that the Middle Riphean-Upper Riphean boundary lies inside the Tungusik Group. However, Shenfil' (1978) recently established the *Inseria tjomusi* Kryl. form typical of the Upper Riphean of the Urals as occurring in the Dzhur Formation in association with the Middle Riphean stromatolites, and he thinks it necessary to revise the age of the Dzhur Formation and to assign it to the Upper Riphean; in his opinion the Middle Riphean and still older stromatolite assemblages are absent in the Yenisei Ridge. This example shows the contradictions the worker in stratigraphy faces when determining age according to stromatolites.

It seems reasonable in this case to attach greater importance to geohistorical criteria in age attributions, for stromatolites cannot be an accurate indicator of age. Applying these criteria the Tungusik Group can be compared with the Middle Riphean or, in other words, it is to be assigned to the upper Neoprotozoic and the unconformities separating this group from the underlying Pogoryuy and the overlying Oslyansk Groups are to be related to the Kibaran and the Avzyan orogenies respectively. The Oslyansk Group, occurring between the Upper Neoprotozoic and Epiprotozoic, can then be regarded as Terminal Neoprotozoic. This conclusion does not contradict the presence of the Upper Riphean *Minjaria nimbifera* Semikh. and *Gymnosolen confragosus* Semikh. stromatolites in both formations. The Nizhneangarsk Formation and the Dashka Formation in part resemble the Zhuya Formation of the Patom Group, but the major portion of the Dashka Formation strikingly resembles the Chencha Formation of the same Patom Group. In particular, the carbonate rocks of both formations are as a rule wholly made up of oncolites. There is nothing surprising in this, for the Neoprotozoic strata of the Yenisei Ridge and of the Patom Highland belong to a single miogeosynclinal belt (zone) that surrounds the Siberian platform from the south (see Fig. 61). At the same time, the Dashka Formation is similar to the Uka Formation of the Southern Urals. In the opinion of some workers, the microphytolites occurring in both formations are also very similar.

In Siberia, in addition to the Riphean geosynclinal strata the synchronous platform sedimentary piles (including the aulacogenic type) are widespread. Their best sequences with abundant phytolites are situated in the Uchur-Maya region of the Siberian platform in the east of the Aldan shield. To the Riphean are assigned the following four groups separated by stratigraphic unconformities, in upward sequence: the Uchur, Aimchan, Kerpil, and Lakhanda, the three latter having earlier been recognized as the Maya Group (Nuzhnov 1967, Komar et al. 1970, Semikhatov 1974). In the area examined all the groups are typically platform, but eastward they pass into the subplatform units bearing the same names. These are the units of transitional to miogeosynclinal type that are developed in the Yudoma-Maya aulacogen; their thickness there increases by two or three times, and the rocks are characterized by higher alteration and more intensive deformation.

The Uchur Group lies unconformably on the Lower Neoprotozoic Birinda Formation examined above, starting with polymictic conglomerates recognized as the Konkulin Formation (up to 400 m); up-section follows with a break the Gonam Formation (80–140 m) made up of red or grey quartz and oligomictic sandstones and gritstones; the sequence of the group finishes with the Omakhta Formation (200 m) composed of dolomites with interbeds of sandstones and siltstones. The dolomites contain stromatolites of the *Omachtenia* group that is close to the *Kussiella* group and also the *Kussiella kussiensis* Kryl., *Gongylina differenciata* Kom., and other groups typical of the Lower Riphean phytolite complex.

The Aimchan Group lies on the Uchur Group or on the Katarchean gneisses of the Aldan shield. It starts with the Talyn sandstone-slate Formation (150 m), then follows the Svetlin Formation (up to 1500 m) of dolomites and argillites with stromatolites *Kussiella kussiensis* Kryl., *Colonella*, *Conophyton*, *Svetliella tottuica* Kom. et Semikh. *S. svetlica* Shap., among which the former is typical of the Lower Riphean and the latter is regarded by Shenfil' as a guiding form of the Middle Riphean.

The Kerpyl Group rests conformably or with a break on the Aimchan Group or directly on the basement gneisses. The Totta Formation (up to 1200 m) of dark-colored sandstones and siltstones is situated at its base, then is successively overlain by the typical Malga Formation (up to 400 m) of variegated fine-bedded limestones and by the Tsipanda Formation (200–600 m) of light silicified massive stromatolitic dolomites. Stromatolites of the Malga Formation are mainly represented by *Malginella malgica* Kom. and those of the Tsipanda Formation by *Malginella zipandica* Kom., *Minjaria sakharica* Kom., *Colonella ulakia* Kom. In the opinion of Semikhatov all these forms belong to the Middle Riphean level, while a form of the *Minjaria* group is registered there, known generally in the Urals only from the Upper Riphean, the remaining forms being "endemic". Shenfil' (1978) thinks that *Inseria tjomusi* Kryl. is also present, together with the forms mentioned above, and accordingly he assigns the Kerpyl assemblage to the Upper Riphean.

The Lakhanda Group lies on the Kerpyl Group conformably as a rule, but in places it occurs with a break and a crust of weathering. The Neryuen Formation (250–800 m), the lower one of the group, is made up of stromatolitic limestones and dolomites with interbeds and bands of variegated argillites and siltstones. Stromatolites are represented here by the following forms: *Minjaria sakharica* Kom., *Colonnella ulakia* Kom. and mainly by *Conophyton cylindricus* Masl., *C. metula* Kir., *Baicalia lacera* Semikh., *B. ingilensis* Nuzhn. Semikhatov regards this assemblage (the "Lakhanda") as typical of the upper part of the Middle Riphean; Shenfil considers it to be Upper Riphean. The Ignikan upper formation of the group (200 m) is made up of dolomites with stromatolites *Inseria tjomusi* Kryl., *I. confragosa* Semikh., *Baicalia maica* Nuzhn., which are assigned to the Upper Riphean by all workers. The Lakhanda Group is unconformably overlain by the Uy Group, which probably belongs to the Epiprotozoic.

All three upper (Riphean) sub-erathems of the Neoprotozoic are easily recognizable in the sequence of the Uchur-Maya region examined above. The lower Uchur Group is naturally compared with the Burzyan Group of the Urals because of its position in the sequence and of the occurrence of the Lower Riphean stromatolite assemblage and belongs thus to the Middle Neoprotozoic. Glauconite from the lowermost part of the Gonam Formation of this group is dated at 1500–1550 m.y. and at 1400 m.y. from the Omakhta Formation by K-Ar method. The glauconites analyzed were collected from almost unaltered platform strata and thus the oldest values obtained may be close to the true age of the rocks. The unconformity separating the Uchur and Aimchan Groups is very likely to have originated as a result of the Kibaran orogeny. The small intrusions of acid magma in the Omna rise are probably related to this orogeny, their K-Ar age being 1400 m.y.

The Aimchan Group overlies the Uchur Group and is mainly characterized by the Middle Riphean stromatolites; it is mostly Upper Neoprotozoic. Glauconite from this group is dated at 1270 m.y., a value which seems to be slightly rejuvenated. The break between this group and the overlying Kerpil Group is possibly related to the Avzyan orogeny that was accompanied by formation of diabase dykes with the K-Ar age of 1250 m.y. in the Omna rise. In the opinion of Semikhatov (1974), these diabases emplaced prior to the deposition of the Kerpil Group.

Finally, the Kerpil and Lakhanda Groups combined are to be attributed to the Terminal Neoprotozoic Suberathem (Upper Riphean). Glauconite from the Lakhanda Group is 680–1000 m.y. old, and does not contradict a certain rejuvenation of the values. This conclusion is also supported by the occurrence of the *Kildinella* acritarchs and the *Nucellosphaeridium* eukaryota form in the Lakhanda Group (Volkova et al. 1980).

In addition to the sequence of the Uchur-Maya region examined above there are other typical and complete sequences of platform strata, though of reduced thickness, in some other parts of Siberia. The strata developed in the north of the Siberian platform, in the area surrounding the Anabar shield, is an example of such a sequence (Tkachenko 1970). At the base of these strata the Mukun Group (up to 640 m) unconformably overlies the Katarchean gneisses, and is made up of red oligomictic sandstones with shallow-water structures (cross-bedding, ripple marks, sun cracks). Upward is situated the Billyakh Group that starts with the thin Ust-Ilinsk (55 m) Formation of siltstones with dolomite interbeds, then follows the Kotuykan Formation (up to 450 m) of variegated dolomites, usually stromatolitic, and finally the Yusmastakh Formation (610 m), which is dolomitic with common stromatolites, occurs. The Starorechenskaya Formation rests on the eroded surface of the rocks of the Billyakh Group, possibly belonging to the Epiprotozoic (Salop 1973 a).

Glauconite from sandstones of the Mukun Group is 1530 m.y. old, and that from the Ust-Ilinsk Formation falls within the 1350–1480 m.y. range. These units seem to belong to the Middle Neoprotozoic judging from these datings (but it cannot be excluded that the Mukun Group belongs to the Lower Neoprotozoic (to the Chaya lithostratigraphic complex to be more exact) because glauconite may be rejuvenated. This conclusion is supported by the occurrence of stromatolites in the Kutuykan Formation of the Billyakh Group that are mostly typical of the Lower Riphean phytolite complex, for instance, *Kussiella kussiensis* Kryl., *Conophyton cylindricus* Masl. etc. The lower part of the Yusmastakh Formation is characterized by stromatolites that are commonly encountered in the Middle Riphean (*Baicalia*, *Tungussia*, etc.) and the upper part contains stromatolites of the Upper Riphean and of the Vendian complex in part (*Nucleella*, *Irregularia* etc.). Glauconite from the lower portion of the Yusmastakh Formation is dated at 1200 m.y. The Billyakh Group is cut by dolerites that do not seem to penetrate into the Starorechenskaya Formation, with an age of 912 m.y. (K-Ar method). This would assign the Yusmastakh Formation to the Upper and Terminal Neoprotozoic Suberathems (or in other words to the Middle and Upper Riphean).

On the Siberian platform the reduced sequences of the Riphean strata are still more widespread, being represented by thin deposits that sometimes belong only to one of the Neoprotozoic sub-erathems. At the same time, these deposits are usually well comparable with the complete sequences of miogeosynclinal and platform types that are examined above. Commonly this correlation is possible by continuous tracing of the deposits from one region to another. For instance, the thick miogeosynclinal strata of the Patom Group developed in the Pri-Lena zone of the Marginal trough gradually wedge out toward the Siberian platform, but their uppermost strata building up the Zhuya and Chencha Formation penetrate into the platform and form the cover represented by thin monotonous dolomites (the Torgo and Tolba Formations) (see Fig. 53).

The platform and subplatform (aulacogenic) strata of the Riphean are also widely distributed in the east of Asia, on the Sino-Korean platform. This includes the Sinian complex in northern China, the best sequences being situated in the vicinity of Peking (Kao Chen-hsi 1962 etc.), and in Korea the strata of the Sanvon and Kore Groups developed within the Phennan trough in the north of the Korean Peninsula. These sequences are shown schematically in Fig. 56, Columns VI and VII. These units are seen to be characterized by alternation of strata (formations) made up of different terrigenous and carbonate rocks; the latter commonly contain phytolites, but they are poorly studied. Some strata contain glauconite of which single datings are available. All these strata are younger than the Mesoprotozoic and older than the sedimentary deposits with tillites that belong to the Epiprotozoic. In Northern China the Sinian complex rocks are overlain by the Lower Cambrian strata (the Manto Formation), but in the south of China the geosynclinal equivalents of the Sinian strata are cut by granites with an age of 900–1000 m.y. and the younger Epiprotozoic strata (comprising tillites) are represented by platform formations. In Northern Korea the Epiprotozoic tillite-bearing deposits (the Kuhen Group) rest directly on the Riphean complex and a great break is ubiquitously observed between them.

According to new geochronologic data (Chung Fu Tao 1977), the lowermost portion of the Lower Sinian (the Chancheng and Chuanlingou Formations) is possibly of Lower Neoprotozoic, as the Pb-Pb whole-rock analysis gave an age of 1920 m.y. for the Chuanlingou slates and an age of 1776 m.y. for the overlying Tanshantsu dolomites. The upper part of the lower Sinian (or the Lower Sinian s. str.) seems to overlie these rocks unconformably, with quartzites at the base. Glauconite from sandstones of the Tahunui Formation is dated at 1640 m.y. by the K-Ar method and Pb-model analysis gave an age of 1430 m.y. for the syngenetic lead mineralization in limestones of the overlying Kaoyuchuang Formation. Taking these datings into consideration, the upper part of the Lower Sinian can be assigned to the Lower Riphean. Glauconite from the uppermost Tieling Formation of the Middle Sinian is dated at 1200 m.y.; as the Sinian rocks are not deformed or altered, the datings of glauconite are suggested to be close to the age of sedimentation. Thus it is possible to state that the Middle Sinian belongs to the Middle Riphean. The Upper Sinian strata unconformably overlying the Middle Sinian are very likely to be of the Upper Riphean.

Correlation of rocks of Northern China and Northern Korea is principally possible by consideration their lithologic aspects and the stratigraphic unconformities separating the units. The lower Sanvon Group is correlative with the Lower Sinian (s. str.) of China and its lower terrigenous Chikhen Formation is also divided into two parts by a break, the lower one being comparable with the Chancheng and Chuanlingou Formations, and is to be assigned to the Lower Neoprotozoic. The overlying rocks of the Sanvon Group are possibly of the Lower Riphean, and the Kore Group, transgressively overlying them and representing the Middle Sinian, is of the Middle Riphean age. The Upper Riphean in Northern Korea seems to have been eroded previous to deposition of the Epiprotozoic Kuhen Group.

The geosynclinal analogs of the Riphean (Sinian) distributed in Southern China are represented by metamorphic strata with poorly known stratigraphy. In the U.S.S.R. corresponding strata are known in Kazakhstan, Central Asia, the Altai-Sayan fold area, in the Far East (in the Burein and Khankay regions) and in the north-east territory (in the Kolyma and Omolon massifs etc.). In most regions they are of typically miogeosynclinal nature and are represented by strongly dislocated rocks altered to a different degree, their composition and structure being different depending on the structural-facial zones in which they occur. All the rocks are composed of polymictic (greywacke), arkose and quartz sandstones, of various slates, siltstones, limestones, and dolomites commonly transformed into marbles; locally the carbonate rocks contain stromatolites that belong to the Middle-Upper Riphean phytolite assemblages. In certain zones of Tien-Shan the basic and acid volcanics are known in subordinate occurrence, and the rocks enclosing them acquire certain aspects of eugeosynclinal formations. A brief survey of the Riphean geosynclinal strata of the U.S.S.R. was given in my book (Salop 1973a, 1977a); they will not be examined here, as they are not sufficiently known, their detailed description is beyond the scope of this work and moreover no additional information is gained toward understanding the geologic history of the era in question.

In the south of Asia, in the Hindustan subcontinent, platform or subplatform strata of the Riphean are fully represented, and form several isolated troughs, grabens or aulacogens in the central and northern parts of the Hindustan platform. The geosynclinal strata are distributed only in the Himalayas, but they are poorly known and thus it is difficult to come to any definite conclusions on their stratigraphy.

The platform strata of Hindustan can easily by divided into two complexes: the lower one Cuddapah, represented by the Cuddapah Group and the upper one Vindhyan, that includes several groups, the Carnool Group in particular.

The Cuddapah Group is developed in the south-eastern part of the Peninsula, to the north of Madras, and forms an arch with is convex side to the west (Fig. 54). The rocks of the group on this convex side are slightly dislocated and altered, but in the zone of the concave side adjoining the Eastern Ghats they are folded and noticeably metamorphosed. The crystalline complex of the Eastern Ghats is made up of old, mostly Katarchean, gneisses; this complex suffered strong diaphthoresis and isotope rejuvenation 1600–1200 m.y. ago, and probably represented the remobilized basement ("roots") of the Neoprotozoic geosyncline, whose strata are now eroded. If this interpretation is accepted, the Cuddapah depresssion is to be regarded as an aulacogen on the Hindustan platform open toward the Eastern Ghats geosyncline.

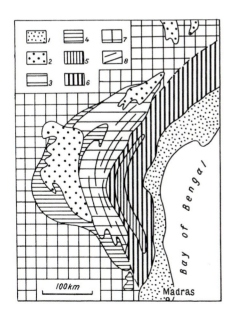

Fig. 54. Cuddapah aulacogen (Hindustan). *1* Quaternary strata, *2* the Carnool Group (the Terminal Neoprotozoic), *3* Upper Cuddapah (the Upper Neoprotozoic), *4* Lower Cuddapah (the Middle Neoprotozoic), *5, 6* crystalline basement of the Neoprotozoic geosynclinal belt of the Eastern Ghats (*5* metamorphic strata of unknown age, Katarchean or Paleoprotozoic?, *6* gneiss-granulite complex, Katarchean), *7* crystalline basement of the Hindustan platform (Katarchean), *8* fold strike in the Cuddapah Group

The Cuddapah Group is of composite structure and great thickness (up to 5500 m). It comprises four rhythmical formations separated by breaks; quartzites with conglomerate inderbeds are situated at their base, upward follow clay shales (or phyllites) and locally carbonates (see Fig. 56, Column VIII). Sills of dolerites and diabases are observed among sedimentary rocks, in the lower formations in particular; some of the sheet bodies of basic rocks of basalt appearance are possibly the covers of volcanics. Many of the rocks are characterized by an original red color, cross-bedding, sun cracks and other aspects of shallow-water environment. Dolomites enclose the *Conophyton* and *Collenia* stromatolites. Microbiota is known from cherts of the upper subformation (Vempalle) of the lower formation and certain unicellular eukaryota forms are registered among it (Schopf and Prasad 1978).

The age of the beginning of sedimentation of the Cuddapah Group is not less than 1450 m.y. judging from K-Ar dating of glauconite from sandstones of the lower formation. The upper age limit of the group falls at 1210 m.y. according to K-Ar dating of dolerites cutting the group (Aswathanarayana 1964). Thus this group can be assigned to the Upper Neoprotozoic that was formed before the Avzyan orogeny, and the dolerite dykes and sills of the Hindustan platform are related to this orogeny. It cannot be excluded, however, that glauconite age is strongly rejuvenated and this means that the lower part of the group (the two lower formations: Papagni and Cheyair or the Lower Cuddapah) belong to the Middle Neoprotozoic.

The Carnool Group is a younger unit of the Neoprotozoic, resting unconformably on different Cuddapah rocks or directly on the crystalline basement of the platform. This group is thin (about 500 m) and is made up of two cyclic formations. The lower one starts with a horizon of diamond-bearing conglomerates (the Banganapalli) with overlying strata of stromatolitic limestones and sheets of clay and calcareous shales. The upper formation is made up of quartzites in the base overlain by bituminous limestones and then by red sandstones with interbeds of limestones. The red beds contain pseudomorphoses after halite, sun cracks and ripple marks.

The Carnool Group can be compared in detail with the Chattisgarh, Bastar, and Bhima Groups and others that form isolated troughs in the central part of Hindustan (Schnitzer 1969). The carbonate rocks of the Carnool Group are characterized by abundant stromatolites, but their classification is different from that accepted by Soviet geologists and thus cannot be compared with stromatolites of the Urals and Siberia. The Carnool Group and its correlatives are very likely to belong to the Terminal Neoprotozoic (the Upper Riphean). Glauconite from sandstones of the lower part of the Chattisgarh Group is dated at 735 and 790 m.y. by K-Ar analysis, but these values seem to be strongly rejuvenated (Kreuzer et al. 1977).

**The Riphean of North America.** The Riphean piles are widespread on this continent and are fully represented. Two formational types are principally known among them: the platform (including subplatform or aulacogen) and miogeosynclinal. The typical eugeosynclinal formations of the Neoprotozoic have not been identified in North America as yet.

The miogeosynclinal strata are mainly developed in the Phanerozoic folded areas surrounding the continent. Their most complete sequences are known in the west of the continent, in the North American Cordilleras. The sequence of the Belt Supergroup, distributed in Montana and Idaho in the U.S.A. has been studied in detail

(Smith and Barnes 1966, Ross 1970, Harrison 1972, etc.). The structural-facial zonation is clearly seen in the sequence of this supergroup as in other geological complexes of the Riphean. In spite of the fact that the same four groups are recognized in these zones, their composition and structure in different zones is obviously different. Upward these groups are: pre-Ravalli, Ravalli, Piegan, and Missoula. A brief characteristic of the central part of distribution of the supergroup (the Alberton zone, the environs of the town of Missoula, Montana will be given here (see Fig. 56, column IX).

The pre-Ravalli or Lower Belt Group is largely built up of siltsones, argillites, and sandstones. Its base is not certain in the zone examined, neither is the major portion of the area in which it is distributed. Only in the extreme south-east of the region, in the Little Belt Mountains, are the crystalline rocks of the basement in the lowermost part of the group unconformably overlain by the Neihart quartzite-like sandstones (230 m) that are in turn overlain by the Chamberlain shales (500 m) and then by dark-grey limestones with slate interbeds that are divided as the Newland strata (up to 700 m). Near the U.S.A.–Canada boundary, in the Glacier National Park (Montana), the Altyn limestones (400–500 m) correspond to the Newland strata. The total thickness of all these piles does not exceed 1500 m and the Prichard Formation only corresponds to these rocks to the west; in the remaining portion of the pre-Ravalli Group distribution this formation represents the whole group, but even in places of incomplete thickness it sometimes exceeds 6000–7000 m. It is built up mainly of terrigenous rocks and is typified by monotonous alternation of siltstones, slates, and sandstones; the upper part of the formation is characterized by dominant fine-grained deposits that are comparable with the carbonate strata in the north and east of the region. In the southern margin of the group distribution the thick La-Hood psephytic strata (up to 3500 m) correspond to the Prichard Formation or to its lower part, these strata having been formed in the margin of the older uplifts composed of metamorphic basement rocks (MacMannis 1964).

The Ravalli Group in the Alberton zone comprises three formations (up-section): Burke (1000–2000 m): grey siltstones, slates and sandstones; Revett (300–600 m): grey sandstones and siltstones; St. Regis (300–500 m): quartzites, sandstones, siltstones, and slates, commonly red colored. In the eastern zones the St. Regis Formation is replaced and partially overlain by the red bed Spokane Formation, principally made up of shales and siltstones with dolomite interbeds.

The Piegan Group or the Middle Belt carbonate, as it is commonly called now, is made up of dolomites and limestones with inderbeds of carbonate sandstones, marls, siltstones, and slates; stromatolites commonly occur in dolomites. In the Alberton zone and in the more westerly zones the group is distinguished as the Wallace Formation and its thickness reaches 3500 m. In the eastern (near-platform) zones the Empire Formation of green-grey dolomites and siltstones and the overlying Helena Formation of grey stromatolitic and oncolitic dolomites and carbonate sandstones correspond to this formation: their total thickness does not exceed 1000–1500 m and locally it decreases down to 200–300 m. The Middle Belt carbonate examined is followed in every zone of the U.S.A. and still farther to the north within British Columbia (Canada) and serves as a perfect level for correlation purposes. For this reason it is recognized as a separate group, but in reality it constitutes simply an upper part of the Ravalli Group and overlies it conformably and even exhibits gradual transition into its upper formation St. Regis (or Spokane).

The Missoula Group is of more composite structure and in the area of its strato-type (the Alberton zone) it is divided into five formations (upward): (1) Miller Peak (up to 1500–2000 m): fine-bedded siltstones and slates with interbeds of carbonate rocks; (2) Bonner (400–600 m): massive quartzite-like sandstones; (3) McNamara (500–700 m): sandstones, siltstones, slates; (4) Garnet Range (up to 2000 m): silt-stones and slates with inderbeds of dolomites; (5) Pilcher (200 m): quartzites. Gener-ally these formations are traced in every zone, the most continuous being the Bonner quartzites. The lower Miller Peak Formation is divided approximately into two halves by a horizon of Purcell lavas (up to 150 m) and abundant carbonate rocks are also observed among siltstones and slates in the area of the U.S.A.–Canada boundary. Many rocks of the Missoula Group are variegated and some contain evaporites as crystal molds after rock salt and gypsum. Glauconite is typical of certain rocks.

The stratiform occurrences or deposits of copper are known from many units of the Belt Supergroup, except the Prichard and Bonner Formations; the Revett, Spokane, and Helena Formations contain very important copper deposits. The rocks of the St. Regis Formation associate with quite large hydrothermal deposits of Pb-Zn, Ag, Au-Wo, and copper ores that are locally accompanied by uranium mineraliza-tion.

The Belt Supergroup rocks are folded and cut by granites with zonal metamor-phism nearby. Additionally, westward and south-westward a general increase of metamorphism is reported up to the highest grade of greenschist facies. In zones of higher alteration the salt-bearing rocks of the Missoula Group are transformed into scapolite-bearing metamorphic schists (Hietanen 1967).

The Belt Supergroup runs continuously from the U.S.A. territory into the British Columbia region (Canada) in the north-west direction, where it is recognized as the Purcell Supergroup (or Group). The equivalents of the Middle Belt carbonate and of the overlying Missoula Group are established there; the lower units are not exposed. Many formations making up the Purcell Supergroup are given local names due to evident facial changes.

The Belt (and Purcell) Supergroup rests on a metamorphic complex of uncertain age that was mobilized and cut by granites at the end of the Mesoprotozoic, about 1900 m.y. ago. The upper age limit of the Supergroup is determined by the uncon-formable occurrence of the Middle Cambrian Flathead quartzite on its different units. In Canada the sedimentary deposits of the Epiprotozoic Windermere Group with tillites unconformably overlie the strata corresponding to the Missoula Group. Thus, the Belt Supergroup and its equivalent, the Purcell Supergroup, belong to the Neopro-tozoic. The thick groups that these comprise are likely to belong to different sub-erathems. The Belt Supergroup is known to reveal a general threefold structure and it is thus possible to suggest that every sedimentary macrorhythm (cycle) in its composition corresponds to one of the sub-erathems. The lower macrorhythm, that is the pre-Ravalli Group, starts with the clastic rocks of the Prichard Formation (and of the La Hood Formation) and finishes with the carbonate-slate rocks of the Altyn Formation. The terrigenous Ravalli Group and the overlying Middle Belt carbonate and also the carbonate-slate Miller Peak Formation of the Missoula Group conform-ably overlying the carbonate strata answer the second macrorhythm. The Missoula

Group without its lower Miller Peak Formation and without its upper Pilcher Formation constitutes the third macrorhythm, starting with the Bonner quartzite and finishing with the carbonate-slate Garnet Range Formation. The Pilcher quartzite starts the fourth sedimentary macrorhythm, the major portion of which was destroyed by erosion, though it cannot be excluded, however, that quartzite corresponds to the regressive part of the third cycle (see Fig. 56, Column IX). Generally the Missoula Group follows the regressive series of the rocks that originated under extreme shallow-water environment.

Proceeding from the cyclic structure of the Belt Supergroup it is possible to draw the boundaries of the groups according to the limits of macrorhythms. American workers state that all the groups demonstrate conformable occurrence, though certain stratigraphic breaks are reported between some of them. A rather sharp change in the type of sedimentation at the boundary of macrorhythms, however, is to reflect manifestations of orogenic processes consisting in a change of directions of oscillating movements.

The isotope datings can be analyzed to elucidate the problem of assignation of every cyclic group to a certain Neoprotozoic erathem. Rb-Sr datings of clay shales and Rb-Sr and K-Ar datings of glauconite from the pre-Ravalli Group rocks give values of the order of 1300–1350 m.y. (Obradovich and Peterman 1968), but these values are undoubtedly rejuvenated, as this group, recognized as the Aldridge Formation in the Purcell Mts, is cut by the diabase sill with an age of 1350–1400 m.y. (K-Ar method) and by a stock of the Hellroaring Creek granophyric granodiorites that yielded the Rb-Sr age of 1335 m.y. (Ryan and Blenkinsop 1971).The same values (1340 m.y.) were obtained for hydrothermal lead mineralization in the Aldridge Formation (Pb-model method on galenite). The diabase and granodiorites do not seem to penetrate higher than the pre-Ravalli Group, and mark the Kibaran orogeny that terminated the Middle Neoprotozoic Subera. At the same time there is an U-Th-Pb dating of 1525 m.y. that was obtained on zircon from the augen gneisses, these latter are regarded as highly altered terrigenous rocks of the group or as granites cutting them (Reid et al. 1970). Many workers, however, suspect that in the zone of high alteration (in Idaho) some remobilized basement rocks occur together with the strongly metamorphosed Belt rocks, and these former, to which the augen gneisses belong, can hardly be distinguished from the latter. Thus it is very probably that the pre-Ravalli Group belongs to the Middle Neoprotozoic, but it cannot be excluded that its lower portion many be Lower Neoprotozoic.

Rocks of the second macrorhythm (the Ravalli and Piegan Groups ant the Miller Peak Formation) are poorly dated; there is only one K-Ar dating of the Purcell basic lavas enclosed in the Miller Peak Formation, where the age of 1150 m.y. obtained seems to be rejuvenated, as glauconite from the rocks of a higher position in the Missoula Group gave the same age. At the same time Semikhatov determined *Collenia frequens* and *Conophyton cylindricus* Masl. stromatolites in the Middle Belt carbonate and its analogs, which occur in the Lower- and Upper Riphean phytolite assemblages. This macrorhythm is probably Upper Neoprotozoic (Middle Riphean).

For the third macrorhythm, datings are available for glauconite from the McNamara Formation: 1150 m.y. and from the base of the Pilcher Formation: 940 m.y. (Obradovich and Peterman 1968). An age of 1000–1050 m.y. was obtained for hydrothermal mineralization in the rocks of the Missoula Group (K-Ar and U-Th-Pb methods); this is the value that reflects the end of sedimentation of the Belt Super-group and it corresponds to the Grenville orogeny. Thus the third macrorhythm belongs to the Terminal Neoprotozoic (Upper Riphean).

It should be noted that abundant stromatolites are determined in carbonate rocks of the Missoula Group, but their classification is different from that applied by Soviet workers and for this reason they cannot be used to correlate this group with the Riphean sequences of the U.S.S.R.

Still southward in the North American Cordillera, the Harrison Group, distrib-uted in the boundary areas of Nevada, Utah, and Idaho, and also the Big Cotton-wood Group developed in Utah are compared with the Belt Supergroup or with its lower part. Both groups are made up of quartzites, sandstones, siltstones, and do-lomites with rare basic lava sheets. They rest on the Katarchean gneisses complex highly altered during the Mesoprotozoic and Neoprotozoic and the Big Cottonwood Group is unconformably overlain by the Epiprotozoic Mineral Fork tillite. In Eastern California the Pahramp Group, excluding its upper Kingston Peak Formation with tillites, is an equivalent of the middle and upper parts of the Belt Supergroup. The Kingston Peak Formation lies unconformably on the underlying rocks and belongs to the Epiprotozoic. Two microrhythms are distinguished in the Pahramp Group, each starting with terrigenous rocks (sandstones and slates) and finishing with stro-matolitic dolomites. Dolomites contain the *Baicalia* sp. and *Jacutophyton* sp. stroma-tolites.

Correlation of the Riphean strata of the Cordillera fold belt and those of its eastern margin within the limits of the North-American platform is of great interest. In the north-west of the Canadian shield, in the area of Great Bear Lake and the Coppermine River situated to the east from the Belt-Purcell belt four platform groups of the Riphean are recognized, up-section they are: the Hornby Bay, Dismal Lake, Coppermine River, and Rae Groups (Baragar and Donaldson 1973).

The Hornby Bay Group overlies the slightly dislocated variegated Lower Neopro-tozoic porphyries (their age is 1750 m.y.) subhorizontally and with a break, and rests on the Mesoprotozoic Cameron Bay Group of strongly folded strata with a signifi-cant angular unconformity. Strata of cross-bedded sandstones with conglomerate lenses (1200 m) are situated in its lower part and upward follow dolomite strata (600 m) with local stromatolites and then strata of fine-grained sandstones, siltstones and slates with ripple marks and sun cracks (500 m). The group is cut by the diabase dykes with K-Ar age of 1150 m.y. and 1350 m.y. This latter value was obtained for the older generation of dykes and indicates the minimal age of the Hornby Bay Group and permits assigning it to the Middle Neoprotozoic, which opens a possibility of correlating it with the pre-Ravalli Group of the Belt Supergroup.

The Dismal Lake Group lies on the underlying rocks with a stratigraphic or faint angular unconformity and starts with the strata (400 m) of quartzite-sandstones, in the upper part alternating with carbonate slates with crystal molds after the rock salt and gypsum. The strata of dolomites of massive and fine-bedded nature with com-monly occurring stromatolites and oncolites (1000 m) overlie these rocks; at different levels intraformational dolomite conglomerates and dolomitic interbeds with cherty concretions are reported.

In the lower part of the group, in cherts among dolomites, certain microbiota is reported, some forms of which resemble those from Mesoprotozoic strata in the

Belcher Islands (Canada), and other microbiota from the Lower Epiprotozoic Bitter Springs Formation (Australia) (Horodyski and Donaldson 1980).

The Coppermine River Group, mainly made up of basalts, rests conformably on dolomites of the Dismal Lake Group and in reality constitutes the upper part of the latter (or probably it is more correct to state that the Dismal Lake Group represents its lower part). The group is divided into two formations. The lower one, Copper Creek (up to 3200 m), consits of approximately 150 basalt sheets with massive lavas in the base of each and amygdaloidal lavas in their upper parts. The sheets are sometimes separated by interbeds of red sandstones with conglomerate lenses. The overlying Husky Creek Formation (up to 1500 m) is typified by the alternation of red cross-bedded sandstones and siltstones with subordinate basalt sheets. Basalts from the lower formation are dated in the range of 770–1250 m.y. by K-Ar analysis and at 1280 m.y. by Rb-Sr analysis; this latter value seems to be very close to the lava outflow. Judging from datings, the Dismal Lake and Coppermine River Groups are to be attributed to the Upper Neoprotozoic and thus are to be compared with the second macrorhythm of the Belt Supergroup (the Ravalli + Piegan + Miller Peak). The occurrence of basalts in the uppermost Belt macrorhythm that correspond to the plateau-basalt of the Coppermine River Group is significant (see Fig. 56).

Earlier it was suggested (Baragar and Donaldson 1973) that during the break separating the Hornby Bay and the Dismal Lake groups the basic magma was emplaced with formation of the Muskox stratified massif of basic-ultrabasic rocks. Recent studies of Hoffman (1980), however, showed that the intrusive rocks of this massif are comagmatic to basalts of the Coppermine River Group and thus were formed simultaneously with the latter. K-Ar datings on micas and whole-rocks from the massif yielded values in the range of 1150–1250 m.y., that is almost the same as those for basalts.

The Riphean sequence is finished with the Rae Group, resting on the Coppermine River Group with a break and in the lower part it is made up of strata (500 m) of cross-bedded sandstones, siltstones and slates, of red color in places, upward strata (up to 900 m) of alternating stromatolitic dolomites, red or variegated shales (with local gypsum inclusions), siltstones, and sandstones are situated. The group is cut by sills and dykes of dolomites that yielded an age of the 650–750 m.y. range. The rocks of the group and dolerites are unconformably overlain by quartzite-like sandstones that are conventionally attributed to the Cambrian. The Rae Group can probably be compared with the upper part of the Missoula Group of the Terminal Neoprotozoic. Young et al. (1979) gave an interesting detailed correlation of the Rae Group and the platform strata of the Shaler Group developed on the Victoria Island (the Canadian Arctic) and also the subplatform deposits of the Mackenzie Supergroup distributed in the mountains of the same name and the geosynclinal piles of the Pinguicula Group (an analog of the Missoula Group) in the Wernecke Mts (the Northern Cordillera) that underlie the Epiprotozoic tillites of the Rapitan Group.

In the Churchill tectonic province of the Canadian shield, two platform groups belong to the Middle Neoprotozoic: the Athabasca in the area of the lake of the same name and the Upper Dubawnt of the more northern areas of the shield in the basin of the Thelon River and Dubawnt Lake (Fraser et al. 1970). Both groups are of similar composition and structure: their lower parts consist of quartzite-sandstones, conglomerates, and siltstones commonly red-colored, lying unconformably on the

Lower Neoprotozoic (the Martin Lake and Lower Dubawnt Group); upward follow dolomites with stromatolites. The Upper Dubawnt Group is cut by diabases with an age of 1415 m.y., with a break it is overlain by basalts that are the correlatives of the corresponding rocks of the Upper Neoprotozoic Coppermine River Group. The basalts are cut by diabase dykes dated at 1150 m.y. Diabases piercing the Athabasca Group revealed an age of 1200–1280 m.y. (K-Ar method).

In the Superior tectonic province the Keeweenawan Supergroup belongs to the Upper and Terminal Neoprotozoic and forms a large superimposed syncline in the area of Lake Superior (Hamblin 1961). The supergroup lies with a sharp unconformity on the older Precambrian formations, including the Mesoprotozoic Animikie rocks. At the base of the supergroup there are red or grey sandstones and siltstones with rare bands of stromatolite dolomites. These strata are recognized as the Lower Keeweenawan Group (or the Sibley Formation); they are rather thin, only about 100–340 m and locally only several tens of meters. The sedimentary rocks are overlain by extremely thick (up to 6000–7000 m) essentially volcanogenic strata consisting of numerous sheets of basalts, of rare acid volcanics alternating with subordinate interbeds and bands of red sandstones, conglomerates and tuffs. Large deposits of native and sulfide copper associate with basalts. The volcanics enclose sills of diabases and a large differentiated Duluth pluton of lopolith-like appearance and formed of gabbroids and granophyres. These volcanogenic strata are usually regarded as the Middle Keeweenawan Group, or they are divided as the Portage Lake Formation. Some workers give it the status of the upper formation of the Lower Keeweenawan Group and it seems more reasonable if the conformable relation with the Sibley Formation is considered. The volcanics are overlain by the Upper Keeweenawan (Oronto) Group with an erosion or slight angular unconformity, the group is a thick pile (up to 3500 m) of red sandstones, siltstones, and conglomerates with local interbeds of cuprous sandstones. Microbiota is registered in cherty slates.

The Upper Keeweenawan Group is unconformably overlain by the red cross-bedded essentially quartz sandstones of the Jacobsville or Bayfield Group (Formation) that are probably Eocambrian; the Upper Cambrian sandstones unconformably overlie them.

Volcanics from the Portage Lake Formation were dated in the range of 1150–1250 m.y. and gabbroids and granophyres of the Duluth massif in the range of 1100–1270 m.y. by K-Ar, Rb-Sr and U-Th-Pb methods. These datings indicate an assignation of the Middle (or Lower and Middle) Keeweenawan Group to the Upper Neoprotozoic and its correlation with the Coppermine River Group of the northwestern part of the Canadian shield, as it greatly resembles this latter. Slates from the Upper Keeweenawan Group (the Nonesuch slates) are dated by Rb-Sr analysis at 1075 m.y. This group seems to belong to the Terminal Neoprotozoic, but at the same time it resembles the upper part of the Coppermine River Group (the Husky Creek Formation).

A very interesting and typical sequence of Neoprotozoic strata is known in the Labrador Peninsula in the Nain tectonic province adjacent to the Grenville belt. This is the Seal Group that occupies a great syncline, the southern limb of which is tilted and ruptured by thrusts in the area adjoining the fracture zone. According to Brummer and Mann (1961), this group includes the following formations (in upward sequence): (1) Bessie Lake (1400–4000 m) – light-grey and pink quartzite that is locally feldspathic, intercalated with amygdaloidal diabases and underlain by conglomerates; (2) Wuchusk Lake (6500 m) – interbedded cherty slate, phyllite, quartzite, and carbonates with stromatolites and oncolites; (3) Whisky Lake (1000 m) – red shale and grey phyllite with subordinate red argillite and quartzite; (4) Salmon Lake (1000 m) – inderbedded red shale and green-grey slate with thick flows of amygdaloidal diabase (basalt); (5) Adeline Lake (450 m) – grey and red shale and quartzite; (6) Upper red quartzite (more than 650 m) – red quartzite with quartz pebbles. All these formations are intruded by diabase and gabbro-diabase sills. Important hydrothermal copper deposits (related to basalts) are present in the Salmon Lake Formation.

The Seal Group resembles platform complexes by its character and paragenesis of the rocks, but it differs in its great thickness (up to 11,000 m), rather strong deformation, and locally in noticeable metamorphism. Thus the group is transitional in character, between platform and miogeosynclinal deposits. Dislocation and alteration grade of rocks increases southward (SSE), approaching the fracture zone at the limit with the Grenville belt. It is suggestive of the fact that earlier the typical geosynclinal strata synchronous to the Seal Group occupied the Grenville belt, but were then uplifted and destroyed by erosion. Certain small outliers of these rocks are known to be preserved only in the tectonic settlings among the crystalline rocks of the basement reactivated during the Grenville orogeny. The Seal Group unconformably overlies porphyry of the Lower Neoproterozoic Letitia Lake Group; its upper boundary is fixed by the Grenville orogeny ($\sim$ 1000 m.y.). Thus the time of its formation falls within the interval of 1600–1000 m.y. The Rb-Sr age of basalts from the Salmon Lake Formation is 1278 m.y. (Smyth et al. 1978), that is they are of the same age as those from the Keeweenawan Supergroup and the Coppermine River Group that are also well-known for their industrial copper mineralization. It seems reasonable, therefore, to attribute the upper part of the Seal Group to the Upper Neoproterozoic, and its lower part to the Middle Neoproterozoic. The boundary of the sub-erathems is problematic, as there are no data on the breaks between the formations. The group is of a cyclic structure and the boundary is probably drawn at the base of the red bed Whisky Formation (see Fig. 56, Column XII).

The Neoproterozoic or Riphean (to be more exact) rocks are also known in Greenland. These are the Thule and Gardar platform Groups and the Eleonora Bay and Hagen Fjord (in part) geosynclinal Groups. The Thule Group extends along the north-western shore of the island and comprises two subgroups separated by a stratigraphic unconformity. The lower subgroup (1150 m) is made up of quartzite-sandstones in the lower part and of shale and dolomite with stromatolites in the upper part, and rests on the metamorphic rocks of the basement cut by diabase dykes with an age of 1620 m.y., but the group itself is pierced by sills (and probable flows) of younger diabases dated at 1220 m.y. by the K-Ar method. The upper subgroup ($\sim$ 800 m) starts with red dolomite and siltstone, then follow strata of grey dolomite and finally strata of red sandstone; the rocks are characterized by pseudomorphs after halite. The upper subgroup is cut by diabase dykes, for one of which an age of 710 m.y. was obtained. Judging from datings the lower subgroup is very likely to belong to the Upper Neoproterozoic and the upper to the Terminal Sub-erathem and even to the Epiprotozoic but the latter is less possible.

The Gardar (Eriksfjord) Group is developed in the southern extremity of Greenland, and lies subhorizontally on the Ketilidian complex of the Mesoprotozoic and on the Lower Neoprotozoic intrusive rocks, including the rapakivi granites, which are 1750 m.y. old; at the same time it is cut by syenites dated at 1300 and 1160 m.y. by the Rb-Sr method (Blaxland et al. 1978). The group is typified by alternation of red sandstone and quartzite with thick basalt flows. In composition the group greatly resembles the Upper Neoprotozoic volcanogenic strata of Canada discussed above, but it may be older – Middle Neoprotozoic.

The Eleonora Bay and Hagen Fjord Groups distributed in different regions of the Eastern-Greenland foldbelt are of miogeosynclinal nature. The first is composed of slightly altered sandstone, quartzite, argillite, and of carbonate rocks with phytolites mainly occurring in the upper part of the sequence. Certain rocks are red-colored, cross-bedded, and exhibit ripple marks. *Inseria* and *Jurusania* stromatolites are determined in the rocks (Bertrand-Sarfati and Caby 1974), and are typical of the Upper Riphean phytolite complex. The base of the group is not certain, but it is definitely younger than the crystalline rocks of the basement that suffered recurrent metamorphism during the Mesoprotozoic; the group is unconformably overlain by tillite-bearing deposits of the Epiprotozoic Merkebjerg Group. On the evidence of stromatolites, the upper part of the group is to be assigned to the Terminal Neoprotozoic; its lower part in this case may be Upper or Middle Neoprotozoic.

The Hagen Fjord Group (or Supergroup, which is more correct) consists of two separate units of the group rank that are separated by an unconformity. Only the lower group belongs to the Riphean, and is composed of terrigenous and carbonate rocks altered to greenschist facies. The base of this group is not known, but it is younger than the so-called Mesoprotozoic (?) pre-Carolinian metamorphic complex; it is overlain unconformably by tillites of the upper Epiprotozoic group. The portion of the group exposed is a possible correlative of the lower part of the Eleonora Bay Group.

The typical sequences of Riphean platform strata are also known in the south of North America: in the Mountains and in the adjacent midcontinental regions in the U.S.A. The most interesting among these is the Grand Canyon Supergroup sequence distributed in the Colorado River basin (Arizona). At the end of last century Wallcott proposed this as a stratotype of the Algonkian. This supergroup lies horizontally and is divided into three groups (up-section): Unkar, Nankoweap, and Chuar separated by unconformities (Ford and Breed 1973). The Unkar Group lies with a sharp angular unconformity on the older Vishnu polymetamorphic complex. The lower part of the group consists of thin (up to 60 m) strata of stromatolite-bearing dolomites and limestones with the basal horizon of red sandstone, but its major upper portion (up to 1800 m) is composed of red sandstone and argillite with flows of basic lava among them; the thickest flow is situated at the top of the group and lies on the erosional surface of sandstones. Diabase sills occur at different levels. The Nankoweap Group is only 100–150 m thick and is largely built up of red cross-bedded sandstones with rare interbeds of argillite; limestone interbeds are also registered in its upper part. In some stratigraphic schemes this group is regarded as the upper formation of the Unkar Group.

The Chuar Group is divided into three formations. The lowest, Galeros (1320 m), consists of massive grey dolomite with interbeds of limestone, shale, and sandstone;

the middle unit, Kwagunt (675 m), is red sandstone and argillite with dolomite interbeds; the upper formation, Sixty Miles, is a thin unit (40 m) composed of brecciated rocks and granule conglomerate with clasts derived from the underlying rocks. The Galeros Formation contains *Gymnosolen* and *Inseria* stromatolites typical of the Upper Riphean phytolite complex; the Kwagunt Formation has *Boxonia* stromatolites mainly occurring in the Eocambrian (Vendian) strata of the U.S.S.R., but this latter form has not been accurately determined. The Galeros and Kwagunt Formations also contain large *Chuaria* acritarchs that are especially characteristic of the Terminal Neoprotozoic. Well-preserved microbiota (filaments and spherical bodies) was reported in cherty oolites from the Kwagunt Formation (Schopf et al. 1973).

The age of the Vishnu complex underlying the Grand Canyon Subgroup is not known, but on the basis of isotope dating its rocks were recurrently metamorphosed at the end of Mesoprotozoic and then underwent isotope rejuvenation some 1350–1400 m.y. ago (K-Ar datings of micas). These values reveal a possible lower limit of the platform strata, that is they are probably younger than Middle Neoprotozoic. Basic lava from the upper part of the Unkar Group gave a K-Ar age of 845 m.y., but many workers think that this value is strongly rejuvenated and that the real age of lava must be about 1200 m.y. (Schopf et al. 1973). Thus the Unkar Group is probably to be attributed to the Upper Neoprotozoic, and the Chuar Group to the Terminal Neoprotozoic.

The Grand Canyon Supergroup is compared with the Apache Group and the Troy strata ("Troy quartzite") developed in a more southerly region of Arizona, in the Mazatzal Mountains (Shride 1967). The Apache Group lies unconformably either on the Mazatzal quartzite (up to 1200 m) of the Middle Neoprotozoic (it is cut by monzonite with an age of 1420 m.y. determined by Rb-Sr method), or on red porphyry of the Lower Neoprotozoic Red Rock Group (1700–1650 m.y.), or directly on metamorphic rocks of the Pinal Group (the Big Back comlex). The group is thin (up to 530 m) and in the lower part is made up of red and grey clastic rocks overlain by stromatolite-bearing dolomite and then by thick basalt flow (up to 120 m). The dolomite contains *Conophyton cylindricus* Masl. and *Tungussia* sp. Semikh. stromatolites (Cloud and Semikhatov 1969), that are common in the Middle Riphean of the U.S.S.R. and also known from different Neoprotozoic levels. The Troy strata (up to 400 m) rest with a break on the Apache Group and are made up of quartzite, arkose, and conglomerate with ventifacts among their clasts. They are cut by diabase with an age of 1150 m.y. (U-Th-Pb method on zircon).

The data examined suggest that the Apache Group is probably Upper Neoprotozoic and the Troy strata – Terminal Neoprotozoic. Basic lava is recorded in the upper part of the Apache Group and is known to be characteristic of the Upper Neoprotozoic of North America.

The Troy strata and the Chuar Group are correlatives of the Uinta Group spread in the mountains of the same name in Utah. This group if formed of thick strata (several thousands of meters) of terrigenous red rocks represented mainly by quartzite and arkose in the lower part (the Mutual Formation) and by shale and siltstones in the upper part (the Red Pine Formation). The shale contains *Chuaria* acritarchs. Rb-Sr age of shale is 950 m.y.

**Riphean of South America.** The Riphean strata in this continent are known in many regions, in Brazil in particular, but their stratigraphy is poorly studied and debatable; even the Riphean age of these rocks is sometimes problematic. The division is complicated because the major portion of the continent, with the most probable Riphean strata, is situated in the zone that was subjected to tectono-thermal activation of the Baikalian (Late Brazilian) orogeny that resulted in isotope rejuvenation of the Precambrian rocks, including the Riphean strata. For this reason the Riphean rocks of South America are poor for studying general regularities of composition and structure of the Neoprotozoic units under examination. It seems reasonable, however, to mention here that mainly terrigenous and carbonate-terrigenous strata in part are represented among the rocks that can be assigned to the Riphean by their structural and stratigraphic situation, and that platform as well as miogeosynclinal formational types are reported among these rocks.

The Cuiaba Group, an example of miogeosynclinal strata, is several thousands of meters thick, and extends in the Paraguay-Araguaia foldbelt that delimits the West-Brazilian (Guapore) craton from the east (Almeida 1968, Almeida et al. 1973). It is made up of rhythmically bedded flyschoid rocks represented by slate and siltstone interbedded with subordinate quartzite and limestone. The rocks are altered to greenschist facies, strongly folded and cut by various granites. The terrigenous piles of the Jangada Group containing tillite overlie unconformably the Guiaba Group, then follow the carbonate rocks (with stromatolites) of the Araraz Group, and finally the red sandstone of the Alta-Paraguay Group; these three groups are probably Epiprotozoic. Granite from the Paraguay-Araguaia belt is dated at 500–600 m.y., but these values seem to give not the true age of granites but the time of their activation.

The Lavras Group is an example of Riphean of the platform type, forming the lower cover in a great superimposed Selitri depression situated in the Bahia state of Brazil. The group varies in composition and thickness in different parts of the depression. In the eastern flange it can be divided into three formations (up-section): Tombador, Paraguazu, and Morou-de-Çapeo (Beurlen 1970). The Tombador Formation (200 m) lies with a sharp angular unconformity on the Katarchean (?) gneiss of the basement and its stratigraphic analogs on the Mesoprotozoic Minas Group or on the granite cutting it. It is made up of inequigranular arkose sandstones and quartzite; conglomerate is registered locally in its base. The Paraguazu Formation (up to 200 m) consists of shale and fine-grained sandstone or siltstone. The Morou-de-Çapeo Formation (100 m) is made up of light quartzite-sandstone. In the central part of the Selitri depression the Lavras Group is unconformably overlain by the Epiprotozoic Bambui Group that is locally underlain by tillites. The Lavras Group seems to be younger than the post-Minas granites with an age of 1300 m.y.; it might be Middle or Upper Riphean.

**The Riphean of Africa.** Different formational types represent the Riphean strata in Africa as in other continents; these are the geosynclinal, platform, and aulacogenic types; sedimentary-volcanogenic strata are also reported, and belong to the taphrogenic type. Many local Riphean units distributed in different parts of the continent are examined in my book dealing with the Precambrian of Africa (Salop 1977b). Here only some few examples will be given, as being the most demonstrative. The Kibara Supergroup, for example, is developed in Equatorial Africa and is commonly regarded as a regional stratotype of Neoprotozoic of the entire continent. This unit will be the first in a brief review of characteristic Riphean units in Africa.

The Kibara Supergroup extends in south-eastern Zaïre, where it forms a major portion of the Kibaran foldbelt that surrounds the older Neo-Kasai platform from the east (see Fig. 62). It represents a thick sequence (up to 12,000 m) of strongly folded sedimentary rocks, exhibiting zonal metamorphism from greenschist to amphibolite facies and locally cut by granite (Cahen 1954, Cahen and Snelling 1966, Cahen and Lepersonne 1967). Four groups are recognized in the supergroup (in upward sequence): Kiaora Mt., Lufira, Hakansson, and Lubudi. The Kiaora Mt. Group rests on the polymetamorphic crystalline rocks of the basement that suffered the youngest thermal processes previous to deposition of the group in question some 1900–1850 m.y. ago. The group is mainly composed of phyllite and micaceous slate, partly of quartzite; acid lava flows and marbled carbonate rocks are also present. The Lufira Group (1300–5500 m) lies unconformably on the previous group and is made up of quartzite, commonly cross-bedded with bands and interbeds of quartz-micaceous slate and phyllite that usually contain graphite; a horizon of conglomerate is situated at its base and thick strata (up to 1100 m) of basic lava with quartzite interbeds at its top. The Hakansson Group (1500–4000 m) is separated by a break from the underlying group and consists of dark graphitic phyllite with light quartzite inderbeds in the lower part that pass into quartz conglomerate in the base of the group. The Lubudi Group (several thousands of meters) rests unconformably on different horizons of the Hakansson and Lufira Groups. Its lower part contains dark arkose with conglomerate lenses that upward passes into black graphitic slate interbedded with sandstone and then into silicified limestone and dolomite with local stromatolites.

The three lower groups are cut by the Mwanza porphyroid granite and by the Bia and Bukena equigranular granites that are dated by the Rb-Sr method at 1385, 1375, and 1300 m.y. respectively. All of them suffered some phases of the world-scale Kibaran orogeny that is mostly evident in the foldbelt of the same name and in other Precambrian belts of Africa. Rocks of the Kibara Supergroup are also cut by tin-bearing muscovite leucogranite and pegmatite with Rb-Sr age falling in the range of 1070–1000 m.y. (Cahen and Snelling 1966). These granites belong to magmatic formations of the Grenville (Late Kibaran) cycle. Unfortunately, the relations of the Kibara granites proper and the Lubudi Group are not known (they exhibit no contact at all). The group rests with a sharp unconformity on the rocks underlying it and it is therefore suggested that the Kibaran granites intruded prior to deposition of the Lubudi Group. If this is so, then the three lower groups are older than the Middle Riphean. These groups may belong to the Lower Riphean or to the Lower Riphean and Akitkanian if the isotope age of the basement rocks is considered. The occurrence of typical Lower Neoprotozoic acid lava in the Kiaora Mt.Group suggests an Akitkanian level for the group. In this case, only the Lufira and Hakansson Groups are Lower Riphean; the uppermost Lubudi Group might be Middle or Upper Riphean.

The adjacent area of Equatorial Africa, situated in Tanzania, also exemplifies Riphean strata and structurally belongs to the Ubendian geosynclinal belt and to the Bukoba near-fracture depression adjoining it; these two structural elements are situated between the Nile platform (including the older Tanganyika craton) and the Bangweulu block of relative stabitily (see Fig. 62). Unlike the typical miogeosynclinal groups of the Kibara belt, these strata are transitional from miogeosynclinal to orogenic; the latter are mostly characteristic of the upper part of the pile. In upward succession these strata include the Kigoma and Itiaso Groups, and the Bukoba Supergroup ("System"). The two lower groups are distributed along the eastern shore of the Lake Tanganyika within the limits of the Ubendian belt and the Bukoba Supergroup to the north-east of these in the depression of the same name situated nearby. A brief description of the units is given below, mainly according to Halligan (1963).

The Kigoma Group (up to 1500 m) is made up of white or lilac quartzite with subordinate interbeds of sandstone and slate; a conglomerate is situated at its base resting with a sharp unconformity on metamorphic rocks of the Mesoprotozoic Ubendian Group. Conglomerates contain pebbles of underlying rocks and porphyries. The latter greatly resemble Lower Neoprotozoic Kate porphyry (1790 m.y.), known from an isolated occurrence in the same eastern shore of Lake Tanganyika. With a stratigraphic break the Itiaso Group overlies the Kigoma quartzite or the older metamorphic rocks with an angular unconformity. The group is divisible into two formations of different thickness. The lower, Mpeta Formation (15–100 m), consists of white and pinkish-lilac quartzite, locally coarse-grained and with feldspar admixture; quartz conglomerate is also reported. Higher in the sequence rests the Makamba Formation (up to 3000 m), which is made up of grey phyllitized and cleaved slates alternating with quartzite, arkose, and greywacke. The Itiaso Group exhibits rather simple folds of north-western strike that accord with a general orientation of structures of the Ubendian belt. A layered gabbroid intrusion is situated at the contact of the Mpeta quartzite and the underlying rocks, apparently younger than the Itiaso Group. An age of 1285 m.y. was obtained on biotite from this gabbro by K-Ar analysis.

The Ukinga Group (more than 3000 m) is a possible stratigraphic and formational equivalent of the Itiaso Group, extending close to the Nyasa Lake in the Bukoba trough, at the junction of the Ubendian and Irumidian belts. This group is made up of phyllite, meta-siltstone, and quartzite. It is cut by the Chimala granite intruded prior to deposition of the Buanji Group. This latter is a satisfactory correlative of the lower part of the Bukoba Supergroup overlying the Itiaso Group. The Chimala granite is of Kibaran age – 1350 m.y. (K-Ar method). The granite age seems to be closer to the true upper age limit of the Itiaso Group than that of gabbroid. It is very probable than the Kigoma and Itiaso (Ukinga) Groups can be assigned to the Lower Riphean. Comparison of the composition of these rocks with the corresponding groups of the Kibara Supergroup also accords well with this conclusion: the Kigoma quartzite correlates fairly well with the Lufira Group, essentially quartzitic in composition, and the Itiaso slate Group with the Hakansson Group, also similar to it in composition (see Fig. 56, Columns XIV and XV). This correlation supports the presumed Lower Neoprotozoic age of the Kiaora Mt.Group of the Kibara Supergroup: this group with a horizon of porphyry is a possible correlative of the Kate porphyry underlying the Kigoma Group.

The Bukoba Supergroup rests unconformably on the Itiaso Group. It is divisible into three groups (up-section): Masontwa, Busondo, and Uha; "Bukoba sandstone" is additionally divided in its upper part. The lower part of the Masontwa Group is composed of green-grey slate and siltstone with subordinate sandstone interbeds (the Mokuba Formation), and the upper part of light-grey or red quartzite (the Mkuyu

Formation). The group is rather thin (up to 300 m) and locally it is missing from the sequence.

The Busondo Group comprises four formations in typical section: Uruwira, Nyanza, Malagarasi, and Kigonero. The Uruwira Formation overlies the Masontwa Group with a break or the Itiaso Group and the Ubendian metamorphic rocks with an angular unconformity. Its lower part is built up of red shale, siltstone, and conglomerate and its upper part of cross-bedded quartzite. Its thickness in the typical section is about 60 m, but in other localities it increases up to 300 m and interbeds of limestone are observed. In the Mwendo Hills area this formation includes also red conglomerate up to 440 m thick with pebble and boulders of red felsite, quartzite, and other rock types. The Nyanza Formation (60–230 m) is built up of grey banded shale and siltstone. The Malagarasi Formation (350 m) starts with coarse-grained white, locally lilac cross-bedded quartzite, upward follow brown shale, then again quartzite of the same aspect as the lower one, but with interbeds of pebbles. The Kigonero Formation (or the "Kigonero flags") is 250 m thick, it lies conformably on the Malagarasi Formation and is made up of distinctly bedded red fine-grained sandstone alternating with shale and quartzite of the same color and with grey dolomitic limestone containing columnar stromatolites. Microphytofossils have been determined in shale of the Masomtwa and Busondo Groups (Bozhko et al. 1974), similar to those of the Upper Riphean of the Southern Urals and of the Russian plate.

The Uha Formation rests on different beds of the Kigonero Formation and starts with the Gagwe amygdaloidal basalt that forms some four flows (with a total thickness of about 700 m) separated by tuff and fine limestone interbeds. K-Ar datings yielded values of the order of 900 m.y. Basalt is overlain with an insignificant break by the strata of Ilagala dolomitic limestone (165 m) with a basalt horizon at the base. Certain varieties of limestone consist of oolites (oncolites?) and contain stromatolites with the *Baicalia lacera* Semikh. form among them (Bozhko et al. 1974). Above dolomite the red beds of the Manyovu Formation (700–1000 m) rest with a stratigraphic unconformity; locally the red beds lie directly on the Gagwe basalt or even on the older Neoprotozoic rocks, including the Itiaso Group. The red beds are made up of fine-grained quartzite and arkose with rare interbeds of shale. Many rocks are typified by cross-bedding and ripple marks. At the base of the formation a thin horizon of conglomerate or coarse-grained quartzite with some indications of copper mineralization is observed.

The normal section of the supergroup finishes with the "Bukoba sandstone": fine- and middle-grained white oligomictic rocks alternating with pure quartzite-sandstone and slate with rare conglomerate interbeds. The grains in the sandstone are well-rounded. The thickness amounts to several hundreds of meters.

At the moment it is not possible to reach any definite conclusion as to the age of the Bukoba Supergroup. The Busondo Group occurring in its lower part closely resembles the Lubudi Group of the Kibaran belt in structure and composition, though it differs in certain aspects, in abundance of red beds for instance. If the microphytofossils established in it and in the underlying thin Masontwa Group are considered, then both these groups might be Upper Riphean, but it must be acknowledged that the paleontological data available are too poor for such a conclusion. The Uha Group contains limestone with stromatolites that are known from the Middle and Upper Riphean strata of the U.S.S.R.; the Gagwe basalt underlying the carbonate strata, however, is younger than the Riphean formations. K-Ar datings of rocks, including basalts, are known to give rejuvenated values. On the other hand, in age determinations according to stromatolites, the more dependent they are on single forms, the less satisfactory they are. If the Gagwe basalt dating is admitted as reflecting the true time of lava outflow, then the Uha Group and the overlying "Bukoba sandstone" are younger than the Riphean rocks and might be attributed to the Epiprotozoic, most probably to its lower part where the Roan Group of the Katangan Supergroup in the adjacent areas of Equatorial Africa belongs (Salop 1977b). Certain indications of syngenetic copper mineralization in the Uha Group are of interest, as these are typical of the Roan Group. It is important that stromatolites known from the Middle–Upper Riphean strata of Eurasia are also established in the Roan Group (and in other Lower Epiprotozoic groups of Africa and Australia). The Uha Group is not likely to be an exception. Finally it should be noted that the unconformity separating the Uha Group from the underlying rocks is the most significant in the Bukoba Supergroup and probably the result of the Grenville orogeny.

Although this presumes Epiprotozoic age for the Uha Group, nonetheless it remains problematic. Additional data are needed to solve this problem, primarily isotopic datings that will provide an uniqualified interpretation.

The Riphean strata of North Western Africa will be examined briefly. They are distributed in the eastern margin of the West African platform and in the zone of the Nigerian geosynclinal belt adjoining it. They are of interest because it is possible to demonstrate the change in the character of strata from typically platform type through aulacogenic and miogeosynclinal to eugeosynclinal type. Platform rocks are represented by the Hombori Group distributed in the south-east of the Taoudeni syneclise, aulacogenic and miogeosynclinal rocks by the Ydouban Group that forms the Gourma aulacogen ("basin") and the adjacent Ansongo miogeosynclinal zone of the Nigerian belt, and eugeosynclinal rocks by the Farusian Supergroup distributed eastward in the Tanezruft-Adrar zone of the same Nigerian belt (Fig. 55). A brief description of the Hombori and Ydouban groups is given largely according to Reichelt (1972) and of the Farusian Group according to Caby (1969, 1972 etc.).

The Hombori Group has the following sequence (up-section): (1) a basal horizon (up to 40 m) quartzite-sandstone with conglomerate lenses; (2) the Beli Formation (300–500 m) – grey shale, commonly calcareous in the lower part, with lenses and interbeds of quartzite-sandstone, conglomerate, siliceous rocks, and sandstone; (3) the Irma Formation (1500 m) – white to dark-grey dolomite with local stromatolites in the lower part, grey to black shale, commonly calcareous and dolomitic, sometimes passing into calcareous-dolomitic breccias; (4) the Hombori-Douentza Formation (up to 150 m) – grey or reddish-grey quartzite-sandstone, middle- to coarse-grained passing into fine-pebbled conglomerate or conglomerate-breccias; (5) the Oualo-Sarniere Formation (300–500 m) – argillite and shale with lenses of quartzite-sandstone, arkose, and stromatolite-bearing dolomite; (6) the Bandiagara-Garemi Formation (600–1000 m) – red and light-grey quartzite-sandstone, locally arkosic and cross-bedded with many horizons of conglomerate in the upper part.

The basal quartzites lie unconformably on the metamorphic rocks of the Birrimian Group of the Mesoprotozoic and on the Eburnean granites cutting them (1900–2000 m.y.). Significant breaks are observed inside the group; these are between the Irma carbonate Formation and the Hombori-Douentza quartzite-conglomerate For-

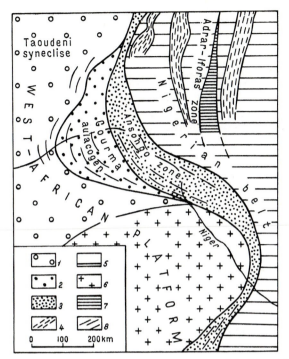

**Fig. 55.** Tectonic scheme of conjugation area of the West African platform and Nigerian folt belt. (After Reichelt 1972 revised).
*1* Riphean platform strata (the Hombori Group), *2* Riphean subplatform strata in the Gourma aulacogen (the Ydouban Group), *3* Riphean miogeosynclinal strata in the Ansongo zone (the Ydouban Group) of the Nigerian belt, *4* Riphean eugeosynclinal strata (the Farusian Group) in the Adrar-Iforas zone, *5* eugeosynclinal zone of the Nigerian belt, *6* metamorphic rocks and granites of the Leone-Liberian massif, *7* the Katarchean In Ouzzal granulite complex – tectonic block in the Adrar-Iforas zone, *8* fold strikes

mation overlying it with an erosion, and also between the Bandiagara-Garemi Formation and the underlying Oualo-Sarniere Formation (Bertrand-Sarfati and Moussine-Pouchkine 1978). There is additionally a small ("dispersed") stratigraphic unconformity at the base of the Oualo-Sarniere Formation. According to Défosser (1963) clastics of the Hombori-Douentza Formation rest on different horizons of the Irma and Beli Formations and in plan they obliquely cut the underlying rocks, suggesting the existence of a structural unconformity. This serves as a basis for recognizing two subgroups (or groups) in the Hombori Group: the lower one and the upper one.

In the Gourma aulacogen the subhorizontally occurring or gently dipping unaltered rocks of the Hombori Group pass into strongly folded and noticeably metamorphosed rocks of the Ydouban Group in the adjacent miogeosynclinal zone of the Nigerian belt. The normal section of this group is best known in the southern and western portions of the aulacogen. According to Reichelt, the following units are recognized here: the basal formation (10–50 m) – quartzite-sandstone and conglomerate; Ib formation or quartzite-argillite formation (1500–2000 m) – shale argillite (up to phyllite) slightly calcareous in places, with horizons and lenses of quartzite sometimes passing into quartz gritstone; II formation or argillite-slate formation (700–1300 m) – shaly argillite, siltstone, and phyllitized clay shale, locally carbonate and ferruginous, locally red-colored in the lower part with a horizon of quartzite in the top; III formation or "heterogenic" formation (more than 1500 m) – interbedding of argillite slate and extremely fine-grained dark quartzite with many horizons or bands of dolomite and limestone; IV formation or sandstone-argillite formation (more than 1500 m) – the lower part consists of interbedding of sandstone, quartzite, phyllitized slate (in the east of the zone down to micaceous schists) and intraformational conglomerate; the upper part of argillite and slate (and their metamorphosed equivalents), commonly passing into sandstone.

In the miogeosynclinal zone the Ydouban Group retains its structure, but its thickness increases somewhat, deformational and metamorphic grade become noticeable higher. Here the group is made up of phyllite, micaceous schist, quartzite with muscovite, marble, and other rock types of greenschist (down to amphibolite) facies. Altered volcanics of intermediate composition appear close to the eugeosynclinal zone.

In the near-platform margin of the aulacogen the Hombori and Ydouban Groups demonstrate a gradual transition and their stratigraphic unity is beyond doubt. Continuous tracing, however, is known only for the lowest parts of both groups and higher units are separated and thus correlation of formations explained differently by different authors. In Fig. 56 (Columns XVI and XVII) the correlation is shown as I understand and accept it (Salop 1977b). The thickness of co-eval units in the aulacogen is seen to increase sufficiently.

Age determinations of both groups can largely be based on isotope datings of the Ydouban Group rocks. This group lies with a sharp unconformity not only on different older rocks of the basement starting from the Katarchean up to Mesoprotozoic but also on dolerites cutting the Birrimian sockle, which are 1395 m.y. old by the K-Ar method (Machens 1973). Thus, it is possible to state that both groups are younger than the Kibaran orogeny and thus, younger than the Lower Riphean (Middle Neoprotozoic). The upper age limit of the Ydouban Group can be defined according to the age of metamorphosed eclogite-like basic rock enclosed in the slate of the Ib formation; values in the range of 865–1100 m.y. were obtained for it by whole-rock K-Ar method and a value of 990 m.y. on muscovite. This latter is the most valuable from the point of view of geochronology, as it is very close to the time of metamorphism of the Grenville (or Late Kibaran) cycle.

It was noted above that the Hombori Group is heterogenous and is separated by an unconformity approximately in the middle of it, at the base of the Hombori-Douentza Formation. The same unconformity (and even an angular type of it) is suggested in the Ydouban Group, at the base of the IV formation that is compared with the above-mentioned formation of the Hombori Group. Granite intrusions known from the lower subgroup of the Ydouban Group are supposed to have been emplaced during formation of this unconformity. A small massif of the Bourre granodiorite is of the age of 1210 m.y. (K-Ar analysis on amphibole, Reichelt 1972), it extends its contact effect on the enclosing rocks of the Ib formation, that is it corresponds to the Avzyan diastrophism of the second order. If this is accepted, the lower sub-groups of both groups examined belong to the Upper Neoprotozoic (Middle Riphean) and their upper sub-groups to the Terminal Neoprotozoic (Upper

Riphean). A still younger (Epiprotozoic) age can be suggested for the uppermost (Bandiagara) formation of the Hombori Group.

In the eugeosynclinal zone that occupies some part of Algeria (the Western Ahaggar) the Farusian sedimentary-volcanogenic Supergroup serves as a correlative of the formations examined above (Reichelt 1972). Direct relations with the Ydouban Group are not observed, as a side band between them is covered with Meso-Cenozoic strata and in some isolated outlying hills the older rocks are strongly dislocated and cut by a large fracture of abyssal type. The occurrence of volcanics in the Ydouban Group close to the eugeosynclinal zone, however, draws both types of these strata nearer together.

The Farusian Supergroup is formed of two large groups that are usually recognized as the Lower and Upper Farusian. In the Tanezruft-Adrar zone the Lower Farusian is commonly divided as the "Série à stromatolites"; it rests here with a sharp unconformity on the Katarchean gneiss and granulite and also on the Mesoprotozoic Tassendjel Group and on the Vallen granite of the Karelian cycle cutting it. The strata of acid, rarely intermediate and basic volcanics, are locally exposed under the "Série à stromatolites"; their thickness varies, but in places it is great and the basal beds of the group overlie these strata with a deep erosion (Lelubre 1969). It is very likely that these strata belong to taphrogenic formations of the Lower Neoprotozoic (Akitkanian).

According to data of Caby (1967, 1971, 1972) the "Série à stromatolites" is divisible into the following units (up the sequence): (1) white or light-grey, massive, fine-grained quartzite with common ripple marks and faint cross-bedding with conglomerate lenses and slate interbeds (thickness is from some hundreds of meters up to 3500 m); (2) limestone and dolomite with columnar stromatolites of *Collenia* and *Conophyton* type with bands of slate, quartzite and jasper (300–1500 m); (3) interbedding of quartzite and slate, calcareous slate and pyroclastic rocks; (4) volcanics of basic and ultrabasic composition with dominant diabase, rare dacite, and porphyry.

The Lower Farusian "Série à stromatolites" is unconformably overlain by the Upper Farusian Verte Group that is divisible into two subgroups closely related: the lower one volcanogenic, and the upper one clastic-volcanogenic. The volcanogenic strata are composed of flows of slightly altered porphyry, of rare andesite, and of their tuffs and tuff-breccias. The clastic-volcanogenic (or Amded) subgroup exhibits the following structure (in upward sequence): (1) conglomerate, arkose and coarse-grained feldspathic quartzite; (2) dolomitic limestone; (3) dolomite interbedding with jasper-like rocks and slate; (4) phyllitized siltstone and greywacke alternating with thick flows of epidotized lava of acid and intermediate composition (mainly dacites) and also with agglomerate tuff and breccias; (5) red fine-grained arkose sandstone. The strata (4) make up the major portion of the Amded Sub-group. Caby thinks the thickness of the Verte Group might amount to some 10–15 km, but it is a rough estimation and seems to be strongly overstated.

Unlike the Lower Farusian, the Upper Farusian is distributed not only in the Tanezruft-Adrar eugeosynclinal zone, but also in Central and Eastern Ahaggar; here the Upper Farusian rocks lie directly on high-grade basement rocks.

The formational aspects of the Farusian Supergroup are quite distinct. The Lower Farusian, with abundant quartzite and carbonate rocks with stromatolites, is of miogeosynclinal nature and even exhibits certain platform aspects; the occurrence of basic and ultrabasic volcanics in its upper part, however, evidences a change in the tectonic environment. The Upper Farusian (Verte) Group might be assigned to the eugeosynclinal formations, though it is uniquely characterized by the dominance of acid and intermediate lavas. The other eugeosynclinal zones of the Neoprotozoic are especially known for such volcanogenic strata. The red sandstone appears in the uppermost part of the Farusian complex and may indicate a termination of geosynclinal regime and its substitution by an orogenic regime.

Farusian rocks are of different grades, starting from the lowest greenschist facies up to epidote-amphibolite, and even amphibolite facies. In the zones of high metamorphism numerous intrusive bodies of granites of different age are located, and some of them penetrate the younger Epiprotozoic strata. Granite datings usually give values corresponding to the latest thermal events in the belt examined: on zircon it is in the range of 590–665 m.y. (U-Th-Pb method) and on micas and whole rock within the 435–745 m.y. interval (K-Ar and Rb-Sr method). The same values are commonly obtained for many pre-Farusian granites. Certainly all these datings are rejuvenated to a certain extent under the influence of the Katangan and Pan-African (Late Precambrian–Early Paleozoic) events that were quite intensive in the Nigerian belt. For this reason isotope methods fail to determine the age of the earliest post-Farusian (Neoprotozoic) granites. However, crystalline basement rocks commonly give values of the order of 1000–1100 m.y. (Rb-Sr method on micas and whole-rock analysis); probably this indicates a strong metamorphic (and isotope) mobilization of the older rocks during the Grenville orogeny. The transgressive occurrence of the Epiprotozoic Ahnet "Série pourprée" with tillites on the Farusian Supergroup favors a supposition that its formation was completed under the influence of the Grenville events. This fact is decisive in determining the upper age limit of the Farusian.

There is sufficient difference in the composition of the Farusian Supergroup and the Ydouban Group to cause problems in detailed (formation-by-formation) correlation. Still, if details are ignored the "Série à stromatolites" is generally similar to the lower subgroup of the Ydouban Group. The upper parts of the large units compared differ greatly in their formational type and naturally they cannot be correlated in their composition. It is, however, important that the Farusian and Ydouban rocks are separated by an unconformity approximately in the middle. It is possible that these unconformities were formed simultaneously. This correlation is supported to some extent by the Pb-model datings of galenite from quartz veins in Lower Farusian rocks, the values obtained falling in the range of 1190–1280 m.y. (Gravelle and Letolle 1963), almost the same as those for Bourre granite in the lower subgroup of the Ydouban Group; it was noted above that the emplacement of the granite is very likely to have taken place during the Avzyan diastrophism. Many workers studying the Ahaggar region indicate an interesting fact: the rocks of the "Série à stromatolites" are of a higher grade than those of the Verte Group. Thus, the break separating the Farusian groups is suggested as being also accompanied by tectonothermal events. Taking into consideration all these facts is seems possible to suppose that the Lower Farusian belongs to the Upper Neoprotozoic, and the Upper Farusian to the Terminal Neoprotozoic.

One more example of Neoprotozoic formations in Africa is examined here to illustrate the peculiar sedimentary-volcanogenic formations of taphrogenic type that extend in Namibia in the south-east of the continent, in the area surrounding the

Transvaal-Rhodesian (Zimbabwan) platform (see Fig. 62). These piles formed after the Karelian orogeny that produced a significant consolidation of the Mesoprotozoic geosynclinal belt but previous to the formation of an extremely thick typically geosynclinal Damara Supergroup.

The Rocks under discussion embrace several larger units separated by unconformities and local granite intrusions (Martin 1964). Recently these units have been regarded as formations of the Sinclair Group (Watters 1977), but it seems more correct to recognize them as groups within the Sinclair Supergroup. According to Watters the following groups (formations) are established in upward sequence: Nagatis, Kunjas, Barby, Guperas, and Auborus.

The Nagatis Group (several thousands of meters) is made up of fluidal felsitic lava, ignimbrite, and agglomerate, usually of reddish hue, with local conglomerate interbedding. It is cut by the Haremub granite and granite-porphyry that are possibly comagmatic to lava. There is no isotope age available for the granite of the group, but it is unconformably overlain by the Lower Riphean strata of the Kunjas and Barby Groups. The stratigraphic situation of the Nagatis Group suggests a younger than Mesoprotozoic age but it is older than the Riphean formations and is thus here attributed to the pre-Riphean Neoprotozoic, or in other words to the Lower Neoprotozoic (Akitkanian). It is included in a general sequence of the Sinclair Supergroup and for this reason is mentioned here.

The Kunjas Group (up to 2500 m) rests with a sharp unconformity on the underlying rocks and starts with basal conglomerate with pebbles of acid volcanics and Haremub granite. Two groups of strata are recognized in the group: the lower strata of conglomerate, arkose, gritstone, and quartzite and the upper strata of phyllite, siltstone slate with limestone lenses.

The Barby Group (up to 8500 m) overlies the Kunjas Group with an insignificant stratigraphic unconformity and commonly starts with a horizon (200 m) of volcanoclastic rocks composed of acid tuff, tuffite, and tuff-conglomerate; up-section in some regions very thick strata of basalt, andesite, trachyandesite, trachybasalt and rhyolite occur in part, and in other regions the strata of basalt of tholeiitic type alternates with felsitic lava or piles of lava of essentially rhyolitic composition with a horizon (up to 400 m) of arkosic sandstone at their base. Volcanics associate with subvolcanic intrusions like trachyandesitic dykes (of shoshonite character in places), small bodies of gabbroid (norite including) and syenite.

Volcanic and subvolcanic rocks of the group are cut by large bodies of Nubib granite, Tumuab granite, etc. Zircon from Nubib granite is dated at $1360 \pm 50$ m.y. (Burger and Coertze 1973). This means the granite had emplaced during the Kibaran orogeny and the Kunjas and Barby Groups are of the Middle Neoprotozoic (Lower Riphean) age.

The Guperas Group rests unconformably on the Barby Group and the Nubib granite. The thickness of its lower part strongly varies (0–3700 m), it is built up of red-brown lithoclastic sandstone and polymictic conglomerate that upwards interbeds with rhyolitic lava and ash tuff; locally at the base of this portion of the group there is a horizon (150 m) of cross-bedded white to red quartzites. The upper part of the group consists of volcanogenic pile (more than 520 m) with basic lava at its base and dominant rhyolitic lava and tuff-lava, and also its tuff and agglomerate at its top. The group is cut by the Rooiberg granite dated at 1270 and 1290 m.y. on zircon

(Watters 1977). These datings correspond to the time of the Avzyan or Late Kibaran orogeny and allows the attribution of the Guperas Group to the Upper Neoprotozoic (Middle Riphean).

The Auborus Group (up to 2800 m) with a sharp unconformity lies on any older groups. A fossil deluvium (regolith) is known at its base and the group itself is made up of bright red cross-bedded arkosic sandstone alternating with conglomerate containing well-rounded boulders and pebbles of granite, gneiss, and basic and acid lavas from the underlying groups. The group is considerably less dislocated than other groups, and in the margin of the platform it occurs almost horizontally and there the Nama platform Group of Eocambrian-Early Cambrian age overlies it. A close equivalent of the Auborus Group is the Tsumis Group that extends along the southern margin of the Damara belt, where it is overlain by the Epiprotozoic strata bearing glacial formations (the latter are younger than the granite with an age of $\sim$ 1000 m.y.).

The Sinclair Supergroup is known to contain all four Neoprotozoic Sub-erathems. They are formed of thick continental volcanogenic and sedimentary strata (the littoral-marine deposits are reported only from the Kunjas Group); these piles had evidently accumulated in deep near-fracture depressions that originated on the basement stabilized as a result of the Karelian folding. By the environment under which they accumulated they greatly resemble the Lower Neoprotozoic (Akitkanian) taphrogenic rocks that are known to be of world wide distribution, and in the supergroup examined they are well exemplified by the Nagatis Group. It is possible, then, to conclude, that the taphrogenic stage lasted during the whole period of the Neoprotozoic in the territory of South Western Africa. It should be noted, however, that in this particular region, as well as in many more regions of the world, the basic lava is dominant in the Riphean volcanogenic strata but in the case of the Akitkanian Nagatis Group the acid lava is a principal component. Notable is the fact that in Equatorial Africa the Early Riphean is characterized by the most intensive outflow of basic lava and that in North America this phenomenon is mostly typical of the Middle Riphean.

**The Riphean of Australia.** In this continent the Riphean is predominantly represented by platform strata that accumulated in extensive depressions ("basins") of enormous depth at places and with inner uplifts and fractures as complicating elements. Some of these depressions developed on the Early Neoprotozoic (Akitkanian) taphrogenic structures and inherit these to some extent. The geosynclinal formations are recorded only in the north-east of the continent in the Riphean Mount Isa foldbelt (see Fig. 63). First a brief description of the well-studied Riphean normal sequences of platform type that are distributed on the Australian craton will be given.

The Riphean strata of the large Bangemall depression situated in the west of the continent between the old Pilbara and Yilgarn shields are represented by two groups: Bresnahan and Bangemall (Daniels 1966, Brakel and Muhling 1976). The Bresnahan Group is locally developed in the northern flank of the depression, where it unconformably overlies the Mesoprotozoic rocks of the Mount Bruce Supergroup that enclose jaspilites. The Cherribuka basal polymictic boulder conglomerate (10–120 m) is recognized at its base, but mainly it is made up of thick (several thousands of meters) piles of light grey and reddish quartz and feldspathic sandstone, arkose, and gritstone, locally cross-bedded and glauconite-bearing, that is the Kanderong strata.

The Bangemall Group rests on the Bresnahan Group with an erosion and on any older rocks with a sharp angular unconformity, the Mesoprotozoic Nabberu Group (an analog of the Hamersley Group?) and including Paleoprotozoic metavolcanics and granites cutting them. It is of wide extension through the entire depression of the same name, but its structure and composition vary from place to place. In the western and central parts of the depression a number of formations are distinguished within it, up the sequence these are: (1) the Tringadee (up to 1000 m), of coarse-grained feldspathic sandstone with conglomerate lenses, with dolomitic interbeds at places in the upper part; (2) the Irregully (up to 1400 m), of dolomite with stromatolites, shale, and argillite with interbeds of sandstone, chert, carbonate conglomerate, and breccia; (3) Kiangi Creek (up to 350 m), of alternating thick-bedded quartz or feldspathic sandstone of the formation; (4) the Jillawarra (1400 m), of grey and black siltstone slate and sandstone with crystal molds after pyrite, halite, and gypsum; (5) Discovery Chert (several tens of meters), of continuous marking horizon consisting of chert with microbiota of spherical shape; (6) Devil Creek (several hundreds of meters), of dolomite, locally detritic and cross-bedded, containing stromatolites and sometimes oncolites; (7) Ullawarra (several hundreds of meters), of shale, argillite, quartz siltstone, and fine-grained sandstone with rare dolomitic interbeds; (8) the Curran (several hundreds of meters), of slate, siltstone and chert; (9) the Coodardoo (several hundreds of meters) of bedded greywacke sandstone with siltstone inderbeds; (10) Fords Creek (hundreds of meters), of green siltstone with quartz sandstone interbeds; the rocks are usually characterized by ripple and current marks, by sliding traces and other aspects of turbidites; (11) Mount Vernon (300 m), of well-sorted middle-grained quartz sandstone with syngenetic clasts of siltstone exhibiting shallow-water structures; (12) the Kurabuka (up to 300 m), of greenish and grey shale and argillite, with cherty slate close to the base of the formation and with fine interbeds of carbonate rocks in its upper part.

In the opinion of Brakel and Muhling, the lower Tringadee Formation was formed under sub-aerial environment and is in part of alluvial origin, the remaining thick sequence of sedimentary rocks of the group having accumulated in the shallow-water marine basin of lagoonal or sometimes stagnant character. In the east of the central part of the Bangemall depression the deposits of the five middle formations (Kiangi Creek, Jillawarra, Discovery Chert, Devil Creek, and Ullawarra) are replaced by thick monotonous strata of shale and siltstone (the Backdoor Formation) that originated probably in the somewhat deeper and better aired portion of the sea basin. The rocks of both groups are of platform type, but they are strongly folded, however, in many places, which is likely to be the result of a fractured basement of the depression.

The following data are available on the age of the rocks examined. A Rb-Sr age of $1080 \pm 80$ m.y. and $1096 \pm 40$ m.y. was obtained for black shale from the middle part of the Bangemall Group and from the rhyolitic neck cutting basal conglomerate of the group respectively (Brakel and Muhling 1976). As the rocks are practically unaltered, these values refer to the time of their formation (or diagenesis), that is to the Upper Riphean. *Conophyton garganicus australe* stromatolites are reported from the Irregully Formation situated in the lower part of the Bangemall Group, and the

*Baicalia carpicornia* from the Devil Creek Formation located in the upper portion of the group sequence (Walter 1972). In the Neoprotozoic of the U.S.S.R. *Conophyton garganicus* is not known above the Middle Riphean and the second form is endemic. Generally the *Baicalia* group occurs in the U.S.S.R. in the Middle Riphean as well as in the Upper Riphean. According to these data the Bangemall Group may be assigned to the Middle and Upper Riphean; the boundary between the sub-groups is very likely to be drawn at the base of quartzite of the Kiangi Creek Formation that unconformably overlies the Irregully Formation. As to the Bresnahan Group, it possibly belongs to the Lower Riphean but it cannot be excluded that it is still older and belongs to the Lower Neoprotozoic (the Chaya complex of Akitkanian).

In Northern Australia the Riphean strata within the limits of the Victoria River depression and in the Fitzmaurice near-fracture mobile zone adjoining it are represented by the Fitzmaurice and Tolmer Groups (Sweet et al. 1974). The Fitzmaurice Group is distributed only in the zone bearing the same name, it is made up of thick (5500 m and possibly even more than 10,000 m) strata of essentially continental clastic rocks with predominant feldspathic sandstone and siltstone, commonly crossbedded, with local conglomerate interbeds and lenses in the lower part of the group (the structure of the group is shown in Fig. 56, Column XXI). The group lies with a sharp unconformity on the Mesoprotozoic rocks of the Finiss River Group. The rocks of the group are intensely deformed, close to the fractures delimiting the Fitzmaurice zone in particular.

The Tolmer Group extends to the east from the zone mentioned, within the limits of a stable part of the Victoria River depression; it is made up of shallow-water marine sandstone, siltstone, and dolomite. The Buldiva sandstone (480–700 m) is recognized in its lower part, this sandstone is white and red-brown, quartz, poorly sorted with interbeds of quartz conglomerate, up-section the sandstone is fine-grained with green and reddish siltstone, with local glauconite. This sandstone is conformably overlain by the Hinde Dolomite, of pink and grey color, fine-bedded alternating with siltstone. The dolomite contains the *Inseria tjomusi* stromatolites typical of the Upper Riphean rocks of the U.S.S.R. The Waterbag Formation (300 m) completes the group; it is built up of fine-bedded reddish-brown sandstone and siltstone, with dolomite interbeds in the upper part; pseudomorphs after halite are common in the rocks. The Lower Cambrian Antrim plateau-basalt lies on the eroded surface of the group.

The Bullita Group (> 600 m) is a possible close analog of the Tolmer Group, distributed in the same region but more southerly. This group is mainly built of dolomite with local stromatolites; its base is not certain, red sandstone and siltstone are, however, known from its lower exposed part. With a stratigraphic unconformity the group is overlain by the Wondoan Hill shale and siltstone (100 m) containing glauconite that is dated by the Rb-Sr method at 1120 m.y. This value and the occurrence of the *Inseria tjomusi* stromatolites in the Tolmer Group allow us to attribute both of the groups compared and the Wondoan Hill beds to the Upper Riphean. The Wondoan Hill siltstone is unconformably overlain by the Epiprotozoic Auvergne (= Duerdin) Group containing tillites.

The Fitzmaurice Group occurs between the Mesoprotozoic and Upper Riphean strata, its age can be within a wide range. It is possible to assume, however, considering its similarity to the Bresnahan Group, that it belongs to the Lower Riphean or to the upper part of the Lower Neoprotozoic.

The Riphean strata are also distributed in the more easterly regions of Northern Australia (the Eastern Kimberley) that belong to the Sturt plate. The following formations are known among these strata (upward): Mount Parker, Bungle Bungle, Wade Creek, and Helicopter (Dow and Gemuts 1969). The Mount Parker Formation (150–300 m) comprises pure quartz sandstone with rare conglomerate lenses. It rests unconformably on the thick Red Rock strata (up to 3000 m) made up of sandstone with siltstone and conglomerate interbeds that are likely to belong to the upper part of the Lower Neoprotozoic (these strata are cut by diabase with an age of 1600–1800 m.y.). The Bungle Bungle Formation (up to 1400 m) lies conformably on the Mount Parker Formation, and is made up of bedded dolomite or dolomite siltstone. Dolomite contains stromatolites, *Collenia frequens* and *Conophyton cylindricus* being determined among them. The first form is known from the rocks of the U.S.S.R. territory in a wide vertical range, the second is commonly known from the Lower and Middle Riphean. The Wade Creek Formation (up to 450 m) lies unconformably on underlying dolomite and is made up of pure quartz sandstone, commonly cross-bedded, with interbeds of siltstone and shale. The Helicopter Formation (up to 250 m) conformably overlies the Wade Creek Formation and is made up of fine-bedded siltstone. With a large break the glacial deposits of the Epiprotozoic Duerdin Group overlie it.

The shale from the Wade Creek Formation is dated at 1130 ± 100 m.y. It permits a comparison between the examined formation and the overlying formation with the Tolmer and Bullita Groups of the Victoria River region and their assignation to the Upper Riphean. The underlying formations separated by an unconformity are very likely to belong to the Middle or even Lower Riphean; by sequence and by composition they are similar to the lower (Middle Riphean) part of the Bangemall Group of Western Australia (see Fig. 56).

In adjacent area the Glidden Group (up to 600 m) is a correlative of the Wade Creek and Helicopter Formations, and comprises sandstone interbedding with shale and siltstone. An age of 1080 m.y. is obtained for the shale by the Rb-Sr method (Dow and Gemuts 1969).

In the Katherine-Darwin region of Northern Australia the Riphean rocks form two platform groups: Mount Rigg and Roper (Randal 1963, Walpole et al. 1968). This region adjoins the Van-Diemen Gulf, where the Pine Creek geosynclinal belt was situated in the Mesoprotozoic. The Mount Rigg Group lies unconformably on the Lower Neoprotozoic Katherine River volcanogenic-sedimentary Group of the Upper (Chaya) lithostratigraphic complex. In the base of the Mount Rigg Group the Margaret Hill Formation (more than 400 m) is situated, and is of polymictic conglomerate and tuffaceous sandstone. The Bone Creek Formation (160 m) of quartz sandstone and oligomictic conglomerate overlies it with a local break, further up the sequence the Dook Creek Formation (330 m) is situated, composed of dolomitic limestone containing stromatolites, locally interbedded with sandstone, chert, shale, siltstone, and oncolitic limestone. The Beswick Creek Formation (100 m) of marl, pink quartz sandstone, siltstone, and basalt completes the sequence.

The Roper Group with a thickness of more than 5000 m is made up of various terrigenous rocks, mostly oligomictic, quartz sandstone, rare polymictic sandstone, siltstone, and shale. Conglomerate, glauconite-bearing sandstone and rare interbeds of limestone and cherty slate occur in the lower part of the group, in its middle part the horizons of sandstone and hematite ore interbeds are observed (the sequence of the group is shown in Fig. 56, Column XXIII),

Glauconite from the middle part of the group (the Crawford Formation) is dated in the range of 1145–1330 m.y. by the K-Ar method and in the range of 1270–1390 m.y. by the Rb-Sr method (McDougall et al. 1965). Dolerite cutting the group is 1160–1280 m.y. old. Judging from these datings, the Roper Group and the underlying Mount Rigg Group belong to the Lower or (and) Middle Riphean (there is a value of 1260 m.y. for glauconite from the Dook Creek Formation, which is certainly rejuvenated).

In Northern Australia, in the McArthur depression situated near the eastern shore of the Gulf of Carpentaria, the normal Riphean sequence is very similar to that examined above (Smith 1964, Rix 1965). There the McArthur Group overlies unconformably the Lower Neoprotozoic Tawallah Group of volcanogeno-sedimentary composition, this latter being cut by the intrusive rocks with an age about 1600 m.y. (a close analog of the Katherine River Group). Five formations are recognized in the McArthur Group (up-section): (1) the Mallapunyah Formation (up to 780 m) of ferruginous sandstone and siltstone with interbeds of dolomite, microbiota-bearing chert and also hematite ore; (2) the Amelia Formation (up to 1900 m) of dolomite that usually contains stromatolites, interbedding with subordinate siltstone, sandstone and chert, with rare hematite ore interbeds; (3) the Top Crossing Formation (up to 1000 m) of massive stromatolitic dolomite with interbeds of cherty breccias, argillite with dolomite interbeds or fine-bedded dolomite intercalating with siltstone and clay shale; (4) the Immerunga Formation (810 m) of massive stromatolitic dolomite, dolomitic and cherty breccias, dolomitic conglomerate, and siltstone; (5) the Billengarra Formation (its thickness varies greatly, but it may reach 1200 m) of massive chert, cherty breccias, sandstone, siltstone, and dolomite.

The McArthur Group is unconformably overlain by the Roper Group, revealing the same sequence and composition as in the Katherine-Darwin region. The Roper Group is overlain here by the Epiprotozoic Wessel Group (790–805 m.y. old on glauconite). The McArthur Group is a correlative of the Mount Rigg Group. The occurrence of the *Conophyton garganicus* in dolomite of the Amelia Formation does not contradict the above conclusion, as in the Precambrian strata of the U.S.S.R. they are commonly reported in the Lower and Middle Riphean.

Riphean platform strata are also known in the south of Australia, in the Eyre Peninsula near the town of Whyalla (the "Stuart shelf"). There the volcanogeno-sedimentary group lies unconformably on the Lower Neoprotozoic sedimentary-volcanogenic strata (the Moonabie rhyolite, the Corruna conglomerate and the Gowler Range rhyolite, all of which are older than 1600 m.y.). This group has the following composition in upward sequence (Thomson 1966): (1) the Roopena porous trachyte and basalt (100 m); (2) the Pandurra red feldspathic sandstone (120 m); (3) the Tregolanna red fine-bedded shale (up to 80 m); and (4) the Tent Hill quartz sandstone (more than 150 m); breaks are observed between every group of strata. The Roopena volcanics are determined at 1345 m.y. old by the Rb-Sr method (Compston et al. 1966); according to this value they belong to the Middle or Lower Riphean. Earlier (Thomson 1966) this sequence was compared with different units of the Epiprotozoic Adelaidean Supergroup, but at the moment this correlation seems to be erroneous in the light of new data available (see Chap. 6); and only the uppermost strata seem to be possible correlatives of certain portions of the Adelaidean sequence.

The taphrogenic sedimentary-volcanogenic strata developed in Central Australia within the limits of the Musgrave block represent a special type of Riphean formation, recognized as the Bentley Supergroup (Daniels 1974). The rocks of the supergroup rest in the sublatitudinal zone of fractures and appear to be in a sharp unconformity with the Katarchean gneiss-granulite complex that later suffered multiple mobilization. The supergroup is divided into four groups: the Pussy Cat, the Tollu, the Cassidy, and the Mission. The Pussy Cat Group (350 m), the lower one, is made up of basic lavas with interbeds of tuff, slate, and quartzite. The Tollu Group is in part coeval with the Pussy Cat Group and consists of thick (several thousands of meters) sequence of different acid and basic volcanics, mainly rhyolite and basalt with subordinate sedimentary rocks. Volcanics are cut by the Giles layered intrusion made up of ultrabasic and basic rocks (up to anorthosite). The Cassidy Group (up to 10,000 m) rests with a break on the Tollu Group and is typified by alternation of acid and basic volcanic flows and of tuff and clastic rock horizons in part. The upper Mission Group, separated by a break from the Cassidy Group, is made up of conglomerate and cross-bedded coarse-grained sandstone (the Gamminah Formation, 500 m) then upward follows fine-bedded dolomite containing stromatolites and interbeds of siltstone and chert (the Frank-Scott Formation, 250 m), then shale interbedding with tuff, basalt, chert, and conglomerate (the Lilian Formation, 1000 m) and finally thick strata of basalt with subordinate interbeds of conglomerate, quartzite, sandstone and slate (the Milesia Formation, 2900 m); basalt is accompanied by copper mineralization. The Mission Group and the older units of the Supergroup and locally the basement gneiss are overlain by the Townsend cross-bedded quartzite and sandstone (225 m); they are conventionally assigned to the Bentley Supergroup. All rocks of the supergroup are transgressively overlain by the Epiprotozoic tillite-bearing sedimentary strata.

Sedimentary and volcanogenic rocks of the Bentley Supergroup (excluding the Townsend quartzite?) are cut by gabbroids and small bodies of hypabyssal granite that are probably genetically related to volcanics.

The sedimentary-volcanogenic complex of the Musgrave block is known to resemble the Lower Neoprotozoic taphrogenic formations, Rb-Sr isotope datings of acid volcanics from the Tollu Group, however, give an age of $1060 \pm 140$ m.y. (Compston and Arriens 1968) and show an Upper or Terminal Neoprotozoic age for the Bentley Supergroup (the Middle or Upper Riphean). At the same time, this supergroup also resembles certain taphrogenic strata of Riphean of SouthWest and Nord West Africa and of the north-western part of the Canadian shield. It is assumed that thick sedimentary-volcanogenic strata of the Bentley Supergroup had accumulated in a narrow but deep near-fracture depression (graben). The fact that the rocks are strongly deformed seems to be explained by differential movements along fractures.

Finally it seems reasonable to characterize briefly the Riphean geosynclinal strata developed in the Mount Isa foldbelt situated in the north-eastern part of Australia (see Fig. 63). It is not possible to give details of these rocks here, as they are of complicated structure due to definite structural-facial and metamorphic zonation. A brief description of the Riphean sequence of only one of the zones that belongs to the western near-platform part of the belt (the Mount Isa region) will be given here. Here

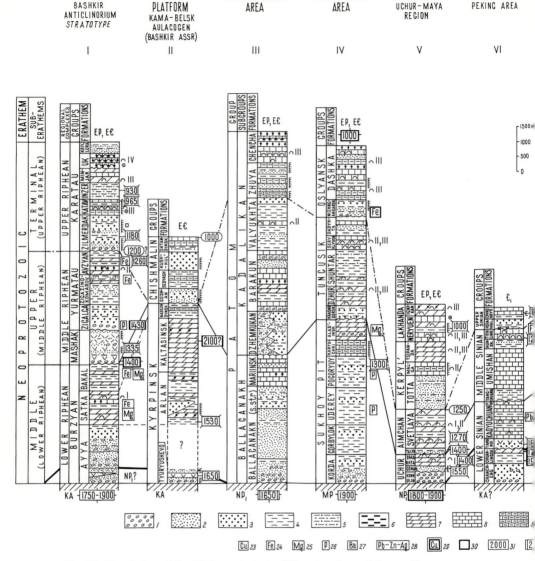

**Fig. 56.** Correlation of the Riphean strata in different regions of the world.
Stratigraphic columns: *I* after Garan' (1946, 1963), Bekker et al. (1979); *II* after Aksenov et al.
(1978); *III* after Salop (1964–1967); *IV* after Kirichenko (1967); Semikhatov (1974) etc.; *V* after
Nuzhnov (1967); Komar et al. (1970), Semikhatov (1974) etc.; *VI* after Kao Chen-hsi (1962) etc.;
*VII* after Salop (1973a) etc.; *VIII* after Krishnan (1960a) etc.; *IX* after Ross (1970), Harrison
(1972) etc.; *X* after Baragar and Donaldson (1973); *XI* after Hamblin (1961) etc.; *XII* after
Brummer and Mann (1971); *XIII* after Ford and Breed (1973); *XIV* after Cahen (1954) etc.;
*XV* after Halligan (1963); *XVI* and *XVII* after Reichelt (1972); *XVIII* after Caby (1969, 1972);
*XIX* after Watters (1977); *XX* after Daniels (1966), Brakel and Muhling (1976); *XXI* after Sweet
et al. (1974); *XXII* after Dow and Gemuts (1969); *XXIII* after Randall (1963), Walpole et al.
(1968); *XXIV* after Smith (1964) and Rix (1965); *XXV* after Carter et al. (1961), Wilson and
Derrick (1976). Only fundamental works are referred to.
Symbols: *1* conglomerate, *2* arkosic and polimictic sandstone, *3* quartz sandstone, *4* shale (and
siltstone), *5* siltstone, *6* chert, *7* dolomite, *8* limestone, *9* clay limestone (marl), *10* oolitic (oncolitic) limestone, *11* sedimentary breccia, *12* basalt, *13* andesite, *14* acid volcanics (rhyolite, ignimbrite etc.), *15* stromatolites (*Roman figures* near the symbol indicate that stromatolites belong to

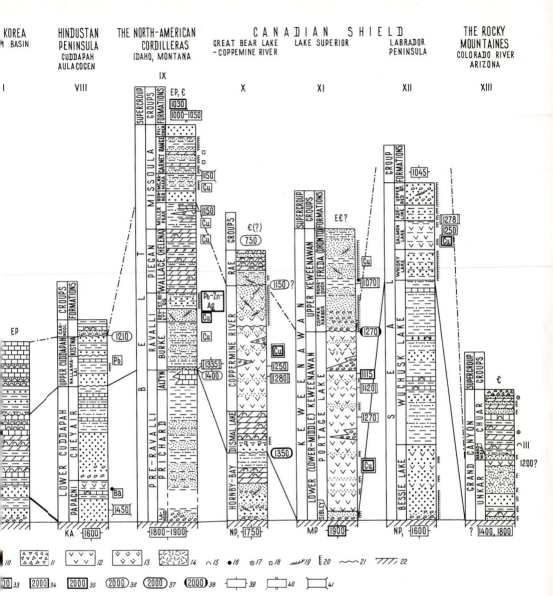

the so-called phytolite complexes: *I* Lower Riphean, *II* Middle Riphean, *III* Upper Riphean);
*16* microbiota, *17* acritarchs from the *Chuaria* group, *18* evaporites and their traces, *19* cross-
bedding, *20* red beds, *21* stratigraphic or slight angular unconformity, *22* angular unconformity
with pre-Neoproterozoic strata, *23, 24* stratiform mineral deposits: *23* copper, *24* iron, *25* magnesium (magnesite), *26* phosphorus (phosphorite), *27* barium (barite), *28* polymetallic ores,
*29* copper deposits associated with basalt, *30* very large deposits; *31–35* datings of sedimentary
and volcanogenic rocks that indicate the time of sedimentation or volcanism (methods: *31* K-Ar,
*32* Rb-Sr, *33* Pb-Pb, *34* Pb-model, *35* U-Th-Pb); *36–38* datings of dabase or dolerite dykes
(methods: *36* K-Ar, *37* U-Th-Pb, *38* Rb-Sr), *39* datings of granitoids and metamorphic rocks
(methods see *31–35*), *40* datings of hydrothermal mineralization (methods see *31, 34, 35*),
*41* K-Ar datings on glauconite.
Indexes: *KA* Katarchean, *MP* Mesoproterozoic, $NP_1$ Lower Neoproterozoic ($NP_1^1$ lower complex,
$NP_1^2$ upper complex), *EP* Epiprotozoic, *ECm* Eocambrian, *C* Cambrian.
Foot-note: In certain columns the strata of the upper lithostratigraphic complex of the Lower
Neoproterozoic (Akitkanian) are given to show the relations of the Riphean and Lower
Neoproterozoic

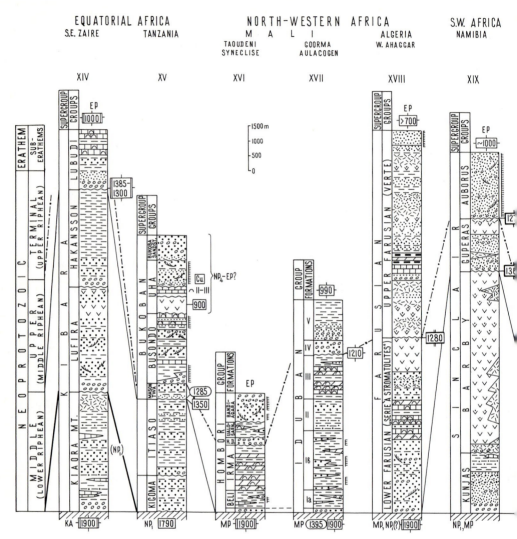

**Fig. 56** (continued)

these rocks are recognized as the Mount Isa Group that rests unconformably on the Lower Neoprotozoic Haslingden Group of sedimentary-volcanogenic composition that is cut by granite with an age of 1650 m.y. The Mount Isa Group is divisible into eight formations in upward sequence: the Warrina Park, the Moondarra, the Breakaway, the Native Bee, the Urquart, the Spear, the Kennedy, and the Magazine (Carter et al. 1961, Wilson and Derrick 1976). The Warrina Park Formation (180–700 m) starts with basal feldspathic sandstone and gritstone with ferruginous matrix, upsection follow carbonate siltstone and sandstone in part and then quartzite and conglomerate that make up a marking horizon. The Moondarra Formation (500–1220 m) consists of red and grey dolomitic siltstone and slate with ferruginous interbeds containing pseudomorphs after pyrite. The Breakaway Formation (260–

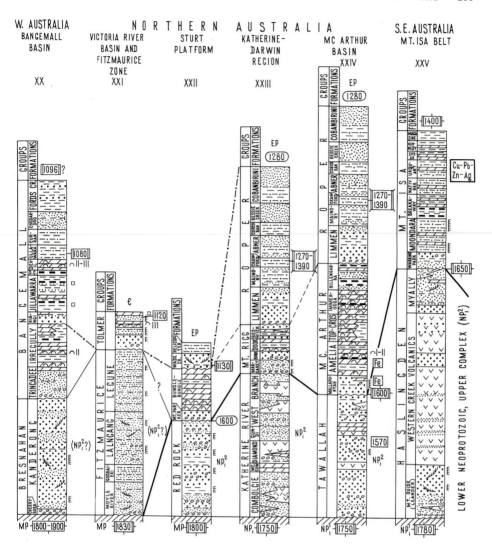

1050 m) is built up of strongly cleaved carbonate and siltstone siliceous slate, of siltstone and chert in part. The Native Bee Formation (300–800 m) comprises well-bedded siltstone that is commonly dolomitic and with interbeds of calcareous slate and rare vitrophyric tuff. The Urquart Formation (250–900 m) is represented by fine-bedded dolomitic slate and siltstone with interbeds of ferruginous and silicified tuffaceous slate; it has a marking horizon of peculiar sedimentary rocks predominantly made up of potassium feldspar, dolomite, and quartz admixture. The carbonate slates of this formation are known to enclose large stratiform deposits of copper, lead, zinc, and silver. The Spear and Kennedy Formations (140–480 m) are very similar, made up also of albitic and dolomitic siltstone, tuffite and sedimentary breccia. The Magazine Formation (up to 215 m) consists of grey to black (reddish-brown on eroded surface) sericite-dolomite-quartz slate with pyrite inclusions. The slate rocks of the group were usually accumulated in a stagnant marine basin that was at times saline.

The three lower formations are correlatives of the Surprise Creek Beds (up to 3000 m) that are distributed to the east of the above-mentioned zone. These beds are represented by two facies: the shelf type, consisting of arkose sandstone and quartz sandstone and also of stromatolitic and oncolitic dolomite, and the other type consisting of principally coarse arkose, feldspathic sandstone and slate, of conglomerate and dolomite in part. Some more formations are also recognizable in other zones of the near-platform portion of the belt, they are generally synchronous to the ones described above.

In the eastern part of the foldbelt the Mount Albert Group and probably the underlying Mary Kathleen Group correspond to a certain extent to the lowermost part of the Mount Isa Group. The sedimentary and volcanogenic rocks of these groups are typified by great thickness, higher deformational grade and much more intensive metamorphism that sometimes reaches the amphibolite facies; the supracrustals there are cut by various granitoids and doleritic dykes. The occurrence of scapolitic slate in some units is reported (for instance in the White Blow Formation of the Mount Albert Group); it contains abundant chloride and is assumed (Ramsay and Davidson 1970) to have been formed as a result of metamorphism of saline deposits.

Granite cutting the metamorphic rocks is dated at 1400 m.y. by the Rb-Sr method. The same age (1350–1450 m.y.) was obtained for micas from the metamorphic rocks by K-Ar analysis (see Wilson and Derrick 1976). All these facts are indicators of intensive Kibaran orogeny in the area and permit assignation of the Mount Isa Group to the Middle Neoprotozoic (Lower Riphean). The Epiprotozoic sedimentary strata with local tillites (the Mount Birnie Beds) overlie the Riphean strata and the granite.

A correlation scheme of principal sequences of the Riphean strata of the world concludes the regional review and illustrates the data discussed in this part of the book (Fig. 56).

### 3. On the Riphean Lithostratigraphic Complexes

In the foregoing review, it was demonstrated that Riphean strata in different regions of the world exhibit many common features and that these are different from any older Precambrian units. Their principal peculiarities were shown in the previous section of the book devoted to general characteristics of the Riphean. It is easily seen that the occurrence of certain rock associations (paragenesis) is related to certain Neoprotozoic sub-erathems and that some regularities can be observed in it, but as a rule they are not on a global scale and are typical only of some regions that can be quite extensive and even embrace an entire continent. For instance, in North America the basic volcanics in the platform (and taphrogenic) strata are abundant only in the Upper Neoprotozoic (Middle Riphean), and in the same strata of Aftrica they occur predominantly in the Middle Neoprotozoic (Lower Riphean); in North Eurasia the "older trapps" are practically lacking in Riphean. In many regions in the upper sub-erathems the abundance of limestones increases in relation to dolomite, but this regularity is not always obvious and is of local character.

A transgressive or transgressive-regressive structure of sequence of every sub-erathem is typical of all regions, and is seen from the occurrence of psammitic and psephytic rocks in the lower part of the normal sequences and of carbonate or carbonate-pelitic rocks in their upper part, in the case of most complete sequences the detractive complexes being registered in their uppermost part. This type of structure certainly shows that the accumulation of rocks of every suberathem occurred during a separate geologic (sedimentary) macrorhytm.

There are no specific Riphean formations that could be regarded as typical of any definite Riphean suberathem. This does not mean, however, that such formations do not exist, but only that they are not recognized and that much more refined analysis of lithologic and chemical composition of rocks is needed and that much more detailed study of their paragenesis must be done. This problem must be solved in the near future, but at the moment it is impossibble to establish any lithostratigraphic complexes in the Riphean, and thus it is impossible to correlate some parts of the Riphean suberathems on a world-wide scale.

# B. Geologic Interpretation of Rock Record

### 1. Physical and Chemical Environment on the Earth's Surface

The average temperature on the Earth's surface some 1300–1200 m.y. ago is supposed to have been within the 40°–50 °C range, according to $^{18}O/^{16}O$ ratio and D/H in hydroxyl silicon of singenetic cherts from carbonate rocks of the Belt Supergroup, the Apache Group, and the Grand Canyon Supergroup that belong to the Middle and Upper Neoprotozoic (Knaut and Epstein 1976). A significant lowering of tempera-ture in comparison with that of the Mesoprotozoic ($\sim 60$ °C) is probably explained by a decrease of concentration of carbon dioxide and water vapor in the atmosphere and a certain consequent lessening of the hot-house effect. Nevertheless, it is clear that the Neoprotozoic average temperature much exceeded the modern one ($\sim 15$ °C).

There are no quantity data available on atmospheric pressure in the Neoproto-zoic, but it is possible to suppose that it was much less than in the Mesoprotozoic and probably exceeded the modern one only slightly. The highest concentration of carbon dioxide in the atmosphere (and in the hydrosphere) was probably characteristic of the Early Neoprotozoic during extensive and intense lava outflow, known to have been a specific feature of the early subera. The intensive precipitation of carbonates during the Riphean must have lowered the partial pressure of $CO_2$ to a great extent. At the same time the abundance of red beds shows a significant increase in free oxygen in the Neoprotozoic atmosphere, that was naturally related to a sharp growth of biomass of photosynthesizing algae and bacteria. In-crease of oxygen concentration must have greatly affected many of the geochemical processes that took place at that time. Probably it also resulted in a change in seawater composition: a higher oxidation of sulfur and hydrogen sulfide must have changed it from chloride-carbonate water depleted of sulfate to chloride-sulfate water, similar in composition to modern water. Probably this process started in the second half of the Mesoprotozoic. An increase of $O_2$ concentration must have transformed the combinations of polyvalent elements, Fe, Mn, Cr, Ni, Co, Cu, V, etc., into the higher oxidized forms of limited solubility, and thus lowered the concentration of these elements in seawater and consequently decreased their geochemical mobility.

The appearance or noticeable intensification of climatic zonation is expected because of the lowering of temperature on the Earth's surface, but in the Neoprotozoic rock record there is no evidence of this. On the contrary, it seems possible to speak of an azonal hot climate at that time, based on the ubiquitous occurrence of red beds and stromatolites in the Neoprotozoic rocks and on the presence of evaporites in many strata situated at different modern latitudes (for instance in polar regions of Canada, Australia, and Africa). It is notable that there are no traces of a fall of temperature, not to speak of glaciations, in the Neoprotozoic strata.

## 2. Life During the Neoprotozoic

Generally the organisms of the Neoprotozoic were very similar to those that originated in the second half of the Mesoprotozoic, but this particular era was typified by a significant increase of biomass and a higher divergence of the groups of organisms that originated earlier. In the Mesoprotozoic numerous algae, bacteria, fungi, and other primitive forms of organisms are known to have existed that are problematic taxons as yet, but in the Neoprotozoic (in the Riphean to be more exact) more varied forms became known and possibly even some new ones, especially among the groups that are usually regarded as microfossils and microbiota. The occurrence of various fossilized microorganisms, which from many aspects may be assigned to the eukaryota, seems quite important and the basis for it is much sounder than in the case of the equivalent Mesoprotozoic forms (Schopf 1974, 1977 etc.). It is not possible to reject a supposition that some peculiar structures (ichnoglypts?) rarely known from the Riphean may prove to be traces of life activity of certain first primitive burrowing or crawling animals, of worms at least.

## 3. Sedimentation Environment

The Neoprotozoic sedimentary rocks are varied, but the most typical and wide-spread rocks are worth being considered here from the point of view of environment, as also the rock types that are of special interest in clarifying processes of past sedimentation.

The terrigenous rocks that make up thick strata at different stratigraphic levels of the Neoprotozoic belong to three principal genetic types according to the sedimentation environment. One of them includes psammites and psephytes occurring in taphrogenic strata of the Lower Neoprotozoic; these are specially characteristic of the Khibelen lithostratigraphic complex and of the lower part of the Chaya complex. These rocks are represented by red-colored, poorly sorted inequigranular arkose, polymictic and lithoclastic sandstones, and also by polymictic conglomerates, accumulated under continental environments, usually in river valleys, lake basins and deltas situated in isolated depressions and grabens. In many cases the supply of clastic material is registered as having occurred from two directions. For instance, during deposition of sedimentary rocks of the Akitkan Group, transportation of terrigenous material was from the Baikalian fold area and from the Siberian platform, and as a result the psammites consist of a mixture of angular clasts of rocks that occur in the parent rock in the east in the highland, and from the well-rounded grains of quartz supplied from the west from a margin of the platform. During the atectonic periods the source of the material was mainly from the platform when in the marginal depression sandstones of essentially quartz and oligomictic type accumulated, their interbeds and bands wedging in the taphrogenic strata. Probably that kind of environment existed also in other places.

Extensive acid lava outflows are known to accompany the formation of taphrogenic strata and in the sedimentary rocks pyroclastic material is quite common. The dissected topography with volcanic structures certainly existed in the areas surrounding the near-fracture depressions during the period when clastic strata were accumulating.

The second type of clastic rocks is characteristic of the Riphean platform and miogeosynclinal strata and of the upper part of the Chaya lithostratigraphic complex of the Akitkanian. The sandstones that comprise it are well-sorted, quartz and oligomictic, sometimes arkosic and polymictic, interbedding with different clay shales and carbonate rocks and commonly containing glauconite. Red beds are locally known among this type of rock, but usually they are monotonous grey or greenish-grey. Judging from different structural aspects from mineral and lithologic associations, these sediments accumulated principally under a marine environment on a stable shelf or in a shallow-water geosynclinal basin.

The third type of clasts is also characteristic of Riphean formations, and is represented by red-colored oligomictic and arkose sandstones of different grade of sorting, with lenses of conglomerate and interbeds of pelitic sediments wedging out. The sandstones are typified by rough parallel- or cross-bedding, usually large-scale and unidirectional, rarely intersecting. In the shales sun cracks and rare traces of rain drops are registered. These rocks are widespread in many regions and were formed under a continental environment, in the extensive alluvial plains (for instance in the East European platform), and also in large intermontane depressions that originated at the final regressive stages of sedimentary macrocycles.

The red beds are abundant in the Neoprotozoic and their distribution can even be compared with that in the continental strata of the Phanerozoic, these latter originating at the time when oxygen concentration in the atmosphere became much higher. This fact seems to be satisfactorily explained by the lack of vegetation on Precambrian land, and as a result ferric iron did not reduce to ferrous iron in the process of sediment burial (as happened later when the concentration of organic matter in soil became significant).

Carbonate rocks forming thick strata in any Riphean sequences accumulated under shallow-water marine conditions; they were deposited on stable shelf as well as in geosynclinal areas. Stromatolites and microphytolites are known almost ubiquitously from these rocks, and these needed good light and a certain hydrodynamic regime typical of shallow-water and shore areas where tidal currents were active. The shallow-water environment is also evidenced by their occurrence in association with quartz sand (quartzite), by admixture of terrigenous minerals, by the presence of cross-bedding of the basin type, by wide distribution of detritic varieties, etc. Syngenetic dolomite is dominant among the carbonate rocks, resulting from a still higher concentration of carbon dioxide in the atmosphere and hydrosphere, from significant mineralization and higher carbonate alkalinity of seawater and its high temperature.

At the same time an increasing abundance of limestones during the Riphean indicated a constant decrease in partial pressure of $CO_2$ and consequently in the alkalinity reserve of seawater. The alkalinity decrease was also favored by a supply of freshwater from extensive continental massifs.

It was noted above that a higher concentration of $O_2$ in the atmosphere and hydrosphere decreased the migration of iron in seawater. As a result, ferruginous formations of jaspilite type (BIF) are not known from Riphean strata. The sheet deposits of hematite and siderite rocks characteristic of Riphean strata were likely to have been formed in shallow-water marine basin that possibly were of semi-closed type. Thus, from the point of view of evolutionary trend, lithogenesis of ferruginous ores migrated from eugeosynclinal areas of the Paleoprotozoic into miogeosynclinal zones of the Mesoprotozoic and then into miogeosynclinal and platform or shelf areas of the Neoprotozoic. Strakhov (1963) established that in the Phanerozoic the process of iron ore formation moved on the platforms entirely. The migrational routes of iron and manganese diverged greatly; in the Riphean these metals usually form separate deposits occurring in different facial environments; iron is registered in miogeosynclinal and platform formations, and manganese mostly in geosynclinal areas.

In the Riphean hematitic ferruginous ores as well as sideritic ores are known to occur, indicative of a peculiar geochemical environment at the time of their deposition. The hematitic ores are known to have originated under an environment rich in oxygen, and sideritic ores, on the contrary, need reduction conditions to be formed. The $O_2$ concentration in the atmosphere and hydrosphere at that time seems not to have been high, and thus some changes in chemical conditions related to facies could have resulted in the deposition of one or the other type of ore. On the other hand, glauconite occurrence in many Riphean sedimentary rocks evidences that seawater contains sufficient oxygen for the formation of this mineral in coastal and shelf areas.

Many other aspects of the Riphean lithogenesis could also be explained by peculiar climatic conditions. Thus, high temperature favored ready solution of silica and its long transportation in seawater. Cherts are widespread in Riphean strata and commonly associate with carbonate rocks, both types seeming to be shallow-water formations. This is also supported by the occurrence of microscopic remains of blue-green algae in these rocks.

Evaporites are more common in the Riphean than in any other older Precambrian units. Usually they are registered as inclusions and crystal molds after gypsum or halite in red beds and carbonate rocks, but locally they form interbeds that are grouped in sufficiently thick horizons or bands. It is of interest that evaporites occur not only in the platform strata, but also in the geosynclinal (miogeosynclinal to be more exact) complexes, for instance in the Missoula Group (the Belt Supergroup) and in the Mount Albert Group (the Mount Isa belt). During the Riphean the climate certainly became more arid, probably due to the formation of extensive land areas.

In the Riphean, different processes of weathering must have evolved intensely due to the existence of extensive areas of peneplanized land, hot humid climate, abundance of carbon dioxide in the atmosphere and also the presence of free oxygen. This is demonstrated by the formation of high-alumina sediments as a result of transportation of products of chemical weathering. An extremely high abundance of high-alumina rocks is observed in the Lower Neoprotozoic (Akitkanian) sedimentary-volcanogenic strata of the Baikal Mountain Land. Their formation seems to be related to intensive volcanic activity that resulted in an abundant supply of carbon dioxide in the atmosphere and ground water (Salop 1964–1967).

One more peculiarity of many of the Riphean sedimentary strata is the occurrence of phosphate bearing rocks and even of some relatively large phosphorite deposits, a feature which seems to be related to a great growth of biomass and to an acceleration of physiological activity in microorganisms. It should be noted that stromatolite structures are quite common in many carbonate rocks and are typified by a high concentration of phosphorus, phosphorite deposits usually occurring in the Middle and especially in the Upper Riphean, whose organic life was mostly rich and distinct.

If Neoprotozoic sedimentary rocks are examined as a whole, it is easily seen that they are similar in many aspects to the Phanerozoic sediments and that many specific ("exotic") formations typical of the older Precambrian units had disappeared. Here we certainly face the fact that the most important and critical changes of geochemical environment on the Earth's surface happened at the beginning of the Neoprotozoic Era and later the evolutionary process in the atmosphere and hydrosphere continued without any evident changes in quality that could have strongly affected the environment of sedimentation.

## 4. Endogenic Processes

When dealing with endogenic processes it is important to note the exceptionally significant global events that are known from an almost the entire duration of the Early Neoprotozoic, that is the large-scale subaerial outflows of acid lavas and the accompanying extrusions and intrusions registered throughout all the continents. The abundance of lava and pyroclastic products extruded at that time is enormous, though is difficult to estimate because a major part of the volcanogenic rocks was destroyed by erosion. It is possible to comprehend the extent of volcanism at that time by knowing a figure of 50,000 km$^3$ of acid lava that accumulated in the Akitkanian near-fracture depression alone (Salop 1964–1967). Magmatic processes of such a nature and on such a scale have never been registered during the Earth's existence, either earlier, or later. The taphrogenic volcanism of the Riphean and the Early Epiprotozoic, similar in some respects to the above, was of much less intensity. There is no doubt that the magmatic processes of the Early Neoprotozoic are explained by a general and rather long (about 300 m.y. duration) rise of the thermal front from the interior of the planet to its surface and by a simultaneous formation of numerous fractures in the Earth's crust that served as magma pathways.

Acid volcanics and comagmatic granitoids seem to have been formed as a result of their selective or complete anatexis of older sialic rocks, judging from their anchieutectic composition (quartz-potassic feldspar-albite), a certain excess of alumina, and common high alkalinity (Salop 1964–1967, Persson 1974 etc.). In many cases the inclusions of sialic rocks not fully melted or some disintegrated remains of dark-colored minerals are registered in volcanics. Anatectic processes were accompanied by the supply of juvenile matter and of volatile components in particular – fluorine and boron and also alkaline metals (K and Na), uranium and possibly thorium. The lowering of pressure in the fracture space could have partially favored melting of the crust rocks.

Basic lavas are much less abundant among the continental strata examined than acid ones, but they are nevertheless present almost ubiquitously, and at places their quantity is quite considerable. They are dominant, however, only in the volcanogenic-sedimentary strata of the upper (Chaya) lithostratigraphic complex of the Lower Neoprotozoic, where they formed under littoral-marine environment. Among these homogenous tholeiitic basalts prevail, andesites are lacking or quite rare. Thus, the Lower Neoprotozoic volcanics generally form a contrast (bimodal) association. Basalts commonly occur, together with dykes and thick sills of dolerite (or gabbro-diabase), that have the same composition as lavas. All the basites are supposed to have been formed from magma rather slightly differentiated and generated in the upper mantle.

The genesis of rapakivi of great interest. All data available support the assumption that the rocks of the gabbro-rapakivi formation generated in the magmatic chambers situated in the lower part of the Earth's crust near the boundary with the upper mantle. This assumption is also supported by the petrographic and petrochemical peculiarities of rapakivi, by their constant association with the earlier intrusive rocks of basic composition (gabbroids and anorthosites), and by their occurrence in acid volcanic strata; they reveal, moreover some regularities in forming composite and multi-phased plutonic bodies and they occupy a certain position in the Earth's crust, and their bodies exhibit a certain shape (flat and laccolith-like types with deep roots). It is suggested that originally the material of the upper mantle was melted as a result of the rise of geoisotherms and then the heat front embraced the sialic shell and the intrusions and extrusions of acid magma appeared. In cases where the heat front was pulsating near the crust–mantle boundary, the intrusions of basic and acid magmas were intermittent in time, and in the border zone the magmatic bodies of mixed composition could originate and syntectic (hybrid) rocks of dioritic, syenitic, and monzonitic type were formed. After the granite melt was crystallized along the newly formed fractures (partially of contraction type) with their roots deep in the interior of the Earth, the basic magma rose again and the dyke series appeared.

If the rocks of the gabbro-rapakivi formation originated under the regime presumed, then many of their specific aspects are easily explained. For instance, a higher initial ratio of strontium isotopes in some rocks indicates a supply of the radiogenic $^{87}$Sr from an older substratum that suffered selective melting. The presence of unequilibrium mineral associations in rapakivi (for example, of "armored" relict minerals) and of homogenic inclusions of basic composition, is probably related to syntexis and contamination phenomena, and the presence of eorroded ovoids of potassic feldspar with plagioclase coatings is likely to be due to pulsations of the heat (and emanation) front and to a simultaneous change in the composition of the residual melts.

This remarkable occurrence of the rapakivi intrusions in the Early Neoprotozoic still remains unclear in many aspects. It can probably also be explained by the exceptional thermal and tectonic regime of that time: (1) rise of thermal front of rather long duration and common for the entire planet so that not only the abyssal mobile zones were affected, but also the under-cratonal areas where the melts that are responsible for the origin of rapakivi and gabbro-anorthosites generated; (2) a relatively quiet tectonic regime during magma crystallization; (3) general expansion of the Earth's crust; (4) the extremely deep situation of the zone of melting (magma-formation) and near-surface formation of major intrusive bodies and a great resulting difference in temperature of solidus and magma crystallization.

All these facts can serve as a possible explanation of the constant occurrence of rapakivi (and anorthosite) in the areas that suffered earlier significant consolidation, recurrences of rapakivi formation during the whole subera, and many petrographic peculiarities of the rocks examined, the massive structure of granites, presence of abundant inclusions, lack of protoclase and of any other traces of stresses, and extensive contraction fractures filled in by the products of basic magma (dolerite dykes) etc. It has been noted above (Chap. 2) that during the formation of layered intrusions of anorthosites, the process of magma crystallization, if it is to occur under quiet conditions and a separation of a crystalline phase, is to occur at the same time. It is notable that rapakivi and anorthosites are lacking among granites and gabbroids of the Lower Neoprotozoic that are situated in the mobile zones, especially in the zones of active abyssal fractures.

The intensive thermal processes recorded in the Early Neoprotozoic are likely to be explained by a sharp increase in differentiation of the mantle matter on a world-wide scale. Brooks and his co-authors (Brooks et al. 1976, Brooks and Hart 1978 etc.) made isotopic studies on Cenozoic tholeiitic basalts of oceanic islands and mid-oceanic ridges and also on the Mesozoic dolerites of tholeiitic composition distributed on different Southern continents. These studies showed that the $^{87}Sr/^{86}Sr$ and Rb/Sr ratios in these rocks exhibit average values, and give a distinct positive correlation on the Rb-Sr isochron diagram, the ratios being determined on numerous samples from different groups of exposures (Fig. 57). The approximated line obtained corresponds to an age of $1620 \pm 55$ m.y. and is reasonably interpreted as the "mantle isochron" by these and many other authors. It gives a time of abyssal differentiation that resulted in the creation and separation of the asthenosphere chemically isolated from the mantle that was formerly of more monotonous composition. The isotope studies of oceanic basalts by the Pb-Pb method done by Tatsumoto (1978) gave an age of 1500–1800 m.y. for the mantle isochron, which also evidences that this dating is valid.

The values obtained on the mantle Rb-Sr and Pb-Pb isochrons are very close and corresond fairly well to the time of global tectono-magmatic events of the Early Neoprotozoic. As the Meso-Cenozoic basalts and dolerites are the products of dif-ferentiation or partial melting of the matter of the new mantle (asthenosphere) that originated 1800–1600 m.y. ago, all other types of basites formed after the events described during the Late Precambrian and Paleozoic are also its derivatives. Thus it is possible to state that the thermal processes of the Early Neoprotozoic reflect a sharp turn in the evolution of the interior of the planet.

Now a brief examination of endogenic events of the Riphean will be given. Three evolutionary stages are recognized in the Riphean (the Early, Middle, and Late) when sedimentary and volcanogenic-sedimentary strata accumulated and three diastrophic cycles terminating these stages occurred (the Kibaran, Avzyan, and Grenville), ac-companied by different intrusions.

During the evolutionary stages of a rather quiet tectonic regime, the endogenic processes principally occurred as local, rather intense lava outflow that was registered on platforms as well as in geosynclines; in different regions the acme of volcanic

activity fell at a different time. Thus, for instance, in the North American platform
the most intensive lava outflow occurred in the Upper Neoprotozoic; and in the
taphrogenic depressions of South Western Africa and Central Australia it was in the
Upper and Terminal Neoprotozoic; in the East European and Siberian platforms the
volcanic processes are hardly distinguishable from the Riphean. In geosynclinal areas
of Central Asia volcanic activity occurred mainly in the Middle Neoprotozoic, in the
geosynclines of Equatorial Africa in the Upper and Terminal Neoprotozoic, and in
North Eastern Africa and Arabia in the Middle Neoprotozoic.

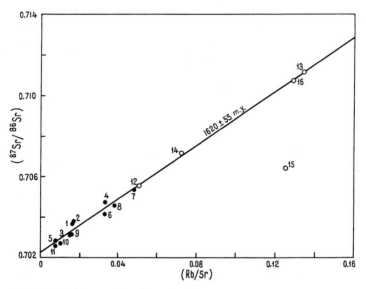

**Fig. 57.** Rb-Sr mantle isochron for the Cenozoic oceanic and Mesozoic continental tholeiites.
(After Brooks and Hart 1978).
*Dots* in the diagram stand for the average values obtained by several analyses of single samples
of a certain region. *Black circles* different oceanic associations, *light circles* different continental
associations.
Numbers of dots: *1–8* tholeiite basalt of different groups of oceanic island, *9–11* tholeiite basalt
of oceanic bottom, *12–16* continental tholeiite (dolerite) from Antarctic, Southern Africa, South
America, and Tasmania

    The following regularity is known to exist: in most cases, in the newly formed
Riphean geosynclinal belts the maximum of volcanic activity fell on the initial stages
of development, as in the platform areas the lava outflow was more characteristic of
the middle (transgressive) stage of an evolutionary cycle when the deepest subsidence
of basin bottom occurred. Naturally, this regularity is clearly observed for the platform strata only in complete (three-fold) sections; in a case where the upper (regressive) portion of the section is eroded, a false impression of occurrence of lava in the
uppermost part of sedimentary macrorhythm is created. It was noted above (Chap. 4)
that such a situation of volcanics was mostly typical of many Mesoprotozoic strata;
it is very probably explained by their poor preservation.

The relationship of volcanic processes and tectonic movements reveals that the basic lava outflow occurred mainly at periods of subsiding crust, and the acid lava (and partly intermediate lava) outflow – at the periods of rising crust. Alternation of lava outflow of different composition is typical of an unstable tectonic regime (frequently occurring change of oscillating movements of different signs). It is assumed that the regularities observed apply not only to endogenic processes of the Riphean, but are of more common significance (as has often been pointed out).

It is of interest that in many geosynclinal belts of the Riphean, initial volcanics of a typical spilitic-keratophyric association are lacking or quite rare (an exception is the geosynclinal belts of Central Asia). This is connected probably with relatively shallow intra-geosynclinal fractures that controlled lava outflow and hyperbasite intrusions and also with the fact that Riphean eugeosynclinal zones are usually overlain by Phanerozoic geosynclinal complexes.

It was noted above that each of the three orogenic cycles of the Riphean was occompanied by intrusions of basic and acid magma, but the scale of plutonic phenomena and the composition of magma products that originated during different cycles were not the same. The Kibaran cycle of the second order was mostly typified by meso- and hypabyssal intrusions of acid magma that were a source of formation of different granitoids. Judging by the mineralogical and chemical composition of the latter, this magma was formed as a result of anatexis of the crust sialic material accompanied by a supply of alkali and volatile components (this especially refers to the formation of certain muscovite rare metal granites and pegmatites in some regions). It seems that some local conditions were favorable for the formation of rapakivi or of rapakivi-like granites that exhibit many aspects of these rocks, though the typical orbicules are faint or even lacking. Basic magma intrusions preceding granite formation were quite rare and on a small scale, and the hypersbasite intrusions were of still less importance in geosynclinal belts. In the interior of platform areas, however, the basaltic magma was generated during the Kibaran cycle, at its very start and also just after its termination, the fracture intrusions of this magma resulting in the creation of a whole system of dolerite dykes. The post-tectonic dykes occur also in geosynclinal belts, where they cut the Kibaran granites.

Endogenic processes accompanying the Avzyan orogeny of the second–third order are mainly represented by intrusions of basic magma. Relatively small massifs of leucocratic and alaskitoid granite, mainly distributed in geosynclinal belts and in the activated portions of platforms, sometimes reveal a lithogenic nature. Only some of the granitoid bodies of granodioritic-dioritic and grano-syenitic composition are likely to have been formed as a result of syntexis of basic hypogenic, and acid lithogenic magmas and partially as a result of contamination. Small bodies of granophyric granite usually occurring in large sills of gabbroids (for instance, in the Duluth massif in the Canadian shield) are probably the extreme products of basic magma differentiation.

During the Grenville orogenic cycle of the first order, a global heating of the planet interior occurred and the most intensive rise of geoisotherms is registered as having been in geosynclinal belts and in the extensive areas of thermal-tectonic activation of the platform basement. There the endogenic processes resulted in the formation of large magmatic chambers and in intensive, though local, metamorphism of the Riphean and pre-Riphean rocks. In the activation zones metamorphism of basement rocks is principally of regressive nature and is easily seen in a partial or complete "isotope rejuvenation" of their age that all isotope methods usually reveal, though to a different extent.

Plutonic processes of the Grenville cycle are recorded as of different intensity in different geosynclinal belts. They are known to be quite strong in the large geosynclines of Central Asia, where at first the pre-tectonic and early tectonic intrusive bodies of gabbroids and hyperbasites were formed, these usually being situated in the zones of abyssal fractures, then large and huge massifs ("batholiths") of syn-tectonic calc-alkaline granites or gneiss-granites mainly occurring in the anticlinorium structures appeared, and finally the late-tectonic small and middle-sized diapir-plutons and stocks of alaskite granite, granodiorite, and granosyenite originated. Granites are usually of lithogenic nature and were formed of anatectic magma that had originated in the abyssal parts of the crust, but were crystallized at a higher level in the meso-abyssal environment. Some of the late tectonic granodiorites and granosyenites are possibly derived from hybridic syntectic magma. The common occurrence of lamprophyric dykes among these rocks is of interest. They formed as a result of filling-in of contraction fractures in hypogenic basic magma contaminated with residual products of granite crystallization.

In the areas of thermal-tectonic activation, the intrusive magmatism was much weaker. Only the late-tectonic granitoids are there in abundance, particularly their alaskitoid and sub-alkaline varieties. In some regions, in Equatorial Africa for instance, the small bodies of tin-bearing granite and pegmatite are very common. In the stable portions of the platforms with unaltered sedimentary cover, the endogenic processes are recorded as having formed numerous dolerite and diabase dykes.

## 5. Tectonic Regime

The beginning of the Neoprotozoic Era was shown to be characterized by fracturing of the Earth's crust in every continent, by formation of near-fracture depressions of different types, and also by intense subaerial outflow of lava. In some parts of the world the fractures and associated volcanic belts are known to form a system of two or more intersecting directions. The taphrogenic structures of Australia are an example of such a system, they cut the older socle of this continent and probably separate the Australian platform from the mobile (geosynclinal?) area that is situated to the east of it (Fig. 58). Problems still exist, however, in reconstructing the systems of fractures and volcanic belts of the Early Neoprotozoic, as in many regions the volcanic structures have been destroyed by erosion and are covered by younger rocks.

The distribution pattern of taphrogenic rift-like structures and peculiarities in sequence, composition, and genesis of sedimentary-volcanogenic strata and of the related comagmatic intrusions can easily be explained by supposing a world-scale fracturing of the Earth's crust as a result of a general expansion of the planet. A question arises; what was this process like in the areas now occupied by oceans? During the Early Neoprotozoic the marine strata did not accumulate at all or are known to be quite insignificant in the territories of modern continents, which prompts the supposition that at that time extensive and deep subsidences must have appeared in the areas of modern oceans and the seawater must have filled them. It is admitted that in that period a new asthenosphere became isolated (under the oceans at least) as a result of differentiation of the mantle matter on a world-wide scale (see above), so that it seems reasonable to suggest that the oceanic basins similar in size to the modern ones originated only at the beginning of Neoprotozoic. It will be shown later that this suggestion fits well in relation to the Pacific ocean.

If we consider the regularity of occurrence of basic volcanism simultaneously with extensive subsidences of the Earth's crust, then it is possible to assume that intensive basalt outflows accompanied formation of oceanic basins in the areas of subsidence.

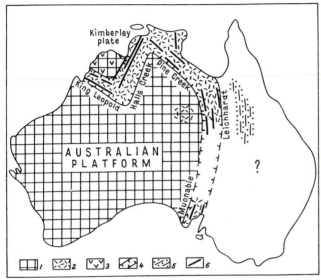

**Fig. 58.** Structural elements of Australia in the Lower Neoprotozoic.
*1* platform areas, *2* taphrogenic belts with acid lava, *3* plateau-basalt of the Kimberley plate, *4* boundaries of taphrogenic volcanic belts, fractures, *5* mobile (geosynclinal) area, *6* some large fractures

The tectonic regime during the Early Neoprotozoic was unstable: fracturing of the Earth's crust and continental taphrogenesis (riftogenesis?) and also the subaerial lava outflows were recorded as happening during the whole subera and were accompanied by folding and different intrusions in the fracture zones. The tectonic pattern shows that these structures originated as a result of vertical movements along the disjunctive dislocations. An evident intensification of tectonic activity in the middle of the Early Neoprotozoic (the Parguazan diastrophic episode) is recorded and mainly concerns the end of the Early Neoprotozoic (the Vyborgian diastrophism of the second-third order), when the most intense intrusions of rapakivi granites occurred.

This Early Neoprotozoic stage was of great importance for the entire subsequent tectonic evolution of the Earth. One great event of this period is the origin of modern oceanic depressions and the other, equally important event found in every continent is a transition of the Earth's crust into a new state, as shortly after its termination, in the Middle Neoprotozoic (Early Riphean), an intensive process of formation of large platforms, similar in extension to the modern ones on the one hand, and a process of generation of new geosynclinal systems on the other is registered, many of these latter continuing to develop during the entire Late Precambrian and Phanerozoic

The situation of major structural elements of the Earth in the Riphean is shown in Fig. 59. This picture is schematic, naturally, and is explained by the small scale of the map and lack of knowledge in some aspects; it is moreover known that outlines of platforms and geosynclines changed during the Riphean and that some small structural elements existed only during one or two suberas of the Neoprotozoic.

In Eurasia three large platforms are distinguished: the East European, Siberian, and Hindustan and other smaller ones: the Chukotkan, North Chinese, and South Chinese, and others. Many small stable blocks of the median-mass type are additionally observed inside the geosynclinal systems that separate the platforms.

The structure of the East-European platform has been studied best, and the geosynclinal belts surrounding it are also well known. In the tectonic scheme (Fig. 60) it is clearly seen that the contours of this platform in the Riphean were very similar to their modern contours, and at the same time they are sharply different from the mosaic outlines of the Mesoprotozoic cratonal elements (cf. Figs. 60 and 43). From the north-west the platform was bounded by a deep narrow trough with thick miogeosynclinal strata of "sparagmite" type of Riphean-Epiprotozoic age. This trough is here called the Eocaledonian geosyncline (Salop 1973 a), as the tectonic movements, favoring this structure origin and folding in it, preceded Caledonian dislocations, and the situation of this trough is very close to that of the Caledonian geosyncline. The Eocaledonian belt continued from Scandinavia through the Finnmark Peninsula, Rybachy, and Kanin Nos Peninsulas south-eastward, into Timan, and there formed an extensive portion of a geosynclinal structure sourrounding the craton that is recognized as the Hyperborean-Timan belt. A small Barents platform cut by the Kolvan aulacogen was situated to the north-east. In the area of the Polyudov Ridge, this belt is included in a large Riphean geosynclinal belt that surrounds the East-European platform from the east. The Riphean strata along this belt are typically miogeosynclinal and metamorphic strata of eugeosynclinal type are known only from the eastern slope of the Urals. Their belonging to the Riphean is not certain but quite probable, as they apparently belong to the inner part of geosyncline. It is seen from this scheme that the Neoprotozoic Riphean belt had almost the same limits as the Hercynian Urals belt.

---------------------------------------------------------------------------►

**Fig. 59.** Principal structural elements of the Earth in Riphean (paleotectonic scheme). Symbols: *1* platforms, *2* platforms and median masses (and their portions) subjected to partial mobilization during the Grenville diastrophism, *3* platform cover, *4* aulacogens, *5* geosynclinal belts, *6* geosynclinal belts where outcrops of basement dominate that suffered strong reworking during the Grenville diastrophism (the "roots of geosynclines").
Platforms: *I* North-American, *II* South American, *III* São-Francisco, *IV* East-European, *V* Siberian, *VI* Chukotkan; *VII* North-Chinese, *VIII* South-Chinese, *IX* Hindustan, *X* West-African, *XI* Nile, *XII* Neo-Kasai, *XIII* Transvaal-Rhodesian, *XIV* Australian.
Geosynclinal belts (figures in circles): *1* Innuitian, *2* Beltian, *3* Carolinian (Eastern-Greenland), *4* Greenville, *5* Old Andean, *6* Paraguay-Araguaia, *7* Eocaledonian, *8* Hyperborean-Timan, *9* Riphean, *10* Mediterranian, *11* Yenisei-Sayan, *12* Baikalian, *13* Central-Asiatic (geosynclinal system), *14* Kolyman, *15* Koryak, *16* Sikhotealin'-South Korean, *17* Old Himalayan, *18* Menaman (Indo-Chinese), *19* Eastern Ghats, *20* Red Sea, *21* Anti-Atlassian, *22* Nigerian, *23* Mauritanian, *24* Kibaran, *25* Mayombe-Kunene, *26* Orange-Natal, *27* Mount Isa

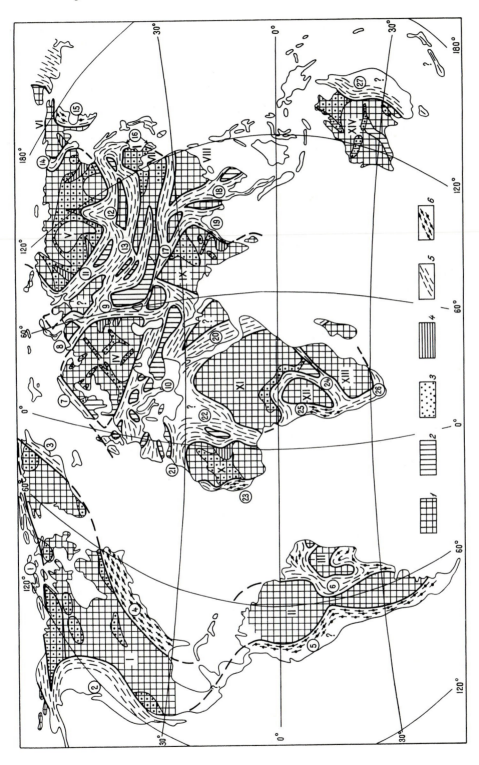

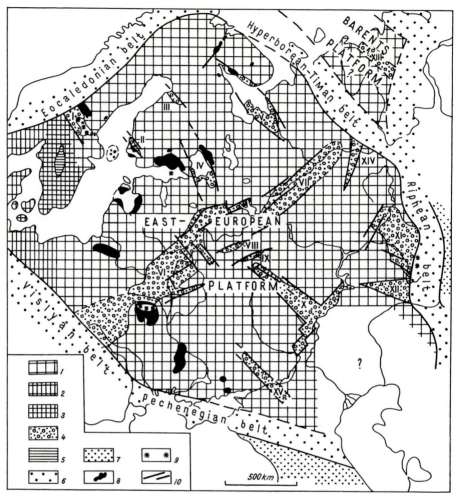

**Fig. 60.** Principal structural elements of Eastern Europe in Neoprotozoic. (After Salop 1973 a, with certain more precise definitions).
*1* platform, *2* a portion of platform activated at the end of the Terminal Neoprotozoic (Grenville-Sveco-Norwegian regeneration), *3* a portion of platform activated at the end of the Middle Neoprotozoic (Kibaran-Gothian regeneration) and the end of the Terminal Neoprotozoic, *4* depressions and aulacogens in the platform with the Riphean strata (*I* Dalarna, *II* Satakunta, *III* Mukhos, *IV* Ladogan, *V* Belomorian, *VI* Volyn-Kreststov, *VII* Soligalich-Yaren, *VIII* Gzhatsk, *IX* Moscowian, *X* Pachelma, *XI* Kama-Belsk, *XII* Radayevo-Abdulin, *XIII* Kolvan, *XIV* Kazhim, *XV* Dnieper-Donets), *5* Terminal Neoprotozoic (Dalslandian) strata deformed during the Sveco-Norwegian regeneration, *6* miogeosynclinal belts, *7* eugeo-synclinal belts, *8* rapakivi granite of the Vyborgian cycle (the Lower Neoprotozoic), *9* the Lower Neoprotozoic subaerial volcanics, *10* fractures

In the south and south-west the Riphean Pechenegian and Vislyan miogeosynclinal belts adjoin the East European platform; their structure is mainly known from drilling and geophysical studies. In the Great Caucasus Range the Riphean strata (the Khasaut Group) are of eugeosynclinal nature. In formational aspect they are correlatives of the Riphean sedimentary-volcanogenic strata of Bulgaria, Czechoslovakia (the Spilitic and pre-Spilitic Groups), and France (the Lower and Middle Brioverian), and at the same time they belong to the northern branch of the Old Tethys geosynclinal system.

Numerous large and small depressions locally surrounded and cut by fractures appeared in the East European platform during the Riphean. These depressions and grabens together formed a system of two perpendicular directions; north-western and north-eastern, approximately parallel to the rectangular platform boundaries and to the orientation of geosynclinal belts surrounding it. In the depressions the red-colored terrigenous or terrigenous-carbonate strata of platform type with local basalt flows accumulated during the whole Riphean (and a part of the Epiprotozoic). The platform depressions situated near the Riphean geosynclinal belt are typical aulacogens. The thickness of the Riphean strata in the Kama-Belsk (Kaltasinsk) aulacogen reaches 8 km.

The western part of the East European platform was activated during the Kibaran (Early Gothian) and Grenville (Sveco-Norwegian) orogenies and resulted in an isotope rejuvenation of the older crystalline basement rocks in Scandinavia, in emplacement of small granite intrusions, and in deformation of the Riphean strata in the Dalsland depression.

A huge Siberian platform appeared in North Asia in the Riphean, exceeding in size the modern platform of the same name (Fig. 61). It was formed as a result of the joining up of the Mesoprotozoic Angara, Chara, and Aldan cratonal blocks (platforms) and of the closing of certain older intra-cratonal geosynclinal belts. The Yenisei-Sayan, Baikal, and Okhotsk geosynclinal belts surrounded the platform from the south-west, south, and south-east, and represented the northern peripheral mobile zones of an extensive Central-Asiatic geosynclinal system. In the east and north-east the platform was bounded by the Kolyman miogeosynclinal belt that separated it from a relatively small Chukotkan platform situated in the north-easternmost part of Asia (see Fig. 59).

Clearly seen differentiation of structural elements exists in the geosynclinal belts surrounding the platform from the south. The mio- and eugeosynclinal zones and inner troughs and uplifs are quite distinct; a narrow and deep marginal trough is situated in the near-platform margin of the Baikal belt (see Figs. 50 and 53). The structural-formational zones and many of the sub-zones were bounded with abyssal fractures. These latter were also sometimes located at the platform boundary. Aulacogens filled in with thick and slightly dislocated subplatform strata of Riphean, deep and wedge-like in plan, branched out from the geosynclinal belts running toward the platform

A major portion of the Siberian platform during the Riphean was occupied by a shallow-water epicontinental sea and in its place are recorded thin terrigenous and carbonate sediments that now cover the basement in many places like a mantle.

The third large platform of Eurasia is the Hindustan platform, situated on the opposite (southern) side of the Central Asiatic geosynclinal system. In the Riphean

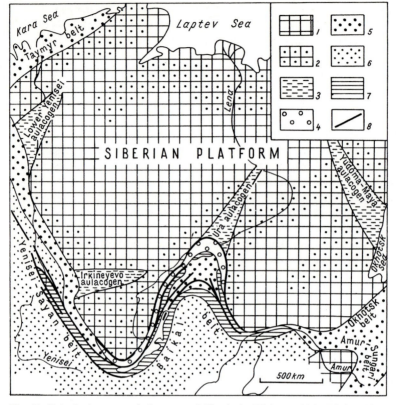

**Fig. 61.** Principal structural elements of Siberia in Riphean. (After Salop 1973a with certain more precise definitions).
*1* platforms, *2* platform cover, *3* aulacogens, *4* marginal troughs, *5* miogeosynclinal belts, *6* eugeosynclinal belts, *7* intrageosynclinal uplifts, *8* abyssal fractures

the old Himalayan belt of the Tethys (Alpine-Himalayan) segment was situated at its border; from the north-west the Afghanian (Afghanian-Baluchistan) geosyncline of Old-Tethys adjoined it; and from the south-east the Eastern Ghats belt bounded it, now only a basement reactivated during the Riphean (the "roots of geosyncline") is preserved. The southern part of the Hindustan platform is covered by the Indian Ocean. The Riphean rocks extensively cover the platform; in the Cuddapah aulacogen, close to the Eastern Ghats belt, these rocks are strongly deformed (see Fig. 54).

The North Chinese platform is situated inside the Central-Asiatic geosynclinal system and divides it into the northern and southern branches. In the west (in Central Asia and Kazakhstan) the semi-cratonal blocks (the median masses) of the basement adjoined it, in Riphean their structure was partially reworked. In the east the platform was surrounded by the Sikhote-Alin'-South Korean geosynclinal belt. Hence, in the eastern margin of the platform, the Riphean cover is widely developed in troughs and aulacogens, and sometimes is quite thick (in Northern China and in Northern Korea).

The tectonic evolution and structure of Africa during the Neoprotozoic is examined in detail in my book devoted to the Precambrian of this continent (Salop

1977 b). The paleotectonic scheme of the Neoprotozoic shown in Fig. 62 is taken from this book. It is seen from this scheme that four large platforms were situated within the limits of this continent, namely the following: West African, Nile, Neo-Kasai, Transvaal-Rhodesian (Zimbabwan), and also a small stable Bangweulu block, all of which were surrounded by geosynclinal belts filled in with dislocated Riphean strata. The Riphean platform cover is mainly developed in the West African platform: in the Taoudeni syneclise and in the Gourma aulacogen (see Fig. 55; in Fig. 62 this aulacogen is arbitrarily included in the Nigerian geosynclinal belt); in other platforms the Riphean cover is of local distribution.

The structural plan of Africa in the Neoprotozoic is different from that of the Mesoprotozoic; its rebuilding is obviously seen (see Fig. 41), and is due to the closing of certain geosynclinal belts during the Karelian orogeny and to a combination of the older cratonal blocks into the earlier mentioned larger platforms as a consequence. The contours and size of geosynclinal belts have also evidently changed. Some of the Mesoprotozoic belts continued to evolve in the Neoprotozoic, but as a rule their boundaries were different. In Equatorial Africa the new belts originated on the base of the older ones, and they partially occupied margins of the older Kasai craton, diminshing its size (the Neo-Kasai platform).

Rebuilding of the structural plan took place also in the Riphean. Thus, for instance, a general inversion of tectonic regime is recorded in the Kibaran, Oubangui-Burundi, Mayombe-Kunene, and partly in the Irumidian geosynclinal belts after the Kibaran orogeny strongly manifested in many parts of Africa, in the Equatorial area in particular; this inversion resulted in a complete or partial stabilization of an extensive part of the continent. The Nile and Neo-Kasai platforms and the Bangweulu block also became combined into a single huge craton – a prototype of the future Epiprotozoic Congo platform (see below). At the same time, a large Nigerian geosynclinal belt appeared in the Middle Riphean, developing in place of the Mesoprotozoic Eburnean and Old Nigerian belts and the movements in the Red Sea geosynclinal belt, inherited from the Mesoprotozoic Old Red Sea, became more intensive. At the beginning of the Middle Riphean a marine transgression started and reached its maximum in the Upper Riphean in the same region of North Western Africa. The transgression embraced not only geosynclinal depressions but also extensive parts of the West African platform.

The Grenville orogeny that completed the Neoprotozoic Era was quite intensive in all of the geosynclinal belts of Africa, but was mostly strong in the inner (eugeosynclinal) zone of the Nigerian belt, where the strained deformation of the Riphean rocks was accompanied by high-grade metamorphism and by different intrusions of basic, ultrabasic, and acid magma; and extensive fields of migmatites and metasomatically altered rocks appeared around large granite intrusions. At the same time, in the geosynclinal belts of Equatorial Africa subjected to inversion during the Kibaran orogeny, the Grenville (Late Kibaran) diatrophism principally resulted in the formation of peculiar fault-fold dislocations and in the emplacement of numerous but small massifs of tin-bearing alaskite granite and granite-porphyry.

To conclude the examination of the tectonic evolution of Africa in the Riphean, it should be stated that the principal trend of this stage of evolution consisted in unification of the older, relatively labile cratonal blocks (protoplatforms) into large stable cratons that possessed all the aspects of true platforms, including the thick

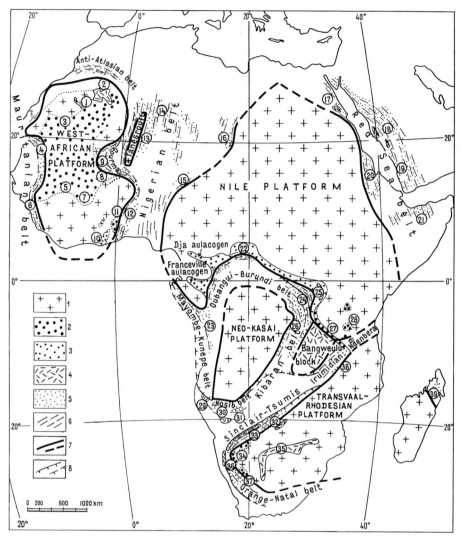

**Fig. 62** Principal structural elements of Africa in Neoprotozoic. (After Salop 1977 b).
*1* basement of cratons (platforms), *2* platform strata of Neoprotozoic (mainly of the Upper
Riphean), *3* the Lower Neoprotozoic subplatform (labile shelf) strata, *4* the Lower Neoprotozoic taphrogenic sedimentary-volcanogenic strata, *5* the Neoprotozoic (Riphean) geosynclinal
strata, *6* fold strikes; boundaries of: *7* platforms, *8* foldbelts stabilized after the Kibaran folding
(*hatching* on the lines run inside the belts).
Figures in circles – names of the Neoprotozoic groups: *1* Aioun-Malek, *2* Guelb el Hadid,
*3* Taoudeni, *4* Selibabi, *5* Segou-Madina Kuta, *6* Rokel River, *7* Bobo, *8* Lower Hombori,
*9* Lower Ydouban, *10* Tarkwa, *11* Morago, *12* Atacora, *13* Farusian, *14* Relaidie, *15* Zinder,
*16* Upper Tibesli, *17* Dokhan, *18* Halaban, Jiddah, *19* Murdama, *20* Tambien, *21* Inda Ad,
*22* Liki-Bembi, Sembe-Ouesso, *23* Mayombe, *24* Burundi, *25* Karagwe-Ankolean, *26* Kibara,
Kalonga, *27* Marungu, *28* Bukoban, Buanji, *29* Oendolongo, *30* Nosib, *31* Khoabendus,
*32* Tsumis, *33* Dorbabis, *34* Auborus, *35* Waterberg, *36* Sinclair, *37* Stinkfontein, *38* Mafingi,
*39* Manambato

cover. Unlike in Eurasia, however, in Africa the process of cratonization was not completed in the Neoprotozoic: the gigantic African platform was formed in the Eocambrian or at the beginning of the Paleozoic.

In Australia during the Riphean only one large platform appeared, its boundaries being very close to those of the modern Australian platform. In the east it was surrounded by the Mount Isa geosynclinal belt (Fig. 63), that probably occupied the entire eastern part of Australia, but its formations are only known from the north-east of the continent (in Queensland), and in the remaining teritory are overlain by a younger geosynclinal complex of the Epiprotozoic (the Adelaide geosyncline) and of the Paleozoic (the Tasmanian geosyncline). It was shown earlier that the Neoprotozoic formations of the Mount Isa belt are represented by taphrogenic strata of the Akitkanian and the overlying Lower Riphean unconformable miogeosynclinal rocks folded and altered during the Kibaran orogeny. It is, however, quite probable that eastward and south-eastward from the exposed part of the belt, the Middle and Upper Riphean geosynclinal strata (the eugeosynclinal formations including) occur under the younger geosynclinal complexes. Granites of the Grenville cycle are known to be present in the Mount Isa belt and thus it is suggested that its geosynclinal evolution was completed only at the end of the Neoprotozoic.

In the northern part of the Australian platform the Riphean cover occurs subhorizontally in deep depressions (the Victoria River, McArthur, and South Nicholson basins), which are inherited to a great extent from the Early Neoprotozoic taphrogenic structures and the Mesoprotozoic geosynclinal structures (cf. Figs. 42, 59, 63). In certain zones (Fitzmaurice and Batten) coinciding with the older abyssal fractures, the Riphean strata are apparently dislocated (near-fracture folding).

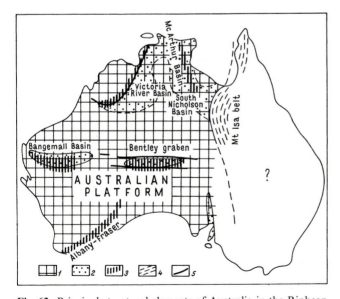

**Fig. 63.** Principal structural elements of Australia in the Riphean.
*1* platform, *2* Riphean platform cover, *3* deformation zones in the platform, *4* geosynclinal belt, *5* fracture zones in the platform

In the central part of the platform two sublatitudinal depressions are situated: the Bangemall and Bentley grabens. The first encloses thick sedimentary strata of all three Riphean groups, and the second still thicker, essentially volcanogenic strata of the Middle (or) Upper Riphean. The Riphean rocks in both depressions are commonly strongly deformed, for which reason they are attributed to geosynclinal formations. However, the strata of the Bangemal graben are typically platform and those of the Bantley graben are typically taphrogenic. It is realistic that the strong deformations of the structures examined are due to their occurrence in the rift-like zone that cuts the platform latitudinally. One more rift-like zone of meridional direction is possibly situated in the south of Australia near Spenser Bay. In the south-west of the platform the Albany-Fraser zone of Grenville activation was situated, within whose limits the older rocks of the basement suffered a strong isotope rejuvenation.

In the Neoprotozoic the tectonic pattern of North America already became close to the modern one on a general scale (Fig. 64). A huge North American platform originated instead of a relatively small cratonal block (protoplatforms) that had existed in the Mesoprotozoic (see Fig. 40). The great Beltian geosynclinal belt bounded it from the west, in its extension coinciding fairly well with the Alpine belt of the North American Cordilleras. Only its eastern near-platform margin is known and its Riphean strata are represented by miogeosynclinal formations. Some aspects of tectonic structure of the belt are suggestive of the possible occurrence of Riphean eugeosynclinal strata under the Phanerozoic folded complexes in the western, near-ocean zone.

The Carolinian geosynclinal belt, initiated in the Mesoprotozoic, adjoined the platform from the north-east, the Grenville belt that also existed in Mesoprotozoic from the south-east, but this latter with contours different from the older belt. It was principally built of basement rocks (from Katarchean to Mesoprotozoic), which had been intensively reworked by the Grenville orogeny. In some localities, however, Riphean metamorphosed rocks are squeezed into the tectonic wedges, and belong there, according to new data to the Flinton Group of Quebec and the Green Head Group of New Brunswick. It should be noted also that in the north-east of the belt, close to the platform, Riphean miogeosynclinal strata (transitional to the subplatform type) are known in the Seal Group, the grade of deformation and metamorphism of these strata evidently increasing south-eastward with increasing distance from the platform margin. On the whole, the Grenville belt represents deeply eroded root portions of the Riphean geosyncline. The so-called Grenville front serves as a boundary line between the foldbelt and the platform, and is the site of intensive differentiated movements and thermal activity registered in the marginal abyssal fractures. During the Paleozoic the Appalachian geosyncline was situated at the site of the Grenville fold belt; this also indicates an inheritance of the Phanerozoic structural plan from that of the Riphean.

The northern (north-western) boundary of the North American platform is not certain because of insufficient data available at the moment. However, the Riphean strata on the Victoria and Baffin Land Islands and in the north of Greenland are known to be of platform nature and on Ellesmere Island they are of miogeosynclinal character. Judging from this, it is possible to suggest that in the Riphean, a geosyncline had already been formed in the place where the Innuitian belt appeared in the Early Paleozoic.

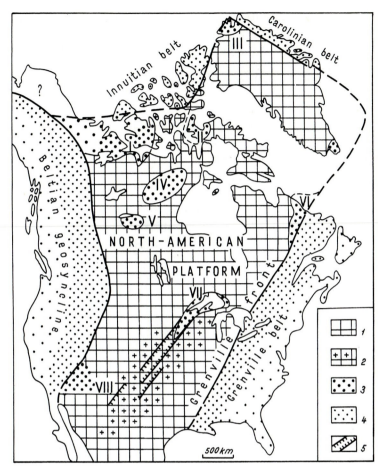

**Fig. 64.** Principal structural elements of North America in the Riphean.
*1* platform (basement), *2* area of Early Riphean tectono-magmatic activation, *3* platform cover, *4* geosynclinal belts (*2–3* rare signs show the areas of their presumed distribution), *5* fractures. Platform depressions (basins of sedimentation and platform volcanism): *I* Coppermine-Bear Lake-Victoria, *II* Borden, *III* Peary Land, *IV* Dubawnt, *V* Athabasca; *VI* Seal, *VII* Keweenaw, *VIII* Colorado

Problems exist as well in establishing the southern boundary of the platform. It is, however, possible to believe that the North American platform was more extensive in the Riphean than nowadays, and was followed further southward from its modern boundary because in the south of the Midcontinent, in the Colorado Plateau, and in adjacent areas of Arizona, the Neoprotozoic strata are of a platform and subplatform nature.

Depressions and grabens existed in the North American platform, as well as being registered in other Riphean platforms, and thick redbeds and volcanogenic strata of the Akitkanian and Riphean accumulated there, and have been examined to a certain extent earlier. Their approximate contours are shown in Fig. 64, a reconstruction of their true outlines is not feasible at the moment.

In the southern part of the platform there is an extensive area of Early Riphean (Kibaran, Mazatzal) tectono-magmatic activation typified by a wide distribution of fractures and rift-like structures associated with small bodies of granites (sometimes of rapakivi-like type), granite-porphyries, gabbroids, syenites, alkaline rocks, and massifs of composite structure made up of gabbro-granites emplaced in the range of 1460–1320 m.y. (Emslie 1978).

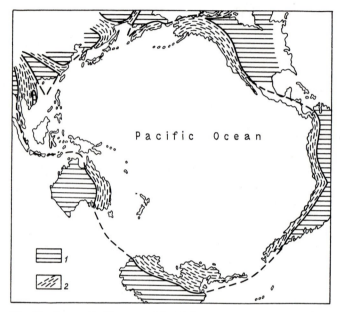

**Fig. 65.** Circum-Pacific area of the Riphean geosynclines. *1* Riphean platforms, *2* Riphean geosynclines (foldbelts)

The structural plan of South America for the Riphean is roughly known. The paleotectonic scheme is shown in Fig. 59 according to data of Almeida and his co-authors (1976). There is seen to have existed in the continent a large South American platform and a part of another craton that these authors recognize as the São Francisco platform. In the west the South American platform is surrounded by the Old Andean extensive geosynclinal belt, where the Precambrian metamorphic complexes and also of the possible Riphean complexes crop out from under the Phanerozoic folded formations; in any case, some of the granites distributed there are 680 m.y. old, which means that the Old Andean geosyncline suffered the tectono-plutonic processes of the Katangan orogeny to a great extent and that this cycle terminated the Epiprotozoic Era. (It is worth mentioning here that the Beltian geosyncline in the western frame of the North American platform also continued to evolve during the Epiprotozoic). From the eastern side the South American platform was surrounded by the Paraguay-Araguaia geosynclinal belt where the Riphean rocks are intensely folded and metamorphosed. The same type of formation surrounds the San Francisco massif (platform), though the principal rocks of the fold belts surrounding the craton are the basement rocks reactivated in the Late Precambrian.

The platform cover in the South American craton is mainly represented by the Lower Neoprotozoic rocks (the Roraima Uatuma Groups, and others), but locally the Riphean younger rocks are also reported (for instance, the Lavras Group of the Selitri depression) and also the Epiprotozoic sedimentary strata (for instance, the Bambui Group of the same depression).

It is exceptionally important that the Old Andean and Beltian geosynclines surround the modern Pacific Ocean from the east along almost the entire extension of both continents, exactly following its boundaries. The Riphean structures of the Beltian geosyncline in the north of the continent curve, forming an arc, and pass into north-eastern Asia and continue there in the Koryak arch-shaped geosynclinal belt that surrounds the ocean from the north-west (Fig. 65). The Western Pacific coast is known to be cut by marginal seas and surrounded by island-arc system that were formed during the Cenozoic as a result of fracturing of the older continental margin of Asia and its submergence when the Earth's crust was expanding (Karig 1971, Kropotkin 1972, Bersenev 1973, Rotman 1975). This is why the Neoprotozoic geosynclinal groups are known there only as fragments: commonly they are covered with thick younger strata or by marginal seas. In Kamchatka the Precambrian sedimentary-volcanogenic strata are known to crop out at some places from under Phanerozoic rocks (the Stenovskaya Group of the Ganalsk block and its analogs in the Khavyven block). These rocks are similar to the Akitkanian taphrogenic strata, but are very likely to belong to the Riphean. Riphean and Epiprotozoic miogeosynclinal formations occur at the base of folded complexes of Sikhote-Alin' (the Khankay massif), of Southern Korea, and possibly of Japan (the Hida zone); the folds exhibit a general submeridional orientation parallel to the western margin of the Ocean. In the south-west of the Circum-Pacific region the Riphean fold structures surrounding it run into Australia and crop out in the Mount Isa belt and probably in the base of the Tasmanian and New Zealand Phanerozoic belts. Finally, in the south, the Circum-Pacific region of the Riphean geosynclines closes with the large Late Precambrian foldbelt of the Antarctic (Ravich and Grikurov 1978).

Thus the foldbelts surrounding the Pacific Ocean originated soon after the Early Neoprotozoic taphrogenic stage (and partly also during it). It is reasonable to suppose that the oceanic depression appeared at the same time, but certainly its limits were different from the modern ones. Here I share the views of many who think that the Circum-Pacific geosynclines were formed at the margins of the older continents and thus belong to the paraliageosynclinal type.

The depressions that had existed in place of the modern Atlantic Ocean are very likely to have widened and deepened during the Riphean. Notable is the occurrence of certain Riphean geosynclinal belts in the margins of this ocean and their parallel orientation to its coasts (see Fig. 59). Thus, it is suggested that an extensive oceanic basin had appeared at the time examined, it was an incipient stage of the Atlantic Ocean. It will be shown below that the Pre-Atlantic Ocean acquired its shape similar to the modern boundaries a little bit later, during Epiprotozoic.

## 6. Principal Stages of Geologic Evolution

In conclusion the sequence of major geologic events of the Neoprotozoic will be briefly examined using the material of the analysis above.

### The Early Neoprotozoic (Akitkanian)

1. $1900-1700 \pm 50$ m.y. Global expansion of the Earth in the continents was accompanied by formation of fracture systems, of grabens and depressions, by intensive outflows of lava principally of acid composition, by intrusions of comagmatic granites and by deposition of thick continental red-colored volcanigenic-sedimentary piles (formation of the Khibelenian taphrogenic lithostratigraphic complex). General sea regression in all the continents and formation of the most extensive land areas. An origin of the large and deep oceanic depressions in the sites of modern oceans (in the Pacific Ocean in particular); a presumed outflow of basic lavas in the bottom of these oceans. The appearance of a new chemically isolated asthenosphere in the upper mantle of the planet.

2. $1700 \pm 50$ m.y. The Parguazan diastrophic episode. Undulating movements, local gentle folding; intrusions of gabbro and anorthosites, of large granite mass, of rapakivi granites in particular.

3. $1700 \pm 50$ m.y.$- \sim 1600$ m.y. Lessening of magmatic and tectonic activity; accumulation of continental red terrigenous or volcanogenic-terrigenous strata that correspond to the lower-taphrogenic part of the Chaya lithostratigraphic complex. At the end of the period: the beginning of sea transgression, accumulation of littoral-marine sediments on the shelves and in the marginal depressions, intensive lava outflow at places, of basaltic composition in most cases (formation of the upper part of the Chayan lithostratigraphic complex).

4. $\sim 1600$ m.y. The Vyborgian diastrophic cycle of the second-third order. Folding (mainly of the near-fracture type), emplacement of basic and acid magma (formation of gabbroid and granite massifs, including that of rapakivi granites.

### The Middle Neoprotozoic (the Early Riphean)

5. $\sim 1600-1400$ m.y. Sea transgression continues in the continents on a greater scale. Initiation or deepening of many geosynclines: of the Riphean, Baikalian, Yenisei-Sayan, Beltian, Grenville, Kibaran, Mount Isa, and others. Some geosynclines evolved in the sites of large near-fracture depressions of the Early Neoprotozoic. Extension of oceanic basins, of the older Pacific Ocean in particular (the "Pre-Pacific"). Deposition of littoral-marine and continental sedimentary piles, commonly red-colored in the shelves and geosynclinal troughs. At places sedimentation was accompanied by basalt outflows. In some near-fracture depressions the sedimentary-volcanogenic strata accumulated and their composition was similar to that of the Lower Neoprotozoic taphrogenic rocks. Considerable evolution of photosynthesizing algae with an accompanying effect of increase of free oxygen concentration in the atmosphere and hydrosphere. The seawater gradually turned from the chloride-carbonate type poor in sulfates into chloride-sulfate type (this process of geochemical evolution is likely to have begun at the end of the Mesoprotozoic).

6. $1400-1350$ m.y. The Kibaran diastrophic cycle of the second order. Folding and granite intrusions in geosynclinal belts. In the platforms there were strong oscillating movements, initiation of fractures, injection of basic and acid magma along the fractures. A partial inversion of the tectonic regime in certain Riphean geosynclinal belts was the result of diastrophism; and some small geosynclines (in Equatorial Africa, for instance) have closed completely.

### The Late Neoprotozoic (the Middle Riphean)

7. 1350–1200 m.y. Shallow-water marine and partially continental sediments accumulated in the platforms, and different marine sediments in the geosynclinal troughs. Carbonate deposits became more abundant in all the basins. In the North American platform basalt outflows were intensive, locally they were accompanied by comagmatic intrusions of basic magma, in other platforms the basalt outflows were of lower-scale. The tectonic differentiation increased sufficiently in many geosynclinal belts; some of the marginal depressions originated, some of them became deeper. Algal biomass became more abundant; eukaryotes were common among microbiota.

8. 1250–1200 m.y. The Avzyan (the Late Gothian) diastrophism of the second-third order. Gentle folding and small intrusions of basic and acid magma in almost every geosynclinal belt. In the platforms the oscillating movements and injection of basic magma along the fractures (the dyke group of diabases and dolerites).

### The Terminal Neoprotozoic (the Late Riphean)

9. 1200–1000 m.y. Deposition of littoral-marine and continental (commonly red-colored) sediments in the platform basins and of different marine sediments locally accompanied by outflows of basic and acid lavas in geosynclinal troughs (sometimes the volcanic processes are quite intensive in the latter). In the mobile belts tectonic differentiation is still increasing and tectonic forms of relief (morphostructures) exhibit a higher grade of contrast. The new geosynclinal belts or the earlier intrageosynclinal troughs become deeper, these processes being accompanied by intensification of volcanic activity (in the Farusian trough of the Nigerian belt, for instance). In the platforms locally near-fracture troughs with psephytic strata and (or) volcanics of taphrogenic type originate. At the end of the stage the uplifts and regressions of sea due to it occur in many geosynclines and in the platforms; the early red-colored molasse starts to form at that time. Further increasing abundance of plant biomass is typical. Stromatolites and microphytolites are at their acme for the whole geologic evolution. Advancing evolution of the eukaryotes of the primitive plant *bios*. Growing concentration of $O_2$ in atmosphere and hydrosphere. Red beds are exceptionally widespread.

10. ~ 1000 m.y. The Grenville diastrophic cycle of the first oder. Intensive folding, gabbroid and granitoid intrusions, thermal-tectonic activation of the older basement rocks, metamorphism of the Riphean rocks in almost every geosyncline. Some geosynclines suffer a complete inversion and close, in some cases a partial inversion of tectonic regime is recorded and they continue to evolve in the Epiprotozoic and Phanerozoic. The areas of the platforms are characterized by oscillating movements, by fracturing and injection of basic magma along the fractures.

# Chapter 6   The Epiprotozoic

## A. Rock Records

### 1. General Characteristics of the Epiprotozoic

**Definition.** The Epiprotozoic consists of supracrustal and plutonic rocks formed after the Grenville orogenic cycle of the first order ($\sim$ 1000 m.y.) but before the termination of the Katangan orogeny of the first order (680–650 m.y.). Thus, the formations examined constitute a separate stage in geologic evolution of the planet and its duration ($\sim$ 350 m.y.) is more than that of the Paleozoic Era (330 m.y.). Nevertheless, this large Precambrian unit was not properly evaluated till recently, and in any of the widely accepted scales it was included in the Upper Proterozoic (the Upper Riphean) or in the lowermost Eocambrian (the Lower Vendian). The Epiprotozoic was recognized as an erathem (era) in the Precambrian in my paper in 1964 (Salop 1964). This erathem could also be called the Katangan or Adelaidean according to the typical strata of Equatorial Africa or Southern Australia.

**Distribution.** The Epiprotozoic rocks are widespread in all continents (they are unknown up to the present in the Antarctic continent alone). It is beyond the scope of this work even to enumerate the larger stratigraphic units of the Epiprotozoic by their proper names, as they are tens and possibly hundreds. Here only examples of the typical units from different continents will be given with the aim of showing what kind of formation is meant. In Europe there are the Hedmark and Vestertana Groups in Norway, the Polarisbreen Group of Spitzbergen, the Lower Dalradian in Great Britain, the Upper Brioverian in France, the Vilchansk Group in Byelorussia, the Biarmian complex of the Middle Urals; in Asia, the Akbulak Group of Kazakhstan, the Chingasan Group of the Yenisei Ridge, the Bodaybo Subgroub of the Patom Highlands, the Kuhen Group in Korea; in North America, a part of the Conception Group of Newfoundland, the Windermere and Pokatello Groups of Cordilleras, the Mount Rogers Group of Appalachians, the Merkebjerg Group of Greenland; in South America, the Jangada and Bambui Goups in Brazil; in Africa, the Katangan Supergroup in Zambia, the Western-Congolese complex of Congo, the Damara Supergroup in Namibia, the "Pourprée serie" Ahnet in Algeria; in Australia, the Adelaidean Supergroup ("system") in the south of the continent, the Kuanidi and Louisa Downs Groups in the north-west, the Duerdin and Albert Edward Groups in the north.

**Brief Characteristics of the Erathem.** The most specific and typical feature of the Epiprotozoic Erathem is the occurrence of glaciogenic rocks known from two stratigraphic levels. In other aspects the Epiprotozoic rocks are similar to their formational analogs of the Neoprotozoic, particularly of the Riphean. The platform and miogeosynclinal strata are widely distributed, but the eugeosynclinal formations seem to be of lower abundance, and they are generally similar to the Riphean ones. An important feature is that the abundance of the red-colored rocks in the platform and miogeosynclinal types is not less and even sometimes higher than in the corresponding Neoprotozoic rocks. At the same time, the sheet siderite ores are quite rare in these rocks, but some sheet hematite and hematite-magnetite deposits are associated with certain glaciogenic rocks. The sedimentary strata contain copper mineralization. Commonly the platform continental strata contain red-colored sandstones that exhibit fine roundness of grains and coarse serial cross-bedding of eolian and stream type. In the lower part of the erathem volcanics of basic, intermediate, and acid composition are abundant in the pre-glacial and lower glacial strata.

The Epiprotozoic strata in any region of the world are commonly divided into two parts by an unconformity that corresponds to the Lufilian orogeny of the second order ($\sim 780$ m.y.). It permits us to distinguish two suberathems (suberas) in Epiprotozoic. The glaciogenic rocks are known to occur at two definite stratigraphic levels and thus for a more detailed divison of the erathem it is possible to use the climate-stratigraphic criterion and to distinguish five global units or horizons: the subglacial, lower glacial, interglacial, upper glacial, and superglacial. Three lower horizons belong to the lower suberathem, and two upper ones to the upper suberathem. It will be shown later that they are quite specific and can thus be recognized as a lithostratigraphic complex.

**Tectonic Structures and Rock Alteration.** The tectonic pattern of the Epiprotozoic is generally the same as of the Riphean formations. This refers also to metamorphic grades of rocks.

**Intrusive Rocks.** During the Epiprotozoic intrusive magmatism is recorded at the beginning of the era, during the formation of volcanic strata of the subglacial horizon, then in the middle of the era, during the Lufilian orogeny and at the end of the era, in particular during the Katangan orogeny of the first order.

The early intrusive rocks are principally represented by subvolcanic and hypabyssal, rarely mesoabyssal granitoids with granite-porphyry, granophyric and medium-grained granites, granodiorite, and granosyenite. As a rule they make up small cutting bodies among volcanogenic strata without any noticeable contact effect. These granitoids, occurring in association with volcanics and being close to these in composition and partially in structure, are possibly comagmatic to the acid and intermediate lavas. It is quite realistic also that some dyke and sheet bodies of gabbro and diabase resting in the volcanogenic strata are related to basic lavas.

In major regions the Lufilian orogeny resulted in the emplacement of dykes and sills of gabbro-diabase and in the manifestation of hydrothermal mineralization (of uranium in particular). In some parts of Africa, however, this orogeny is related to numerous small intrusions of the so-called red granites.

The plutonic processes of the Katangan orogeny completing the era were mostly intensive in the Epiprotozoic. In many geosynclinal belts they resulted in intrusions of small masses of basic and partly of ultrabasic magma at first and in large and small intrusions of granitic magma that became a source of formation of different calc-alkaline granites. In the Baikal geosynclinal system, for instance, the muscovite-bearing granite-pegmatite and pegmatite veins of the Mama mica-bearing field and also pink leucocratic granites of wider distribution belong to the Katangan cycle. Thermal processes of the Katangan orogeny are additionally responsible for isotope rejuvenation of granites of the Barguzin complex of the Upper Riphean in some cases (data of Mirkina, VSEGEI, 1981) and for a higher metamorphism of Epiprotozoic rocks. In the platforms, the Katangan events were of local effect, in the emplacements of ultrabasic-alkaline magma with a subsequent formation of small ring intrusive bodies of composite structure (the Ingilin massif in the Aldan shield serves as an example of it).

**Organic Remains.** Carbonate rocks of the subglacial horizon mostly contain stromatolites with the forms established from the Middle Riphean (*Baicalia*), Upper Riphean (*Inseria, Katavia, Minjaria, Gymnosolen*, etc.) and Eocambrian (*Linella, Boxonia*, etc.) in the territory of the U.S.S.R. and also many "endemic" forms that are not known in Eurasia. Acritarchs are reported among microfossils, including those of the group *Chuaria*, their representatives being mostly typical of the Upper Riphean. It is quite realistic to suggest that further studies of this group (and of the related group *Kildinella*) will enable forms typical only of the Epiprotozoic to be established within it. Abundant eukaryota are reported among microbiota together with prokaryota.

## 2. The Epiprotozoic Stratotype and Its Correlatives in Different Parts of the World

The Katangan Supergroup of Equatorial Africa and the Adelaidean Supergroup ("system") of Southern Australia are worth being proposed as a stratotype of the Epiprotozoic Erathem. Both these larger units similarly answer the requirements of a world stratotype, for which reason it is difficult to give preference to either of the two. Here the Katangan Supergroup is only arbitrarily taken as a stratotype (Salop 1972b, 1977b).

**Stratotype: the Katangan Supergroup and Its Correlatives in Other Parts of Equatorial Africa.** The Katangan miogeosynclinal Supergroup is distributed in the southern part of Zaïre (Katanga) and in adjacent areas of Zambia. Many workers have studied its stratigraphy, but Cahen and his co-workers made a very important contribution to it in many papers on the problems of the isotope geochronology of this supergroup (Cahen 1954, Cahen and Lepersonne 1967, etc.). Mendelson (1961) and also Drysdall and his co-workers (1972) gave details on the sequence structure of the supergroup in the Copper belt of Zambia.

The Roan Group constitutes the lower part of the Katangan Supergroup, which in the Copper belt is divisible into two parts or subgroups, the Lower and Upper Roans. The Lower Roan (600–1200 m) starts with the basal conglomerate resting unconformably on different pre-Katangan (Kibaran) rocks, up-section follow the thick strata of quartzite, feldspathic quartzite, phyllitized slate, greywacke, and arkose with dolomite interbeds in the upper portion. The Upper Roan Subgroup is composed of slate with sandstone and quartzite interbeds in the lower part (300–1500 m) and of dolomite interbedding with slate, quartzite, and of some horizons of sedimentary breccias (300–600 m) in the upper part. In Katanga the lower part of the Roan sequence does not crop out, as it was not affected by erosion. The exposed portion of the sequence, with a thickness of more than 1500 m, starts with cloritized sandstone with dolomitic cement, upward appear dolomites with sandstone and slate (phyllite) interbeds, then follow chloritic and talc slate with interbeds of sandstone and cherty dolomite, and finally the strata of dolomitic limestone with slate interbeds; stromatolites are locally known from limestone and near the top this limestone is wholly built up of oncolites. Syngenetic copper mineralization of exceptional economic importance occurs in the horizons of clay dolomite, carbonate slate, sandstone, and argillite in the Roan Group (in the Lower Roan mainly).

The overlying Mwashya Group (up to 800 m) rests with a break on the Roan Group. In Katanga it is made up of dark slate, oligomictic or feldspathic sandstone, and local arkose and puddingstone (the "Mwashya conglomerate") that are sometimes regarded as tillites, but it is not certain that they are glaciogenic rocks. In the Copper belt this group is mostly built up of fine-bedded calcareous slate and subordinate interbeds of cherty dolomite and quartzite.

The Mwashya Group is overlain by the so-called Great Conglomerate with a thickness of up to 330 m. Locally it rests on the older rocks of the group and even on the Kibaran formations due to erosion. The Great Conglomerate is a typical tillite with bands and lenses of fluvioglacial conglomerate and also with interbeds of quartzite and slate (phyllite).

The upper units of the Katangan Supergroup are combined into the Kundelungu Group. In the Copper belt this group is divided into three parts, and in Katanga into two parts. The lower part is separated from the overlying parts by a significant unconformity that according to Cahen corresponds to the early (Lufili) phase of the Katangan folding. That is why it is more reasonable to distinguish two separate groups: the Lower and Upper Kundelungu. In the Copper belt the middle and upper parts of the group (s. lato) belong to the Upper Kudelungu.

The Lower Kundelungu Group rests conformably on the Great Conglomerate; in Katanga the conglomerate is locally known at its base, limestone and dolomite appear upward, dolomite sometimes contains stromatolites, then follows slate and finally sandstone with interbeds of sandy limestone. In the south of Katanga the thickness of the group reaches 2500 m, but northward it decreases and even wedges out completely near the Bangweulu block. In the Copper belt the group is mainly built up of dolomite and slate.

The Upper Kundelungu Group overlies different units of the Katangan Supergroup, and directly the Kibaran and pre-Kibaran rocks in the north of Katanga. The so-called Little Conglomerate is situated at the base of the group, in the north of Katanga it represents a typical tillite and in the south it consists of marine-glacial sandstone and slate with disseminated pebbles (diamictite). The thickness of glaciogenic rocks varies from 50 to 80 m. In Katanga above the Little Conglomerate the following formations are situated in successive order: (a) the Kalule Formation (150–300 m): sandstone, calcareous slate, and pink stromatolitic limestone or dolomite; (b) the Kyubo Formation (1500–2000 m): red and grey sandstone, slate, calcareous slate; (c) the Plateau Formation (1500 m): red-colored arkose, gritstone, and slate. Some features of erosion are observed between the formations and common conglomerate or coarse-grained sandstone occurs at the base of the formations. The most significant unconformity is known at the base of the Plateau red beds that are almost unfolded. They probably originated after the Katangan orogeny and thus should be assigned to the Eocambrian formations.

The rocks making up the Katanga Supergroup are folded and the folds are combined into an arch-like system with its convex side to the north (see Fig. 72). This Katangan or Lufilian fold arc is a natural continuation of the Damara geosynclinal belt of Southern Africa. The intensity of deformation and metamorphic grade within this arc increase southward, that is toward the inner part of the fold belt. In the north, at its boundary with the Bangweulu block, the Kundelungu aulacogen branches out from it; here the platform or subplatform strata are distributed and the analogs of the upper units of the supergroup are observed, of the Upper Kundelungu Group in particular, but of reduced thickness.

The lower age boundary of the Katangan Supergroup is drawn according to its occurrence on the post-Kibaran post-tectonic tin-bearing granites that are dated at 1000 m.y. and thus belong to the Grenville (Late Kibaran) orogeny. The emplacement of the Lusaka granite is confined to a break between the Lower and Upper Kundelungu, and it forms small bodies in the Roan Group of the Copper belt in Zambia. This group is also cut by pegmatite veins. The intrusive bodies are lacking in the younger rocks of the Katangan Supergroup (the post Katangan granites and pegmatites are not known from Katanga). The Rb-Sr age of Lusaka red granite is 780 m.y. (Cahen and Snelling 1966) and its U-Th-Pb age on zircon is 860 m.y. (Barr et al. 1978). This scattering of values is great and the choice of the true value according to radiological (analytical) data is problematic. The figures close to the first value were, however, obtained for Lufilian granite from other African areas, including the adjacent regions. Thus the time of the Lufilian orogeny is suggested to be close to the Rb-Sr dating. The age of the earliest epigenetic uranium mineralization in the Katangan Supergroup of 714 m.y. is also close to this dating (Cahen and Snelling 1966). The Rb-Sr age of the Musoshi pegmatite (Cahen et al. 1970) that cuts the Roan Group is not reliable (see Salop 1977b).

It is notable that the Lufilian orogeny, fixed by the thermal events and by significant break in sedimentation, seems not to have been accompanied by intensive folding; in any case all rocks of the Katangan Supergroup are conformably folded and any obvious angular unconformities between its units are quite rare.

The Katangan orogeny, completing formation of the Katangan Supergroup, is dated in the range of 620–670 m.y. by the U-Th-Pb method on uranite from the epigenetic ores of the Shinkolobwe and other deposits (Cahen and Snelling 1966). The formation of hydrothermal uranium deposits is known to take a long time and thus the oldest value is supposed to be most suitable for the time of the early thermal events and the lower values characterize different stages of uranium redeposition. The thermal events of the Katangan belt seem to have ended during the Baikalian orogeny (520–550 m.y. ago), as is evidenced by dating of the remobilized granite of the basement and of the latest occurrences of uranium mineralization in redeposited hydrothermal deposits.

The Bushimaie Group, distributed in the Kasai craton and in the Neoprotozoic Kibaran belt to the north-west of the Katangan arc, is a platform analog of the lower part of the Katangan Supergroup or the Roan Group, to be more exact. Two successive assemblages of stromatolites are determined in the Bushimaie dolomites; among the first are the following stromatolites: *Conophyton ressoti* Mensh., *Tungusia, Baicalia mauritanica* B.-S., *B. aff. anastomosa* B.-S., etc.; the second is represented by *Gymnosolen* aff. *ramsayi* Steinm., *G.* (*Minjaria*) *uralica* Kryl., *Conophyton* sp., *Baicalia* (Bertrand-Sarfati 1972). Bertrand-Safati considers both these assemblages to be typical of the Upper Riphean. In the opinion of Raaben (see Semikhatov 1974) stromatolites from the Roan Group of Katanga "are of Middle Riphean appearance". In the previous chapter of this book it was shown that the Riphean is older than 1000 m.y., and the Roan and Bushmaie Groups are younger and thus stromatolites of Middle or Upper Riphean appearance occur in the younger strata in Equatorial Africa. It will be demonstrated later that the same situation is also typical of other Epiprotozoic strata.

The Lindi Supergroup is one more platform analog of the Katangan Supergroup, and occupies an extensive area in the north-eastern part of Zaïre, in the Oubangui-Congo interfluvial area. Details of its stratigraphy are known from the Verbeek's monograph (1970) and correlation of this group units and those of the Katangan Supergroup is given in Salop's book (1977b). The lower, Ituri quartzite-slate carbonate Group (~200 m), is naturally compared with the Roan and (or) Mwashya Groups, the overlying, Akwokwo tillite, with the Great Conglomerate, and the overlying rocks of the Lokoma Group (1100 m) and the Aruwimi Group (1700 m) with the Lower and Upper Kundelungu Groups respectively. Equivalents of the upper tillite (the "Little Conglomerate") of Katanga are lacking in north-eastern Zaïre. A break between the Lokoma and Aruwimi Groups corresponds chronologically to these rocks. The Aruwimi Group, principally composed of red-colored cross-bedded arkose and quartzite and of clay shale is in part a close analog of the molasse strata of the Upper Kundelungu. An age of 755 m.y. was obtained by the Pb-model method on galenite from the quartz vein cutting the Akwokwo (Mt. Homa) tillite (Cahen and Snelling 1966). This value probably corresponds to the thermal events of the Lufilian orogeny. A later episode of mineralization revealing the Katangan cycle gives the age of galenite (690 m.y.) from the carbonate vein in dolomite in the middle part of the supergroup.

The sequence of the Epiprotozoic miogeosynclinal strata in the Western Congolese belt is similar to the stratotype. These rocks are developed in the Atlantic coast of Equatorial Africa and are recognized as the Western Congolese complex (Cahen 1954, Cahen and Snelling 1966). The lower unit of the complex is the Sansikwa or Bamba Group (up to 2000 m), is made up of limestone, dolomite, different slates (or phyllites), sandstone, and quartzite. It is underlain by basal conglomerate and arkose that occur on the Lower Neoprotozoic Mayombe Group and the granite cutting it with unconformity; the youngest of the granites are dated at about 1100 m.y.

The Sansikwa Group is overlain with an erosion by the glaciogenic rocks divided as the "Lower Tillite" or the "Mamba Tillite". In Zaïre the tillite associates with basalt; in Gabon the tillite is represented by schistose mixtite with a thickness of up to 150 m. The glacial horizon reaches 400 m in other sites. The tillite is overlain by argillite, calcareous slate, calcareous sandstone, sandy limestone, and arkose with a thickness from 600 to 1000 m. These strata are recognized under different names in different parts of the foldbelt, but commonly they are called the Louila or Haut Shiloango Group.

Up-section the "Upper Tillite" or the "Niari Tillite" rests with a stratigraphic unconformity, and consists of unsorted puddingstone-like conglomerate with interbeds of schistose argillite and varved clay shale. The thickness of glacial strata varies from 5 to 150 m; locally tillite is missing in the section. Still further up in the sequence of the complex the "slate-limestone group" (1000–1250 m) rests, typified by alternation of bands of dolomite and limestone commonly silicified and oncolitic with stromatolites at places, with bands of slate, argillite, marl and sandstone, some of the rocks being red-colored.

The Mpioka Group (up to 1000 m) lies on the "slate-limestone group" with a break, and is made of red terrigenous rocks; quartzite-sandstone, cross-bedded arkose and conglomerate and argillite and clay shale that mainly occur in the upper part of the group.

The uppermost unit of the complex is the Inkisi Group (1500 m). Its rests unconformably on different horizons of the Mpioka Group and sometimes even on the "slate-limestone group" and in this case an angular unconformity is observed in its base. Conglomerate is recorded in the bottom of the group, which is generally composed of red arkose and quartzite.

The syngenetic sheet bodies of copper-lead-zinc ores are known from the "slate-limestone group". Galenite from the ores is dated at $740 \pm 50$ m.y. by the Pb-model method; probably it is the time of deposition of rocks of this group. The time of tectonic movements that completed formation of the Western-Congolese complex is 655 m.y. by the Rb-Sr method on mica from Boma muscovite granite and pegmatite (Cahen and Snelling 1966).

There are no difficulties in correlation of units of the Western-Congolese complex and of the Katangan Supergroup (see Fig. 67, Columns 1 and 2). The Sansikwa Group is a reliable correlative of the Roan and Mwashya Groups, that are similar to it in composition and are of the same stratigraphic situation. A natural correlation can be made between the "Lower Tillite" and the Great Conglomerate, the "slate-limestone group" and the Lower Kundelungu Group, and the "Upper Tillite" (Niari) and the Little Conglomerate of Katanga respectively. In this case the "slate-limestone group"corresponds to phyllite and slate resting on the Little Conglomerate in Zambia and there they are sometimes regarded as the Middle Kundelungu. The Mpioka and Inkisi red-colored molasse Groups may be correlated with the Upper Kundelungu red-colored Group; the first of these is compared with the Kalule and Kyubo Formations (the lower molasse), and the second one with the Plateau thick Formation (the upper molasse). An unconformity between the Mpioka and Inkisi Groups reflects the Katangan orogeny.

The Epiprotozoic tillite-bearing strata are also known from other parts of Equatorial Africa. It is the Mangbei Group of Cameroun, the Bembe (Chella) Group of Angola, the upper part of the Bukoba Supergroup of Tanzania, the Bunyoro Group of the Uganda and others (see Salop 1977 b). The Epiprotozoic glaciogenic rocks also became known in the east of Zaïre recently, in the Eastern Kivu province, where they make up the Chibangou Group (2000 m) consisting of tilloids, sandstone, and slate (Cahen et al. 1979). This group lies with an angular unconformity on the Riphean Nia-Kasiba Group and the granite cutting it, this latter being about 1000 m.y. old.

Granite cutting the Chibangou Group is dated at 775 ± 40 m.y. (Rb-Sr method); it permits the suggestion of a Lower Epiprotozoic age for the group examined.

**Epiprotozoic of Southern Africa.** The Epiprotozoic formations are mostly widespread in the north of Namibia, in the region of Damaraland, and the rocks are divisible here as the Damara Supergroup. This supergroup is a thick sequence of geosynclinal strata of the Damaran geosynclinal system that extends along the Atlantic Ocean and is followed in the north-eastern direction into the Katangan belt of Zambia and Zaïre (see Fig. 72). The inner structure of the supergroup is extremely composite due to facial changes in different directions. According to Martin (1965) two main facies are distinguished: the Outjo miogeosynclinal facies and the Swakop eugeosynclinal facies. The first is present along the northern margin of the geosynclinal system near the Congo platform, the second occupies the inner part of the geosyncline. A common feature of both is the presence of glaciogenic rocks. A brief generalized characteristic of the supergroup according to Martin is given below, a detailed subdivision of the supergroup is to be looked for in Hedberg (1979).

The miogeosynclinal Outjo facies is divided into the Otavi Group and the overlying Mulden Group. The Otavi Group starts with the Abenab Formation (1900–2700 m) of grey clay, sometimes phyllitized slate interbedding with dolomite and limestone with common stromatolites and oncolites. It contains additionally greywacke, gritstone, quartzite, and arkose. Large deposits of polymetallic ores (Pb-Zn-V-Cu) occur in the carbonate rocks of the Abenab Formation, probably of primary sedimentary origin. Pb-model dating of galenite from these ores gave 790 m.y. The overlying Tsumeb Formation (2300–3000 m) in the lower part is made up of tillite with interbeds of conglomerate, quartzite, and slate, and with lenses of sedimentary hematite ores (the "Tsumeb tillite"); then up-section follow thick strata of grey dolomite, limestone, shale, and quartzite. Tillite lies on the underlying rocks with a break; its thickness (together with the alternating rocks) varies from 210 to 600 m. Semikhatov determined the *Baicalia* aff. *rara* stromatolite from dolomites of the Abenab Formation, a form commonly encountered in the Middle-Upper Riphean strata of Siberia. In this particular case, however, it occurs in the younger rocks.

The Mulden Group rests unconformably on the Otavi Group, with a great break, but without any conspicuous angular unconformity. In the Otavi Mountains it comprises quartzite, arkose, and greywacke with a horizon of red sandy slate in the upper part; its thickness is 700 m. In the Northern Kakoveld massive gritstone and sandstone are dominant with bands and interbeds of shale and conglomerate, fanglomerate, greywacke, and quartzite; here its thickness is much greater and reaches 2200 m.

Eugeosynclinal piles of the Swakop facies are divisible into the Hakos Group – a correlative of the Otavi Group and the Khomas Group arbitrarily compared with the Mulden Group. The Hakos Group rests unconformably on the Upper Riphean Nosib Group and on the Mesoprotozoic metamorphic rocks and granites. Two formations are recognized in the Hakos Group: the Lower and Upper Hakos. The Lower Hakos Formation consists of marble interbedding with biotite schist, quartzite, chert, and amphibolite or amphibole schist (metavolcanics); its thickness reaches 1600 m. In the north-western margin of the eugeosynclinal belt, near the Kamanjab median mass, very thick strata (up to 8000 m) are situated at the base of the Hakos Group, these strata consist of altered acid volcanics of taphrogenic type, mainly ignimbrite, porphyry and tuff with interbeds of metasedimentary rocks in the upper part, probably these rocks substitute to some extent the metasedimentary rocks of the Lower Hakos Formation. Usually these strata are regarded as the Naauwpoort Formation (Miller 1980). Ubiquitously the Upper Hakos Formation starts with tillites with calcareous-slate cement interbedding with phyllite, quartzite, conglomerate, and containing lenses of magnetite-hematite ores (the "Chuos tillite"); the thickness of this glaciogenic pile amounts to 600 m. The tillite is overlain by strata (500–700 m) of quartzite, micaceous and amphibole schist, and marble; in some localities quartzite and schist is dominant, in others marble.

The Khomas Group overlies the Upper Hakos Formation without any break. It is made up of biotitic or garnet-biotitic schist, orthoamphibolite, and marble. Its thickness varies from 3000 to 8000 m, but the true thickness is difficult to estimate because of complicated folding. It is obvious that the Khomas Group has little similarity with the Mulden Group in its composition, but usually these two are compared. Unlike the latter, that rests unconformably everywhere, it is also closely related to the underlying rocks. It seems probable that this group belongs to the upper part of the Hakos Group, but in the miogeosynclonal zone its analogs were destroyed by erosion that occurred prior to deposition of the molasse strata of the Mulden Group.

Eugeosynclinal piles of the Damara Supergroup are of a higher alteration and of more intense folding than its miogeosynclinal equivalents. Many granite massifs affecting the wall-rocks are known among them; local migmatites are recorded in the contact aureole. Granites are dated by the U-Th-Pb method on different radioactive minerals and by the K-Ar method on micas, the values obtained (440–570 m.y.) reveal the time of the Damaran (Baikalian) folding, but it is possible that intrusions of the Katangan cycle also occur among various granites.

A great similarity is seen between the lower units of the Damara Supergroup: the Abenab Group (and the Lower Hakos Formation) and the Roan and Mwashya Groups, and the Tsumeb (and Chuos) tillite and the Great Conglomerate of Katanga when comparing piles of Damaraland with these of the stratotype of the erathem. The Upper Hakos Formation and overlying Khomas Group are very likely to be the correlatives of the interglacial Lower Kundelungu Group, and the Mulden molasse-like Group occurring unconformably greatly resembles the Upper Kundelungu Group similar to it in composition. Thus, the analogs of the upper glacial horizon (the Little Conglomerate) are lacking in Damaraland; probably they were destroyed by erosion prior to deposition of molasse.

In southern Namibia (Namaqualand) the Gariep Group, developed near the western margin of the Kalahari platform, is an analog of the Damara Supergroup. Opinions differ in interpreting the inner structure of this group. The stratigraphic scheme of Kröner (1971) seems to be well reasoned, but the older Upper Riphean Stinkfontein Formation (Group to be more exact) is wrongly included in the Gariep Group; it is cut by the granite (920 m.y.) that does not penetrate into the Gariep Group (s. str.). If this correction is accepted, the Gariep Group consists of two formations, Hilda and Numees.

The Hilda Formation (up to 4000 m) unconformably overlies the Stinkfontein Group and is made up of gritstone, arkose, and fine-pebbled conglomerate with bands of limestone and dolomite and acid volcanic flows. The age of the latter is

720 m.y. by Rb-Sr analysis. The Numees Formation rests on the Hilda Formation conformably or with an erosion. According to Martin (1965) at places it lies directly on the Stinkfontein Group. The upper limit of the formation (and of the whole Gariep Group) is difined by the transgressive occurrence on it of subhorizontal platform strata belonging to the Eocambrian-Lower Cambrian Nama Group.

In the Numees area thick strata (up to 600 m) of the Numees tillite with interbeds of quartzite are situated at the base of the formation, up-section appear the strata (1000 m) of stromatolite-bearing dolomite. The Numees tillite is of different composition: it can be typical continental fossil moraines or marine-glaciogenic rocks. The phyllitized slate with huge (up to 11 m) blocks and boulders of dolomite and gneiss are quite common among the tillites. The glacial origin of these rocks is favored by the presence of faceted and striated boulders, dropstones, and by the occurrence of related varve-like slate. A typical feature is also the presence of a horizon of magnetite-hematite ores. This very aspect permits a reliable correlation of the Numees tillite and the Tsumeb and Chuos tillites of the Damara Supergroup and the Great Conglomerate of the Katangan Supergroup. From this, the underlying Hilda Formation should correspond to the Abenab Formation (or the Lower Hakos Formation) of Damaraland and to the Roan and Mwashya Groups of Katanga, and the dolomite strata overlying the Numees tillite to the dolomite strata of Damaraland and to the Lower Kundelungu of Katanga respectively.

In Southern Africa the following groups (besides those described above) belong to the Epiprotozoic: the Buschmannsklippe Group (or Formation) of southern Damaraland, the Sijarira Group of Zimbabwe, and the Malmesbery and Kango Groups of the Cape belt; all of them contain glacial deposits.

**Epiprotozoic of Western and North-Western Africa.** The Epiprotozoic strata are exceptionally widespread in this part of Africa and are usually distributed in the areas where the Riphean rocks are known to exist, but they occur unconformably on the latter everywhere. They are characterized by platform and geosynclinal formations. The first type of rocks, together with the Riphean strata of the same formational type, fill in a huge Taoudeni syneclise and the Volta depression in the West-African platform, the second type of rocks occurs in the Nigerian and Anti-Atlas geosynclinal belts (see Fig. 72).

Two types of the platform sequence are recognized in the Volta depression, they correspond to different stratigraphic levels of the Epiprotozoic. One of these belongs to the Lower Epiprotozoic and these strata are developed in the north of the depression, in Togo and Benin. This is the Buem Group that is divisible into three formations. The lower formation (600 m) is made up of tillite, quartzite, and argillite with limestone interbeds. It is characterized by the occurrence of sheet hematite ores. The association of tillite and sedimentary iron ores is known to be a peculiar aspect of the lower glacial horizon of Equatorial and Southern Africa; later it will be shown that this regularity is on a world-wide scale. The middle formation (1200 – 1300 m) comprises quartz-sandstone and arkose, variegated argillite and siltstone. The upper formation (more than 600 m) consists of volcanogenic rocks: basalt, andesite, porphyry, trachyte, agglomerate lava, and tuff. The rocks are of extremely low grade and rest subhorizontally on the Upper Riphean Bassa (Kande) Group or on the Eburnean granite of the basement.

The other type of the Epiprotozoic sequence in the Volta depression is destributed in Ghana. This is the Oti Group that also consists of three formations, but they are of different composition (Bozhko 1975). The Byupe tillite (up to 200 m) occurs in the lower part of the group and rests unconformably on the Upper Riphean Morago and Akwapim Group. Unlike the Buem tillite, these glacial strata do not contain quartzite and limestone interbeds and iron ores are absent in the tillite. The tillite is overlain by the Prang Formation (650 m) made up of argillite with common interbeds of sandstone, limestone, and dolomite. Carbonate rocks contain stromatolites known from the Upper Riphean and Vendian of the U.S.S.R. Glauconite from the sandstone is dated at 620 m.y. by the K-Ar method. Dolomite contains barite that will be shown to be a typical feature in the carbonate rocks overlying the upper tillite in the Taoudeni syneclise. The Nasia Formation (700 m) completes the sequence, it is built up of green sandstone and argillite with fine limestone interbeds.

The Oti Group is overlain by the Obosum Group (400 m) with an erosion, and consists of red and green sandstone, argillite, and conglomerate; up in the section the Gambaga quartz sandstone (400 m) unconformably rests, and is arbitrarily compared with the deposits of the Taoudeni syneclise that are close in composition and contain arganic remains of Ordovician or Cambro(?)-Ordovician.

Thus, in the Volta depression two horizons of glaciogenic rocks of different age are known, corresponding to the Lower und Upper tillites of the stratotype. Naturally, the rocks separating them are compared with the interglacial sedimentary strata of Equatorial and Southern Africa, and the overlying rocks with the superglacial complex of these regions. The age of the Obosum Group is not quite certain. This group is very likely to be a correlative of red beds of Equatorial Africa (the Upper Kundelungu Group) that constitute an early molasse of the Katangan cycle, but it cannot be ruled out that it may be Eocambrian, and in this case could be regarded as a later molasse of the same cycle.

In the Taoudeni syneclise the Epiprotozoic strata consist of sedimentary rocks with tillites that in most cases belong to the Upper Suberathem. In different regions they are given different names, but generally the sequences are of similar composition and structure.

In Mauritania, in the northern limb of the syneclise the Epiprotozoic Kayes or Kiffa Group is distributed. It rests unconformably and sometimes with an angular unconformity on different units of the Riphean Taoudeni Supergroup or on the basement crystalline rocks of the platform. A horizon of the Jbeilat tillite is traced through the base of the group, pebbles and angular clasts with glacial striation are sometimes enclosed in carbonate-clayish cement. In the stratotypical sequence (the Afole anticline) the group has the following structure: (1) the Jbeilat tillite (0–60 m); (2) the "Lower dolomitic horizon": dolomitic limestone and dolomite with barite containing the *Collenia* and *Conophyton* stromatolites (50 m); (3) variegated clay shale with cherty interbeds (100–270 m); (4) sandstone and siltstone (100–1000 m; thickness increases toward the Mauritanian belt); (5) the "Upper dolomitic horizon" – sugar-like dolomite (25–60 m); (6) variegated clay shale with sandstone interbeds (50–150 m) that are substituted by red sandstone with pebbles along the strike. The "Upper dolomitic horizon" at places lies with a break on the underlying rocks and is probably comparable with the lower part of the Eocambrian (?) Dhar Group.

Clay shale from the Jbeilat tillite is 680 m.y. old according to Rb-Sr analysis (unpubl. data of J. Bock and L. Kloeft from the town of Strasburg; pers. comm. of N. Chumakov 1978). These data permit us to assign this tillite to the upper glacial of the Epiprotozoic.

In the southern limb of the syneclise, in Mali, the Nara Group overlies unconformably the Bandiagara red cross-bedded sandstone that makes up the Upper part of the Riphean Hombori Group (see Chap. 5). The Nara Group starts with tillite (several tens of meters), then upward it passes into a band (up to 80 m) of cream-colored sandstone and argillite with inderbeds of jasper-like rocks and dolomite with inclusions and veinlets of barite and then into variegated strata (up to 500 m) of argillite or clay shale alternating with red sandstone.

The Mali Group, distributed in Senegal and in the north-west of Mali, has a similar sequence, as also some of its correlatives. The cherty-carbonate rocks with a higher concentration of barium are a typical pile overlying tillite in every unit. Later it will be shown that this association is characteristic not only of the Taoudeni syneclise, but also of the corresponding Upper Epiprotozoic strata in some other parts of the world.

In the Nigerian belt the "série pourprée" Ahnet belongs to the Epiprotozoic geosynclinal formations. In the western Ahaggar it consists of thick sedimentary-volcanogenic strata with tillites. Essentially this group consists of two or three groups separated by breaks (Caby 1969, 1971, Caby and Moussu 1968) and thus it may rather be regarded as a supergroup.

The lower group in the north-western part of Ahaggar is divided as the Tagengant Group and in Adrar-Iforas as the Nigritian Group. In both regions it is principally composed of red acid lava and clastic rocks of strongly varying thickness – from some tens of meters up to 3–4 km, and locally the group is absent in the sequence. The Nigritian Group (up to 2000 m) consists of porphyry (rhyolite), ignimbrite, dacite, and andesite and also of tuff and breccias interbedding with arkose; rare limestone interbeds and bands are also recorded among them. There is a pile of red conglomerate at the base of the group, which rests with a sharp unconformity on the Riphean Farusian Supergroup. The Tagengant Group is of the same composition as the Nigritian Group, but three conglomerate bands occur among the red lava arkose, the upper one of which near the top is evidently of glacial origin (the "lower tillite"). Certain copper mineralization is registered in red arkose; copper-bearing dolomite and siltstone are also known.

The Vallen-In-Semmen Group lies on the Tagengant Group with a break. Conglomerate strata occur at the base of it, upward follow strata of purple and variegated sandstones with local interbeds of conglomerate and then the flysch-like strata of green clay shale rhythmically alternating with sandstone. The upper ("Main") tillite (100–500 m) rests on these rocks or directly on the Tagengant Group unconformably and contains mainly blocks of fine-grained arkose characteristic of the lower part of the Tagengant Group. A band of red arkose is situated above the tillite. It seems reasonable to recognize the upper part of the group starting from the "Main" tillite as a separate group. The total thickness of the Vallen-In-Semmen Group is 3000 m.

Deposits of the "série pourprée" (supergroup) accumulated in isolated intermontane depressions formed during the inversional stage of evolution of the Nigerian geosynclinal system. Lava outflow and sedimentation occurred mainly under continental environments that changed into lagoonal conditions for short periods of time. The supergroup can be divided into the following parts in relation to the tillite horizons (up-section): (1) the sedimentary-volcanogenic sub-tillite portion; (2) the lower tillite portion; (3) the sedimentary inter-tillite portion; (4) the upper tillite portion, and (5) the super-tillite portion. The three lower probably belong to the lower suberathem, and the upper two separated from the lower ones by a great break to the Upper Epiprotozoic Suberathem. The sedimentary copper mineralization is known to occur in the pre-tillite portion of the group and is also typical of the corresponding level in the stratotype of the erathem.

Geochronologic studies show that the Riphean granites piercing the Farusian rocks suffered strong isotope mobilization some 680–650 m.y. ago during the Katangan cycle. It is probable that the Katangan granites proper are present among the Faruzian and older rocks of Ahaggar (for instance, the Tihoiiarene, Anfeg, and Nasobir granites).

In other parts of the Nigerian belt the Proch Tanere Group of the Eastern Ahaggar (Niger) and the Birnin-Gwari Group of Nigeria belong to the Epiprotozoic. Both contain tillite, but their stratigraphy is poorly known because of intensive folding (and in the case of the Birnin-Gwari Group also because of high alteration).

In the Anti-Atlas Mountains (Morocco) the Epiprotozoic strata embrace two tillite levels. The Siroua-Sarhro Group and the synchronous Tidilin-Lizat Group distributed in different parts of the Anti-Atlas belt belong there to the Lower Epiprotozoic. The sub-tillite parts of these groups are principally built up of acid or basic lavas; above the lower tillite the flysch-like strata of sandstone are situated. Porphyry from the base of the Tidilin-Lizat Group yielded an age of 960 m.y. by the Rb-Sr method (Choubert and Faure-Muret 1972); this value gives the time of lava outflow corresponding to the beginning of the Epiprotozoic Era.

Both synchronous groups rest unconformably on the so-called "série de calcaires et quartzites" of the Middle or Upper Riphean and are cut by small granite massifs. At the same time the Upper Epiprotozoic Ouarzazate sedimentary-volcanogenic Group, made up of almost unaltered porphyry (rhyolite) and andesite with subordinate interbeds of terrigenous rocks and rare limestones, overlies these groups and the granites. According to Choubert tillite is observed in the group in the Ougarta Mts. The Ouarzazate Group is unconformably overlain by the Adoudou Supergroup, the lower part of which belongs to the Eocambrian (see Chap. 7).

**Epiprotozoic of Australia.** The rocks of this erathem are exceptionally widespread in this continent. They include platform and miogeosynclinal formations; the first type strata fill in extensive depressions ("basins") in the Australian platform, the second build the Adelaidean foldbelt surrounding the platform from the south-east. Their sequences are as complete as those of the Epiprotozoic strata of Africa, and their geology is as well known, in particular the piles of the Adelaide geosyncline of Southern Australia that are regarded by the Australia workers as a stratotype of the Adelaidean "system", or in other words of the Epiprotozoic of Australia. It was pointed out earlier that these rocks can also be proposed as a world stratotype and can complete the rocks of the Katangan Supergroup as such.

The Adelaidean "system" is divisible into four chronostratigraphic units (up-section): Willouran, Torrensian, Sturtian, Marinoan, but they have no distinct boundaries and cannot be mapped in the field. Rocks of the Adelaide geosyncline proper can be recognized as the Adelaidean Supergroup. Recent division of the supergroup based on accepted lithostratigraphic principles was done by Thomson and his co-authors in some papers (1964, 1976 etc.) Preiss and Krylov published a brief summary of stratigraphic data on the supergroup in 1980.

The Adelaidean Supergroup is divided into four groups (in upward succession): Callanna, Burra, Umberatana, and Wilpena.

The Callanna Group (or the Callanna Beds) is distributed in isolated localities and the rocks making it up are strongly dislocated; thus its sequence is problematic. Its more or less complete sequence is known in an area where it is well exposed. There the metamorphic basement rocks are overlain by basal feldspathic quartzite and conglomerate (up to 1300 m), then upward follow dolomites (up to 150 m), then basalt and andesite piles (more than 700 m) and finally red sandstones (up to 2000 m) with interbeds of argillite with crystal molds after rock salt and gypsum. At some places the syngenetic stratiform deposits (and occurrences) of copper are related with the Callanna Group.

The Burra Group rests unconformably on the basement rocks as a rule and sometimes on the Callanna beds, but its relations with the latter are usually not clear. The structure of this group, as well as of other groups of the Adelaidean sequence, are rather complicated due to strong facial changes (Fig. 66). In the most complete sections the group starts with the basal arkose (up to 1200 m) that upward passes into siltstone, fine-grained sandstone, and dolomite (up to 700 m); in succession these are overlain by quartz-feldspathic sandstone (up to 500 m), by dolomite and magnesite with interbeds of sandstone and slate (locally reaching a thickness up to 4000 m), siltstone (up to 300 m), then again by quartz-feldspathic sandstone (up to 200–600 m), dolomite with siltstone interbeds (up to 600 m), and siltstone with sandstone and dolomite interbeds (up to 4000 m). The group finishes with thick strata (up to 3600 m) typified by alternation of coarse-grained feldspathic sandstone and dark siltstone and slate. The total thickness of the sequence just described amounts to 15 km, but this value is obtained by summing up the maximal thicknesses of its units; the true thickness, however, seems not to exceed 6 km, and near the western flange of geosynclinal depression it decreases to 600 m and even to zero at places.

The Umberatana Group, with a total thickness of about 7000 m, is made up mostly of terrigenous rocks and of carbonate rocks in part; horizons of glacial rocks are registered in its base and top. The lower glacial horizon is mostly distinct and is divided as the Sturtian unit. Its tillites are given different names in different places (Sturt, Bibliando etc.). In the east of the trough the tillites are separated into two sub-horizons by a pile of marine-glacial sediments; total thickness of glaciogenic and related rocks here can be exceptionally great – up to 3000 m. Locally the basic lava outflows occur among the glacial deposits; sheet bodies of iron ores associate with tillite here and there. Tillite of the lower horizon is characterized by many aspects of buried moraines (striated pebbles, glacier bed, exotic clasts, etc.); the marine-glacial deposits are characterized by fine varved lamination and presence of "natant" pebbles and boulders (dropstones).

The lower glacial horizon rests unconformably on the basement rocks that crop out in the inner uplifts or on different rocks of the Burra Group. The overlying thick interglacial deposits make up the middle part of the Umberatana Group (the Farina

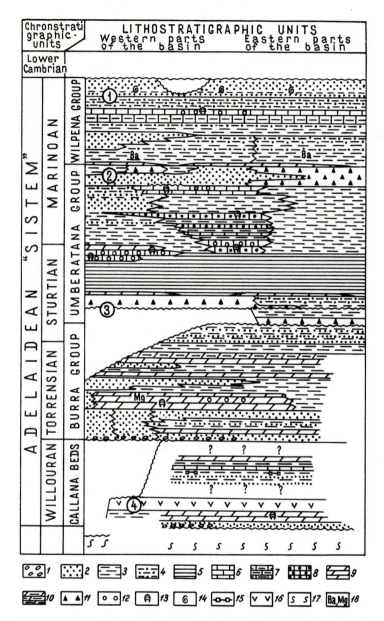

**Fig. 66.** Epiprotozoic and Eocambrian strata of the Adelaide miogeosyncline. (After Preiss and Krylov 1980).

*1* conglomerate, *2* sandstone, quartzite, *3* siltstone, argillite, *4* sandy siltstone, *5* fine-bedded siltstone, *6* limestone, *7* silty limestone, *8* sandy limestone, *9* dolomite, *10* silty dolomite, *11* tillite and related glaciogenic formations, *12* oncolites and oolites, *13* stromatolites, *14* nonskeletal multicellular organisms (the "Ediacara fauna"), *15* crystal moulds after salt and gypsum, *16* basic volcanics, *17* metamorphic rocks of basement, *18* Mg-magnesite, Ba-barite.

Figures in the sequence (in circles): *1* the Pound quartzite (Eocambrian), *2* the Yerilinian glacial horizon (the "upper tillite"), *3* the Sturtian glacial horizon (the "lower tillite"), *4* the Wooltana, Depot Creek, Beda volcanics and others (age is $\sim$ 1000 m.y.)

Subgroup); they principally consist of grey and greenish-grey, rare red or variegated fine-bedded siltstone, argillite, and slate with bands of sandstone and terrigenous-carbonate rocks with stromatolite- and oncolite-bearing limestones. Sun-cracks are locally observed in slate.

The upper glacial horizon is called the Yerilinian or Marinoan and is made up of tillite, marine-glacial pink feldspathic sandstone, and siltstone with disseminated boulders of different rocks and with lenses of red tillite with a thickness from 60 to 1000 m. It lies on the underlying rocks with an erosion.

The Wilpena Group (up to 3500 m) starts with a thin marking horizon of pink dolomite, then appear strata of red siltstone and argillite overlain and partially replaced by light quartzite typified by cross-bedding, sun-cracks and other aspects of shallow-water environment. Dolomite and argillite exhibit higher concentration of barium and sometimes enclose even the barite deposits. The higher portion of the section again consists of red siltstone and argillite, with bands of silty limestone containing stromatolites and oncolites.

The Pound quartzite completes the sequence of the Wilpena Group and contains casts of nonskeletal organisms, the so-called Ediarara fauna, and the enclosing rocks are thus assigned to the Eocambrian. Some details on these strata will be given in the next chapter. It should be mentioned here, however, that the Pound quartzite is overlain by sedimentary rocks containing the Lower Cambrian fauna. These relations clearly define the upper age limit of all units discussed; it is evident that the Adelaidean Supergroup is older than Eocambrian (except the Pound quartzite).

The lower limit of the supergroup is drawn according to its unconformable occurrence on the metamorphic basement rocks that are older than 1900 m.y. and also on the Lower Neoprotozoic taphrogenic sedimentary-volcanogenic rocks with an age of about 1600 m.y. and finally, on the Riphean platform strata with the Roopena basic volcanics among them that are dated at 1345 m.y. The Beda basic volcanics occurring in the lower part of the Callanna Goup (or strata) are 1100 m.y. old ($1076 \pm 34$ according to the constant $\lambda^{87}\mathrm{Rb} = 1.42 \cdot 10^{-11}$, see Preiss and Krylov 1980). A similar age (950 m.y.) was also obtained for clay shale from the same group. Thus, sediments of the Adelaidean Supergroup started to accumulate soon after the Grenville orogeny or even during it.

In many units of the supergroup carbonate rocks are stromatolite-bearing. The preliminary determinations done by Preiss on these structures permitted the Soviet workers (Semikhatov 1974, etc.) to assign the Burra Group to the Middle Riphean and the overlying groups to the Upper Riphean and Vendian. At the same time, Preiss (1977, etc.) stresses many times that almost all of the forms of stromatolites from the Adelaidean are local ("endemic") and have no analogous forms described by the Soviet workers. Again it was confirmed in a joint work by Preiss and Krylov, who pointed out that "there are no stromatolites in Southern Australia fully identical with the Riphean forms from the U.S.S.R. except *Conophyton garganicus* Kor." (1980, p. 69).

At the same time, the dominating stromatolites from carbonate rocks of the Burra Group are from the *Tungussia* and *Baicalia* groups that are so abundant in the Middle Riphean of the U.S.S.R., and from the Umberatana Group they are from the *Inseria*, *Gymnosolen*, *Katavia*, *Jurusania*, and *Tungussia* typical of the Upper Riphean, but the *Patomia*, *Boxonia*, and *Linella* groups are also reported as being common in the

Eocambrian (Vendian). In the opinion of Preiss and Krylov, stromatolites indicate an Upper Riphean and Vendian age for the Adelaidean Supergroup (stromatolites form the Burra Group are thought not to be useful in age determinations because of their specific character). Still, the Riphean age of the supergroup is strongly contradicted by geologic data and by isotope datings in particular. It will be shown later that the Epiprotozoic strata of the Urals analogous to the Adelaidean rocks rest unconformably on the Upper Riphean piles. It seems quite probable that phytolite assemblages enclosing the Upper Riphean and Eocambrian stromatolites are typical not only of the youngest Upper Riphean rocks but also of the Epiprotozoic rocks.

The Adelaidean Supergroup exhibits a great similarity with the stratotype, the Katangan Supergroup of Africa. In the base of both the terrigenous and terrigenous-carbonate rocks (with stromatolites) are situated, they contain horizons with syngenetic copper mineralization of the cuprous sandstone type. Over these strata the lower glacial horizons occur in both supergroups and the tillites are characterized by association with sheet iron ores. Up-section the thick terrigenous or terrigenous-carbonate strata rest under the upper glacial horizon with a break. In Australia, as well as in some parts of Africa, the upper tillite is directly overlain by dolomite and slate rich in barium. The upper portions of both supergroups are made up of terrigenous red beds.

The Epiprotozoic platform strata widespread in the Australian craton, in the Ngalia and Officer basins, and in the Amadeus aulacogen greatly resemble the strata of the Adelaide miogeosyncline in structure and composition, but the subglacial rocks are lacking there and so the sequence starts with the lower (Sturtian) glacial horizon. Preiss and his co-authors (1978) made a correlation of these strata and of the Adelaidean Supergroup rocks. Here only general characteristics of platform rocks will be given as an example; they are developed in the north-west of the continent in the Kimberley region and belong to two groups: the lower, Kuanidi, and the upper, Louisa Downs.

At the base of the Kuanidi Group the Landrigian tillite (or "lower tillite") lies directly on the crystalline basement, its thickness is up to 330 m. In upward sequence follows the Stein Formation made up of red high-ferruginous quartz subgreywackes with local interbeds of dolomitic sandstone, then the Wirrara Formation consisting of monotonous green dolomitic siltstone that passes upward into pink cross-bedded sandstone of the Mt. Bertram Formation, crowning the group. The total thickness of all these rocks usually does not exceed 1000 m, but sometimes it may be up to 2000 m.

The Louisa Downs Group lies with an erosion on the underlying rocks and starts with the Egan Formation (30–200 m) of tillite, interbedding with dolomite, limestone, arkose, and siltstone in places. It is overlain by the Yurabi Formation (30–210 m) of quartz and feldspathic sandstone with cross-bedding and ripple marks, then follows the McAlly Formation (1500 m) of dark and greenish clay shale, then the Tean Formation (120 m) of feldspathic sandstone and subgreywacke and finally the Lubbok Formation (more than 1800 m) of grey, green, and pink siltstone, subgreywacke, and feldspathic sandstone. The Lower Cambrian rocks unconformably overlie these formations and are represented by the Antrim plateau-basalt.

The age of the lower tillite (Landrigan) is 780 m.y. by the Rb-Sr method, but this value is no longer considered valid, as it was done on the slate samples from different formations. The age of a slate from an adjacent area that is compared with the McAlly slate is 685 m.y., but even this value is to be viewed with caution (Coats and Preiss, in press).

There are no problems in correlating the Epiprotozoic rocks of the Kimberley region and those of the Adelaidean Supergroup, shown in Fig. 67.

**Epiprotozoic of North America.** The Epiprotozoic rocks of this continent are mainly represented by miogeosynclinal formations, and build up the Riphean-Phanerozoic fold belts surrounding the North American platform.

In the North American Cordilleras the Epiprotozoic tillite-bearing rocks are known along the whole extension of these mountains from Alaska to California. In the northern Cordilleras the Rapitan Group is a well-known Epiprotozoic unit. It lies unconformably on carbonate rocks of the Riphean Purcell (Belt-Purcell) Supergroup and in the Mackenzie Mts area (Eisbacher 1978) it starts with the Sayunei Formation (0–50 m), built up of fine-bedded reddish siltstone and argillite with local inclusions of different rock clasts and lenses of coarse-grained material. The strata seem to be of marine-glacial origin judging from occurrences of dropstones in argillite. In the upper part of the formation argillite is commonly ferruginous and contains sheet hematite ores of economic value. The Shezal Formation (10–250 m) of tillite and marine-glacial deposits overlies the Sayunei Formation and locally rests directly on the Riphean rocks. Tillite is composed of rounded boulders and pebbles of dolomite, limestone, quartzite, metamorphic and crystalline rocks enclosed in argillite cement. The surface of certain boulders is faceted and striated. Tillite of the lower part of the formation contains interbeds of hematite ores or hematite clasts, and due to this the tillite is red-colored.

The Twitya Formation (600–1170 m) overlies tillite and is made up of clay shale with sandstone and siltstone interbeds; the rocks are commonly characterized by fine-serial cross-bedding and submarine slumping structure. The group finishes up with the Keele Formation (300–600 m) of rhythmically alternating thick-bedded sandstone, dolomite, limestone, and siltstone. The Sheepbed Formation (760 m) rests in upward succession, though its relations with the Keele Formation are conformable, the Canadian workers do not attribute it to the Rapitan Group. It is made up of siltstone with subordinate interbeds of sandstone and dolomite, and is overlain by the Eocambrian strata with a buried unconformity.

The Rapitan Group (with the Sheepbed Formation as a possible unit of it) represents the Lower Epiprotozoic Suberathem. It was pointed out, and will be shown later, that tillites associate with iron ores only in the lower glacial level of the Epiprotozoic.

The Windermere Group, developed in the more southerly area of British Columbia is a close analog of the Rapitan Group. In the Salmo area (Salmo 1965) it overlies unconformably the Purcell Group cut by a diabase with an age of 1200 m.y. and is unconformably overlain by the Eocambrian (?) Three Sisters Formation. It consists of the following units in upward sequence: the Toby Formation (20–600 m) of tillite with interbeds of sandstone and siltstone; The Irene Formation (0–2500 m) of basic volcanics alternating with argillite, conglomerate and sandstone, rare limestone; and the Monk Formation or Horsethief Creek Formation (up to 2000 m) that rests with a local erosion and is principally composed of argillite, siltstone, and sandstone with conglomerate and limestone interbeds; additionally, it includes a horizon of tilloid rocks that possibly belongs to the upper glacial level of the Epiprotozoic. Basic volcanics from the Irene Formation are dated at 825–900 m.y. by the K-Ar method, but these values are only approximate because of secondary alterations of the rocks (Miller et al. 1973).

In southern portions of the North American Cordillera, the Epiprotozoic miogeosynclinal strata are known in Utah, Idaho, California, and in the boundary areas of Nevada. In Utah (the area of the Great Salt Lake, Wasatch Range) the Mineral Fork Formation with tillite belongs to these strata; it rests with an erosion on the Riphean Big Cottonwood Group and is unconformably overlain by quartzite of the Eocambrian Mutual Formation. In Idaho the Pokatello Group is Epiprotozoic; in it the tillite interbeds with greenstone volcanics and quartzite, slate, and conglomerate in part. In the lower exposed part of the formation slate and argillite is situated over and under the tillite and in the uppermost part of it limestone occurs. The Pokatello Group is overlain by the Brigham thick quartzite that belongs to the Eocambrian (conformably overlain by the Cambrian rocks).

In California (Death Valley) the Epiprotozoic strata are represented by three successively lying formations, resting with a break, namely the Kingston Peak, Noonday, and Johnnie Formations (Steward 1970). The Kingston Peak Formation (300–600 m) is of unconformable occurrence on the Riphean Pahramp Group and is made up of sandstone with interbeds of argillite and siltstone that pass upward into tillite containing striated and faceted boulders of different rocks; sandstone encloses disseminated ("natant") pebbles of quartz and granite. The overlying Noonday Formation (400–660 m) consists of grey massive dolomite; the uppermost Johnnie Formation (600–1300 m) is made up of sandstone, siltstone, and dolomite with stromatolites: *Linnella ukka* Kryl., and *Boxonia gracilis* Korol., mainly known in the U.S.S.R. from the Eocambrian and Epiprotozoic. All these strata are unconformably overlain by the Eocambrian Stirling quartzite, which in its turn is overlain by the Lower Cambrian fossiliferous rocks. It is highly probable that all the rocks described above belong to the Lower Epiprotozoic Suberathem.

Epiprotozoic strata are also widespread in the Appalachian foldbelt. In their north-eastern flange, in Newfoundland (the Avalon Peninsula) the major portion of the Conception Group and the underlying Harbour Main Group belong to this erathem (Wiliams and King 1979).

The Harbour Main Group (more than 1500 m) is made up of red or pink rhyolite and ignimbrite and of their tuff and agglomerate, partly also basalt. The base of the group is not known. The Conception Group starts with the Malle Bay Formation (up to 800 m) that is typified by interbedding of pink and grey quartzite, volcanoclastic rocks, sandstone, and argillite. It exhibits a conformable occurrence on the Harbour Main volcanics and is conformably overlain by the Gaskiers Formation (up to 300 m), composed of tillite interbedding with subordinate conglomerate, tuff, and volcanoclastic rocks. In the opinion of Anderson (1977), the accumulation of sedimentary strata (including the glacial rocks) was accompanied by volcanic processes and the Harbour Main Group is in essence the lower part of the Conception Group.

The Gaskiers tillite is unconformably overlain by the Drook Formation (1500 m) of grey and pink sandstone and siltstone with gritstone interbeds; up-section the Briskal Formation is situated (up to 1200 m) consisting of grey coarse-bedded sandstone with interbeds of pink and green slate and conglomerate lenses.

It is possible that deposition of the Briskal Formation took place after intrusion of the Whalesback gabbro that cuts sandstone of the Drook Formation, but both formations exhibit conformable occurrence (Williams and King 1979). The group sequence finishes with the Mistaken Point Formation (up to 400–600 m), this latter being typified by alternation of pink sandstone and clay shale. Numerous casts of soft-bodied organisms of the Ediacara type (mainly of medusoids) are reported from the clay shale, indicating the Eocambrian age of the enclosing rocks. A contact of this formation with the underlying rocks is apparently conformable. In upward succession some thick strata (several thousands of meters) are situated, but they also belong to the Eocambrian, as they are overlain by rocks containing fossils of the Lower Cambrian.

Volcanics from the Harbour Main Group and the Holyrood comagmatic granite yielded an age of about 600 m.y. by Rb-Sr analysis. These values seem, however, to be strongly rejuvenated due to multi-fold effects of Caledonian and Early Hercynian movements and thermal processes, and taking into account an extremely great thickness of the Precambrian strata and their complex structure. An age of $795 \pm 80$ m.y. by the Rb-Sr method is to be mentioned here, as it was obtained for volcanics of the Coldbrook Formation, a close analog of the Harbour Main Group, that rests on the Riphean Green Head Group in New Brunswick.

In the U.S.A. the Epiprotozoic strata are distributed in the more southerly areas of the Appalachians. In North Carolina the Mount Rogers (Grandfather, Ashne) Group is made up of metamorphosed sedimentary rocks: greywacke, arkose, and slate alternating with acid and basic metavolcanics. In the lower part of the group there are volcanic strata of arkose and unsorted boulder-pebbled conglomerate that seems to be a tillite in many respects and can probably be compared with the Gaskiers tillite. The group lies with a sharp unconformity on the Cranberry gneiss and is unconformably overlain by the Lower Cambrian strata. Acid volcanics gave an age of 820 and 850 m.y. by the U-Th-Pb method. According to these datings the Mount Rogers Group claims a Lower Neoproterozoic age, at the same time it is the best unit in revealing the age of volcano-glaciogenic strata of the Appalachian belt. In Virginia (the Blue Ridge) the Glenarm metamorphic complex is a stratigraphic analog of the Mount Rogers Group, and includes the Lynchburg metasedimentary Formation and the overlying Catoctine Formation made up of basic metavolcanics.

In the Eastern Greenland Caledonides the upper part of the Hagen Fjord Formation of Kronprins Christians Land and the Merkebjerg Group of the Kong Oskars Fjord belongs to the Epiprotozoic. The lower part of both units is built up of tillite and marine-glacial strata and up-section appear thick slate-limestone-dolomitic strata. Tillite lies with a sharp unconformity on the Riphean rocks and the groups are generally unconformably overlain by the Lower Cambrian. It is very probable that both groups belong to the Lower Epiprotozoic, but any reliable data are lacking.

In northern Greenland, in Peary Land, the platform Epiprotozoic strata are developed, but this is the only place in North America where this formational type is known. At the base of these strata tillite (5–10 m) lies unconformably on the Riphean Midsommer sandstone (cut by a diabase with an age of 1050 m.y.), up-section follow strata (up to 720 m) of sandstone, shale, and dolomite. The latter contains phytolites and microbiota that greatly resemble the ones established in the Lower Epiprotozoic (?) Bitter Spring Formation of Central Australia.

**Epiprotozoic of South America.** The rocks of this age in this continent are known for sure in Brazil and possibly in other regions. Their stratigraphy, however, is poorly studied and the little we known about it adds hardly anything valuable to characteristics of the erathem discussed. Here it will be recalled only that the Jangada Group of miogeosynclinal terrigenous strata with tillite developed in the Paraguay-Araguaia belt and the Bambui Group of platform terrigenous-carbonate strata with tillite distributed in the Selitri depression ("basin") belong to the Epiprotozoic; they were mentioned briefly in the previous chapter.

**Epiprotozoic of Europe.** The Epiprotozoic rocks are sufficiently extensive in this continent and are represented by platform and geosynclinal formations. The platform strata are known from drilling in many depression and grabens in the Russian plate where they lie unconformably on different Riphean rocks. They are best studied in Byelorussia and Volyn'. Their detailed characteristics are to be found in many papers, including Chumakov (1978 a).

The Epiprotozoic rocks of Byelorussia are represented by the Vilchansk Group that is divisible in the stratotypical section of the Orshansk depression into the Blonsk Formation (up to 240 m) of red sandstone with some bands of tillite and the overlying Gluska Formation (up to 300 m) of different glaciogenic rocks: tillite, varved clay, sand and silt; tillite forms from one to 3–6 thick bands separated by lacustrine-glacial and fluvio-glacial deposits. The Vilchansk Group overlies with a certain erosion the Lapichi Formation (up to 80 m) of siltstone, clay shale, and dolomite, this latter in its turn resting unconformably on the Middle-Upper Riphean red sandstone. Dolomite of the Lapichi Formation contains microphytolites of the so-called Vendian complex that are also known from the upper part of the Upper Riphean of the Southern Urals. At the same time it cannot be excluded that the Lapichi Formation belongs to the sub-tillite horizon of the Epiprotozoic. The Vilchansk Group is unconformably overlain by Eocambrian strata represented by the Rataichitsy (Svisloch) Formation of the Valday Group.

The Epiprotozoic strata of the Volyn' depression form the Volyn' Group. The Brody Formation (up to 45 m) of gritstone and sandstone with bands of tillite and varved clay shale is situated at its base, it lies unconformably on the Riphean Polessian Group and is overlain by the Gorbashevsk Formation (5–60 m) of sandstone, partially siltstone and argillite; the rocks commonly are characterized by an admixture of pyroclastic material. The group sequence finishes with the Berestovets Formation (up to 500 m) of basalt and its tuff. The basalt is dated by the K-Ar method and the values obtained fall within the 640–850 m.y. range (there is also one value of about 1000 m.y.). The Volyn' Group is overlain with a break by the Eocambrian Valday Group. Chumakov supposes that tillite of the Volyn' and Vilchansk Groups belongs to the Laplandian glacial horizon of the lowermost Upper Epiprotozoic (the "upper tillite"). However, if the oldest datings of the Volyn' basalt are accepted as its true age, then this supposition must be revised. Unfortunately, more reliable datings of volcanics are not available at the moment. Many workers correlate the Volyn' and Vilchansk Groups, but is cannot be excluded that they are of different age, the first being older than the second; thus both glacial horizons may be present in the Russian plate.

In addition to the above-mentioned platform tillite-bearing rocks, other localities of these are known from drilling in the northern and central parts of the Russian plate, in the Pachelma depression (aulacogen) in particular.

The Epiprotozoic geosynclinal rocks crop out in many Paleozoic fold belts of Europe; they are reported from Caledonides of Spitsbergen, Norway, and Scotland, from the Variscan of France and Czechoslovakia, and from the Hercynian belt of the Urals.

In the Spitsbergen Archipelago two groups belong to the erathem examined: Polarisbreen distributed in the Western Spitsbergen Island, and Gothian distributed in the North-East Land (Krasil'shchikov 1973). The Polarisbreen Group consists of two formations. The lower, Wilsonbreen Formation (140–240 m), is made up of two tillite horizons with a thickness of 100 and 60 m separated by terrigenous-carbonate rocks, which also overlie the upper tillite. The upper, Drakoisen Formation (280 m), composed of variegated shale, siltstone, argillite, and dolomite that encloses microphytolites of the Vendian complex. The Gothian Group is of the same structure. In the base of it the Sveanor Formation (50–130 m) is situated, made up of quartzite-sandstone that upward passes into tillite; the Klakberg Formation (250 m) overlies the first one conformably and consists of dolomite interbedding with quartz sandstone. Both groups are overlain with a break by Lower Cambrian rocks containing *Salterella*, *Volbortella*, and other remains of skeletal fauna. The data available are not sufficient to attribute these groups to any definite part of the Epiprotozoic; usually these tillites are assigned to the upper glacial horizon without any reliable and valid facts.

The Epiprotozoic sequences studied in detail are situated in the Norwegian Caledonides. Two of them will be discussed here in brief, one is typical of the north of Norway and the other of its southern part. In Southern Norway, in the so-called Sparagmitic trough the Hedmark Group is recognized and is divisible into two subgroups: Lillehammer and Rena, which are separated by a break (Bjørlykke et al. 1976). In the base of the group the Brøttum Formation (up to 2000 m) unconformably lies on different older rocks and consists of greywacke sandstone (sparagmite) with interbeds and bands of conglomerate and slate; it is overlain by the Biskopäs conglomerate (up to 200 m) that resembles the fluvio-glacial rocks in the opinion of many workers; up-section the Biri limestone is situated (up to 700 m) and encloses oncolites that are similar to the structures in the Vendian complex of microphytolites.

The Ring Formation (40–60 m) starts the Rena Supergroup, rests unconformably on the Biri limestone, and consists of conglomerate and sandstone; in upward succession the Muelv tillite (15–30 m) appears, representing buried moraine and marinoglacial deposits. Tillite is overlain by the Ekre green or red shales (30–40 m), and these are overlain by the Vangsås Formation (up to 230 m) with a break, which is made up of quartzite-sandstone and quartz gritstone. In the upper part of this formation (the Ringsaker quartzite) there are imprints of worms.

In the sequence of the Hedmark Group described above the Ring Formation, the Muelv tillite, and the overlying Ekre shale that is the lower part of the Rena Subgroup certainly belong to the Epiprotozoic or to its Upper Suberathem, to be more exact. The uppermost-Vangsås Formation is most probably Eocambrian. The age of the Lillehammer Subgroup is more problematic. If the Biskopäs conglomerate is really fluvioglacial, then it belongs to the lower glacial horizon of the Epiprotozoic, and correspondingly the Biri limestone to the interglacial horizon, and the unconformity between the Lillehammer and Rena Subgroups is due to the Lufilian orogeny. If this is accepted, the Brøttum sparagmite then belongs to the lower pre-glacial part of Epiprotozoic and the Hedmark Group thus embraces the whole Epiprotozoic Erathem. Assignation of the Lillehammer Subgroup to the Upper Riphean is an alternative concept. This problem could be solved by geochronologic (isotope) studies.

In Northern Norway, in the Finnmark Peninsula, the Epiprotozoic rocks form the Vestertana Group resting unconformably on the Riphean thick essentially terrigenous strata. This group comprises three units: (1) the Smallfjord or "lower tillite" (30–60 m) represented by green and red-brown silty-clay rocks with boulders and pebbles of dolomite, sandstone and granitoid; (2) the Nyborg Formation (up to 400 m) of brown and red sandstone interbedding with red and greenish silt and slate; and (3) the "upper tillite" (60 m) similar to the lower one in composition and structure; in places it is overlain by silt and sandstone with slate interbeds. The Vestertana Group is unconformably overlain by sandstone of the Eocambrian Stappugiedde Formation that is in its turn overlain by sedimentary rocks with organic remains of the lowermost Lower Cambrian.

Rb-Sr dating of slate from the Nyborg Formation gave values of the order of 665–680 m.y. and that of slate from the Stappugiedde Formation – 530 m.y. (Pringle 1973). It is possible that both datings are rejuvenated, from which it can be concluded that the Stappugiedde Formation must be older than 570 m.y., the lower age boundary of Cambrian, and that the age obtained reflects the time of the Early Caledonian rejuvenation. For this reason these values cannot characterize the true age of these rocks.

It is probable that the lower and upper tillite horizons of Finnmark belong to different Epiprotozoic Suberathems, but at the same time it cannot be ruled out that both horizons reflect different stages of a single, later glacial epoch, in which case the whole Vestertana Group must be assigned to the Upper Epiprotozoic.

In the Scottish Caledonides the lower and middle parts of the Dalradian Group (Supergroup is more exactly) belong to the Epiprotozoic. The lower part of the group or the Lower Dalradian is made up of limestone-slate strata resting unconformably on the Riphean Moine and Torridonian Groups or directly on the Katarchean gneiss. The boulder conglomerates are situated above it (the Shichallion, Port-Askaig, Fanad, and other conglomerates), and exhibit many aspects of a tillite. The Middle Dalradian thick strata overlie tillite with a break, these piles are of composite structure consisting of quartzite and slate with dolomite and limestone, they also include tillite. The upper half of these strata (above the Scarba conglomerate) is possibly Eocambrian, for up-section it gradually passes into the Upper Dalradian rocks with fossil fauna of the Lower Cambrian. The Lower Dalradian is very probably Lower Epiprotozoic, and the lower half of the Middle Dalradian – Upper Epiprotozoic.

In France, in the Armorican massif, the Upper Brioverian Group is Epiprotozoic, it rests unconformably on the Riphean Lower Brioverian Group and is overlain with an angular unconformity by the Lower Cambrian archaeocyathean-bearing rocks. The Granville tillite is situated in the lower part of this group, up-section it passes into fine-bedded (varved) siltstone and coarse-grained sandstone rhythmically interbedding with slate. The "post-Spilitic Group" (2600 m) in the Krušne Mountains of Czechoslovakia is a close analog of the Upper Brioverian. It consists of slate and sandstone with tillite bands, lies unconformably on the eugeosynclinal sedimentary-volcanogenic rocks of the Neoprotozoic "Spilitic Group", and is overlain by the Lower Cambrian strata.

The Epiprotozoic rocks of the Middle and Southern Urals are of special interest, for in this area it is possible to establish clear relations of these rocks with the Riphean stratotype. The most complete sequence crops out in the Middle Urals, where the Epiprotozoic rocks make up a major portion of the western slope of the range. According to data of Mladshikh and Ablizyn (1967), and pers. comm. of Mladshikh (1980), four groups there belong to the Upper Precambrian: the Kedrov, Basega, Serebryanka, and Sylvitsa Groups. The lowermost Kedrov Group embraces two units (up-section): (1) the Sinegorsk Formation (2800–3900 m) of quartzite or quartzite-sandstone with thick bands of phyllite in the middle and upper parts; and (2) the Klyktan Formation (630–1000 m) of phyllite, chlorite-carbonate slate, marbled limestone, and dolomite; there are interbeds of altered volcanics at its base. Carbonate rocks contain stromatolites *Tungussia* sp., *Gymnosolen* sp. and *Linella ukka* Kryl. and also various microphytolites. The first two forms are typical of the Upper Riphean and the third of the Eocambrian (Vendian). In the Riphean stratotypical sequence in the Southern Urals this kind of mixed phytolite assemblage is characteristic of the Uka Formation that lies in the upper part of the Karatau Group or, in other words, of the uppermost part of the Upper Riphean that has recently been divided under the name Kudash.

The Basega Group overlies unconformably the Kedrov Group. It consists in upward succession of the following units: (1) the Oslyansk Formation (110–300 m) of quartzite or quartzite-sandstone with ripple marks; (2) the Shchegrovit Formation (50–900 m) of basalt, porphyrite, keratophyre, quartz porphyry, and orthophyre and also of tuff and tuff-breccias and subordinate quartzite-sandstone and chloritic slate; (3) the Fedotovo Formation (200–1100 m) of dark carbonaceous pyritized phyllite, slate and quartzite-sandstone with a band of limestone; (4) the Usva Formation (200–1100 m) of quartzite-sandstone with subordinate interbeds of phyllitoid slate or siltstone.

The Serebryanka Group rests unconformably on the underlying rocks and consists of five related formations (up-section): the Taninsk, Garevsk, Koyva, Buton, and Kernos. The Taninsk Formation (350–500 m) is principally built up of glaciogenic rocks: tillite and marino-glacial deposits; sandstone, silty slate, altered basic volcanics and limestone are present as sharply subordinate components. Microphytolites of the "Vendian type" are known from the limestone interbeds. In the eastern extension of the group the tillite (belonging to the Vilva Formation) associates with magnetite-hematite ores of economic importance. The Garevsk Formation (550–750 m) is made up of pelite-silty slate and sandstone with limestone interbeds. The Koyva Formation (225–700 m) exhibits varying composition and colors of rocks. It consists of silty and clay slate, sandstone, alkaline basalt, marl, and limestone. Red tilloid conglomerate is locally reported, possibly formed as a result of erosion and redeposition of the Taninsk (Vilva) tillite and hematite rocks on the slopes of the paleouplifts. The Buton Formation (150–350 m) is of dark carbonaceous ("coaly") silty slate, phosphatized locally. The Kernos Formation (250–1100 m) comprises feldspathic-quartz and arkose sandstone and siltstone, partly slate, dolomite, and limestone, with local thick flows of basaltoid and picrite.

The Sylvitsa Group overlies the Kernos Formation of the Serebryanka Group and lies with a break and a local slight unconformity and consists in upward sequence of four units: the Staropechnaya, Perevolok, Cherny Kamen', and Ust'-Sylvitsa Formations. The Staropechnaya Formation (450–500 m) is made up of siltstone and argillite with interbeds of polymictic sandstone; at its base there is a horizon of tillite (15–80 m) with boulders and pebbles of various rocks plunged into sandy-siltstone cement. The Perevolok Formation (180–270 m) consists of dark-grey slate with a higher concentration of phosphorus. Acritarchs are reported from slate, they resemble the ones from the Valday Group of the Russian plate. The Cherny Kamen' Formation (1700 m) is built up of grey polymictic sandstone with interbeds of siltstone and argillite. Sandstone at the base of the formation is characterized by sun cracks, current traces, and mechanoglyphs. The imprints of *Tirasiana* – a soft-bodied organism closely resembling the Ediacara *Cyclomedusa* form – is determined in the lower part of the formation; because of it the Cherny Kamen' Formation must be assigned to Eocambrian. The Ust'-Sylvitsa Formation finishes the sequence of the group, with a thickness of 570–610 m, and is made up of cross-bedded polymictic sandstone that is locally variegated.

In the Middle Urals the Basega, Serebryanka Groups and two lower formations of the Sylvitsa Group belong to the Epiprotozoic. Here, as in other complete sequences of the erathem, the following strata are distinguished: the pre-glacial deposits (the Basega Group), the lower glacial deposits (the Taninsk Formation), the interglacial deposits (the Garèvsk, Koyva, Burton, and Kernos Formations), the upper glacial deposits (the Staropechnaya Formation) and superglacial deposits (the Perevolok Formation). I proposed (Salop 1979a) calling these units the Biarmian complex (after the name of an ancient state: Biarmia = Permian the Great). A break between the Serebryanka and Sylvitsa Groups seems to correspond to the Lufilian orogeny.

In the stratotypical area of the Riphean rock distribution, that is in the Bashkir anticlinorium of the Southern Urals, the Kurgashli (Kaltyagay) Formation and the Bakeevo Formation, its analog, lie unconformably on the Uka and Krivaya Luka Formations of the Karatau Group of the Upper Riphean, latter containing Vendian phytolites. These formations are from 100 to 300 m thick and made up of sandstone, siltstone, and clay shale; a horizon of tillite occurs in the base of the Kurgashli Formation. These formations are overlain with a stratigraphic unconformity by the Eocambrian Asha Group that starts up with conglomerate and arkose of the Uryuk Formation. This group is continuously traced along the western slope of the Southern Urals and is widespread in the Urals foreland (depression) of the Russian plate. In the Middle Urals the upper part of the Sylvitsa Group combining the Cherny Kamen' and Ust'-Sylvitsa Formation corresponds to this group. Tillite of the Staropechnaya Formation corresponds to tillite of the Kurgashli Formation. Thus, in the Southern Urals equivalents to the Basega and Serebryanka Groups are lacking, that is the major portion of the Biarmian complex of the Lower Epiprotozoic is absent (see Table 5). Their absence in the sequence is possibly related to the Lufilian orogeny.

**Epiprotozoic of Asia.** The Epiprotozoic rocks with tillites are mostly known from Kazakhstan and Central Asia. The most complete sequence of this erathem is situated in the Baykonur synclinorium area (Zaytsev and Filatova 1971). There the Akbulak Group (up to 1500 m) of conglomerate and sandstone, commonly tuffaceous, overlies

**Table 5.** Correlation of the Upper Precambrian sequences of the Southern and Middle Urals

| General scale | Epiprotozoic horizons | Southern Urals | | | Middle Urals |
|---|---|---|---|---|---|
| Eocambrian | | Asha Group | Zigan Form. Kukkarauk Form. Basa Form. Uryuk Form. | Sylvitsa Group | Ust'-Sylvitsa Form Cherny Kamen' Form |
| Epiprotozoic — Upper | Super glacial | | | | Perevolok Form |
| | Upper Glacial | | Kurgashli Form. | | Staropechnaya Form |
| | Interglacial | | | Serebryanka Group | Kernos Form Buton Form Koyva Form Garevsky Form |
| Epiprotozoic — Lower | | | | | Taninsk (Vilva) Form |
| | Lower Glacial | | | | |
| | Subglacial | | | Basega Group | Usva Form Fedotovo Form Shchegrovit Form Oslyansk Form |
| Neoprotozoic — Upper Riphean | | Uka and Krivaya Luka Formations of Karatau Group ("Kudash") | | | Klyktan Formation of Kedrov Group ("Kudash") |

the Upper Riphean Koksuy unconformably, upward the Ulutau Group is situated with a break, and divided into the following units: (1) Zhaltau Formation – quartz sandstone and granule conglomerate that passes upward into phosphate-bearing shale interbedded with limestone; (2) Satan Formation – tillite, sandstone, and shale (the lower tillite horizon); (3) Bozigen Formation – dolomite; (4) Kurayli Formation – variegated shale and limestone; (5) Baykonur Formation – tillite and conglomerate (the upper tillite horizon). The Lower Cambrian strata lie on the eroded surface of the Ulutau Group.

In Central Asia Epiprotozoic strata enclosing tillite are reported from a number of areas of Tien-Shan, and have been studied in detail by Zubtsov (1972), according to whose data four correlative horizons can be identified within this sequence. These are (up-section): (1) Kichitaldysuy or the sub-tillite (100–200 m), made up of conglomerate, arkose, and siltstone with dolomite interbeds in many areas; (2) Dzhetym or the lower tillite (400–2500 m), made up of terrigenous rocks with tillite associated with finely bedded shale; in the Dzhetymtau Ridge hematitic and magnetitic rocks are associated with glacial deposits; (3) Dzhakbolot or the intra-tillitic sequence (200–500 m and even more), consisting of terrigenous and carbonate rocks; and (4) Baykonur or the upper tillite (15–100 m and even more), largely composed of tillite. These four units were found by Zubtsov, from Kazakhstan to the Kuruktag Ridge in China

(Sinkiang). It is very likely that the Akbulak sedimentary-tuffaceous Group of Ulutau is also to be assigned to the lower sub-tillite level of the Epiprotozoic.

The Epiprotozoic glaciogenic rocks are widespread also in south-eastern Asia. In Southern China, in Hupeh and Yunnan provinces, platform strata of the Liangtou, Nantou, and Toushantou Formations which the Chinese geologists (Lü Hung-yün and Sha Tzuan 1965) attribute to the Sinian "system" may be Epiprotozoic, but in the stratotypical area of Northern China this system is of an older – Neoprotozoic age (see Chap. 5). The Liangtou Formation (50–500 m) consists of red coarse-grained arkosic sandstone interbedded with argillite, siltstone, and conglomerate. These rocks lie sub-horizontally on Neoprotozoic (?) metamorphic rocks cut by granite dated at 900 m.y. (K-Ar method). The Nantou Formation (up to 200 m) is composed of tillite. It overlies with an erosion the Liangtou sandstone or the metamorphic basement. The younger Toushantou Formation (80–420 m) is dark argillite interbedded with limestone, dolomite, and sandstone. Glauconite from this formation is dated at 620–670 m.y., but this age is probably slightly rejuvenated. The Toushantou Formation is overlain by the Taning Formation (dolomite and limestone with phosphorite) which contains organic remains resembling hyolites in its upper part. Higher up the section there are Lower Cambrian strata with the trilobite *Redlichia*. The Taning Formation may be Eocambrian or lowermost Lower Cambrian.

In North Korea the Kuhen Group is Epiprotozoic; it amounts to 1200 m and is distributed in the Phennam trough where it rests unconformably on the Riphean Sanvon Group and is unconformably overlain by the Lower Cambrian strata. The group consists of phyllitized slate with rare beds of limestone and calcareous sandstone. It includes one or two tillite units in its lower part.

In the Hindustan Peninsula the Epiprotozoic interval is not known for certain. The rocks earlier accepted by Indian geologists as Late Precambrian (Vendian) glacial strata are really of the Mesoprotozoic Bijawar Group, as was shown by later studies. However, it cannot be excluded that the Melany Group of Rajasthan belongs there, being made up of acid volcanics and lying subhorizontally on the Mesoprotozoic Aravalli Group. Rhyolite from this group is dated at 745 m.y. by the Rb-Sr method.

In a vast territory of Northern Eurasia the Epiprotozoic Erathem is known from many places, but the tillite-bearing rocks are reported only from the Yenisei Ridge and in the north-east of the U.S.S.R. in the Kolyma River basin; at the same time, in many regions rocks are known which originated under extremely cold climatic conditions, judging from their peculiar features.

In the Yenisei Ridge the Chingasan Group and the lower part of the Chapa Group, developed in the north of the Teyan superimposed trough belong to the Epiprotozoic. The Chingasan Group lies inconformably on the Tungusik Group of the Middle Riphean and consists of three units (up-section): the Lopatino, Kar'yernaya, and Suktal'ma Formations. The Lopatino Formation (0–500 m) is made up of red conglomerate, gritstone, and silty-clay shale rhythmically interbedding in the lower part and of the grey dolomite, oncolitic limestone interbedded with shale and siltstone in the upper part. The Kar'yernaya Formation (up to 1000 m) is composed of grey quartz sandstone which upward passes into silty-clay shale with beds of quartz sandstone and dolomite. Glauconite from sandstone is dated in the range of 747–815 m.y. (K-Ar method, several determinations from different places). The Suktal'ma (Chivida) Formation (up to 800 m) rests with an erosion on the older formations; it consists of tillite, sandstone, argillite, and shale, locally interbedding with tuffaceous rocks and volcanics. Tillites are preserved as disrupted or densely packed bolders (up to 1.5 m across) and pebbles or polymictic conglomerate enclosed in argillitic or inequigranular cement. The supply of clastic material was mainly from the east during deposition of tillite. The following acrytarchs are determined in argillite: *Kildinella hyperboreica* Tim., *K. sinica* Tim., *Protosphaeridium paleaceum* Tim. etc.

The Chapa Group lies on the eroded surface of the underlying rocks. It starts with a locally developed Suvorov Formation (0–150 m) – red sandstone, gritstone, and siltstone, with dolomite interbeds in the upper part. Higher up the section the Podyemskaya Formation (up to 300 m) lies–sandstone and dolomite. Glauconite from sandstone is dated by the K-Ar method, the values obtained for different samples fall within the range of 560–810 m.y. The Podyemskaya Formation is overlain with a break by the Nemchanskaya (Uglovskaya) Formation of siltstone, sandstone, and gritstone, with dolomitic interbeds in the upper part. The rocks from this formation yielded an age of 635 m.y. on glauconite. In the opinion of workers studying the Yenisei Ridge, the Nemchanskaya Formation is Vendian (Eocambrian); it is overlain by the Lebyazh'ya Formation of the Lower Cambrian.

The Chingasan Group belongs to the Lower Epiprotozoic judging from glauconite datings (probably they are rejuvenated), but the Suktal'ma tillite may well belong to either the lower or upper glacial level. The Suvorov and Podyemskaya Groups are probably of the Upper Suberathem. The extremely great scattering of age values obtained on glauconite prevents us from relying on these datings.

The tillite-bearing rocks of the north-eastern part of the U.S.S.R., in the Kolyma region, are very poorly known and the data available are discrepant. According to Furduy (1968) they rest on the Upper Riphean rocks and are made up of dolomite and limestone with phytolites of the Vendian type in the lower part, and higher up they are overlain with a possible erosion by the strata (up to 400 m) of red sandstone, shale, and conglomerate that enclose a band of tillite (up to 130 m) preserved as unsorted pebble-block-bolder rocks with striated bolders. These strata are overlain by the Eocambrian Korkodon Formation of dolomite containing Vendian phytolites that in turn is overlain by Lower Cambrian variegated fossiliferous rocks.

In Eastern Siberia Epiprotozoic rocks are reported from the Sayan, Western Baikal, and Patom Highland regions. In the Sayan region the Oselochnaya Group is Epiprotozoic; it rests unconformably on the Riphean Karagas Group cut by the diabase dykes with an age of 1200 m.y., and is also unconformably overlain by the Eocambrian-Lower Cambrian Ust'-Tagul Formation. The Marna Formation (100–400 m) is at the base of the Oselochnaya Group, and is made up of conglomerate, polymictic sandstone, and siltstone, with rare limestone beds in the upper part; higher up the section, with a possible break, the Uda Formation (300–420 m) is situated, of sandstone and siltstone with interbeds of limestone enclosing the Vendian microphytolites and the sequence of the group finishes up with the Aysa Formation (200–1000 m) of gritstone, sandstone, and siltstone. Glauconite from the Marna sandstone is dated at 737 m.y. by the K-Ar method. In the Western Baikal region the Ushakovka Formation (up to 1300 m) of dark-grey and greenish-grey greywacke sandstone, gritstone, and conglomerate belongs to the Epiprotozoic; it overlies with an erosion the Riphean Baikal Group and is transgressively overlain by the Eocambrian-Lower Cambrian Mota Formation. Conglomerate occurring mainly in the lower part of the

Ushakovka Formation locally pass into puddingstone (or mixtites) that are similar to glaciogenic rocks. The bolder tilloid conglomerate developed as an isolated outcrop among granites along the Buguldeyka River possibly belongs to the formation examined.

According to data of O. Sennov (1980, pers. comm.) whose special study was the lithology of the Upper Precambrian strata of the Sayan and Western Baikal region, the Marna Formation sandstone of the Oselochnaya Group and that of the Ushakovka Formation differ greatly from other psammitic rocks developed in these areas by the occurrence of clasts of quite unstable minerals (of pyrrhotite, for instance) and the absence of any evident changes. In the opinion of this worker they were formed under an extremely cold climate, possibly even close to the glacial type climate.

In the Patom Highland of the Baikal Mountain land the Bodaybo miogeosynclinal Subgroup of the Patom Group belongs to the Epiprotozoic and lies without any apparant unconformity on the Middle-Upper Riphean Kadalikan Subgroup (see Chap. 5). In the base of the Bodaybo Subgroup is the Aunakit Formation (200–1200 m) of quartz metamorphosed sandstone with interbeds of altered siltstone; higher up the section the Vacha Formation (80–700 m) occurs, consisting of black carbonaceous siltstone and slate with quartzite interbeds; then the Anangro Formation (up to 2000 m) of coarse-bedded greywacke sandstone with gritstone and conglomerate lenses mainly occurring in the lower and middle parts of the formation; the sequence finishes with the Iligir Formation (up to 1000 m), typified by interbedding of calcareous sandstone and phyllitoid slate with rare dolomitic horizons (dolomite locally bears poorly preserved phytolites). The Anangro Formation is compared with the Ushakovka Formation of the Western Baikal region (Salop 1964–1967). As in the latter its greywacke sandstone contains abundant clasts of metavolcanics and dark-colored minerals, but many aspects of the original composition and structure of rocks are masked due to a higher grade of alteration. It is possible that the Anangro Formation was also deposited during a period of a sharp cooling of climate, but these effects of climatic changes are hardly observed because of quick burial of sediments. The Bodaybo Subgroup is cut by the pegmatoid micaceous granite that is likely to have been formed during the Katangan orogeny.

In the Eastern Siberia the Uy Group belongs to the Epiprotozoic, and is distributed in the eastern portion of the Aldan shield, where it rests with a break on the upper Riphean Lakhanda Group (see Chap. 5). The group is built up of green-grey rarely red sandstone, siltstone, and argillite. In the shield area the group is typically platform and its thickness does not exceed 500 m, but eastward, in the Yudoma-Maya aulacogen, it acquires certain miogeosynclinal aspects and becomes noticeably dislocated, its thickness increasing up to 3000–4000 m (Nuzhnov 1967). Acrytarchs are reported from the lower, Kandyk Formation (certain indications as to the presence of medusoid imprints were not corroborated); glauconite from the upper, Ust'Kirba Formation is dated in the range from 610 to 720 m.y. by the K-Ar method on different samples. The Uy Group is cut by ultrabasic and alkaline rocks of the Ingili massif with an age of 610–680 m.y. by the K-Ar method, but the most reliable datings, falling at $650 \pm 20$ m.y. (Tugarinov and Voytkevich 1970), were obtained by the U-Th-Pb method on different minerals, and correspond to the time of the Katangan orogeny. The lithologo-mineralogical aspect of the rocks of the Uy Group is poorly known as yet, and thus is cannot be excluded that the Kandyk Formation sandstone appears to be close by some characters and genesis to the similar sandstone of the Marna and Ushakovka Formations of the Sayan and Baikal regions.

## 3. The Lithostratigraphic (Climate-Stratigraphic) Complexes of the Epiprotozoic

In the previous pages of the book the Epiprotozoic strata of different continents were examined and correlation of these strata and stratotype was given. In Fig. 67 the structure and correlation of some Epiprotozoic sequences is shown. Two glacial horizons are seen to be recognized in the most complete sequences of the erathem, the lower one being formed prior to the Lufilian orogeny, that is prior to 780 m.y., and the second one prior to be Katangan orogeny, that is prior to 680–650 m.y. Later it will be shown that during the Epiprotozoic Era the climatic conditions were of the kind that short and exceptionally abrupt climatic changes on a world-wide scale had to occur in order to cause a glaciation. Thus it is possible to accept that the glaciogenic rocks originated all over the world simultaneously; and both glacial horizons can serve as wonderful markers.

Isotope datings of sedimentary and volcanogenic rocks enclosing glacial deposits are not suitable for determining the time of glaciations because the values obtained are commonly discrepant for different reasons. However, the geologic and geochronologic (radiometric) analysis done is suggestive of the 850(880?)–820 m.y. interval for early glaciation and the 760(780?)–750 m.y. interval for late glaciation. A principal task for future work is a more exact definition of the true time of glaciations.

It was pointed out that the Epiprotozoic Erathem, in relation to the glacial levels, is divisible into five horizons: subglacial, lower glacial, intraglacial, upper glacial, and supperglacial. The lithostratigraphic (or climate-stratigraphic) complexes correspond to these horizons, their brief characteristics will be given below.

*The Subglacial complex* in major cases is built up of terrigenous-carbonate strata: shale, sandstone, quartzite-sandstone, conglomerate, limestone, dolomite, and marl. Abundance of carbonate rocks varies greatly, in some places these are rare or absolutely lacking. Sandstone and marl is commonly characterized by inclusions of gypsum and crystal molds after rock salt. Various phytolites are reported from carbonate rocks; the Upper Riphean forms are typical of stromatolites, but the Vendian forms are also known. In some regions the strata are built up of two groups separated by stratigraphic unconformities; their total thickness may reach several thousands of meters. The terrigenous-carbonate rocks of the subglacial complex are mostly known for occurrences of stratiform copper deposits (together with Co, Zn, U, V, Cd, and Ge), and are sometimes large, as for instance in the Copper belt of Zambia and in Katanga.

The other type of rocks of this complex is the thick sedimentary-volcanogenic strata mainly composed of acid (rarely basic) volcanics and their pyroclastic products, developed in the Appalachian, Nigerian, and Red Sea foldbelts, but also registered in other belts on different continents. Sedimentary-terrigenous and carbonate rocks are subordinate among them, and they exhibit many features typical of the first type of rocks.

The Roan and Mwashya Groups of Katanga are the stratotypes of terrigenous-carbonate rocks of the complex, and the Tagengant Group of Algeria of volcanogenic

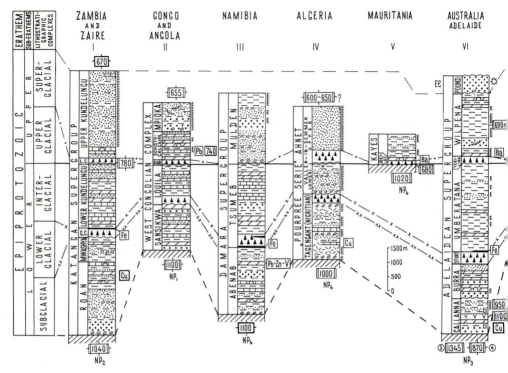

**Fig. 67.** Correlation of the Epiprotozoic strata of different regions of the world.
Stratigraphic columns: *I, II* after Cahen and Lepersonne (1967), *III* after Martin (1965), *IV* after Monod (1952), *V* after Caby (1971), Caby and Moussu (1968), *VI* after Thomson et al. (1964), *VII* after Dow and Gemuts (1969), *VIII* after Eisbacher (1978), *IX* after Williams and King (1979), *X* after Bjørlykke et al. (1976), *XI* after Chumakov (1978a), *XII* after Mladshikh and Ablizin (1967), *XIII* after Zubtsov (1972). Only fundamental works are referred to, additional information (isotope datings etc.) is taken from different sources.
Symbols: *1* conglomerate, *2* sedimentary breccia, *3* tillite, *4* arkosic and polymictic sandstone, *5* quartzite-sandstone (quartzite), *6* tuffaceous sandstone, *7* shale and siltstone, *8* dolomite, *9* limestone, *10* ferruginous rocks, *11* basic volcanics, *12* acid volcanics, *13* subvolcanic granite, *14* stromatolites, *15* fossil remains of the Ediacara fauna, *16* traces of rock salt or gypsum, *17* cross-bedding; *18* stratigraphic unconformities (*a* slight, *b* large, locally angular), *19* angular unconformity with the pre-Epiprotozoic strata, *20* red beds, *21* sedimentary iron ore mineraliza-

rocks. Their numerous analogs were examined in the regional review of the erathem. The time of formation of the subglacial complex falls within the range of 1000–850 m.y.

*The Lower Glacial complex* is of wider extension than the first and sometimes starts the sequence of the erathem. It is made up of tillite and related glaciogenic rocks (fluvioglacial, marino-glacial deposits) and also of the rocks formed as a result of washing out of continental moraines. The thickness of this complex varies from some tens to some hundreds of meters, but usually it is not great and commonly is less than the thickness of underlying and overlying complexes.

An association of the Epiprotozoic lower tillite with the sedimentary iron ores is typical, the deposits are of sheet or lens-like type occurring as a rule near the base of

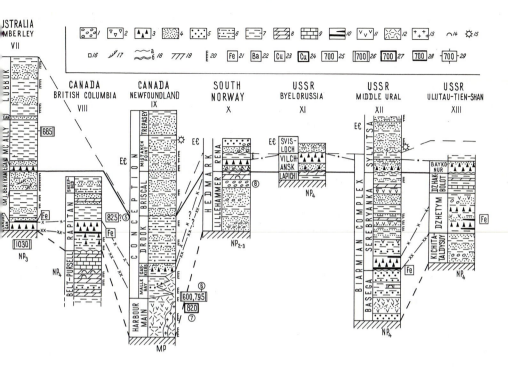

tion or deposits, *22* barite mineralization or deposits, *23* sedimentary copper mineralization and deposits, *24* very large sedimentary copper deposits, *25–28* isotope datings of sedimentary and volcanic rocks – time of sedimentation or volcanism (*25* K-Ar, *26* Rb-Sr, *27* U-Th-Pb, *28* Pb-model), *29* isotope datings of granitioids and metamorphic rocks (method see *25–27*).

Foot-note (figures in circles): *1–7* datings: *1* shale, *2* the Beda basic volcanics, *3* the Neoprotozoic Roopena porphyry, *4* the Houghton metamorphite, underlying the Adelaide Supergroup (it is possibly rejuvenated), *5* the Windermere Group volcanics overlying the Toby tillite, a correlative of tillite of the Rapitan Group, *6* the Coldbrook volcanics, New Brunswick, a reliable correlative of the Harbour Main volcanics, *7* the Mount Rogers Group volcanics of Appalachians, associated with tillite, *8* conglomerate of presumed fluvioglacial origin.

Indexes: *MP* Mesoprotozoic, $NP_{1,2,3,4}$, different Neoprotozoic suberathems, *EC* Eocambrian, *GC* the "Great Conglomerate" tillite of Katanga, *LC* the "Little Conglomerate" tillite of Katanga

the complex. Such formations have now been reported from many parts of the world. This empiric regularity is of great stratigraphic value, as it can be used when identifying the lower tillite and is of especial significance when the upper tillite is absent. Iron ores in some regions are of economic importance (British Columbia, Middle Urals, etc.). Volcanogenic rocks are one more typical feature of the Lower glacial complex (though their occurrence is not ubiquitous). The Great Conglomerate of Katanga or the Sturtian horizon of the Adelaidean may be chosen as a stratotype of the complex.

*The Interglacial complex* rests, as a rule, conformably on the lower tillite and is formed of thick strata (up to 3000 m) of terrigenous or terrigenous-carbonate rocks that locally interbed with basic and acid volcanics. Sedimentary rocks of this complex generally are the same as those of the Subglacial complex, but they do not alternate and form isolated bands or strata (formations). In many parts the red arkosic sandstone is abundant, but it rarely encloses evaporite or cuprous rocks, these latter being absent as a rule. Phytolites in limestone are the same as those of the Subglacial complex, but the Vendian forms seem to be more common. The Lower Kundelungu Group of Katanga or the middle (major) part of the Umberatana Group of the Adelaidean is a stratotype of the complex. The time of formation of the complex falls within the range of 820–780 m.y.

*The Upper Glacial complex* lies unconformably on the underlying rocks. It is exceptionally widespread in every continent and corresponds to the later Laplandian (Varanger) Glaciation that is expressed most evidently in Europe. Like the Lower Glacial complex it is composed of different facial types of rocks, the thickness of which varies from some tens of meters to a thousand and more meters. In cases where the thickness is great, the proportion of glaciogenic rocks proper (tillite etc.) is low; its major portion is then made up of deposits originated under alluvial, proluvial, and basin environments, with the products of washing out of the older moraines also constituting an essential part. The upper tillite, unlike the lower one, never associates with ferruginous rocks, but in a number of places in Africa (Ghana, Senegal, Mali, Mauritania), of Southern Australia (the Adelaide region) and of Asia (Kazakhstan), dolomites or carbonate siltstones with a higher concentration of barium are known to directly overlie it, sometimes enclosing barium deposits. Probably this is also peculiar to synchronous strata in other regions, but up to the present, no attention has been paid to it. It seems useful in recognizing the Upper Glacial complex and in its correlation. The Little Conglomerate of Katanga or the Yerilinian glacial horizon of the Adelaidean is a stratotype of the complex.

*The Superglacial complex* of the Epiprotozoic is less extensive: as it is the uppermost unit in the erathem, it was eroded in many places prior to deposition of Eocambrian sediments. In formational aspect it corresponds to the lower molasse of the Katangan orogeny. In some regions (in Western and North Western Africa, for instance) the detractive deposits are known to overlie it unconformably, probably representing the upper molasse of the same cycle. The red clastic rocks are mostly typical of the complex: sandstone, gritstone, conglomerate, shale, locally phytolite-bearing carbonate rocks. In certain localities, in its lower part the strata of dolomite and siltstone or argillite are situated on the upper tillite and the earlier mentioned primary barite and local copper-lead mineralizations associate with these strata. The thickness of the Superglacial complex might reach 4–5 thousands of meters. The complex was deposited in the range of 750–670 m.y. Its stratotype is the Upper Kundelungu Group of Katanga or the Wilpena Group of Adelaide.

# B. Geologic Interpretation of Rock Record

## 1. Physical and Chemical Environment on the Earth's Surface

Isotope paleothermometric determinations for Epiprotozoic rocks are lacking, but these are available for the cherty rocks of the Neoprotozoic and Lower Paleozoic; they indicate a temperature of the order of 40°–50 °C for the seawater in the Middle-

Upper Riphean and about 35 °C in Cambrian (Knaut and Epstein 1976). Thus it is possible to suggest that during Epiprotozoic, except in its glacial periods, the seawater temperature was of an average value ($\sim$ 40 °C) and in comparison with the modern one was still very high. Whether the climate at that time was warm or hot is also evidenced by geologic data, by wide distribution of red beds, evaporites, dolomites, and various phytolites in the Epiprotozoic strata in particular. An abrupt lowering of temperature during the glacial periods is sure to have been an "abnormal" phenomenon and does not correspond to the climate of the era as a whole. The possible causes of glaciations will be discussed below. Here it is noted only that in the rock record of the Epiprotozoic, as well as of the Neoprotozoic, there is no reliable evidence indicating the existence of climatic zonation, but if this zonation was faintly expressed (and that is very probable) then it could quite easily have been missed (see below).

The chemical environment near the Earth's surface differed slightly from that of the end of the Neoprotozoic, but the concentration of oxygen in the atmosphere and hydrosphere seems to have increased to a certain extent. It can be concluded from the disappearance, or at least, from a sharp decrease of sideritic rocks; the Epiprotozoic sedimentary ferruginous ores are commonly composed of oxide minerals, hematite and magnetite.

## 2. Origin of the Older Glaciations

The problem of glaciations refers to the Epiprotozoic Era as well as to the Mesoprotozoic and to the Phanerozoic, the extreme cooling of climate resulting in glaciations is known to have occurred during all these periods. One would think this problem might be solved by the study of the youngest, Quaternary glaciations, for these are best known. As a matter of fact, this is not so; the Quaternary glaciations occurred when an average annual temperature of the Earth's surface decreased strongly and under conditions of distinctly expressed zonal climate, only a slight further decrease of it in the middle latitude was needed (for some 4°–5 °C less) to cause a glaciation, and this lowering could have been caused by different local facts, for instance, by increase of land of continents, by building up of higher mountains, by manifestations of volcanism, etc. Thus, there is no unanimous opinion as to the causes of glaciations among geologists studying the Quaternary period.

The older glaciations, Paleozoic and Precambrian in particular, took place under rather unfavorable conditions, when the climate was very hot and azonal (or faintly zonal). Nevertheless, abrupt cooling of climate embraced the entire planet. For instance, evident traces of Epiprotozoic glaciation are reported from absolutely different regions, irrespective of their geographical latitude, either modern or older, for in any paleogeographic reconstructions (also in palinspastic reconstructions) the outcrops of the older tillite appear to be at a different distance from the poles, independent of their paleo-situation (Fig. 68).

The fact that glaciogenic rocks commonly have direct contact with hot climate strata enclosing red beds and dolomite and limestone with phytolites seems to be very important for clarifying the problem. Modern phytolites are known to form only in warm mineralized water in the seashore areas or in lagoons in the tropical zone only (or in the hot springs). Additionally, the rocks enclosing tillite commonly contain evaporite, caolinitic sandstone, and even bauxite. A quick and even sudden (in the geologic sense) beginning and termination of glacial epochs is indicated. The glacial formations are usually thin in comparison with the thickness of the rocks in which they are enclosed, and also evidence a relatively short duration of glacial periods. It must also be taken into consideration that the glacial deposits accumulated very quickly and that their thickness does not thus characterize the longevity of sedimentation. Then, as noted above, the glaciogenic rocks associate with rocks that originated as a result of redeposition of moraine material after termination of a glacial period.

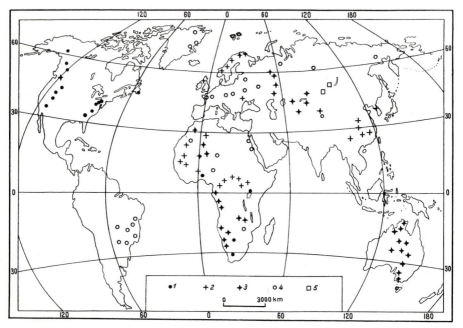

**Fig. 68.** Exposures of Epiprotozoic glaciogenic rocks.
*1* lower horizon of tillite ($EP_1$), *2* upper horizon of tillite ($EP_2$), *3* lower and upper horizons of tillite, *4* tillites that are not assigned to any definite Epiprotozoic level, *5* nonglacial clastic deposits of cold climate

What were the causes of great glaciations? These global-scale phenomena could have occurred in the Precambrian if the temperature lowered by several tens of degrees. Certain paleogeographic changes as formation of extensive montane uplifts could not be the cause, for they could not have caused such a great lowering of temperature all over the whole planet. Moreover, high mountains are not known in the Precambrian and the older glaciations do not correlate with the principal orogenic cycles. A hypothesis stating a relation between glaciations and intensive volcanic processes resulting in outbursts of abundant gases and pyroclastic material in the atmosphere is either satisfactory. There is no correlation between the time of volcanic epochs and glaciations. For instance, the volcanic eruptions occurring during the whole Early Neoprotozoic that were the largest and longest in duration in geologic history were never accompanied by glaciations. Generally, intensive volcanic processes are registered as having occurred quite often, while glaciations occurred only a few times during the whole evolution of the Earth. It is possible that in some cases the volcanic eruptions favored intensive development of glaciation, but they could not have been the original cause of it.

The hypothesis that states that the Precambrian glaciations originated as a result of a lessening of the hothouse effect due to decrease of carbon dioxide concentration in the atmosphere while carbonates were precipitating also fails to explain this strong lowering of temperature. The "antihothouse" hypothesis is not valid because of lack of correlation between tillites and carbonate strata; accumulation of thick strata of carbonates occurred recurrently and was not accompanied by glaciations.

Generally it is difficult, or rather it is impossible, to present any terrestrial cause of glaciations in the Precambrian if one proceeds from the evolution of the outer shells of the planet only. The cause must be nonterrestrial, and must be related to certain phenomena in the Cosmos. The most natural explanation of the oldest glaciations seems to lie in a change of intensity of the Sun's heat reaching the Earth's surface. In the Phanerozoic, when the average annual temperature decreased, terrestrial causes could have caused a global cooling of climate to a greater extent. It is very likely that the Phanerozoic glaciations and the temperature minima could be the result of interference of terrestrial and cosmic phenomena.

Now it is impossible to state substantially what cosmic phenomena were responsible for the abrupt cooling of climate of the Earth. At first sight the simplest and most natural explanation lies in a change of intensity of Sun radiation. Studies of Sun activity reveal that variations of the Sun constant do not exceed 3%, which is not sufficient to cause a glaciation, even more so in the Precambrian. Many astrophysicians think the Sun is a stationary star with a stable thermal regime. However, lately a great deficiency of the Sun neutrinos has been recorded and new Sun models are being elaborated with the Sun again regarded as a variable star. According to a well-known model of Fawler (1972), the high temperatures necessary for nuclear processes appear in the inner parts of a star at definite intervals of the order of 200–300 m.y. Due to convection instability the heated material then rises and mixes with the relatively cold material of the surface; in this process the insolation decreases by some 35% and the temperature on the Earth by 30 °C and more, a situation which continues for about 10 m.y. One would think that this represents a satisfactory explanation of terrestrial glaciations, but in 1980, Soviet physicians established that neutrinos had a remaining mass, which, in the opinion of some workers (Pontekorvo, etc.), can lead to the following result: a certain portion of neutrinos irradiated by the Sun may be transformed in a way that can no longer be recorded by known methods. If this is so, then there is no necessity to revise the old established model of the Sun.

A well-known astronomic hypothesis of Milankovich that connects glaciations with a change in inclination of the Earth's ecliptics either fits an explanation of Precambrian glaciations. These changes in inclination cannot cause a total lowering of temperature for several tens of grades. At the same time, it is possible to admit that during the Quaternary period the glaciation stages could have been the result of this phenomenon, but the principal cause of development of glaciations still seems to have been of a different nature.

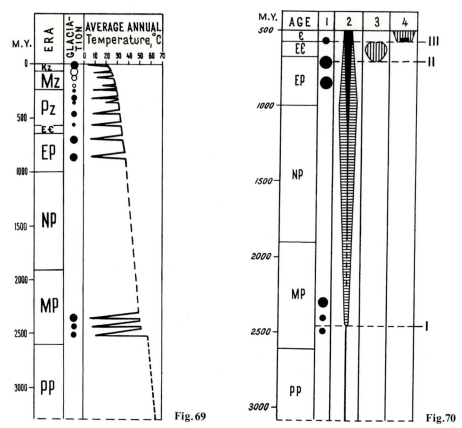

**Fig. 69.** Glaciations and temperature minima in geologic history of the Earth. (After Salop 1977c.) *Black circles* glaciations, *light circles* temperature minima. In diagram to the *right*, a general trend in change of an average annual temperature of the Earth's surface is shown

**Fig. 70.** Glaciations and the epochs of acceleration in biologic evolution during the Precambrian and Cambrian.
*1* glaciations, *2* phytolites (*black inner contour* eukaryotes), *3* Ediacara fauna, *4* skeletal fauna (*black* archaeocyatheans).
*I* beginning of a stage of rapid reproduction of blue-green algae and bacteria *II* appearance of the Ediacara soft-bodied fauna at the Epiprotozoic-Eocambrian boundary, *III* appearance of skeletal fauna, appearance and extinction of archaeocyatheans

Some more hypotheses (of G. Tamrasyan, G. Steiner, and E. Grillmair) tried to explain the cyclicity of glaciations by cosmic factors. But the factual data we possess contradict even the idea of periodicity: they evidence that glaciations in the Earth's history are episodic phenomena occurring at different time intervals. The interval between the last Mesoprotozoic and the early Epiprotozoic glaciations is thus about 1500 m.y., the duration of the interglacial epoch of the Epiprotozoic is of the order of 80–100 m.y., and the intervals between the Phanerozoic glaciations (and the temperature minima – "failed glaciations") vary from 40 to 125 m.y. (Fig. 69).

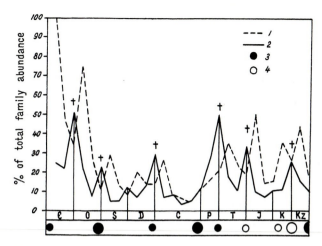

**Fig. 71.** Epochs of extinction and flourishing among animals. (After Newell 1967, with additional data on glacial epochs and temperature minima in the lower part of the figure.) The plot shows the percentage of first and last appearance of animal families during geologic time. *1* first recorded appearance, *2* last recorded appearance, *3* glaciations, *4* temperature minima (cooling of climate that did not cause glaciations)

Some years ago I called attention to the fact that the time of glaciations coincides with time of quick mass extinction of different groups of organisms and their simultaneous and subseqent outburst (Figs. 70 and 71). In a paper specially devoted to this problem (Salop 1977c), it was shown that glaciations could not be the cause of biological crises, which are very likely to be due to episodic increase in intensity of rigid ionizing cosmic radiation when mutation rates jump quickly. On studying the problem in detail, I proposed a hypothesis that a common cause exists to explain glaciations and biologic revolutions, and that is the outbursts of Supernovae in the environment of the Sun system which result in intensification of cosmic radiation and in shielding of the Sun system by gas-dust nebulae that appear during the cosmic outbursts.

## 3. Life During the Epiprotozoic

The organic world of Epiprotozoic differs only slightly from that which existed during the Middle and Late Riphean, but during the examined era some new prokaryotic and eukaryotic forms of microorganisms (mainly algae) probably appeared, which determined the origin of new groups of phytolites widely distributed in the Eocambrian. Great changes, probably, also took place in the composition of microphytofossils (and in the composition of microbiota on the whole), but they have not been well studied. However, the appearance of variety of nonskeletal animals at the begining of the Eocambrian indicates that already at the end of the Epiprotozoic important biological premises originated for this greatest event in the history of Life.

## 4. Sedimentation Environment

All Epiprotozoic sedimentary deposits, except glaciogenic rocks, were formed under an environment very similar to that of the Late Neoprotozoic. This naturally results in a great similarity of rocks of both erathems. Certain peculiarities of sedimentary iron ores of the Epiprotozoic related to an increase of $O_2$ concentration in the atmosphere were mentioned earlier. It is notable that cuprous sandstones also acquired a wider distribution at the beginning of Epiprotozoic. During geologic evolution these formations are known to occur recurrently, but they were related in each case to a special facial and tectonic environment. Conditions favorable for their formation are reported to have existed also in the Epiprotozoic, but there is no adequate explanation of the fact at present, why many and rather large deposits occur in the rocks formed at the beginning of this erathem. The solution of this problem, however, greatly depends on knowing the source of copper and concentration of it in the cuprous sandstone formations.

Now the environment of glaciogenic rock deposition will be examined. It was noted that these rocks are quite widespread; but the limits of the areas embraced by glaciations remain problematic. The principal difficulty is the burial of Epiprotozoic strata under younger sedimentary strata or their destruction by erosion. Additionally, the continental glaciogenic rocks, of the montane valley type in particular, have but little chance of being preserved in the sequences of the older strata. In many cases these rocks appear to be secondary in occurrence (commonly even in a strange facial environment) for they were redeposited by rivers, mudflows, subaqueous slumps, and turbidity currents. In spite of the fact that in a number of cases their primary origin can be recognized with a certain reliability, few criteria could have helped in reconstructing the type of glaciation.

It is, however, accepted that in some regions, for instance in Eastern Europe, in a major portion of Africa, in South-Eastern Asia (China and Korea), and also in Central Australia, the Epiprotozoic glaciation was of an ice-cover nature and the glaciers were very extensive. All the regions enumerated used to be platforms in the Epiprotozoic and naturally the relief was smooth or slightly undulating, and is very likely to have been buried under the ice. The nature of glacial deposits (with common marine-glacial type of rocks among them) indicates that certain portions of the land were at times covered by drift ice. Probably this type of glaciation embraced the South American platform. In the North American platform, with the exception of its northernmost margin (North Greenland), Epiprotozoic glaciogenic rocks are not known. They could have been destroyed by erosion, or possibly they are not recognized under the younger sedimentary strata. The glacial deposits are widespread in the geosynclinal belts surrounding the North American platform and so it is difficult to suggest that they have never been present there. Glacial formations are also not known in the Hindustan platform, but the Epiprotozoic strata are not certain there; meanwhile, the Epiprotozoic tillite is widespread in the geosynclinal belts surrounding this platform from the north. The Epiprotozoic strata of the Hindustan platform seem to have been quite thin and later to have been destroyed by denudation.

The situation in the Siberian platform is different; the Epiprotozoic strata there and in the surrounding geosynclinal belts are quite extensive, but the glacial deposits are known only near the south-western margin of the platform, in the Yenisei Ridge.

In the Baikal and Sayan regions the terrigenous strata (the Ushakovka and Marna Formations) are present instead of glacial deposits; they do not contain tillite, but were formed under extremely cold climate conditions (see above). A similar environment is suggested for the synchronous strata developed in the Patom Highland and in the Aldan shield. It seems that glaciation did not embrace the whole territory of Eastern Siberia. It is difficult to clarify the reasons for it, but possibly it was the climatic zonation or some other local conditions.

The Epiprotozoic geosynclinal belts are also characterized by glacial deposits in the same way as the platforms, and here also they are quite distinct and complete. Buried continental moraines are known among them, but the marino-glacial strata and various marine deposits with enclosed redeposited glacial material are of wider distribution; the latter are mostly typical of slopes of the intra-geosynclinal uplifts. It can be suggested that glaciation embraced the island mountains as well as the adjoining water areas that were covered by consolidated or drift ice.

A wider distribution of glacial deposits is typical of the areas surrounding the Pacific and Atlantic Oceans (with the exception of the western shore of South America, but Epiprotozoic strata are either known there). Is it not possible to relate the distribution of glaciers in the margins of continents with the higher moisture of the atmosphere in the oceanic shore areas? It was shown in the previous chapter that the Pacific Ocean existed in the Riphean within limits close to the modern ones. For the Atlantic Ocean, it will be shown below that its "opening" began in the Early Epiprotozoic, that is prior to the first glaciation.

The sub-latitudinal zone of Tethys following from Western Europe to Middle and Central Asia, then the meridional belt of the Urals and submeridional Adelaide belt of Eastern Australia (this latter belonging to the areas surrounding the Pacific Ocean) are the areas of wide distribution of glaciation in the Epiprotozoic geosynclinal belts. High moisture in these belts probably favored the development of glaciations. If these suggestions are true, then it is possible to claim the existence of climatic zonation in the Epiprotozoic, though it is thought to have been not on a world-wide scale depending only on geographic latitude, but on a local scale related to distribution of land and oceans.

The environment of formation and the problems of recognition of glaciogenic rocks in the Precambrian are known in detail from many specialized works (for instance Chumakov 1978a) and will not be dealt with here. It should, however, be pointed out that certain aspects of the older glacial deposits have no adequate explanation. In particular, it is not clear why the sedimentary iron ores occur in the Epiprotozoic lower tillite, and the barite-bearing rocks in the upper part of the erathem above the upper tillite. Only general suppositions can be made on this phenomenon. A common occurrence of volcanics is known among rocks of the lower glacial horizon and thus it is possible to suggest that the first association is due to transportation of iron by the volcanic sources and to its further migration in the sea or groundwater; the glacial deposits were only a suitable environment for infiltration of solutions and precipitation of iron oxides over the water-resisting beds. In this case the association of iron ores and volcanics is not a necessary condition: the volcanic sources could have been situated aside. The accumulation of barium in the shallow-marine or lagoonal dolomite and argillite overlying the upper glacial horizon in some localities may be related to the climate becoming more arid, which is known to occur after glaciations according to data on the Quaternary period.

A significant change in the oceanic level must have occurred after these extensive and thick ice covers were degraded. In the periglacial zones, in the drained sea-shores in particular, extensive outwash plains probably existed and their lithified strata, preserved as arkosic and quartz sandstone, commonly occur in sequences of both Epiprotozoic glacial horizons. Sandstone of eolian origin is widespread among interglacial and superglacial deposits, probably due to recurrent fanning and redeposition of sands of the outwash plains.

## 5. Tectonic Regime

Epiprotozoic strata as a rule occur in the tectonic elements that were formed in the Riphean. They are separated, however, from the Riphean strata by unconformities, but generally their dislocation is conformable and so distribution of major paleotectonic elements of both erathems coincides closely in many parts of the world, in the northern continents in particular. Thus, in North America, in Europe, and in Asia the Epiprotozoic geosynclinal strata occur in the same mobile belts as the Riphean strata and in some regions they are even closely related (as for instance, in the north of the Baikal geosynclinal belt). In the East European platform the Riphean and Epiprotozoic strata commonly occur in the same depressions (grabens). Certain more evident changes of a structural pattern are observed in Africa and Australia in part, which incidentally may also be related to better exposures of Epiprotozoic strata on these continents. A paleotectonic scheme of Africa is shown in Fig. 72; it is seen that the Western African platform was almost of the same shape in the Epiprotozoic, but in the west a new foldbelt adjoined it, occupying the place of the former Riphean Mauritanian geosynclinal belt and a subplatform regime established in this foldbelt during the era examined (later tectonic movements became effective again in this belt). A major portion of the continent used to be occupied by a huge Congo platform, its outlines becoming quite distinct only at the end of the Riphean (cf. Figs. 72 and 62). The Kalahari platform, situated in the southernmost part of the continent, was inherited from the Riphean Transvaal-Rhodesian (Zimbabwan) platform, but a peculiar block originated between it and the Kalahari platform, and as a result these two became combined. The tectonic regime in the area of this block (the "Bangweulu Isthmus"), however, was not stable; its basement is dislocated by numerous fractures and subsided deeply at places, for instance, in the Kundelungu aulacogen. The platform cover (including the glacial deposits) is extensive in every platform, in the Taoudeni syneclise and in the Congo syneclise of the Western African platform in particular.

During the Epiprotozoic the Nigerian, Red Sea, and Anti-Atlas geosynclinal belts existed within their earlier boundaries, but the tectonic regime there suffered great changes: the red sedimentary strata enclosing volcanics of the rhyolite-andesite formation started to accumulate in these belts instead of eugeosynclinal formations with greenstone volcanics typical of the Riphean; the beds were characteristic of the later or orogenic stages of evolution of mobile belts.

The tectonic evolution of geosynclinal belts of Equatorial Africa and Southern Africa is reported to occur in a different fashion. There, at the beginning of the

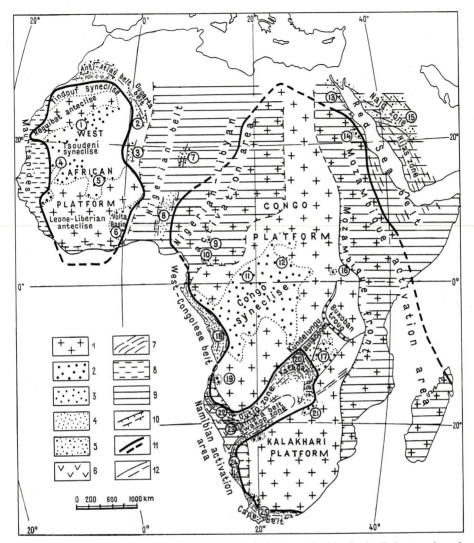

**Fig. 72.** Scheme of situation of principal structural elements of Africa in the Epiprotozoic and of areas of Early Paleozoic activation. (After Salop 1977b).

*1* basement of platforms, *2* platform strata, *3* aulacogenic strata, *4* miogeosynclinal strata, *5* red sedimentary-volcanogenic strata of the late stage in evolution of geosynclines, *6* eugeosynclinal strata, *7* fold strikes, *8* area of the Late Riphean stabilization (the Mauritanian foldbelt), *9* areas of the Early Paleoozoic (Pan-African, Baikalian) activation, *10* boundaries of the Early Paleozoic activation, *11* craton boundaries, *12* fractures.

Figures in circles – names of the Epiprotozoic groups or formations: *1* Kayes, *2* Ahnet, *3* Nigritian, *4* Kiffa, *5* Nara, *6* Buem, Oti, *7* Proch Tanere, *8* Birnin Gwari, *9* Mangbei, *10* Dja, *11* Oubangui, *12* Lindi, *13* Hammamat, *14* Awat, *15* Shammar, *16* Bilati, *17* Luapula, *18* West Congolese, *19* Bembe, *20* Katangan, *21* Sijarira, *22* Otavi, *23* Hakos, *24* Gariep, Holgat, *25* Malmesbery

Epiprotozoic, a new geosynclinal system of Katangides originated and the older foldbelts consolidated in the Riphean served as a basis; this new system followed mainly along the Atlantic shore and pierces inland only between the Congo and Kalahari platforms. It consists of the following geosynclinal belts: West-Congolese, Damaran, Katangan, and Cape. In the north the West-Congolese belt was probably connected with the Birnin-Gwari zone of the Nigerian belt that is close to the Katangides strata in structure and composition. Very thick miogeosynclinal strata had accumulated in every belt mentioned above and in the inner part of the Damaran belt (in the Swakop zone) the eugeosynclinal strata containing greenstone volcanics were also deposited. Folds of Katangides are of a parallel orientation with the contours of the platform; in the Katangan belt proper they form an arch with its convex side to the "Bangweulu isthmus" and to the Kundelungu aulacogen situated within the former. This type of junction of fold arches and aulacogens opening toward them is a characteristic feature of many fold systems of different ages.

The general plan of tectonic structure of Australia did not change greatly during the Epiprotozoic, as compared with the Riphean time (Fig. 73): earlier the major portion of the continent was occupied by a platform and bounded by a geosynclinal belt in the east. This latter, however, moved far southward in comparison with the situation of the Riphean Mount Isa belt, and probably was partially formed at the expense of fracturing of a margin of the Riphean platform, the fragments of which it encloses as median masses (cf. Figs. 73 and 63). This belt, usually called the Adelaide geosyncline, is exposed only in its smaller near-platform part where the Epiprotozoic strata are represented by miogeosynclinal facies only. Its inner zone is suggested to have been situated eastward, under the Phanerozoic rocks of the Great basin and the Tasmanian folded area, and probably was of an eugeosynclinal character.

In the Australian platform a vast syneclise appeared during the Epiprotozoic, consisting of several depressions ("basins"), including the Amadeus aulacogen, which was situated at the site of the Riphean Bentley graben. This aulacogen is opened toward the Adelaide geosyncline and the Epiprotozoic platform strata developed there exhibit great thickness and rather strong folding (probably it is of the near-fracture type).

Generally, the tectonic regime of the Epiprotozoic is typified by further stabilization of platforms, by less extension of eugeosynclinal zones, by the appearance of local and general inversions in many geosynclines that resulted in their temporary transition into an orogenic stage and in some cases in their complete closing. At the same time, the tectonic regime at the beginning of the Epiprotozoic was quite peculiar. New geosynclinal belts originated in some parts of the world at that time and rift depressions or taphrogeosynclines appeared within the Riphean foldbelts and became filled with lavas of rhyolite-andesitic or rhyolite-basaltic formation.

It is important that many newly formed geosynclines and taphrogeosynclines were situated in the margins of the continents. Most significant is the fact that a very extensive volcanic belt of the Early Epiprotozoic in North America ran along the modern Atlantic coast. Its thick volcanogenic or sedimentary-volcanogenic strata are traced along the whole extension of the Appalachian Mts from Newfoundland down to the Gulf of Mexico for a distance of more than 3000 km. These strata in different regions are given different names: the Harbour Main, Coldbrook, Lynchburg, Catoctine, Mount Rogers (or Grandfather) Groups. All of these strata lie with a sharp unconformity on the basement rocks and start the Eocambrian-Paleozoic complex of the Appalachian geosynclinal belt; it allows the suggestion that this latter started to form on the activated south-eastern margin of Grenvillides only at the beginning of the Epiprotozoic Era.

This situation of the volcanic belt in the margin of the continent and the occurrence of taphrogenic volcanics in it and also clastic continental and littoral-marine deposits may evidence that the formation (or "opening") of the Atlantic Ocean started at the time discussed and was accompanied by an origin of marginal abyssal fractures, grabens, or rifts (Bird 1975, King 1977, Schwab 1977 etc).

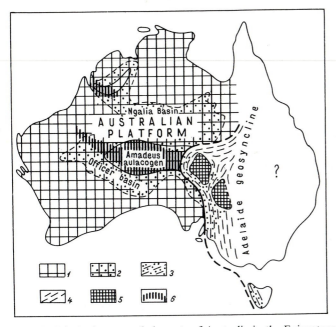

**Fig. 73.** Principal structural elements of Australia in the Epiprotozoic.
*1* platform basement, *2* platform strata, *3* geosynclinal exposed strata, *4* geosynclinal strata covered with Phanerozoic rocks, *5* median masses, *6* deformation zones in the platform

The situation of the principal belt of Katangides of Africa and the Eocaledonian ("Sparagmitic") belt of Scandinavia along the eastern shore of the Atlantic Ocean seems to be related to this problem. It is very probable that these belts, like the Appalachian, originated on the site occupied earlier by paraliageosynclines in the margin of the Pre-Atlantic. It is important that the evolution of the near-Atlantic Swakop zone of the Damaran belt of Katangides, as well as of the Appalachian belt, started with an intense outflow of acid lava of taphrogenic formation (the Naauwpoort Formation). This all seems to indicate that contours of the Atlantic Ocean resembling modern ones had already appeared at the beginning of the Epiprotozoic.

The principal Epiprotozoic lineaments of Africa run parallel, on one side the principal belt of Katangides and the Nigerian belt situated on its continuation, and on the other the Red Sea belt of North-Eastern Africa. From these two sides they form the framework of the continent, omitting only a relatively small West African craton (platform) that is also bounded by large fractures and in addition, by belts of volcanics of rhyolite-andesite formation from the east and north. The same volcanics are also known to be widely spread in the Red Sea belt. The direct relationship of the principal belt of Katangides and the Nigerian belt suggests the possibility that an extensive area situated within the latter represents the "failed ocean". The same can be stated about the Red Sea belt. The north-eastern branch of Katangides (the Damaran and Katangan belts) and also the sublatitudinal Anti-Atlas belt impart an orthogonal character to the lineaments of Africa (see Fig. 72).

It seems reasonable to suggest the same phenomena of destruction and subsidence of larger portions of the Earth's crust in other parts of the world, in Northern Eurasia in particular. The occurrence of taphrogenic volcanics in the Lower Epiprotozoic strata of the Urals and Yenisei belts that are situated on the opposite flanges of a vast geosynclinal system of Western Siberia may also indicate that this latter suffered intensive submergence at the beginning of the era discussed and could have turned into an oceanic basin.

## 6. Principal Stages of Geologic Evolution

Principal aspects of geologic evolution during the Epiprotozoic are summed up in the scheme given below, which reflects the succession of major events.

### The Early Epiprotozoic

1. 1000–~850 m.y. Formation of a system of large fractures of the Earth's crust. Initiation of new (partially inherited) geosynclinal troughs within the Riphean geosynclinal or fold belts. Deposition of terrigenous and carbonate sediments in geosynclinal troughs and platform basins. Accumulation of thick sedimentary-volcanogenic strata that enclose lava of taphrogenic rhyolite-andesite or rhyolite-basalt (rarely alkaline-basalt) formation in the near-fracture troughs in the margins of certain geosynclinal belts. The beginning of formation ("opening") of the Pre-Atlantic Ocean.

2. A range of 850–820 m.y. The first glacial period in the Epiprotozoic. Glaciation embraced almost every part of the world, irrespective of its geographic latitude. Formation of the Lower complex of glaciogenic rocks. Lava outflows continued in the areas where they were reported earlier.

3. 820–780 m.y. Deposition of terrigenous and carbonate (commonly red-colored) rocks under hot climate typical of the Precambrian (formation of the Interglacial lithostratigraphic complex). Obvious lessening or absolute ceasing of volcanism.

4. ~780 m.y. The Lufilian orogeny of the second or third order. Oscillating movements mostly of positive sign. Local gentle folding and small intrusions of basic and acid magma.

### The Late Epiprotozoic

5. A range of 760–750 m.y. The second glacial period in the Epiprotozoic. The same global-scale distribution of glaciation as in the Early Epiprotozoic. Formation of the Upper complex of glacial deposits.

6. 750–670 m.y. Accumulation of red principally terrigenous, sometimes carbonate-terrigenous strata. At the beginning of the period the climate became somewhat more arid soon after the end of glaciation. In the second half of the period deposition of thick red terrigenous strata of molasse type (the early molasse of the Katangan cycle) in many geosynclines and in the margins of the platforms adjoining them.

7. 680–650 m.y. The Katangan orogeny of the first order. In a number of geosynclines the intensive folding, small pre-tectonic intrusions of basic and ultrabasic magma, large syn-tectonic intrusions of granitoid magma, metamorphism of wallrocks (of the greenschist facies mainly). The upper red molasse is formed at the end of the cycle and partially at the beginning of the Eocambrian.

# Chapter 7  The Eocambrian (Vendian sensu stricto)

## A. Rock Records

### 1. General Characteristics of the Eocambrian

**Definition.** Different geologic formations that originated after the Katangan orogeny are here attributed to the Eocambrian, the time of their formation being after 680–650 m.y. and prior to the beginning of the Cambrian period, that is earlier than 570 m.y. The duration of the Eocambrian is thus 100–80 m.y., which is much shorter than that of other principal units in a general scale of the Precambrian. Moreover, unlike other Precambrian eras and suberas, separated by diastrophic cycles, the Eocambrian is marked by a large orogeny only at its beginning, and its boundary with the Cambrian shows no traces of any important tectonic events.

The problem of the stratigraphic rank of the Eocambrian and its position in a general scale will be discussed later, after its examination here. The informal term "complex" is now used to designate the rocks forming it, a term that will be used here too. Some workers think the name Eocambrian unhappy, as it was proposed originally for the Sparagmitic complex of Scandinavia that is now known to embrace strata of different age. However, the scope of this term, as of other old stratigraphic terms, changed with time. Lately many workers (including those from Scandinavia) apply Eocambrian widely for the youngest Precambrian strata underlying the Cambrian, but there is still no unanimous opinion as to the volume of this unit.

One more term applied for the rocks discussed is Vendian or the Vendian complex, proposed by Sokolov (1952) for the first time for the sedimentary strata of the platform cover of the Russian plate underlying the Cambrian. At first Sokolov assigned the Valday Group rocks to Vendian that here are regarded as and serve as a good stratotype of Eocambrian, but later Sokolov and others added other older strata of the platform cover enclosing tillites to Vendian, which are here attributed to the Epiprotozoic (the Volyn Group and its equivalents). Some workers, for instance Keller and Semikhatov, used microphytolites and stromatolites of the so-called Vendian or fourth phytolite complex for dividing the Vendian strata. As a result, more and more old strata (older than 1000 m.y.), including the Upper Riphean (Kudash), were supposedly assigned to the Vendian. At the same time, the studies of Jacobson (1966 etc.) and Solontsov and Aksenov (1970) proved that the Valday Group (Vendian sensu stricto) rocks in the Russian plate form a separate structural stage and that they overlie the Volyn rocks (and other old rocks) with a significant unconformity. The same can be stated about the Siberian platform (Salop 1973a). Naturally, it is not valid to combine them as Vendian structural complexes, whose sequences are different in composition and structure.

The fact that the Eocambrian or Vendian proper (Vendian s. str.) strata contain specific organic remains, typical only of these rocks, is of great importance as these strata contain fossils that are not known from underlying or overlying rocks. It permits division of the Eocambrian using biostratigraphic criterium. This fact also helped some workers to revise their opinion on the "Great Vendian", and they of necessity came to limit the Vendian volume to the Valday Group (Keller 1979, Semikhatov 1979). Nevertheless, the "Great Vendian" still exists with its stratotype – the Volyn and Valday Groups in a new stratigraphic scale of the Precambrian of the U.S.S.R. accepted by the Interdepartamental Committee on Stratigraphy.

**On Geologic Boundaries of the Complex.** The age boundaries were stated when a definition of the Eocambrian was given, but certain difficulties arise in establishing and following these geologic boundaries of the complex. This concerns in particular its upper age boundary and is explained by the fact that the lower age boundary of the Cambrian is not clearly defined. Usually it is stated that Cambrian strata are recognized by the appearance of remains of the skeletal fauna, but it is well-known that these remains with hard outer skeleton or with certain traces of biomineralization of tissues are also encountered in the Eocambrian, in the upper part of the complex in particular (see below). A more precise understanding of the term "skeletal fauna" is needed, be even if it arbitrarily chosen. Workers now accept drawing the Eocambrian (Vendian)-Cambrian boundary at the base of the so-called Tommotian stage that is recognized by the presence of the remains of the oldest hyolithids, brachiopods, gastropods, kamenids, and archaeocyatheans (Cowie and Glaessner 1975, Rozanov and Sokolov 1980, etc.). At the same time it is quite usual that the first faunal remains in geologic sequences occur in the strata of thick monotonous unfossiliferous rocks related to Eocambrian rocks. In such cases the Precambrian-Cambrian boundary line is drawn arbitrarily.

The lower boundary of the Eocambrian is clearer as a rule, as the Eocambrian strata usually rest transgressively and are separated from the underlying rocks by an unconformity (of an angular or stratigraphic type). Sometimes, however, this unconformity is masked, as happens commonly in miogeosynclinal belts where the Katangan orogeny was feeble, and then it is expressed only in a change of nature of sedimentation and in formation of a new transgressive macrocycle with coarser terrigenous rocks in its lower part (examples are the Conception Group of Newfoundland or the Sylvitsa Group of the Middle Urals described in the previous chapter). In the platform these masked unconformities are traced during geologic mapping when extensive basal beds of the Eocambrian complex are followed. Sometimes two or three unconformities (breaks) are observed in the Epiprotozoic-Eocambrian boundary beds and it is difficult to give preference to any of them when drawing the boundary of the units. This problem is, however, quite common in stratigraphy when geologic formations of different age are being studied.

Thus the boundaries of the Eocambrian can be drawn according to its occurrence over the youngest Epiprotozoic strata and below the oldest Cambrian rocks, according to the organic remains enclosed, and finally according to isotope datings of syngenetic minerals and rocks. It should be noted here that many K-Ar datings on glauconite and clay minerals can be considered as reliable because in the platforms the Eocambrian rocks are unaltered and usually are high in the sequence of the older strata.

**Distribution.** The Eocambrian strata extend in every continent. They are most complete and specific in the East-European and Siberian platforms; they are of less distribution in the Australian, African, North American, and other platforms. Miogeosynclinal formations are reported from many foldbelts: the Urals, Eastern-Sayan, Baikalian, Cordilleran, Appalachian, and Adelaide, in Variscides and Caledonides of Western Europe in particular. The Eocambrian eugeosynclinal formations are supposed to occur in some Phanerozoic foldbelts, but it is difficult to distinguish them from the Cambrian sedimentary-volcanogenic strata; paleontological methods can hardly be applied in this case, as imprints of soft-bodied organisms are rarely preserved in highly altered and strongly deformed rocks.

The Valday Group of the Russian plate is accepted as a stratotype of the Eocambrian complex as was pointed out above. Its well-known stratigraphic analogs in Europe are the Stappugiedde sandstone in Northern Norway, the Nekse sandstone and the Balka quartzite in Sweden, the Verkhnebavlinsk Group of the Urals region, the upper parts of the Sylvitsa Group of the Middle Urals and the Asha Group of the Southern Urals; in Asia, the Yudoma Formation of Aldan, the Zherba and Tinnaya Formations of the Patom Highland, the Kirbakta and Berkuta Formations of Tien-Shan; in North America, the Stirling Formation and the lower part of the Wood Canyon Formation of California, the Jacobsville (and Bayfield) Formation in the Great Lakes area, the upper part of the Conception Group together with the Hodgewater Group of Newfoundland, the Great Smoky Group in the Appalachians; in Africa, the Lower and Middle Adoudou Groups in Morocco, the lower part of the Nama Group in Namibia; in Australia, the Pound quartzite in the south of the continent and the Arumbera Formation in the north. These Eocambrian units and others will be examined below.

**Brief Characteristics of the Complex.** It is very difficult to give general characteristics of Eocambrian strata, as they are not so marked with the exception of certain specific organic remains, and do not differ greatly from the Riphean, Epiprotozoic, or Cambrian rocks. It is possible to state only some aspects in this connection, though even these are not obvious.

The platform strata are mainly represented by shallow-water marine sedimentary rocks. The continental strata are subordinate and in the case of their forming separate units, their belonging to the Eocambrian remains as a rule doubtful. In the East European, North American and Australian platforms, the Eocambrian strata are principally made up of terrigenous rocks as in the Siberian platform the carbonate rocks are dominant and it is only the lower thin portion of the complex that is composed of well-sorted quartzite-sandstone. In the African platform terrigenous as well as terrigenous-carbonate or carbonate epicontinental rocks are reported. The sandstone-slate strata are obviously dominant among miogeosynclinal deposits. The red-colored or variegated rocks are common among all types of strata, of the platform ones in particular, but still they are not so abundant as in the Neoprotozoic or Epiprotozoic strata. This is certainly due to the fact that marine sedimentation was dominant. In the red beds local inclusions (or crystal molds) of gypsum and rock salt are observed. Certain carbonate rocks contain bituminous matter.

It is notable that basic volcanics ("traps") are lacking or quite rare in the Eocambrian platform strata, where in the Riphean rocks they were typical. At the same time, in these same strata, mostly in their lower part, acid volcanics and (or) their tuffs occur.

The folded areas are typified as a rule by terrigenous formations of orogenic type that terminate the geosynclinal evolution of mobile belts. These are, for instance, thick conglomerate-sandstone strata of the inner parts of the Baikal foldbelt, of Eastern Sayan, and the western slopes of the Urals. Some of them can be assigned to the late molasse of the Katangan cycle. Glacial deposits are absent in the Eocambrian complex proper, but in the beds transitional to Cambrian strata, tillites or rocks formed under extremely cold climate are observed in some regions (South Western Africa, Southern Norway, Southern Sweden, Scotland, etc.).

Transgressive structure is typical of almost every Eocambrian sequence, as is seen from an upward change of coarse terrigenous deposits into more fine-grained (purely pelitic) and carbonate rocks. All of the marine sedimentary strata are of extreme shallow-water environment. The deep-sea sediments are lacking not only in the platform but also in the miogeosynclinal complexes.

The thickness of the Eocambrian platform strata is not great: it amounts to some tens or several hundreds of meters and in rare cases reaches some thousands of meters. The thickness of miogeosynclinal formations (including the orogenic ones) is commonly several thousands of meters.

The Eocambrian platform strata are absolutely unaltered and rest horizontally with the exception of zones of near-fracture dislocations and margins of foldbelts. Geosynclinal rocks of the complex are of the same grade as the overlying Cambrian rocks usually of a zonal character and not exceeding greenschist facies. Intrusive rocks of Eocambrian age are not known for certain. The plutonic bodies cutting Eocambrian strata belong to the Baikalian (Damaran) orogeny of the Upper Cambrian or to the Caledonian and still younger tectono-magmatic events.

Mineral deposits related to the Eocambrian sedimentary strata are generally not numerous. It is worth mentioning the following deposits: stratiform mineralization and occurrences of lead and polymetallic (copper-lead-zinc) ores in carbonate strata (Eastern Siberia, Anti-Atlas etc.), a small bauxite deposit (Eastern Sayan) and deposits of phosphorite (Eastern and Western Siberia, Mongolia, etc.). Certain manifestations of oil and gas in the Eocambrian platform terrigenous-carbonate strata in Eastern Siberia; of great scientific interest and probably of certain economic importance, are the Mota and Tolba Formations that underlie beds with the oldest Cambrian fossils.

**Organic Remains.** The most striking feature of the Eocambrian is the presence of remains of organisms that are lacking or not known in the older Precambrian units. The Eocambrian is typified by the occurrence of imprints of nonskeletal organisms that are given the name Ediacara fauna. Earlier this fauna was known from single localities, the Ediacara hill area in Southern Australia being especially rich in it, but now there are many sites in different continents where it is registered and more and more localities containing it are being reported. Some of them are not inferior in richness to the famous Ediacara biota. Twenty five species of soft-bodied invertebrates belonging to 21 genera were recently reported from deposits on the Onega Peninsula in the White Sea (Fedonkin 1980).

The Ediacara fauna consists mainly of coelenterates: single polypoid medusoids, hydroid and scyphoid medusa that comprise up to 80% of all forms in large biotas. Additionally, the imprints of the following organisms are registered: annelids, flat-worms (*Dickinsonia*), arthropods (*Vendia*, *Praecambridium*), echinoderms (*Tribrachi-dium*); many forms (*Pteridinium*, *Rangea*, *Parvankornia*, *Charnia* etc.) cannot be assigned with certainty to any taxon or even type or class, but some are of cosmopolitic distribution. Imprints of the Ediacara fauna sometimes associate with fossil traces of multi-cellular organisms, mainly of worms. Various ichnocoenoses are common as separate assemblages.

In the upper part of the Eocambrian, in the boundary beds with the Cambrian strata, the remains of organisms with hard (biomineralized) outer skeleton are known, commonly tube-shaped. These tube-like remains of problematic classification include the following organisms: *Anabarites*, *Sabelliditides*, and *Platysolenites*. Their skeleton is built up of chitin, silica, phosphate-carbonate material. There is no doubt that these remains are organogenic, but it cannot be excluded that some structures greatly resembling these tubes are mechanoglyphs or ichnoglyphs.

The Eocambrian carbonate rocks contain stromatolites and microphytolites of the so-called Vendian phytolite complex that are also typical of the older strata, the Terminal Riphean and Epiprotozoic. But the calc-secreting algae *Renalcia* and others appear only in the Eocambrian. Thallus of brown algae *Vendotaenia*, films *Laminari-tes*, accumulations of different trichoms, various elements of lichen and fungi are also very typical.

Acritarchs are common, mostly of the *Leiosphaeridia* group, some of them probably occurring only in this complex (Timofeev 1979). Acritarchs of the *Kildinella* group widely distributed in the Upper (and partly in Middle) Riphean are either not known from the Eocambrian or are quite rare, *Chuaria* is completely absent.

In cherts (including syngenetic flints of carbonate rocks) various microbiota is known, but no specific Eocambrian forms have been recognized as yet. A peculiar spiral form has been determined for the first time among the spherical and filamentous microremains from the Eocambrian Tina Formation of the Siberian platform, it was given the name *Obruchevella* and is now being compared with modern blue-green algae *Spirulina*. It was considered that *Obruchevella* forms have appeared only since the Eocambrian (Vendian) and then passed through up to the Ordovician. Recently, however, they were found in the Riphean strata of the Yenisei Ridge and in the Altai-Sayan area. Their stratigraphic importance thus appears to be less.

## 2. The Eocambrian Stratotype and Its Correlatives in Different Parts of the World

**The Eocambrian Stratotype – the Valday Group of the Russian Plate and Some of Its Correlatives in Europe.** The strata known under the name Valday Group were distinguished in 1950–1952 for the first time by Sokolov in the north-west of the Russian plate as the Gdov and "laminaritic" beds. Further works helped by numerous drillings showed that the Valday Group builds up the base of Paleozoic cover for a major portion of the East European platform. Jacobson (1966) and later Solontsov and Aksenov (1970) established that the Valday Group (or Vendian s. str.) unconformably mantles all the older Precambrian strata, including the Riphean and Epiprotozoic rocks and these latter, unlike the Valday rocks, occur only in isolated depressions-grabens in the basement of the platform (Fig. 74).

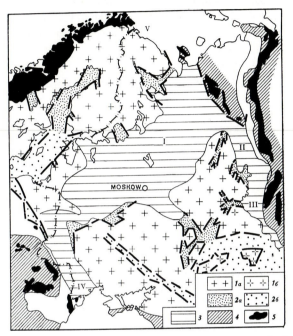

**Fig. 74.** Eocambrian structure of the East-European platform. (After Aksenov et al. 1978). *1* shields, massifs, anteclises: *a* know, *b* presumed; *2* Riphean graben-like depressions: *a* known, *b* presumed; *3* Eocambrian (Vendian s.str.) syneclises and depressions: *I* Moscow syneclise, *II* Upper Kamian, *III* Shkapovo-Shikhan, *IV* Lvov-Kishenev, *V* Barents Sea marginal depressions; *4* foldbelts, *5* exposures of the Upper Precambrian in the areas surrounding the platform

In different regions of the Russian plate the strata discussed are divisible into formations and units of lower rank that are given many local names. The most complete sequence of the Valday Group is known from drilling in the near-axial part of the Moscow syneclise. Four formations are distinguished within it there (up-section): Pletenevo, Ust'-Pinega, Lyubim, and Reshma, the two lower ones being sometimes combined into the Redkino Subgroup, and the two upper into the Povarovo Subgroup. Their general characteristics are given according to Aksenov et al. (1978).

The Pletenevo Formation (up to 80 m) is distributed in the pre-Valday peneplain and is made up of dark siltstone and argillite underlain by a band of conglomerate and gritstone. The Ust'-Pinega Formation (up to 350 m) is composed of greenish-grey, dark-grey and brown argillite with interbeds and bands of siltstone and sandstone. There are some three horizons of tuffite, ash tuff, and montmorillonite clay developed after them among these rocks. All the occurrences of imprints of nonskeletal organisms of the Ediacara type are situated in the Russian plate in the rocks of this formation. A K-Ar age of about 600 m.y. is obtained on glauconite and vitroclastic tuff, but is probably slightly rejuvenated. The Lyubim Formation (up to 490 m) rests on different horizons of the Ust'-Pinega Formation and directly on the crystalline basement in the western margin of the syneclise. It comprises dark greenish-grey, variegated upward, siltstone, argillite, and sandstone. An age of 600 m.y. is obtained on glauconite by K-Ar analysis. The Reshma Formation (up to 500 m) is built up of red polymictic and quartz sandstone, siltstone, and argillite. It is conformably overlain by the Lontova horizon with remains of the oldest skeletal fauna of the Lower Cambrian (hyolithids, gastropods, etc.).

In the north-western regions of the Moscow syneclise the Gdov Formation of sandstone and the Kotlin Formation of argillite-siltstone are recognized within the Valday Group, and are likely to correspond to the Lyubim Formation of the central part of the Moscow syneclise. The thickness of both formations is not more than 300 m. In the western margin of the syneclise, in the environs of the town of Gdov and westward, in the Baltic area, the Vendian s. str. is represented only by the Kotlin Formation, or in places the overlying strata of quartz sandstone and siltstone are added to it (the Voronkovo Formation, up to 30 m), which possibly correspond to the Reshma Formation of the stratotype. The Valday rocks of the Baltic area contain films of the *Laminarites* algae, the shells of *Sabellidites*, acritarchs, ichnoglyphs etc.

In Byelorussia, as well, only the lower part of the Vendian complex (s. str.) is developed. The Svisloch' (Rataychin) Formation of tuff, argillite, and sandstone (up to 60 m) is recognized at the base of it, resting unconformably on the Epiprotozoic Vilchansk tillite-bearing Group or on the basement rocks. Upward the Liozno Formation of sandstone, gritstone, and siltstone is situated (up to 110 m).

In the south-west of the Russian plate, in the Dniester area, the following formations correspond to the Valday Group (up-section): (1) the Mogilev Formation (up to 80 m) of sandstone and siltstone with rare imprints of medusoids; (2) the Yaryshev Formation (up to 100 m) of argillite and sandstone, with horizons of tuffite and tuff in the upper part, it contains numerous imprints of soft-bodied organisms, of medusoids mainly; (3) the Nagoryany Formation (75 m) of sandstone and conglomerate that upward passes into bituminous argillite with phosphorite nodules; (4) the Kanilov Formation (up to 200 m) of variegated and green siltstone and sandstone that lies with an erosion on the underlying rocks. All these strata are usually correlated with the Redkino Subgroup of the Moscow syneclise.

On the western slopes of the Southern Urals the Asha molasse-like Group is assigned to the Eocambrian. It starts with the Uryuk Formation (100–350 m) of arkosic sandstone, gritstone, and conglomerate that unconformably overlies various Riphean strata and the Upper Epiprotozoic tillite (the Kurgashli Formation). In the upward sequence follow the Basa Formation (600–900 m) of quartz sandstone, siltstone, and argillite, the Kukkarauk Formation (350 m) of red sandstone and conglomerate, the Zigan Formation (up to 480 m) of quartz sandstone interbedding with argillite and siltstone. Glauconite from the Uryuk Formation is dated at 600 m.y. by the K-Ar method.

In the western slope of the Middle Urals the upper part of the Sylvitsa Group, examined in the previous chapter, corresponds to the Asha Group. The imprints of coelenterates of the cyclomedusa group are known to be determined from sandstone of this group.

In Northern Europe Eocambrian strata are reported from several localities of Norway and Sweden. In Northern Norway, in the east of the Finnmark the Stappugiedde Formation is Eocambrian, being built up of slate and sandstone with problematic organic remains. It overlies unconformably the tillite-bearing Epiprotozoic Vestertana Group and is conformably overlain by siltstone of the Breyvik Formation that contains *Platysolenites antiquissimus*. The overlying strata contain trilobites *Holmia* sp. typical of the lower "trilobite beds" of Cambrian. In the western portion of the Finnmark the lower part of the Dividal terrigenous strata is a correlative of the Stappugiedde Formation, it contains imprints of medusoids and worms; upward these beds gradually pass into rocks characterized by the oldest Cambrian skeletal fauna (hyolithids, brachiopods, gastropods). In Southern Norway the Vardal (Vangsos) sandstone and the overlying Ringsack quartzite with worm tracks (*Scolithus, Monocraterion*, and *Diplocraterion*) are possible Eocambrian; sandstone lies unconformably on the Muelv tillite and the Ekre shale of Epiprotozoic (see Chap. 6), and quartzite underlies the Lower Cambrian strata with *Holmia*. In Norway all the rocks mentioned are of miogeosynclinal nature.

In Southern Sweden (the Bornholm Island) the Nekse sandstone and the Balka quartzite with worm tracks (*Scolithus* and *Diplocraterion*) can be attributed to the Eocambrian; these are overlain by Lower Cambrian siltstone with hyolithids, brachiopods, and gastropods. Rocks of the Bornholm Island are typically platform and lie on the peneplained surface of crystalline rocks of the basement.

In the British Isles the complex under discussion is represented by strongly dislocated geosynclinal rocks. In the Scottish Caledonides the Middle Dalradian Group, with a thickness up to 2000 m, is likely to belong to the Eocambrian, and rests with a break on the Epiprotozoic tillite (the Lower Dalradian). In the lower part it is made up of quartzite with local conglomerate interbeds, upward it passes into rhythmically bedded carbonaceous slate, this latter is conformably overlain by the Upper Dalradian thick strata with the Middle Cambrian faunal remains in the upper part. Thus in the Dalradian sequence the Eocambrian-Cambrian boundary is roughly drawn.

In Central England the geosynclinal rocks of the Uriconian Supergroup (Shropshire) and the Charnian Supergroup (Leicestershire) belong to the Eocambrian. The Charnian Supergroup rocks are of special interest as they contain well-preserved imprints of nonskeletal organisms of the Ediaraca type. The Supergroup is divided into three groups (up-section): Blackbrook, Maplewell, and Brand, with a total thickness up to 4000 m. The two lower groups constitute the major portion of the supergroup and are composed of volcanogenic-sedimentary rocks: tuff, welded tuff, and lava-breccias are dominant, then of lesser abundance are acid, intermediate, and partly basic volcanics, subgreywackes, slates, and tuffites; the upper group (360 m) lies with a supposed break on the Maplewell Group and starts with conglomerate passing upward into quartzite-sandstone and slate alternating with sandstone. Imprints of *Charnia masoni, Charnodiscus concentricus*, and *Cyclomedusa davidi* are reported from the upper part of the Maplewell Group (Ford 1979).

Charnian porphyroids are dated at 684 m.y. (K-Ar analysis) and diorite cutting the supergroup at 547 m.y. (K-Ar analysis) and at $552 \pm 58$ m.y. (Rb-Sr analysis). These datings accord well with the Eocambrian age of the rocks in Leicestershire, but their interpretation is not reliable, as the Charnian rocks suffered alteration under the influence of later thermal events. Geologic methods can also give no certain upper age boundary of the Charnian rocks, as they are overlain only by the Triassic strata.

The Eocambrian formations of Central England mentioned above are commonly attributed to eugeosynclinal rocks. However, this opinion does not seem reasonable, as acid volcanics and their tuffs are dominant among them, and clastic rocks with monomictic quartzite-sandstone as well as greywacke are also reported. It is highly probable that they can be regarded as taphrogeosynclinal rocks.

**The Eocambrian of Asia.** In this continent Eocambrian strata are principally distributed in Siberia, where they build up an extensive cover of the Siberian platform and thick detractive (orogenic) strata in the foldbelts surrounding it. This cover comprises many units, but the Yudoma Formation (or Group) is the best known among them, being developed in the Aldan shield. It is reasonably exposed, contains various phytolites and exhibits distinct geologic boundaries with the underlying and overlying rocks. For this reason some workers were inclined to contrast it to the Valday Group of the Russian plate and to propose it as a stratotype of the whole complex, suggesting instead of the name Vendian the name Yudomian or the term combined of the two names – Vendomian. However, the Vendian seems to have been now as a stratotype definitely and finally accepted on the establishment of rich occurrences of Ediacara fauna in the Russian plate.

The Yudoma Formation is rather simple and continuous in structure: in the lower part it consists of light-grey, rarely dark-grey quartz sandstone and argillite that upward pass into carbonate rocks: dolomite and dolomitic limestone that make up the major portion of the formation. Locally the terrigenous bed is quite thin, even wedging out completely, and carbonate strata are characterized by interbeds and bands of argillite at different levels, mostly of their lower part. Dolomites contain phytolites of the fourth (Vendian) phytolite complex, these are mostly oncolites and katagraphs, rare stromatolites (*Boxonia*, *Yurusania*, *Gongilina* etc.); in the uppermost part of the formation anabarites tubes are registered. Dolomites are sometimes slightly bituminous. Shallow-water structures are typical of all rocks. The thickness of the formation varies from 150 to 250 m, but in some sites (depressions) it reaches 800 m.

The Yudoma Formation lies transgressively on the peneplained surface of different Precambrian rocks: from the Katarchean gneisses to the Epiprotozoic Uy sandstone Group; it rests also on the ultrabasic and alkaline rocks of the Ingili pluton with an age of 680–650 m.y. that cuts the latter group. The age of the lower part of the formation is determined on glauconite by several K-Ar datings in the range of 635–650 m.y.

The upper age limit of the formation is defined by the fact that the carbonate-terrigenous variegated Yuedey Formation overlies dolomite; it contains in its base remains of skeletal fauna of the Tommotian stage of the Cambrian (small hyolithids, gastropods, brachiopods, camenids etc.) and somewhat up-section appear archaeo-cyatheans and fragments of trilobites of the Protolenidae family.

The Yudoma Formation continues with certain facial changes till the northern near-platform margin of the Baikal Mountain Land, where the Zherba Formation (50–60 m) of quartzite-sandstone and the overlying Tinnaya Formation (100–600 m) of dolomite and cherty limestone with phytolites and microbiota (including *Obruche-vella*) corresponds to it. The Tinnaya Formation, like the Yudoma, is overlain by the Lower Cambrian "variegated" formation. Glauconite from the base of the Zherba Formation is dated at 610 m.y., but this value may be rejuvenated.

In the western margin of the Baikal Mountain Land the lower part of the Mota Formation (up to 300 m) belongs to the Eocambrian. It is mainly made up of grey and red sandstone with dolomite interbeds in the upper part. The Mota Formation rests with a break on the Ushakovka Formation of the Epiprotozoic or on older Precambrian rocks. Sokolov determined the imprints of *Pteridinium* sp., the Ediacara form, the tracks of mud-eaters (*Cylindrichnus*), and *Vendotaenia* algae. The upper part of the Mota Formation, composed of carbonate rocks with rare anhydrite interbeds, belongs to the Lower Cambrian, as it contains arcaeocyatheans of the Sunnagin horizon. It should be noted here that some workers call the Mota Formation the essentially carbonate strata of the Irkutsk amphitheater of the Siberian platform, characterized by skeletal fauna and acritarch remains typical of the lower horizons (Rovno and Lontova) of the Cambrian of the Russian plate. In my opinion they correspond only to the uppermost part of the Mota Formation of the Baikal region.

Analogs of the Yudoma Formation are also known along the western region framing the Siberian platform: in the Yenisei Ridge, in the Khantay-Rybinsk and Chadobets uplifts. They are all composed of terrigenous-carbonate strata that are usually thin (200–300 m) and contain phytolites (mainly microphytolites) of the Vendian type, local sabellidite remains and in the Ostrov Formation of the Yenisei Ridge the imprints of cyclomedusa are also reported. These strata lie transgressively on the Upper Riphean and Epiprotozoic (in the area of the Yensei Ridge) rocks and are conformably overlain by the Lower Cambrian fossiliferous strata.

The Eocambrian strata are also quite abundant in the north of the Siberian platform. In the Anabar anteclise and in the Udzhin uplift the lower part of the Eocambrian rocks is represented by the Starorechenskaya Formation (up to 200 m) with dolomite and interbeds of sandstone and argillite at the base and strata of limestone and finally bituminous limestone. The rocks contain stromatolites (*Paniscollenia, Colleniella, Boxonia*) and microphytolites. The formation lies with a great break on the platform rocks of the Riphean Billyakh Group. Glauconite from the lower or middle part of the formation is dated at 670 m.y. and from the upper part – at 620 m.y. by the K-Ar method.

The Starorechenskaya Formation is overlain with a break by the Nemakit-Daldin (Manykay) Formation. Along the Kotuykan River (the Anabar anteclise) the major lower portion of the formation (60 m) consists of limestone interbedded with polymictic sanstone; its upper part (24 m) is made up of massive stromatolitic limestone. The lower part of the formation belongs only to the Eocambrian, and contains *Anabarites trisulcatus, Sabellidites*, acritarchs, and vendotaenides. In the upper carbonate portion of the sequence there are gastropods (*Helcionella*), then conodont-like structures, algae, and acritarchs and ten meters up the sequence there is a rich complex of skeletal fauna of the Lower Cambrian Tommotian stage, belonging to the *Oelandiella Korobkovi-Anabarella plana* zone (Savitsky et al. 1980). Thus the Nemakit-Daldin Formation is a striking example of rocks transitional from Eocambrian to Cambrian.

Eocambrian orogenic formations, the Eastern Sayan and Baikal foldbelts, are developed in the areas south of the Siberian platform, and are represented by thick molasse clastic rocks, were formed in intermontane depressions soon after termination of intensive tectonic movements of the Katangan cycle. The Mamakan Group of the inner part of the Baikal foldbelt (the North-Muya Range) represents a standard type of such strata. This group embraces the Gukit, Lower Padrokan, and Upper Padrokan Formations. The rocks composing them occur in narrow but deep depressions (troughs-grabens) that are situated in the zones of abyssal fractures. The rocks are red and greenish-grey polymictic sandstone and conglomerate, their thickness varies greatly from some tens of meters in the slopes of uplifts up to 4500 m in the central parts of depressions. These rocks are conformably overlain by the Sidel'te carbonate-terrigenous Formation (300–600 m) mainly composed of quartzite-sandstone, marl, and dolomite, and containing, together with microphytolites, some rare remains of small inarticulate brachiopods. Up-section the Yangud Superformation of carbonate rocks rests, with a thickness up to 3500 m, and characterized by remains of archaeocyathids and trilobites of the Lower Cambrian Lenan stage; in its uppermost part there are trilobites of the Amgan stage of the Middle Cambrian (Salop 1964–1967).

In the Verkhneangarsk Range the Kholodnaya Formation, an analog of the Eocambrian formations of the Mamakan Group, has a thickness up to 3500 m and is of conglomerate-sandstone composition. The Tukalomi Formation of siltstone-slate overlying it conformably corresponds to the Lower Cambrian Sidel'te Formation; it contains acritarchs typical of the Valday Group of the Russian plate, and worm burrows. It is overlain by the same Lower and Middle Cambrian carbonate rocks as in the area of the North-Muya Range. All the Eocambrian and Cambrian rocks of the Baikal foldbelt mentioned exhibit large symmetrical folds and locally are cut by granites and alkaline syenites. Unlike the Riphean and Epiprotozoic strata developed in the same regions of the Baikal Mountain Land, they did not suffer regional metamorphism and are altered only in the contact aureoles of the Paleozoic intrusive massifs.

In the Eastern Sayan foldbelt, the Eocambrian orogenic formations are developed in its western part, in the Uda-Derba zone (the Anastas'ina and Angul Formations); in addition to them the co-eval miogeosynclinal rocks are also known from the Bokson-Sarkhoy synclinorium, being represented by two lower members of the Bokson Group. This group lies unconformably on the volcanogenic-terrigenous Sarkhoy Formation of the Upper Riphean or Epiprotozoic, and is divisible into three formations. The lower, Zabit (Gorkhon) Formation (500 m), starts with basal conglomerate, but is mainly composed of dark bituminous limestone and grey dolomite with phytolites of "Upper Riphean and Vendian appearance"; phosphorite horizons are reported from two levels of it. The middle, Tabirzurta Formation (more than 1000 m), is principally built up of dolomite. A bauxite-bearing horizon is known in its lowermost part, the medusoid imprints are observed there; some 400 m higher up from the bottom of the formation the *Renalcis* algae are determined in dolomite, in its uppermost part the remains of the Lower Cambrian archaeocyathids and trilobites are known. The upper-Khuzhirtay Formation contains abundant remains of fauna of the Aldanian stage. Thus, in the area examined the Eocambrian-Cambrian boundary runs inside the monotonous dolomite strata of the Tabirzurta Formation.

The Eocambrian strata are also developed in other foldbelts of Asia, in the northeast and east of the continent in particular, in the foldbelts of Central Asia, but their age is not certain there. In Kazakhstan many different units of the Upper Precambrian are usually assigned to the Vendian, but as a rule these strata are tillite-bearing and here they are regarded as Epiprotozoic. The Maly Karatau Ridge is the only place where the Eocambrian formations are certainly established. There the thin strata occurring unconformably on the Upper Riphean or Epiprotozoic Malokaroy Group and belonging to the lowermost part of the Tamda Group are Eocambrian. They start with variegated carbonate rocks interbedded with glauconite-bearing sandstone (0–170 m), up the sequence a band (up to 30 m) mainly of dolomite is situated, containing microphytolites of the Vendian complex and protoconodonts (?) (*Protohertzina*) and also anabarites (*Anabarites trisulcatus*). The dolomite is overlain with an insignificant break by the Chulaktaus Formation of phosphorite-bearing slate that is characterized by abundant and various remains of hyolithids, inarticulate brachiopods, gastropods, and kamenids that are the same as in the Lower Cambrian Tommotian stage in Siberia. Still up-section there are beds that contain trilobites of the Lower Cambrian.

**The Eocambrian of North America.** The Eocambrian of this continent is represented by three types of rocks extending in isolated regions. The rocks of one of these types occur mainly in the foldbelt of the North American Cordillera, and were formed at a late stage in the evolution of the Riphean-Epiprotozoic Beltian geosyncline. These rocks are very similar to the platform type, but at the same time they differ by intensive folding and local low-grade alteration. Another type of Eocambrian rocks is widespread in the Appalachian foldbelt, and is represented by thick miogeosynclinal essentially terrigenous strata that sometimes exhibit taphrogenic (or riftogenic) aspects. Finally, the thin clastic rocks developed in single sites of the North American platform belong to the third type. Below several examples of these types of rocks are given.

In Northern Cordillera, in the Mackenzie Mts of British Columbia, the strata of oligomictic and essentially quartz sandstone interbedded with bands of limestone and dolomite recognized as the Benyon Range Formation (more than 1300 m) rest with a possible break above the Rapitan Group (with tillite) and the Sheepbed Formation overlying it conformably. In the upper part of the Benyon Range Formation there is a band of interbedded slate, siltstone and dolomite with remains of conodont-like structures *Protohertzina anabarica* at its base, and some 370 m up-section are known imprints of *Gyrolithes polonicus*, *Didychnus*, and *Phycoides* sp. and still upward there are imprints of *Tretichnus*, *Planolites* and *Neonorites*. In the uppermost part of this strata there are remains of hyolithids and brachiopods together with imprints which are given the names *Bergaueria*, *Teichichnus*, *Risophycus*, and *Scolicia*. The Benyon Range Formation is conformably overlain by the carbonate Sekwi Formation with archaeocyathids and trilobites belonging to *Parafallotaspis*, *Nevadella*, *Salterella*, *Olenellus* genera and others. Thus, in the Mackenzie Mts the Eocambrian-Cambrian boundary is in the upper part of the Benyon Range Formation (Fritz 1980).

Further examples of Eocambrian strata of this type will be given from the more southern regions of Cordillera situated in the U.S.A. The Stirling and a part of the Wood Canyon Formations belong to the Eocambrian in the boundary areas of Nevada and California. The first lies transgressively on the Johnnie Formation that completes the Epiprotozoic sequence, and is made up of red sandstone up to 1000 m thick. The second is of conformable occurrence; it rests on sandstone and is more composite, including quartzite-sandstone, siltstone, shale, and limestone in part (up to 700 m). It is overlain conformably or with an insignificant break by the Zabriskie quartzite (70 m), which is in turn overlain by the Carrara carbonate Formation without any break; in its lower part there are remains of the Lower Cambrian fauna, and in its upper part the lowermost Middle Cambrian fauna. The first faunal imprints represented by worm burrows (*Scolithus*) are registered at 100–300 m below the top of the Wood Canyon Formation, and near the top there are archaeocyathids and trilobites of the Lower Cambrian. Organic remains typical of the oldest Cambrian beds are not known as yet, which makes the drawing of the Eocambrian-Cambrian boundary in this region problematic. As a rule it is accepted as lying within the Wood Canyon Formation by the evidence of *Scolithus*.

In Idaho the limestone completing the Epiprotozoic Pokatello Group (with two horizons of tillite) seems to be overlain with a masked stratigraphic unconformity by an extremely thick pile (up to 6000 m) of pink and light-grey quartzite, of sandstone in part, with rare interbeds of argillite, slate, and limestone. This pile is called the Brigham Quartzite and is divided into several formations similar in compositon. In its divided upper part the Camelback Formation is recognized, made up of quartzite and separated from the underlying quartzite by a horizon of quartz conglomerate. The Gibson Jack Formation (up to 700 m) of slate and argillite interbedded with sandstone and quartzite is situated in the uppermost part of the pile; *Naraoia* arthropods are determined from it (known from different levels of the Cambrian) and the *Olenellus* trilobites typical only of the Lower Cambrian. The Brigham Quartzite is overlain by limestone with different trilobites of the Middle Cambrian. Obviously the oldest Cambrian fauna is not determined in the sequence examined. American workers (Trimble 1976 etc.) suggest drawing the Precambrian-Cambrian boundary inside the unfossiliferous Camelback Formation.

Eocambrian miogeosynclinal rocks are spread in different portions of the Appalachian belt. The earlier-mentioned strata of Newfoundland extending in the north of the belt belong to this type of rocks. In the previous chapter it was noted that the terrigenous strata making up the Briscall and Mistaken Point Formations with a thickness of up to 1600 m are situated in the upper part of the Conception Group, and rest on Epiprotozoic rocks that constitute their lower part with a masked unconformity, and that intrusions of basic rocks are likely to have occurred during the time this unconformity was coming into existence. The Mistaken Point Formation has abundant imprints of Ediacara fauna, principally of medusoids. Up the sequence there are two more Eocambrian Groups: St. John's and Signal Hill. The first, with a thickness of up to 2000 m, consists of grey bedded sandstone that encloses in its central part thick strata of dark-grey to black slate with sandstone lenses; it lies conformably on the Conception Group. The Signal Hill Group (more than 1400 m) is separated from the underlying rocks by a local break; it is mainly built up of coarse-bedded grey, rarely red sandstone with subordinate siltstone interbeds, and it encloses a band of pink slate and quartz gritstone (Williams and King 1979).

At other sites of the Avalon Peninsula the Hodgewater Group (more than 4500 m) and the coeval Musgravetown Group, with a thickness of up to 7000 m, are probably the equivalents of the two upper groups mentioned above (McCartney 1967). Both these groups are composed of grey and red arkosic sandstone, siltstone, and conglomerate; acid and basic lavas are also known from the lower part of the Musgravetown Group. In the uppermost part of the Hodgewater Group, just below the Random quartzite overlying it with an erosion there are cephalopods *Volbortella*, hyolithids, and algae *Epiphyton* sp., and also the imprints of *Cruziana* type in slate, fossils typical of the lower horizons of the Lower Cambrian (Greene and Williams 1974). Thus in Newfoundland the Eocambrian rocks are represented by extremely thick (up to 10,000 m) terrigenous strata, the upper part of which is typical of orogenic or taphrogenic formations.

In the south of the Appalachian belt the thick Ocoee Supergroup (more than 7000 m) belongs to the Eocambrian, and is developed in the eastern part of Tennessee (U.S.A.). It consists of three groups separated by breaks: Walden Creek, Snowbird, and Great Smoky, all of which are made up of different clastic rocks, mainly of sandstone, subgreywacke, quartzite, siltstone, and slate. The upper, Great Smoky Group, accumulated in deep trough at the margin of a continental slope and there turbidites with conspicuously expressed graded bedding are quite abundant. The Chilhowee Group (~ 1000 m) overlies the Ocoee Supergroup with a stratigraphic unconformity; it starts with conglomerate, but is principally composed of sandstone, siltstone, and slate. At different levels it contains imprints of *Scolithus* worms and fragments of *Olenellus* trilobites near its top. In the opinion of American geologists, this group seems to include all the Lower Cambrian, but the possibility cannot be excluded that its lower part is Eocambrian.

**The Eocambrian of Africa.** Reliable Eocambrian strata in this extensive continent are only the platform terrigenous-carbonate rocks distributed in the north-west and south-west of it, in Morocco and Namibia. It is probable, however, that the essentially continental clastic strata resting unconformably on the Epiprotozoic rocks in some regions of Equatorial, Western, and North Western Africa also belong to the complex examined (see previous chapter). Locally distributed red sandstone is usually regarded as an upper molasse of the Katangan cycle. It is unfossiliferous and is overlain only by Ordovician or Carboniferous and Cretaceous rocks.

In Morocco the lower part of the Adoudou Supergroup extending along the whole Anti-Atlas foldbelt belongs to the Eocambrian. It lies unconformably on different Precambrian units, including the Upper Epiprotozoic Ouarzazate Group, and surrounds the anticlinal uplifts ("boutonnières") of the older rocks. Its rocks occur almost everywhere subhorizontally and form composite folds only at places, especially in the margins of the "boutonnières", these folds probably being of gravitational origin. According to Choubert and Faure-Muret (1973), this supergroup is divisible into three groups. The Lower Adoudou Group or the group of lower dolomite reflects the beginning of a great marine transgression that embraced the major portion of the Anti-Atlas. The basal strata (100–400 m) are recognized at the base of the group; they start with a dolomite bed or a conglomerate horizon and are overlain by green slate exhibiting sun-cracks and ripple marks. Up the sequence the dolomite strata occur, with local bands of coarse-grained arkosic sandstone with carbonate matrix. The stratiform copper deposits are locally known in dolomite. The thickness of the group is from some hundreds of meters to 2000 m in the west of the Anti-Atlas; in the extreme east of the belt the group wedges out almost absolutely. The medusoid imprints are reported from shales of the basal beds of the group (Hauzay 1979).

The Middle Adoudou Group or "Lie de vin série" is made up in the east of the belt of continental strata, mostly slate, siltstone and sandstone with sun-cracks and molds after halite. In the west of the belt the continental strata are replaced partially or completely by marine rocks, mainly limestones with stromatolites, and the thickness of the group increases from 250 m up to 1200 m. The Upper Adoudou Group or the group of upper limestone reflects a new marine transgression that embraced the whole territory of the Anti-Atlas. Its limestones and dolomites contain the stromatolites *Linella*, *Tungussia*, and *Tifounkeia*. The thickness of the group varies, but is not great, amounting to some 200–500 m. In this group (sometimes regarded as the upper part of the "Lie de vin série") there are various archaeocyathids of the Atdaban horizon and also the trilobites *Eofallotaspis*, *Hupetina*, *Bigotinops*, ets. at the level 100 m below its top. Some of them are very similar with the ones recently determined in the "variegated formation" of Aldan, these are the new forms belonging to the Protolenidae, therefore characterizing the oldest Cambrian strata belonging to the Tommotian stage that was earlier considered "pre-trilobite" (Fedorov et al. 1979). Trilobites of a higher level of the Lower Cambrian corresponding to the *Fallotaspis* zone are known from the slate-limestone group that is of conformable occurrence on the Upper Adoudou Group (Sdzuy 1978). Thus, in the Anti-Atlas the Precambrian-Cambrian boundary can be drawn somewhere inside the Upper Adoudou Group.

In South Western Africa the rocks discussed make up the Nama Group ("System") that occupies a large part of south Namibia. The group composes the lower part of cover of the Kalahari platform and rests subhorizontally on different older rocks from the Archaean to the Epiprotozoic. Near and within the Damaran belt of Katangides, the group reveals significant deformation together with the Epiprotozoic rocks. The group is divided into four formations (up-section): Kuibis, Schwarzkalk, Schwarzrand, and Fish River. In the stratotypical area the following aspects are characteristic of the formations named above (Martin 1965, Germs 1972, 1974 etc.). The Kuibis Formation (30–200 m) lies on the pre-Nama peneplain, which is hilly with differences in elevation marks up to 300 m. Fine pebbled conglomerate makes its basal beds, but principally the formation consists of coarse-bedded pink feldspathic with common cross-bedding and ripple marks. The Schwarzkalk Formation comprises fine-bedded dark to black bituminous limestone; its thickness is up to 300 m close to the Damaran belt, but eastward and southward it decreases down to 15 m and limestone is replaced by dolomite and sandy rocks. The Schwarzrand Formation (up to 400 m) is composed of grey-green clay shale with two sandstone horizons; at different levels purple siltstone interbeds occur and there is a horizon of black limestone with abundant scutiform stromatolites in the middle–upper part of the formation. In the Little Koras Mts there are peculiar polygons, flattened and striated surfaces of beds in the rocks of the formation and faceted and striated boulders in shale. Many workers (Germs, Kröner etc.) think that these are traces of an older glaciation. Their arguments seem quite sound, but some (Martin, for instance) suggest that these aspects could be a result of turbid current activity. The Fish River Formation (up to 700 m) lies with an erosion on the underlying rocks. In the lower part it is formed of two layers of quartzite and quartz conglomerate separated by a band of shale; up the sequence appears purple or red sandstone, then again quartzite and finally dark purple shale and siltstone interbedded with quartzite-sandstone. Cross-bedding and ripple marks are common in the rocks. The Nama Group is unconformably overlain by the Dwyka tillite of the Upper Carboniferous.

Abundant fossil remains of the oldest soft-bodied organisms are registered in the Nama Group. The imprints of *Rangea schneiderhöni, Pteridinium simplex, Paramedusium africans, Cyclomedusa, Orthogonium, Nasepia, Ernietta,* etc. are determined in the Kuibis quartzite and in sandstone of the Schwarzrand Formation. Rocks of all formations have traces of life activity of worms and of other crawling organisms (*Scolithus, Scolicia, Taenidium, Phycoides pedum, Helmintoidichnites,* etc.) that are known from the Cambrian as well as from the Eocambrian.

The skeletal tube-like structures found in the Kuibis, Schwarzkalk, and Schwarzrand Formations are of special interest. Germs (1972) assigned them to cribrocyathids (*Cloudina* nov. gen.). All the occurrences of cribrocyathids are known only from the Lower Cambrian; in Namibia, however, they occur in association with the typical Eocambrian Ediacara fauna never registered in Cambrian. But not all workers agree with this classification of the fossils; Glaessner (1978) suggests that *Cloudina* is more likely to be a representative of an extinct polychete order, but this opinion seems to be debatable. A question arises: which faunal group is to be given preference in evaluating the age of the group? The answer is far from being easy. It is probably reasonable to suggest that the Nama Group belongs to the Eocambrian-Lower Cambrian transitional strata. Data of isotope age of granite and syenite cutting the Nama Group do not contradict this conclusion. In the Lower Orange River there is a group of small massifs (stocks) built up of granite, granite-porphyry, aplite, and syenite that associate with dykes of different types. One of these massifs – the Bremen massif – is dated at $553 \pm 13$ m.y. by Rb-Sr and U-Th-Pb methods and another – the Kuboos massif – at $550 \pm 30$ m.y. (on zircon). That means that granitoids were emplaced in the Early Cambrian or probably in the Middle Cambrian.

Some geologists (Kröner, Van Eeden, etc.) are inclined to suggest a correlation between the Nama Group and the Damara Supergroup. The common arguments for this are the occurrence of tillites in both units, certain similarity in rock composition and in sequence structure, similar age of some granites cutting the units. In the light of reliable data, however, this concept seems to be erroneous (Germs 1974, Salop 1977a). The main point is that the Nama Group and its equivalents lie unconformably on the members of the Gariep Group and other Epiprotozoic groups of South Western Africa that enclose tillites. At the same time the Numees tillite of the Gariep Group is certainly an analog of the Chuos tillite and the Tsumeb tillite of the Damara Supergroup (all of them exhibit a typical association of glaciogenic rocks and sedimentary iron ores). The similarity between the Nama Group and the Damara Supergroup is relative; thus, the first contains no rocks that could have been compared with the rocks of the lower and middle units of the second. Notable also is the difference in grade of alteration of rocks of the Nama Group and the rocks of the adjacent Damara and Gariep Groups (situated in the margin of the Kalahari platform). Finally, the strata mentioned and many other Epiprotozoic groups of South Western Africa are unfossiliferous; they are characterized only by the common Upper Precambrian phytolites. It seems reasonable to acknowledge the existence of a separate glacial horizon in the lowermost Lower Cambrian, because of the occurrence of rocks resembling tillite in the Nama Group, and there are no data to refute there glacial origin. It was noted above that this type of formation is also known from this level in other regions of the world. It is quite true, however, that the tectonic movements occurred in the Damaran (Damaran-Cape) belt not only after the Epiprotozoic but also during the Cambrian (the Damaran or Baikalian epoch of activation). For this reason the radiometric datings of the "younger" granites of South Western Africa are very similar. At the same time Germs suggests that the Nama Group may correspond to the Mulden miogeosynclinal Group resting with a break on other units of the Damara Supergroup, a supposition which seems quite realistic.

**The Eocambrian of Australia.** Here only the Pound Quartzite and its analogs distributed in the Adelaide region and the Arumbera Formation of the Amadeus depression (basin) are reliable Eocambrian.

It was noted above (see Chap. 6) that Eocambrian strata form the uppermost part of the Adelaidean miogeosynclinal complex (supergroup) that is known under the name "Pound Quartzite". This unit consists of thick (up to 3000 m) terrigenous strata divisible into two parts. The lower (the Bonney Sandstone) is made up of red sandstone with sun-cracks and the upper (the Rownsley Quartzite) of cross-bedded quartzite-sandstone with thin interbeds of siltstone. In the Flinders Ranges, in the rocks of the upper part, a little above its base, Ediacara fauna was determined for the first time (Glaessner 1961, etc.).

The Pound Quartzite seems to rest conformably on the underlying limestone of the Epiprotozoic Wonoka Formation (the Wilpena Group), but the type of sedimentation is quite different at this level and so it is possible to suggest the existence of a masked unconformity there. Ubiquitously the Cambrian strata overlie the quartzite with an evident unconformity. It starts with the Uratanna Formation (up to 500 m) made up of green micaceous siltstone and fine-grained sandstone with *Scolithus*; in the uppermost part of the formation there is a horizon of dark red to white sandstone with sun-cracks; it contains abundant imprints of *Rusophicus*, *Didymaulichus*, *Phycoides pedum*, etc. The Uratanna Formation or the Pound Quartzite is unconformably overlain by the Parachilna Formation that is composed of sandstone and siltstone; the basal beds of the formation are characterized by different bioturbations, including the burrows of *Diplocraterion*. Slightly above the base there are casts of *Bemella* gastropods. In the overlying Ajax limestone Formation there are archaeocyathids, those typical of the Atdaban horizons of the Aldanian stage in the lower part of the formation, and those typical of the Lenan stage in its upper part.

The unconformity separating the Eocambrian strata in the Adelaide belt from the Lower Cambrian rocks seems to be insignificant; it is evidenced, for instance, by the fact that the U-shaped burrows of *Diplocraterion* pierce not only the basalt beds of the Parachilna Formation, but also the Pound Quartzite adjacent to them, this latter apparently not having been properly consolidated at the time bioturbation occurred (Daily 1976).

The Arumbera Formation of Central Australia rests unconformably on the Pertatataka Formation that makes up the upper part of Epiprotozoic (with tillite) in the Amadeus depression. Its thickness is up to 900 m (commonly it is 350 m) and mainly consists of brown-red median-to-coarse-grained arkosic sandstone and siltstone with local pellets of shale and pebbles of quartzite or chert in the lower part. There are strata in the upper part of the formation that are built up of siltstone and clay shale with subordinate interbeds of sandstone and dolomite. The traces of *Rangea arborea* are known from the lower part of the formation and arthropods tracks and burrows of *Scolithus* from the upper part, that claims a Lower Cambrian age for a part of the

Arumbera Formation. The rocks examined pass upward into carbonate strata (the Todd River, Chandler Formations, and others) that contain archaeocyathids, brachiopods, hyolithids, gastropods, and other remains of skeletal fauna. Some Australian geologists consider the Arumbera Formation to be at the base of the Pertaoortta Group that embraces the strata up to the Upper Cambrian (Wells et al. 1970).

## 3. On the Stratigraphic Rank of the Eocambrian

Typical Eocambrian strata from different parts of the world have been examined and now it is time to discuss the rank of the unit and its place in a general stratigraphic scale.

It is to be noted that the Eocambrian is the only unit among the Precambrian units that can be recognized according to paleontological data. The older units are usually divided and recognized mainly by tectonic or geohistorical criteria; their biosedimentary structures (phytolites) or the remains of primitive plant biota are principally used for correlation (in company with lithologic characteristics). Fossil remains of multi-cellular organisms and of multi-cellular plants of higher organization appear for the first time in the Eocambrian. The typical Eocambrian metazoa remains (imprints) of Ediacara fauna are lacking in the older as well as in the younger strata. Earlier statements on some single occurrences of this fauna in the Epiprotozoic have been refuted, now it is known that the rocks where these occurrences were registered are true Eocambrian. Certainly, it is reasonable to expect that some forms of soft-bodied organisms will be determined in the Upper Epiprotozoic beds, for it is unbelievable that this varying Ediacara biota had no ancestors. This does not need to be confusing, however, for exactly in such cases there is a necessity to establish some older forms. It is an important fact for biostratigraphy that Ediacara fauna is absent in the Cambrian and some of its representatives are known only from the transitional beds conventionally assigned to the lowermost Cambrian (the Nama Group).

The Eocambrian, like the Phanerozoic, is divided according to a common principle (paleontological method). There is no doubt that the Eocambrian belongs to the Phanerozoic ("evident life") as it is characterized by remains of Metazoa and Metaphyta seen with the naked eye. The fact that the Eocambrian is more closely related to the Cambrian than to the Epiprotozoic in its geohistorical aspects is important in solving the problem under examination; accumulation of its rocks reveals the beginning of a great marine transgression that reached its maximum at the end of the Early Cambrian to the beginning of the Middle Cambrian. That is why the Eocambrian can be attributed to the Paleozoic. It is also significant that the duration of the Eocambrian (80–100 m.y.) accords well with other Paleozoic periods, with the Cambrian period (70–75 m.y.) in particular. The Eocambrian can thus be regarded as the first Paleozoic period (system).

Many workers have already raised the question as to the status of Eocambrian strata (Vendian s. str.), suggesting them to be regarded as a separate system, which seems to be quite clear, only needs to be presented at some representative International Geological Forum. Sokolov (1980) suggested the name Vendian System for the unit under discussion, a proposal which it seems reasonable to accept as the best in this case. At the same time the necessity arises of replacing the unfortunate, though commonly used, term Precambrian by some new term that is yet to be found, as the names used earlier, like Pre-Paleozoic or Cryptozoic are also unsatisfactory.

# B. Geologic Interpretation of Rock Record

### 1. Physical and Chemical Environment on the Earth's Surface

Judging from rock records, physical and chemical conditions during the Eocambrian differed slightly from those at the end of the Epiprotozoic or during the Cambrian. Paleotemperature isotope studies show that the average annual temperature of the surface seawater was of the order of $35°-40\,°C$, that means much hotter than now. Concentration of free oxygen in the atmosphere and hydrosphere due to increase of plant biomass must have grown, but it could not have become very high. Many geologists, after Berkner and Marshall (1967), suggest that the appearance of Metazoa in the Eocambrian-Cambrian was related to the increase of atmospheric $O_2$ up to 0.01 of its modern concentration (the "Pasteur level"). The facts available, however, indicate that the red beds and highly oxidized iron ores acquired wide distribution in the Neoprotozoic and Epiprotozoic strata, and favor the supposition that the Pasteur level already existed at the beginning of Riphean and probably in the second half of the Mesoprotozoic. According to Schidlowski (1975) the oxygen pressure exceeded the Pasteur level some $1500 \pm 300$ m.y. ago, for at that time abundant eukaryotic organisms appeared possessing oxygen metabolism or, in other words, breathing. In the opinion of this author a still higher $O_2$ concentration was necessary for the oldest metazoa to appear. Budyko and Ronov (1979) suggest that some 500 m.y. ago the oxygen content in the atmosphere was 1/3 of the modern level $(1.2 \cdot 10^{21}\ g)$.

At the very start of the Eocambrian a faint climatic zonation probably appeared due to the existence of vast continental massifs, but the greatest marine transgression that soon occurred must have allayed the climatic differences and tempered the arid conditions existing in some continents at the end of the Epiprotozoic.

### 2. Life During the Eocambrian

An extremely quick and abrupt evolution of life is a striking feature of the Eocambrian, as the rate of evolution at that time was much quicker than during any older chronostratigraphic periods. For the entire geologic history of more than 3500 m.y. a very slow, hardly noticeable transformation of uni-cellular prokaryotic algae into their eukaryotic forms (partially they developed into multicellular at the end of this period) was observed, and during a rather short period in the Eocambrian various metazoa and metaphyta of complex organization appeared. It was by no means a single, sudden creative action. Naturally, it was favored by the whole earlier biologic and partly geologic evolution. An increase of $O_2$ concentration is one of the important facts that could have influenced this evolution, actively affecting many many bioenergetic processes related to metabolism. The increased abundance of microphytoplankton supplying food for the organism was also important. Certain geographic factors could also have favored this evolutionary process, for instance the appearance in the Eocambrian of vast shallow-water shelves that were suitable environment for organisms. Particularly during the Epiprotozoic glacial epochs the climatic contrasts grew, the relief of marine basin bottoms became more complicated, the tidal currents became more intensive, all factors which could stimulate the adaptability of organisms and thus favor their evolution. A fact of exceptional importance is a striving of all living creatures to complexity and regulation of their inner organization and to a maximal increase of populations.

It is still not possible to explain the striking acceleration of evolution and some puzzling biological phenomena that are registered during the Eocambrian and Early Cambrian. The problem exists why the Ediacara fauna that appeared or at least flourished at the very start of the Eocambrian then suddenly (in the scale of geologic time) became extinct before the beginning of the Cambrian. It is hardly probable that this extinction was due only to severe competition with more active representatives of Cambrian fauna.

There is no explanation of a "biological outburst" at the very beginning of the Cambrian period, when extremely various and rich cryptogenic fauna of trilobites, brachiopods, mollusks, echinoderms, sponges, archaeocyathids, and radiolaria appeared, a fauna amounting to thousands of species and represented by highly specialized forms. Meanwhile, there are no fossils of that kind in the underlying Eocambrian rocks that are either hardly altered or if they are altered retain the same grade as the Cambrian rocks, both forming complexes that reveal the same geohistorical aspects. Up to now there is no single occurrence of the primitive forms of Cambrian organisms that could have been considered the direct ancestors of this fauna. They may be expected to be found in some thin transitional beds. This should underline that the Ediacara fauna had no direct continuation in the Cambrian fauna from the point of view of its organization, but was certainly an accessory branch of evolution. There is no doubt that the Eocambrian-Cambrian boundary has no equivalents in all geologic history in its striking and abrupt change in fossils.

Some authors are inclined to see the change in organisms at this boundary only (or principally) in skeleton formation. Thus, Lowenstam (1980), for instance, states reasonably that processes of sclerotization (biomineralization) of surface tissues already started in the soft-bodied organisms of the Ediacara type (Petalonamae group), but he uses this as a basis to neglect the existence of an abrupt biological limit between the Eocambrian and Cambrian and regards this concept as "a historical relict of opinions". This problem, however, concerns not only the appearance of skeletons, but principally a sharp change in the inner organization of animals and the origin of a new type of animals evidently philogenetically related to modern ones.

The true causes (or a cause) of these biological revolutions and of many more that occurred later during the Phanerozoic are not known. Many hypotheses have been proposed to explain these phenomena, but up to now no single one could have provided a satisfactory explanation of the well-known facts. The most complicated problem is a rapid mass flourishing and subsequent extinction of many varying groups of organisms and the causes of an abrupt philogenetic divergence. In my work (Salop 1977c) devoted to this problem I came to a conclusion that the most valuable hypothesis proposed to explain these facts is that which suggests a sporadic radiation of the planet surface called out by certain cosmic phenomena, for instance, by outbursts of Supernovae in the environs of the Sun system.

Rigid radiation is known to cause a strong mutagenic effect and increases by many times the frequency of mutations. Mutagenesis by radiation is not directed, and the genetic variations are haphazard, registered in a large population. In highly organized animals this mutation usually leads to harmful consequences, in organisms of lower organization and in plants it could appear to be favorable for their evolution. The radiogenic mutations can either increase or decrease the rate of reproduction. The importance of radiogenic mutation lies also in the fact that, when accumulated and accompanied by genetic isolation of species, it results in formation of new species and of higher taxa. It seems there is no other reason for philogenetic divergence.

It can be supposed that the cosmic impulses resulting in intensification of rigid radiation could stimulate creation (or evolution) and subsequent extinction of Ediacara fauna at the beginning of the Cambrian. The same factors could have caused the "biologic outburst" at the beginning of the Cambrian. It is significant that the Late Epiprotozoic glaciation preceded the appearance of the Ediacara fauna and that the appearance and accelerated evolution of the Cambrian fauna coincided with a sharp Early Cambrian cooling of climate and local glaciation (see Fig. 71). It was noted earlier (Chap. 6) that these biological and climatic phenomena probably have a common cause.

### 3. Tectonic Regime

It is known that tectonic events were not abundant during the Eocambrian (Vendian) period, the length of which was relatively short. It has been noted that a marine transgression started at that time in many parts of the world and embraced major portions of platforms and also of certain geosynclines. There are no data available to state definitely the scale of the event, still it is possible to state that the transgression was quite extensive, but not on a world-wide scale. It is clearly seen in the platforms of Northern Eurasia: the East Siberian and Siberian platforms were almost entirely embraced by a shallow-water epicontinental sea. At the same time, only a minor portion of some platforms (African, North American and Australian, for instance) was involved in the transgression, and a geocratic regime governed their major territory.

The tectonic regime was distinct in different geosynclinal belts. In the intracontinental geosynclinal belts of Eurasia (Baikalian, East-Sayan, and the Urals) and of Africa (Red Sea and Katangan), which suffered a major inversion of tectonic regime or even a complete closing at the end of the Epiprotozoic during the Katangan orogeny, deep intermontane and (or) marginal depressions originated during the Eocambrian and there the thick clastic (detractive) strata accumulated; these rocks represent the late molasse.

In the geosynclinal belts surrounding the Pacific Ocean (the Cordilleran geosyncline of North America and the Adelaide geosyncline of Australia) where the Katangan orogeny did not greatly affect the tectonic pattern, the thick sub-platform or even platform terrigenous or terrigenous-carbonate strata were deposited during the

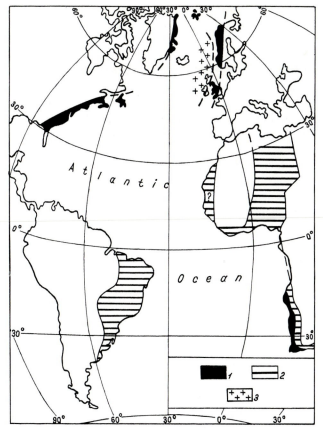

**Fig. 75.** Scheme of structural elements of the Baikalian cycle in the margins of the Atlantic Ocean.
*1* geosynclinal belts (paraliageosynclines and taphrogeosynclines), *2* activation areas, *3* the Katarchean gneiss-granulite basement in the oceanic bottom and in the adjoining land areas (islands)

Eocambrian and were later deformed together with the Cambrian strata conformably overlying them.

In the geosynclinal belts surrounding the Atlantic Ocean (in the Appalachian, Caledonian, West Congolese, Damaran, and Cape belts) the Eocambrian (and Cambrian) rocks are commonly of miogeosynclinal type transitional to taphrogenic (or riftogenic), i.e. like the underlying Epiprotozoic strata. It is thus suggested that the Pre-Atlantic ocean was at that time bounded by paraliogeosynclines and marginal taphrogeosynclines (or rifts) as it had been earlier; otherwise, the process of "opening" of this ocean that started in the Epiprotozoic continued in the Eocambrian-Cambrian (Fig. 75).

Eocambrian eugeosynclinal formations are not known for certain as yet, but it is possible to suggest that if the Lower Cambrian strata of some belts (or zones) are represented by eugeosynclinal formations, the underlying rocks of presumed Eocambrian age are of the same nature, as the type of geologic evolution in the Cambrian was inherited from the previous period. Such belts or zones are registered in some geosynclinal systems of Central Asia and of Western Europa in part.

When examining the tectonic regime on the whole it must be stressed that the period in question reflects only the beginning of a new geologic cycle, its evolutionary stage is characterized by dominant subsidence and sedimentation. This cycle terminated with intensive tectono-magmatic processes that took place in the middle and in the end of Cambrian, in the range of 540–520 m.y. during the Baikalian orogeny. This latter resulted in folding and significant intrusions of magma in the geosynclinal belts and in thermal-tectonic activation in the platforms that led to isotope rejuvenation of the older rocks of basement, to formation of large uplifts, depressions, fracture systems, near-fracture dislocations, and small intrusions of granite, syenite, and alkaline rocks. The Baikalian orogeny was on a global scale like many other orogenies; in Siberia it is called the Baikalian, in Western Europe the Cadomian or Assynthian, in North America the Avalonian, in South America the Brazilian (Late Brazilian to be more exact) or Pan-American, in Africa the Damaran, Pan-African, or Mozambique, etc.

Manifestations of thermal-tectonic processes of this cycle in the African platform are of special interest; there it became possible to establish several areas of activation due to excellent exposures of older rocks, the largest among them being Mozambique, Nigerian-Libyan, and Damaran (see Fig. 72). The first occupies an extensive territory in the east of the continent. Its eastern limit is not known, as it is occupied by the Indian Ocean, but Madagascar and the Seichel Islands certainly belong to it. The so-called Mozambique front, a rather wide zone of fracture and diaphthorites serves as its western boundary. Within the Mozambique area radiometric datings of rocks usually give values rejuvenated to a certain extent, irrespective of their true age, commonly revealing the time of the Baikalian (Mozambique) orogeny. In many aspects this area greatly resembles the Grenville province of North America: both are situated in the continental margins, Katarchean and some younger Precambrian rocks are widespread there, and the rocks are known to have suffered tectono-magmatic activation; the older rocks are deeply eroded, both run along the zones of intensive dislocations (the Mozambique and Grenville fronts respectively) and there adjoin the areas that were not subjected to the activation processes. The main difference in this case is that this activation occurred in the Mozambique area in the Cambrian during the Baikalian orogeny, and in the Grenville province at the end of the Neoprotozoic during the Grenville orogeny.

The Nigerian-Libyan activation area occupies a major portion of the Riphean-Epiprotozoic Nigerian foldbelt and of the Epiprotozoic Congo platform. It includes moreover the sub-latitudinal zone situated in the southern margin of the West African platform along the coast of the Bay of Guinea. Its general orientation, as in the Mozambique area, is meridional. The western limit of the area coincides with a zone of abyssal fracture bounding from the east the Tanezrouft-Adrar block and with a zone of smaller dislocations in the southern part of the Nigerian belt. The eastern limit of the area is not certain; it is situated within the Congo platform and probably crosses different older structural elements. In the south the Nigerian-Libyan area is buried under the Atlantic Ocean, in the north it is covered by the Hercynian-Alpine fold system of Mediterranean (Tethys).

The Namibian (Damaran) area of activation is situated in South-Western Africa, it occupies the eastern part of the Damaran foldbelt and the western margin of the Kalahari platform. It is quite possible that the Namibian and the Nigerian-Libyan areas constitute a single activation region that is divided by the ocean. In this case it is quite comparable with the Mozambique area in extension and both represent wide submeridional belts that divide the African platform into huge segments.

The association of these areas (belts) with the Epiprotozoic foldbelts, Red Sea, West-Congolese and Damaran, is notable. However, the Early Paleozoic activation is not a result of posthumous folding of the Katangan orogeny in these belts, as it occurred after a long atectonic pause. It is clearly evidenced by the fact that this activation brought deformation of typical platform strata of the Nama Group of Eocambrian-Lower Cambrian age that build up the cover of the Kalahari platform, and also by the fact that these strata are intruded by hypabyssal granite and syenite with an age of 550 m.y. (granite of the "Kuboos type"). The same is observed in the north-east of Africa; judging from datings of the so-called "young" granite, the activation there took place in the platform after the Early Cambrian, that is after the moment the platform cover started to form in some parts, for instance, in the northernmost portion of the Red Sea belt (the "Sik sandstone").

In Africa the Baikalian activation is likely to have taken place also within the Mauritanian and Cape foldbelts in addition to the areas mentioned above. This suggests that the modern contours of the continent were predetermined to a certain extend by the belts of Early Paleozoic activation. One more fact is also related to the discussion: on the opposite side of the Atlantic Ocean, in South America, the area of intensive Baikalian activation occurs in a wide belt extending along the Brazilian coast. All the facts seem to favor the idea of a very early start in formation ("opening") of the Atlantic Ocean.

The causes of activation suggested are many and various. Two conceptions are best known: according to one the thermal-tectonic activation in the platforms is due to the processes that occurred in the adjacent geosynclinal belts, while the second states that these two phenomena are quite separate and that geosynclinal processes have nothing to do with it (so-called autonomous activation). However, with the facts referring to the Precambrian and some younger areas of activation in different continents at hand, it seems more reasonable to suggest that this activation was due to a global rise of the heat front from the planet interior, but it led to different changes depending on the tectonic regime: in geosynclines the rocks were folded and metamorphosed, numerous intrusions of magma occurred and many more tectono-plutonic processes typical of these areas were registered; in platforms and in the stabilized older foldbelts the changes mentioned above occurred, which belong to the activation proper. These changes are of select character, the way they are expressed and their intensity depending on the grade of consolidation of the cratonal areas. Naturally, they are most intensive in the portions where the basement is built up of the youngest fold structures and/or is cut by numerous dislocations. According to a general isotope rejuvenation of the older rocks, the thermal processes were quite intensive during the activation, in any case, the temperature of the basement rocks must have exceeded $300° - 350 °C$ at a relatively shallow depth; this is the threshold of isotope activation, critical for K-Ar dating.

Morphologically the activation belts of different age usually represent huge arch-like uplifts complicated with fractures and near-fracture depressions. After orogeny, as a rule, during the beginning of a new geologic cycle, when extension of the Earth's crust becomes a governing process, large rift zones with typical basaltic volcanism are formed in the axial portion of these arch-like uplifts. It thus cannot be excluded that contours of the African and some other continents, with certain aspects of their larger morphostructures, were predetermined to some degree by the older rifts that appeared in the arch-like uplifts of the Baikalian cycle of activation.

### 4. Principal Stages of Geologic Evolution

The Eocambrian (Vendian) period is simply an initial stage of a geologic cycle that terminated with the Baikalian orogeny in the second half of the Cambrian. Geologic evolution during the Cambrian is beyond the scope of this work; however, to complete the picture of this large cycle, the principal aspects of all its stages will be given below:

1. 650–570 m.y. – Eocambrian. The beginning of a large marine transgression, evidently expressed in the platforms of Northern Eurasia. Sedimentary cover is formed in all platforms. Thick sedimentary strata of different fomational types were accumulated in geosynclinal belts. In taphrogeosynclines surrounding the Pre-Atlantic Ocean the sedimentary-volcanogenic strata were deposited with abundant acid volcanics. Formation of late molasse in the belts stabilized by the Katangan folding. Appearance of multi-cellular soft-bodied organisms and multi-cellular algae at the beginning of the Eocambrian. Appearance of organisms with outer tabular skeleton at the end of Eocambrian.

2. 570–540 m.y. – the Early Cambrian and partially the Middle Cambrian. Maximal extension of marine transgression. Deposition of different sediments mostly of carbonate type. Accumulation of thick salt piles in certain platforms in the Early Cambrian. At the very beginning of the Cambrian a short but intensive cooling of climate with local glaciations. Appearance of various and abundant skeletal fauna and simultaneous extinction of the Ediacara fauna. Extinction of archaeocyathids at the end of the Early Cambrian.

3. 540–520 m.y. – the Baikalian diastrophic cycle. Folding, intrusions of basic and acid magma, zonal metamorphism of rocks in geosynclinal belts; intensive activation processes in platforms.

# Chapter 8    Geologic Synthesis

In the previous chapters the principal aspects of composition and structure of larger Precambrian units established were examined and certain peculiarities of geologic evolution of each of these units were pointed out. Some general problems of geologic (mainly tectonic) evolution ensuing from the analysis will be briefly outlined in this chapter. In the first place these are the cyclicity, tendency, and continuity (inheritance) of geologic processes.

## I. Evolution and Cyclicity of Geologic Processes

### 1. Geologic Cycles and Megacycles

It was shown that Precambrian subdivision is based on recognition of natural stages of geologic evolution or, in other words, of large geologic cycles each consisting of a long period of relative tectonic quietness and a subsequent shorter period of orogeny. During the first period the dominant movements of the Earth's crust are of submergence type accompanied by fracturing and stretching; sedimentary strata accumulate in the submerged portions of land, lava outflows and associated comagmatic intrusions are recorded in many places. A gradual extension of marine trangression is observed and a certain regression is marked at the end of the period; as a result the strata formed exhibit a general transgressive-regressive structure of the sequence. Simultaneously in the mobile areas (geosynclinal belts) the volcanic products change their composition from basic (and ultrabasic) to acid. During the orogenic period the rising of the Earth's crust dominates, accompanied by intensive differential movements; in the mobile belts (geosynclines) deformation and metamorphic transformation of rocks occur, a large mass of basic magma and of acid magma in particular is emplaced. Magmatism becomes of homodromic character: basic magma intrusions give place to calc-sodium granitoids and then to granites rich in potassium, up to alaskites.

This succession of events is generally typical of all geologic cycles irrespective of their age. The history of geologic evolution is, however, not a multifold repetition of identical cycles. This becomes absolutely evident when examining Precambrian history that embraces the major portion (85%) of the existence of the planet. The facts dealt with in this book and the conclusions made undoubtedly show that this cyclicity happens against the background of a directed and irreversible evolution of the Earth. They also show that during certain cycles, the nature and significance of global events brought about fundamental changes in geologic history.

At least 12 complete geologic cycles are established in the Precambrian, corresponding to an equal number of large chronostratigraphic units, eras and suberas. Additionally, at the very end of the Precambrian there is one more (13th) incomplete cycle, the Eocambrian or Vendian, that can probably be assigned to the Phanerozoic. These cycles are grouped into cycles of a higher rank – megacycles that correspond to five Precambrian eras: Katarchean, Paleoprotozoic (Archeoprotozoic), Mesoprotozoic, Neoprotozoic, and Epiprotozoic; each terminates with a global orogeny of the first order. The concise characteristics of these eras, and their most typical aspects since the earliest time of the Earth as a planet will be given below.

The "pre-geologic" stage in the Earth's evolution, of relatively short duration (100–200 m.y.), preceded the Katarchean Era; there are no rock records of this period, but indirect evidence is at our disposal to judge its peculiarities. The Earth's crust at that time was still thin, but it already contained some sialic material that originated as a result of quick initial differentiation of the matter from the planet interior and of its layering into shells of different composition. It was the period of exceptionally intensive "bombardment" of the Earth's surface by large cosmic bodies (planetosimals) that fell from sufficiently dense protoplanetary nebula. The impacts must have caused the additional heating of the Earth's crust, formation of heterogeneities within it and in the upper mantle and accumulation of regolith material on the Earth's surface.

The Katarchean Era is characterized by a high thermal regime of the interior of the Earth, by thick and dense essentially vapor-water and carbon dioxide atmosphere lacking oxygen (the secondary atmosphere in relation to the "pre-geologic" one). The Earth's surface during a long time interval was almost completely covered by the primary hot ocean (Panthalassa). The temperature regime on the Earth's surface was mainly governed by the hothouse effect. During most of the era intensive subaqueous outflows of ultrabasic and basic lavas occurred, at the end of the era acid lava outflows were added. Sedimentation was quite specific and mostly of chemogenic nature. The oldest sediments exposed are represented by chemogenic quartzites that were likely to have been formed at the expense of the decomposition and solution of primary sialic rocks of the crust. The banded ferruginous rocks of the Azovian type closely associated with basic volcanics are also typical among the chemogenic formations. Abundant pelitic and carbonate rocks appear only in the second half of the sedimentation period, and psephytic rocks are sporadically known at the end of the megacycle. Structural-facial zonation is lacking. The formation succession of different supracrustal rocks was roughly the same everywhere and was determined by the tectonic and geochemical evolution of environment on a world-wide scale.

Diastrophic events during the Katarchean megacycle are presumed to have happened recurrently, but only two of them are known, one occurring ∼ 4000 m.y. ago (the Godthaabian orogeny) and the second at the end of megacycle, 3750–3500 m.y. ago (the Saamian orogeny of the first order). At that time supracrustals had been ubiquitously altered under granulite and amphibolite facies, migmatized and cut by gneissoid diorites and tonalites. During orogenies large concentrical fold systems (the "gneiss ovals") appeared in places where large planetosimals had fallen in the "pre-geologic" time. Generally the tectonic regime of the Katarchean was of a permobile nature, typified by a total mobility of the Earth's crust. At the end of the megacycle, after the Saamian orogeny, a gneiss-granulite crust (of the order of 20–25 km) was formed.

The Paleoprotozoic megacycle differed strongly and in many respects from the Katarchean. At its very start some portions of the Earth's crust with a different tectonic regime became isolated, namely, the labile protoplatform blocks and the protogeosynclinal belts separating them; these latter consisted of several greenstone belts occurring in the zones of stretching (abyssal fractures). As a result the structural-facial zonation became conspicuously visible; volcanic outflows and sediment accumulation were principally in the near-fracture protogeosynclinal depressions. Komatiites are typical among various volcanics, and were formed under high-temperature melting of ultrabasic material of the mantle. Sedimentary strata are represented by different rocks, but these are mostly of clastic type and poorly sorted; monomictic rocks are present only among the deposits accumulated at the end of the sedimentation period. Jaspilites of the Algoma type associated with basic and acid volcanics are mostly typical among the chemogenic formations. The cherts rarely contain microscopic remains of prokaryotes and carbonate rocks – stromatolites. With the exception of the youngest Paleoprotozoic strata (the Moodies complex) the greenstone rocks dominating in this erathem are similar to the eugeosynclinal formations. During the megacycle examined three or probably four cycles are distinguished, terminating with diastrophic periods (the two strongest ones: Swazilandian of the second order and Kenoran of the first order, can be used for global stratigraphic subdivision). The tectonic structures that appeared during these orogenic events were of linear character, but are almost everywhere complicated with isometric mantled gneiss domes, many of these being commonly formed as a result of modification of synchronous astroblems. Magmatism during orogenies was expressed in early intrusions of basic and ultrabasic magma and in later intrusions of acid magma, mostly of tonalitic composition.

The Mesoprotozoic megacycle consists of three cycles regarded as suberas, and these in turn consist of two cycles of lower rank corresponding to groups (or lithostratigraphic complexes). Each is characterized by specific sedimenatation and magmatism. However, they exhibit many common features as well. The main difference between this megacycle and the previous ones, however, lies in a greater consolidation of the Earth's crust. During the Mesoprotozoic the first true platforms originated on the base of the older semicratonal blocks, basement and cover are conspicuously seen in them, but these platforms, in comparison with the later ones, were relatively small and labile. For the first time in geologic history miogeosynclines were isolated as separate belts or as peripheral zones of eugeosynclines. The platform and miogeosynclinal strata are widespread with abundant mature clastic rocks, products of chemical weathering, and also dolomites.

Sedimentation of the Early Mesoprotozoic took place in an atmosphere lacking oxygen, so that the gold-pyrite-uraninite placers could have originated at that time. In the Middle Mesoprotozoic the formation of gold-uranium conglomerates stopped because of a notable increase of abundance of blue-green algae that resulted in a higher concentration of oxygen in the atmosphere and hydrosphere. At that time the ferruginous-cherty formations of the Krivoy Rog type and then of the Lake Superior type started to accumulate intensely. Red beds appeared for the first time in the Middle Mesoprotozoic. The first eukaryotic algae also probably appeared at that time. In the Mesoprotozoic four glacial epochs of short duration are registered: three during the Early Mesoprotozoic and one during the Late Mesoprotozoic; the climate of the remaining periods was, however, hot.

The terminal Karelian orogeny among the three diastrophic periods of the Mesoprotozoic is the longest and the tectono-magmatic processes then were the most intensive. It is also characterized by strong folding of linear type, by emplacement of a great mass of calc-alkaline (essentially potassic) granites and by zonal metamorphism from greenschist to amphibolite facies. Structures of the type of mantled domes are not ubiquitous and in rare cases the traces of impact phenomena are preserved in them.

The Neoprotozoic megacycle consists of four principal cycles each usually embracing two or more cycles of lower rank. All the cycles are successively related; during the whole period of the megacycle the process of consolidation and growth of large platforms was steady, but certain destructive events also took place against this background. The Akitkanian cycle – the earliest one – is of greater significance for advancing geologic evolution, as during this cycle numerous fracture zones and taphrogenic (rift) structures appeared as a result of global stretching of the Earth's crust accompanied by a general rise of the heat front; mass subaerial outflows of lavas of acid and rare basic composition and large intrusions of comagmatic granite, mostly rapakivi, occurred in these regions. At that time a new chemically isolated asthenosphere was formed in the upper mantle. It is probable that a huge depression of the Pacific Ocean also started to form at the same time.

The three remaining Neoprotozoic (Riphean) cycles are characterized by much more abundant marine and continental strata in platforms than in the Mesoprotozoic period; psephytic and carbonate rocks in miogeosynclines also became notably abundant. The role of dolomite decreases and that of limestone increases with time. Red beds are also more common than in the Mesoprotozoic. Sedimentary iron ores are represented by sheet deposits situated only in platform and miogeosynclinal strata. Ferruginous-cherty formations are no longer formed. Volcanics are of less abundance than in the older formations. Basalts are dominant among the platform rocks. The average annual temperature of the Earth's surface lowered a little as compared to Mesoprotozoic temperature, but the climate remained hot during the whole era. Biomass increased greatly; the eukaryotic algae became abundant and first metaphytes appeared among them.

During the Early Riphean many large geosynclinal belts (and systems) originated, then they developed for a long time also in the Phanerozoic, the paraliageosynclines of the Circum-Pacific orogenic belt being among them.

Four diastrophic periods are registered during the Neoprotozoic, the most important among them – Grenville – terminating the whole megacycle. With every orogeny the tectonic morphostructures became more complicated. In geosynclines numerous inner uplifts appeared and structural-facial zonation became more distinct. In the platform margins, close to geosynclinal belts, foredeeps were formed and aulacogens appeared in many platforms. The Neoprotozoic fold structures are of linear type only and their orientation, shape, and vergence were mainly governed by a local tectonic regime existing in intrageosynclinal troughs and uplifts of mobile areas. Mantled domes and astroblems are small and rare. Metamorphism is of a zonal type and is known only in geosynclinal belts. Rocks of the platform cover are effected by diagenesis only. Granitoids are obviously dominant over basites among the plutonic rocks; normal calc-alkaline granites rich in potassium are mostly typical.

The Epiprotozoic megacycle embraces two geologic cycles and resembles the Neoprotozoic in many respects. Its most striking and specific peculiarity is the two global glacial epochs, the first in the middle of the early cycle, and the second at the beginning of the late cycle. The rocks, except for those of glaciogenic type, are generally similar to those of the Riphean formations of the same type. The platform and miogeosynclinal strata are dominant, but among the latter the piles of terminating stages of geosynclinal cycle are abundant, the orogenic (detractive) strata in particular. Except for short glacial epochs, the Epiprotozoic climate was usually hot, but certain climatic zonation became traceable. The philogenetic aspect of life did not suffer a great change in comparison with Riphean life, but at the very end of Epiprotozoic the conditions became suitable for the appearance of metazoa at the beginning of the Eocambrian.

With the start of the Epiprotozoic megacycle a strong fracturing and fragmenting of the Earth's crust and associated taphrogenic volcanism occurred. This period is also characterized by the formation of some new geosynclinal belts, by the beginning of the "opening" of the Atlantic Ocean and by the formation of paraliageosynclines, taphrogeosynclines, and rifts at its margins. Of the two diastrophic periods of this megacycle, the terminal Katangan orogeny was most intensive in its tectono-plutonic manifestations and resulted in the closing of many geosynclines.

## 2. Common Trends in Geologic Evolution

Geologic evolution is only one aspect of evolution of the Earth as a cosmic body. However, only few cosmic or planetary phenomena affecting geologic processes are known for certain or with a course grade of reliability from the point of view of their directed change over the cause of time. The slowing down of growth of the Earth's mass because of exhaustion of matter of the protoplanetary nebula (and the resulting decrease in size and number of astroblems), dimunition of gravitational energy with lessening of intensity in differentiation of matter in the interior of the planet, decrease in supply of radiogenic heat into a total energy balance of the interior part of the Earth – these are some of the phenomena mentioned above. Many of the cosmic and planetary phenomena are of periodic or quaziperiodic (sporadic) or variable (reversible) character. This means that the position of the Earth (the solar system) in the Galaxy is changing, as also the velocity of the Earth's rotation, the emission of the Sun's light, the position of the magnetic and geographic poles, the inversion of the magnetic field, the intensity of cosmic radiation, the parameters of the Earth-Moon system and other phenomena. The existence of some of these is not as yet proven, but can be of exceptionally great geologic significance (for instance, the variation of the gravitiy constant).

It is naturally impossible to neglect any of the phenomena while clarifying general regularities of geologic evolution, but here only those trends that the rock records help us to conclude upon will be discussed. For convenience, the principal trends of evolution will be examined separately for tectonic, magmatic, exogenic geochemical, climatic, biologic, and sedimentation processes, though many of them, if not all, are closely related.

*The directed changes in tectonic events* that are registered during the Pecambrian and continued to a certain extent in the Phanerozoic are as follows:

- the heterogeneity of the Earth's crust was increasing;
- the thickness and rigidity of the Earth's crust was also increasing;
- the role of abyssal fractures in the structure of the Earth's crust became greater;
- the cratonal elements of the Earth's crust were formed, their size and stability increased (the evolutionary row: protoplatform blocks, labile platforms, stable platforms);
- the tectonic differentiation of the Earth's crust intensified;
- geosynclines were formed, their type varying with time (the evolutionary sequence: "greenstone belts", initial eugeosynclines and miogeosynclines, "secondary" and relict geosynclines);
- the inner structure of geosynclines became more complicated;
- the vertical (oscillating) movements and morphostructures acquired greater amplitude and contrast or gradient;
- the rate of downwarping of geosynclines increased;
- acceleration of tectonic (diastrophic) processes and decrease in duration of "atectonic" periods in geologic megacycles;
- the life period of certain geosynclines became shorter;
- the concentric structures (gneiss ovals and gneiss domes) formed in places by astroblemes less in number and size;

Almost all trends of tectonic development are freely drawn from data discussed in different chapters of the book, still certain explanations are to be added to some of them. Thus, increase in amplitude and gradient of oscillating movements proceeds from an increase in thickness of strata filling depressions in the Earth's crust over a period of time and also from increasing psephytic rock with time. It is reflected, for instance, in a greater role of detractive (orogenic) formations. In the Early Precambrian they are quite insignificant (in the Katarchean they are absent). Molasses that can be compared with the Phanerozoic ones appeared only in the Late Riphean or even in the Epiprotozoic, and their thickness is not more than 4 km. According to Kay (1956), the maximal thickness of the Caledonian molasse is 6 km, of the Hercynian 12.5 km, of the Cenozoic 20 km (this latter value seems to be overstated). These data are absolute evidence of the growing altitude of mountain structures and deepening of associated marginal and intermontane depressions that appeared during the orogenic stage of evolution of the mobile belts.

This gradual increase of amplitude of oscillating movements and of rate of downwarping of geosynclinal troughs with time is also evidenced by the growing maximal thickness of sediments deposited for an equal period of time during different geologic periods. If the thickest strata of different Precambrian eras are examined for an equal time interval, for instance, for 100 m.y., then a certain increase of maximal thickness related to 1 m.y. is observed: from 80–100 m in Mesoprotozoic, to 130 m in the Neoprotozoic and 140 m in the Epiprotozoic. According to Gilluly (1949) the maximal rate of sediment accumulation was already 150 m in 1 m.y. in the Cambrian, and grew continuously to 500 m in the Pliocene. Figure 76 shows in diagram form that the rate of downwarping in the Phanerozoic greatly increased, and became most rapid at the beginning of the Paleozoic and Jurassic or Cretaceous periods. The same trend will be revealed if another method of estimation is applied. It is well known that the maximal thickness of strata accumulated during one complete sedimentation period of a geologic cycle, that is between the orogenies, does not as a rule exceed 15 km in any region of the world and irrespective of the age of the strata, a thickness that is likely to be explained by the maximal possible depth of downwarping of the Earth's crust governed by processes of isostatic compensation. The atectonic periods become shorter with time (from 300–200 m.y. in the Early Precambrian down to 40–20 m.y. in the Cenozoic), so that, naturally, the maximum subsidence rate ("sedimentation rate") is automatically different.

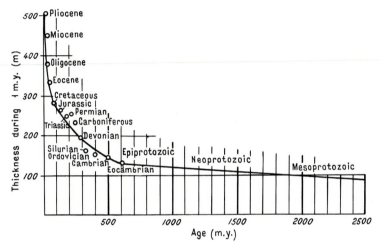

**Fig. 76.** Acceleration of the rate of geosynclinal subsidence with the time according to data of maximal thickness of strata accumulated during one m.y. [After Salop (1964, 1973a), with certain more precise definitions. Gilluly's data (1949) are used for the Phanerozoic]

A shortening of existence of geosynclines is evidently seen during the Precambrian, but it becomes still more obvious if this trend is traced during the Phanerozoic. Thus, the geosynclinal troughs, from their initial stage to their complete closing, existed during the Paleoprotozoic and Mesoprotozoic for the whole megacycle, that is 900 or 800 m.y.; the duration of existence of many Riphean and Epiprotozoic geosynclines is only one-two geologic cycles, that is up to 200–400 m.y. Many of the Phanerozoic geosynclines also evolved during one geologic cycle, but with time a gradual shortening of their existence is seen from 150–100 m.y. to 30–20 m.y.

It was noted in the previous chapters (see also Salop 1973a, 1977b) that at the end of the Riphean or in the Epiprotozoic, the evolution of many Early Precambrian orthogeosynclines terminated by an inversion of the tectonic regime and by formation of relic troughs in their place that were filled in with detractive deposits. However, the next generation of geosynclines that originated in the Neoprotozoic or Epiprotozoic became activated at the beginning of the Paleozoic and continued to develop rapidly, passing through every stage of geosynclinal evolution. A new Phanerozoic megacycle of evolution of the mobile belts started, but was not continuous, as certain turning points are also registered in its course, the most important being in the Jurassic period when a sharp acceleration in the global stretching of the Earth's crust and horizontal movements of its separate parts are known to have occurred. Thus, the tectonic evolution of the Earth can be presented in the following stages, starting from the earliest:

1. the protoplanetary ("pre-geologic") stage: $\sim$ 4500 (5000?)– $\sim$ 4400 m.y.;
2. the permobile stage, the Katarchean megacycle: $\sim$ 4400–3500 m.y.;
3. the protoplatform-protogeosynclinal stage (or the stage of the "greenstone belts"); the Paleoprotozoic megacycle: 3500–2600 m.y.;
4. the platform-geosynclinal initial stage (or the stage of labile platforms); the Meso-protozoic megacycle: 2600–1900 m.y.;
5. the platform-geosynclinal mature stage (or the stage of stable platforms); the Neoprotozoic and Epiprotozoic megacycle: 1900–650 m.y.;
6. the platform-geosynclinal acceleration stage characterized by a sharp increase of rate and amplitude of tectonic movements and by a strongly increased heterogeneity in the Earth's crust structure; the Phanerozoic megacycle: 650 m.y. up to now.

Periodicity and acceleration of diastrophic events with time will be discussed later; here some regularities in *evolution of magmatic and metamorphic processes* known from the Precambrian will be enumerated. Variations in scale and character of granitoid plutonism are quite conspicuous. A steady decrease is observed in the role of autochtonous syntectonic (palingenetic and metasomatic?) granitoids and a relative increase of allochtonous magmatic syntectonic and in particular late-tectonic granitoids. The strongest change in evolution of these types of granitoids is registered as having occurred after the Saamian orogeny, and was probably related to an essential decrease of energy supplied from endogenic sources at that time. In the Katarchean, granitization (sensu lato) of supracrustal rocks was known to occur almost ubiquitously in every continent and since the Paleoprotozoic it has became localized to narrower zones. At the same time it cannot be stated that during the Protozoic an advancing strong decrease in abundance of all kinds of granitoids (irrespective of their genesis) took place, as some very large plutonic bodies of granites were formed in mobile belts even at the end of the Epiprotozoic during the Katangan orogeny.

The proportion of plagiogranite-tonalite is becoming notably less and the part of "normal" potassic calc-alkaline granite is increasing (it is of interest that this trend is observed during the evolution of a plutonic process and during a single diastrophic cycle). Since the end of the Mesoprotozoic the "normal" granites are dominant. Enderbites and charnockites occur in the Katarchean only (with the exception of palingenetic charnokites or paracharnokites that may be of different age).

In the Precambrian geosynclinal belts the abundance of basic and ultrabasic intrusive and volcanogenic rocks decreases with every following megacycle and the role of acid volcanics becomes greater (this trend is known to be typical of development of volcanic processes during a single evolutionary period). This trend, however, typical of the Precambrian, cannot be applied in the case of the mobile belts of the Phanerozoic megacycle. Magmatic phenomena in the platform areas were identical in the Precambrian and in the Phanerozoic: outflows and intrusions of basic magma, on the contrary, became more intensive, but this trend was not of a regular nature. In the oceanic depressions an exceptional increase in volume of basic volcanics was noted

(Ronov et al. 1979). The growing abundance of platform and postorogenic intrusions of alkaline and nepheline syenites and also of alkaline gabbroids is known.

Geosynclinal basites decreased during the Precambrian, which is likely to be related to the generel tectonic evolution of the older geosynclines mentioned above and, to a certain extent, to the filling in of the abyssal fractures by granite bodies. The growing proportion of platform basites is probably related to a general "sclerotization" of cratonal blocks that favored formation of deep fractures of the Earth's crust and also to a general growing and stretching of the lithosphere (see below).

A sharp decrease in ouflow of basic and ultrabasic komatiitic lavas at the end of the Paleoprotozoic was caused by a greater thickness of the Earth's crust that resulted in shifting high-thermal processes to a greater depth, and also by a general evolution of the mantle composition.

The exceptional occurrence of large intrusive bodies of anorthosites in the Precambrian is probably explained by the special tectonic regime during their formation that existed only in the early stages of tectonic evolution of the planet (see Chap. 2), and the occurrence of rapakivi granite intrusions and associated mass outflows of acid lavas in the beginning of the Neoprotozoic only is related to a quite specific endogenic regime during the Akitkanian – a turning-point in geologic history.

The variation in thermal regime of the Earth interior with time is also reflected in a general lessening of intensity of metamorphic processes that became located in the narrower zones.

*The chemical and physical evolution* of the outer shells of the Earth resulted in:

- a change in atmosphere composition; decrease in concentration of carbon dioxide and "acid smokes", and growing concentration of free oxygen;
- a change of composition of hydrosphere from sulfate-lacking chloride through sulfate-poor chloride and chloride-carbonate to chloride-sulfate;
- a gradual lowering of temperature of the Earth's surface (from $\sim 100\,°C$ in the Katarchean to $70\,°C$ in the Paleoprotozoic, $60\,°C$ in the Mesoprotozoic, $50\,°C$ in the Neoprotozoic, $40\,°C$ in the Epiprotozoic, $35\,°C$ in the Eocambrian-Cambrian and down to $14° – 16\,°C$ at present (with the exception of short periods of cooling during glacial epochs), is was mainly determined by lessening of the hothouse effect;
- a decrease of density of atmosphere and its growing transparence for penetration of the Sun's light;
- a lowering of UV and cosmic radiation;
- change of the azonal climate (typical of almost the whole Precambrian) into a locally faint zonal climate in the Epiprotozoic-Eocambrian.

*Evolution of life* during the Precambrian was as follows:

- transition of anaerobic and autotrophic organisms to aerobic and heterotrophic forms;
- the inner structure of organisms became more complicated and metabolism processes increased;
- philogenetic divergence grew;
- total biomass became more abundant;
- change of cellular division into sexual reproduction (appearance of eukaryotes in the second half of the Mesoprotozoic and their rapid evolution since the Riphean);

- appearance of Metaphyta in the Riphean;
- appearance of Metazoa at the beginning of the Eocambrian;
- appearance of various and abundant skeletal fauna at the beginning of the Cambrian.

A combination of all these changes in atmosphere, hydrosphere and biosphere and also of the tectonic regime of the planet resulted in the *evolution of the Precambrian sedimentary formations*, as seen from the following:

- heterogeneity of strata increased;
- role of miogeosynclinal, platform, and orogenic (detractive) formations became greater;
- role of clastic rocks (of the psephytic type in particular) became greater;
- red beds became more abundant;
- carbonate piles grew thicker (and at the end of the Precambrian limestones were formed at the expense of dolomites);
- evaporites became more widespread to a certain extent;
- types of ferruginous formations changed; the ferruginous-cherty banded (BIF) formations typical of the Early Precambrian, changed into sheet deposits of massive iron ores of the Urals type in Neoprotozoic; simultaneous migration of sedimentary iron ore formation from eugeosynclines into miogeosynclines and in platforms took place. Evolution of ferruginous-cherty formations is revealed in a successive change of the following types: Azovian (Katarchean), Algoman (Paleoprotozoic), Krivoy Rog, and Lake Superior (Mesoprotozoic).

Formation of gold-uranium conglomerates existed exceptionally in the Early Mesoprotozoic (the environment was not favorable for its origin either earlier or later).

### 3. Continuity (Inheritance) of Evolutionary Changes

The Precambrian subdivision accepted reflects the natural stages in geologic evolution, but still the geologic cycles and megacycles or, in other words, the cyclic occurrences of global tectonic events are its basis. Many of the evolutionary changes mentioned above are registered, however, as having occurred at different time intervals and their boundaries usually do not coincide with the boundaries of tectonic events. Nevertheless, all these changes merit to be used in giving a more complete characteristic of geologic cycles and in making peculiarities of larger stratigraphic units more distinct.

When examining different aspects of evolution, it becomes evident that a definite continuity exists in the development of all or almost all geologic events, despite the occurrence of sharply recognizable turning points. This continuity is seen from the fact that every subsequent geologic cycle is characterized by certain elements of the previous one. It is supported by the existence of "transitional" units and formations. For instance, the Isuan lithostratigraphic complex, the youngest in the Katarchean, greatly resembles some of the Paleoprotozoic complexes; the Moodies complex, the youngest in the Paleoprotozoic, reveals many aspects in common with the Mesoprotozoic formations and on the contrary, some Lower Mesoprotozoic strata resemble the Paleoprotozoic ones; the Vepsian, the uppermost complex of the Mesoprotozoic, is similar to the strata of the lower part of the Neoprotozoic, etc.

This continuity is clearly exemplified by many formations, by the ferruginous-cherty formations in particular: thus, the Algoma-type jaspilite typical of the Paleoprotozoic starts with the end of the Katarchean, but at the beginning of the Mesoprotozoic it still exists, though faintly expressed; the Krivoy Rog-type jaspilite typical of the Mesoprotozoic appears at the end of the Paleoprotozoic, the "degenerated", quite specific ferruginous-cherty formation is known from the lowermost Neoprotozoic (the Franceville Group, Western Africa).

Many more examples of the kind can be cited, but not here, as some of them have already been mentioned earlier and others can be thought out by the attentive reader. The inheritance and continuity in tectonic evolution of larger structural elements is well known and will here be recalled again. Generally speaking, it is possible to state that continuity is an obligatory element of evolution, or in other words that there is no evolution without continuity. I pay attention to this well-known fact only with the aim of avoiding an erroneous impression from geologic history as if it were some chain of unrelated events.

## 4. Periodicity of Orogenies

Periodicity in tectonogenesis is one of the principal problems in geology. It has been discussed with animation for many years, but even now there is no unanimous opinion on any aspect of the problem. Are the tectonic events synchronous or asynchronous, do the global tectonic cycles exist, is there a strict periodicity in these cycles, how can it be revealed, does it change with time and in what nanner? These are only some of the questions whose solution is acutely necessary. Up to now an attempt has been made to answer these questions using data of studies of Phanerozoic rocks. However, tectonic events in the Phanerozoic were frequent, and it is problematic to fix the time of their occurrence, as the resolving power of paleontological and isotope methods is insufficient as a rule for this purpose. In the Precambrian the situation is quite different: the diastrophic epochs were rarer then and separated by rather long periods of relative quietness, which is an advantage in the application of isotope methods for dating these events. Besides, among the Precambrian formations there are various intrusive and metamorphic rocks – the objects in dating diastrophic events. Thus, studies of Precambrian formations, paradoxal as this is, appear to be more useful in clarifying the problem as a whole.

Earlier and even sometimes now, histogram method of processing of isotope dating of rocks from separate regions and for the whole planet was widely accepted. This type of analysis gives certain general regularities in the distribution of datings, however, the peaks on diagrams do not always reveal the time of tectonic events, and commonly they characterize the time of termination of thermal events or the time of uplift of abyssal rocks above the critical level where migration of radiogenic isotopes stops. In the areas of polycyclic evolution (that constitute the major proportion) only the single, so-called relict datings give the true age values of the early tectono-plutonic events. Thus it seems more reasonable to interpret every dating in combination with geologic data to obtain more valid conclusions.

The analysis of geologic and geochronologic data on the Precambrian of different parts of the world, some results of which are given in this book, has shown that several global diastrophic periods can be recognized in the Precambrian; they are usually called the diastrophic (tectonic) cycles. (It was noted above that the diastrophic period is a part of a geologic cycle. It is only a tradition to use the term "cycle" for it. However, the essence of the tectonic events examined also answers at least one of the meanings of this term). It must be emphasized that the diastrophic cycles were recognized exceptionally on the basis of geologic data; isotope datings permitted only their allocation in the geochronologic scheme, an interregional correlation and an evaluation of their global importance.

It seems appropriate to explain the essence of the "global diastrophic cycles" as understood here. Diastrophic events are known to be established on the basis of unconformities between strata and on relations of the latter with the intrusive bodies. However, it is impossible to presume (and it would not accord with well-known facts) that the whole surface of the Earth became dewatered as a result of diastrophism; naturally, in some regions, subsidence of basins took place and sedimentation occurred. In such regions the sedimentary sequences would have been continuous, but even there observation reveals certain traces of tectonic events (variations in type of sedimentation, erosion, intraformational disturbances, local intrusions, etc.). At the same time, the unconformities separating the oldest Precambrian erathems (megacycles) that are due to orogenies of the first order are followed for exceptionally great distances, sometimes for the whole continent. It is very probable that this is only a result of a long duration of Precambrian diastrophic cycles and is caused by integration of several local unconformities formed during separate tectonic impulses (phases) of these cycles in different parts of mobile belts in the process of migration of zones of folding (see Salop 1973a, 1977a).

Isotope datings of various plutonic and metamorphic rocks revealing the time of a certain diastrophism are usually of a wide range of values, as the Early Precambrian tectonic cycles of the order of 200–250 m.y. and even more. The dispersion in this case related to accuracy of methods and to laboratory errors is almost negligible. Commonly this scattering of values is explained by natural causes: long duration of tectono-plutonic phenomena and extremely slow extinction of abyssal thermal processes that lasts for many millions of years. This is evidenced by isotope datings of rocks of different phases of a plutonic cycle and by field observations of a composite pattern of intersection of intrusive bodies formed during different phases of a single cycle, mostly typical of the Katarchean.

Thus, the global scale of diastrophic cycles is to be understood in so far as they occur approximately at the same time at relatively large intervals all over the world, but in different parts of it their mode and intensity may also be different. It is important that within the diastrophic cycle the climax of tectonic activity may fall for a different time in different regions; in some places intensive folding and plutonism may be registered and in others these processes were already extinct and the sedimentary strata of the next supracrustal complex ("transitional formations") started to accumulate.

The occurrence in time of the Precambrian diastrophic cycles starting from the oldest is examined below.

In the Katarchean the terminal Saamian cycle of the first order is certain, and is established in every place where the oldest formations are exposed; and a significant structural unconformity is known everywhere between the Katarchean gneisses and the overlying Paleoprotozoic greenstone strata. For the cycle discussed there are relict isotope datings in the range of 3750–3500 m.y. As these datings correspond to all ranges, an impression of a continuous nature of tectonic events is created. It is possible, however, that it is a result of different secondary processes and that the tectonic events were discontinuous. The most reliable datings are suggestive of the maximum tectono-plutonic activity having been about 3750 m.y. ago, and the thermal events that took place 3600–3500 m.y. ago characterize a separate, Late Saamian diastrophism of the possible earliest Paleoprotozoic.

The earlier Godthaabian diastrophic cycle of the Katarchean is recognized with less reliability; it occurred prior to deposition of the Isuan lithostratigraphic complex, approximately 4060–4000 m.y. ago. It has been established only in Western Greenland, but a similar geologic situation seems to exist in some other parts of the world. In any case close relict isotope datings of Katarchean plutonic and metamorphic rocks are known from several regions of different continents (see Table 3).

Some workers suggest unconformities among the still older formations of the Katarchean: at the base of the Slyudyankan complex (the Baikal area, Stanovoy Range, Greenville province?), between the Sutamian and Fedorovian complexes (the Aldan shield, White Sea area, Madagascar), between the Fedorovian and Ungran complexes (the Aldan shield, Azovian massif, Eastern Africa, Madagascar). All these unconformities are far from being certain but we known for sure that at these boundaries a change in composition of supracrustal strata is registered and that horizons of alumina-rich gneisses and local quartzites are present. It was noted above that in the Early Katarchean sedimentation probably took place in the boundless oceanic basin (Panthalassa) and it seems problematic that under such conditions the tectonic events could have resulted in common unconformities. Meanwhile, the structural analysis of the tectonic pattern of gneiss complexes is suggestive of the possibility of several epochs of deformation during the Katarchean. Moreover, these Katarchean gneiss complexes quite often enclose deformed dykes of basites and the granitoid bodies are inter-penetrating, which also probably shows the multifold character of tectono-plutonic phenomena.

Four orogenies of different grades of intensity are distinguished during the Paleoprotozoic. The earliest, and probably the weakest, is the Belingwean diastrophic episode (3400–3370 m.y.), which took place between the formation of the Komatian and Keewatinian lithostratigraphic complexes; the next Swazilandian diastrophism of the second-third order (3200–3150 m.y.) preceded deposition of rocks of the Timiskaming complex; and a relatively weak Barbertonian diastrophic episode (3000–2950 m.y.) occurred prior to deposition of sedimentary strata of the Moodies complex. Finally, the Kenoran diastrophism of the first order (2800–2600 m.y.) terminated the whole Paleoprotozoic megacycle.

All these orogenies are recognized and well-based geologically in many parts of the world, but their geochronology is not adequately known. Isotope methods gave the true time of the Kenoran cycle, but still the question remains; does it not consist of two separate cycles? This is also true for the Saamian cycle. The data available permit the assumption that the climax of orogeny took place at the beginning of the cycle

(2800–2750 m.y.) and partially at the end of it (2650–2600 m.y.). The remaining three diastrophic epochs are dated in some regions only and the age values obtained are not certain. It is expected, therefore, that further works will bring some corrections in estimation of their age. Correlation of lithostratigraphic complexes (Chap. 3), however, shows that the unconformities separating these epochs are also on a global scale. The high intensity and world-wide character of the Kenoran cycle is suggested by the numerous local names it is given: in different parts of Africa it is recognized as Zimbabwan (Rhodesian), Shamvaian, Aruan, Liberian, Louizan, pre-Birrimian; in Europe: Scourian, Icartian, Kuhmoidan, Belomorian, Rebolan; in Asia: Muyan; in South America: Guianan etc.

The Mesoprotozoic Era embraces six diastrophic epochs of different importance, which divide the geologic megacycle into six lithostratigraphic complexes (groups, units). Three of them are well-expressed and accompanied by intrusive events, the Seletskian epoch (2450–2400 m.y.), the Ladogan (2200–2160 m.y.) and the Karelian (2000–1900 m.y.), they are useful in dividing the Mesoprotozoic into suberas. The remaining diastrophic epochs or episodes terminate the geologic cycles of lower rank, they have no proper names, though in places they result not only in stratigraphic but also in angular unconformities. The Karelian diastrophic cycle of the first order terminating the era was mostly intensive in all regions. Its wide distribution can be judged by the great number of its regional synonyms: in Europe: Svecofennian, Svecokarelian, Inverian, or Laxfordian, Lihouan; in North America: Hudsonian, Penokean, Arizonian; in South America: Transamazonian, Surinamian, post-Baraman; in Africa: Birrimian, Mayombe, Suggarian, Usagaran, Ubendian, Mirian, Ruvensorian, Berberan, Tihaman; in Asia: Liuliang, post-Udokan; in Australia: Olarian, Williaman etc.

The three principal orogenies are well dated by isotope methods and their synchroneity in different continents is proved geochronologically and geologically. Geologic data show that the climax of tectono-plutonic activity during the Karelian diastrophic cycle fell in the range of 1950–1900 m.y. and only in Equatorial Africa are these processes likely to have been intensive in the first half of the cycle (~ 2000 m.y. ago).

The Neoprotozoic Era is characterized by four diastrophic cycles of different strength, used for dividing it into suberas, and by at least four smaller diastrophic episodes that terminated the geologic cycles of lower rank. The Vyborgian orogeny is the earliest among the first group, having occurred 1650–1600 m.y. ago, but it was preceded by certain thermal-tectonic events that are registered during the whole Early Neoprotozoic (Akitkanian), in particular prior to deposition of the Chayan lithostratigraphic complex (Parguazan or Korostenian episode). The synonyms of this cycle are: Bakalian (the Urals), Early Gothian or Smoland (Sweden), Sanerutian (Greenland), post-Tarkwan (Western Africa), Oftalmian (Australia).

Kibaran, the next large diastrophism that terminated the Middle Neoprotozoic (the Lower Riphean), occurred in the range of 1400–1300 m.y.; in different countries it is recognized as follows: Prikamian (the Urals region), Late Gothian (Sweden), Early Gardarian (Greenland), Elsonian (Canada), Mazatzal (U.S.A.), Irumidian (Mozambique), Hijaz (Arabia), Espinhaço or Uruaçuano (Brazil). The Avzyan orogeny terminated the Upper Neoprotozoic (the Middle Riphean), and occurred 1250–1200 m.y. ago. It is undoubtedly a separate cycle of thermal-tectonic activity, but it is weaker than the previous one, which is probably the reason why it is commonly combined with the Kibaran or Gothian (Late Gothian). In the Northern Cordilleras it is distinguished as the Nadalian orogeny.

The Grenville orogeny that completed the Neoprotozoic megacycle 1100–1000 m.y. ago is one of the largest in the Precambrian and thus belongs to diastrophic cycles of the first order. In different parts it is commonly given this name, but regional synonyms exist for it as well, for instance, it is called Sveco-norwegian (Scandinavian), Late Garder (Greenland), Late Kibaran (Equatorial Africa), Farusian (NW Africa), Late Irumidian (E Africa), Gattar (NE Africa), Bishah (Arabia), Raclan or Haynook (Northern Cordilleras, Canada), Issedonian (Kazakhstan), Minas-Gerais, or Minas-Uruaçuano (Brazil) etc.

All four Neoprotozoic diastrophic cycles enumerated are geologically documented and isotopically dated; their global scale is beyond doubt. Diastrophic cycles terminating the lower rank geologic cycles have no proper names, they are not dated and their world importance is problematic.

The Epiprotozoic Era is approximately divided in its middle section by the Lufilian orogeny of the second order; it took place about 780(760–780) m.y. ago, but as pointed out (Chap. 6), this dating should be refined. It is likely that one further weak diastrophism occurred in the Early Epiprotozoic, corresponding to the unconformity between the Subglacial and Lower Glacial lithostratigraphic complexes; its time is estimated to be at 900 m.y. according to datings of pegmatite veins from the Roan Group, but these do not seem quite satisfactory. The Epiprotozoic megacycle was terminated by the Katangan diastrophic cycle of the first order, and occurred ubiquitously in the range of 680–650 m.y., but it is quite probable that it was preceded by an outburst of tectonic activity (diastrophic episode) that is revealed in local unconformities and in deposition of the early red molasse (of the Mpioka Group type). Katanga is a typical tectonic area of this orogeny, in other parts of Africa it is called the West Congolese or Late Gattar, as Najd in the Arabian Peninsula, as Caririanian in South America, as Delamerian in Australia, but commonly it is erroneously combined with younger Baikalian orogeny.

The Baikalian diastrophism of the first order was the first notable tectonic event in the Phanerozoic megacycle (see Chap. 7). It is striking that it occurred at 100–140 m.y. after the Katangan orogeny, as in the Precambrian the time interval separating diastrophic cycles of the first order was much longer. After the Baikalian cycle a real "tornado" of large tectonic events followed: from 15 to 20 diastrophic epochs during the remaining 560 m.y. of Phanerozoic time, that is on the average every 30–35 m.y. and virtually with an ever greater acceleration: from 75 to 10 (or even 5) m.y. at the end of the Cenozoic. These epochs are usually called phases, but in reality some of them can be compared with the Precambrian diastrophic cycles by their intensity and independent occurrence. A combination of groups of such phases into tectonic cycles of a higher rank, for instance, in the Caledonian or Hercynian orogeny, is quite conventional, as the anorogenic periods separating these cycles are commensurable with the anorogenic periods between the tectonic "phases" of every cycle. Considering this acceleration of tectonic activity with time, it seems more reasonable and correct to regard the principal tectonic "phases" of Phanerozoic as separate diastrophic epochs or cycles, the more so as many of them are divided into several phases (up to 5 and even more at the end of the Alpine cycle).

The principal diastrophic cycles of the Phanerozoic were also shown to be on a world scale in some fundamental works of Pronin (1969a, b, 1973a, b). Kunin and Sardonnikov (1976) came to the same conclusion on the basis of statistical data given by Pronin, and still earlier this was done by Rubinshtein (1967) by analysis of isotope (mostly K-Ar) datings of Meso-Cenozoic magmatic rocks. Thus at the moment it is possible to state that the rhythm of all diastrophic events is uni-planetary.

The distribution of diastrophic epochs during geologic history is shown in Fig. 77. An evident shortening of intervals is clearly seen there between the Precambrian diastrophic cycles of the highest order with time, these cycles being accompanied by strong granite intrusions. As a result the geologic megacycles also became more frequent with time: the duration of the oldest Katarchean magacycle is not known, but it is likely to have been not less than 900 m.y.; the Paleoprotozoic megacycle lasted about 900 m.y.; the Mesoprotozoic 800 m.y., the Neoprotozoic, consisting of two large cycles (megacycles in reality): the Akitkanian and Riphean separated by long (multiphase) intensive granitoid magmatism lasted 400 and 600 m.y. respectively; the Epiprotozoic for 300–350 m.y. This regularity would appear still more strict if the Kibaran cycle and not the termination of the Vyborgian cycle is taken for the upper age boundary of the Akitkanian or, in other words, if the Lower Riphean is assigned to the Akitkanian (megacycle?), as certain geologic data suggest (see Chap. 5). It is seen in the picture (see Fig. 77) then, that since the Eocambrian (Vendian), an obviously accelerated of evolution in the tectonic history of the planet has started. It is striking that interchange of the Precambrian megacycles of different orders is generally uniform, with intervals of about 200–250 m.y. (a great Akitkanian cycle is an exception, for the tectono-plutonic events then took place recurrently). The same cyclicity may be suggested for the Phanerozoic diastrophic cycles, but this assumption needs further checking (see Kunin and Sardonnikov 1976). Finally, a shortening is noted in the duration of diastrophic cycles, of the first order in particular, with time. These latter seem to have been of bi-modal or more composite character in the Early Precambrian, and commonly embrace an interval corresponding in time to that between two diastrophic cycles of a lower rank, represented by the doubled and reinforced "common" cycles.

The causes of tectonic cyclicity have long been discussed by many workers, but it must be stated that they are still not certain, though some of many hypotheses proposed may be true or close to reality. I am inclined to take the hypothesis relating cyclicity with periodical changes in gravitational potential as valid, as it is almost beyond doubt that the principal source of tectonic events is gravitational energy. At the same time, the assumption of many endogenic phenomena regulating the global cyclicity of tectonic processes seems unsatisfactory. I share the opinion of many workers who see the cause of this cyclicity in more general physical and cosmic phenomena, most probably in the change of gravitational potential due to the effect of mass in the Universe (Kropotkin 1964, 1970, 1980, Machado 1967 etc.).

A coincidence of duration of a geologic cycle (200–250 m.y.) and of a Galactic year estimated to be of 220–180 m.y. by many workers is striking, and in the opinion of some authors this latter shortens or varies in this or a slightly longer period of time. It is possible to suggest that the solar system, while passing along the Galactic orbit, crosses the domain with a great, relatively dense mass of matter that increases the

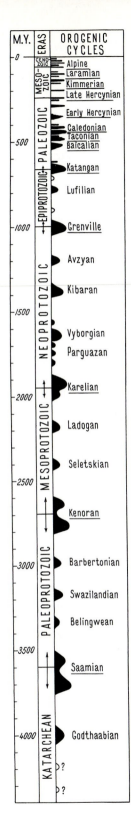

**Fig. 77.** Distribution of diastrophic (orogenic) cycles or epochs during the history of the Earth

gravity potential of the Earth. As a result the Earth and other bodies of the solar system suffer compression or expansion in different parts of the Galactic orbit.

One more important problem is the cause of acceleration of diastrophic processes, that becomes mostly evident at the end of the Precambrian and in the Phanerozoic. It must be emphasized that this problem is not at all clear. However, in my opinion (Salop 1977a) a possible cause in acceleration of tectonic movements is a transportation of radioactive elements at higher levels in the planet interior with time. I proceed from the fact that radioactivity is the second important source of energy responsible for thermal-tectonic processes. The rate of heat conduction through a 100 km pile of rocks is known to be about 300 m.y. (Verhoogen et al. 1970). It is possible that the duration of periods between the diastrophic cycles of the highest order (when the most intensive processes of selective melting of the interior occur accompanied by granite formation and subtraction of light and radioactive elements) corresponds to the time necessary for transportation of energy from the heat-generating layer rich in radioactive elements to the Earth's surface, where the heat flow is consumed for deformation and metamorphism, for magma formation, and is partially emitted in space.

The length of the Paleoprotozoic anorogenic period is about 900 m.y., which infers that at the beginning of this era the heat-generating layer was at a depth of about 300 km. The length of the anorogenic period of the Mesoprotozoic Era is about 600–700 m.y., thus it is possible to conclude that this generating layer at the beginning of this era (after the Kenoran orogeny) appeared to be at 200–230 km beneath the day surface. The following Precambrian periods between the diastrophic cycles of the first and second orders lasted about 250–350 m.y. and thus, at the beginning of any of these cycles, this generating layer was situated at approximately 100 km from the surface. Thus, at the end of Mesoprotozoic, this layer rose by 100–150 km in comparison with its previous position. This extremely great rise is probably due to intensive migration of radioactive elements during the exceptionally strong Karelian orogeny that was accompanied by formation of a huge mass of potassium-rich granite. An insignificant rise of the generating layer during the Precambrian is probably explained by relatively low granite formation in the Riphean. Judging from the relatively short duration of anorogenic Phanerozoic periods, it is possible to assume the generating layer to have a moved from a depth of 25–30 km at the beginning of the Paleozoic to a depth of 10 km at the end of the Mesozoic. In the Cenozoic, the major portion of radioactive elements appeared to be concentrated in the "granite layer" of the Earth's crust and thus the heat generation took place at a shallow depth: in the continents it was practically at the level of tectonic and magmatic processes.

The ideas discussed above accord well with geochemical data that reveal migration of radioactive elements (K, Rb, U, Th) from the interior of the planet to its surface during differentiation of matter of the Earth's shells, during metamorphism and formation of granites. It must be emphasized that the depths of the generating layer given are only preliminary, as they were evaluated without taking transportation of heat by emission and convection into consideration. The conductive transportation of heat in the thermal processes of the Earth, however, seems to be of principal significance, though in the zones of higher penetration (for instance in the zones of abyssal fractures) convection may play the leading part.

Emission of radioactive heat is continuous, but its tectonic effect may be discontinuous, for diastrophic events can occur only after accumulation of heat up to a certain critical value sufficient to trigger tectonic and plutonic processes. The accumulation of heat energy usually occurs below the isolating cover of sedimentary or volcanogenic-sedimentary rocks that exhibit much lower heat-conductivity than the underlying crystalline rocks. Moreover, loss of heat during orogenies and its emission in space results in a "freezing" of the heat system for a certain period of time.

In the course of time the growing thickness of the sedimentary cover on the Earth may have provided better thermal insulation of the interior and resulted in greater tectonic effects (increase in rate and amplitude of movements). Upward movement of the generating layer led to a reduction of the area involved in active tectogenesis and metamorphism, or in other word, to a narrowing of mobile belts. The deeper the "disturbing source", the broader the surface structure caused by it. The upward movement of the heat-generating layer would have caused a shortening of orogenic cycles with time, for the closer the layer is to the Earth's surface, the more rapidly is the radiogenic heat consumed.

Thus it is evident that there are two sources of thermal-tectonic processes: gravitational compression and radioactive decay; and in the upper solid shell of the Earth, in the tectonosphere, the significance of the first phenomenon seems to decrease with time and that of the second increases due to upward transportation of radioactive elements (in spite of a total decrease of intensity of radioactive heat flow with time). The strongest and longest thermal-tectonic events that are identified as diastrophic cycles of the first order very probably originated as a result of interference of heat flow fronts due to gravitation and radioactivity at the level of the tectonosphere.

In conclusion, it is necessary to state that all these facts by no means signify an increase in intensity of thermal-tectonic processes with time. This increase concerns only their morphostructural expression and grade of differentiation of tectonic elements due to growing thickness and rigidity of the Earth's crust (and also of the altitude of arch rises), to the growing importance of abyssal fractures and to a general increase in heterogeneity of the crust that is seen in its division into belts and blocks of different age and structure.

## II. Global Tectonics in the Light of Precambrian Geology

While reading this book, the reader has certainly come to the conclusion that I have avoided a conception of lithospheric plate tectonics, its terms and methods when dealing with the Precambrian tectonics. Many data that became available in studying the older formations contradict certain aspects of this conception, at least in its widely accepted orthodox version. At the same time, it can be seen that different geodynamic ideas of the "classical" geosynclinal theory are insufficient to interpret some important facts.

In my opinion we have no general tectonic theory at the moment that would correspond to the principal well-known facts and that could be regarded as having no conflicting claims, but it is very probable that such a theory, when created, will combine many aspects of conceptions of fixism and mobilism (Khain 1978). The geodynamic concept of fixism well suits an explanation of the concentration of folding and different types of magmatism in relatively narrow and deep depressions in the Earth's crust in various parts of continents, relations of different shapes and vergence of folds and the shape of depressions of sedimentation, succession in tectonic and magmatic phenomena, structural-facial (formational) zonation, inheritance of tectonic evolution and other well-known empiric regularities. At the same time geodynamic concepts of fixism provide insufficient evidence to explain such phenomena as expansion of the Earth's crust, formation of oceanic basins and tectonic structure of continental margins, processes of tectonic-magmatic activation, and other events.

The conception of plate tectonics is mainly based on the recent abundant data on the morphology of the oceanic bottom, on the structure of mid-oceanic ridges, rifts and continental margins, on data of deep underwater drilling and geophysics (seismic and magnetic studies in particular). However, workers give different, sometimes debatable, interpretations of geophysical data, because there is no reliable geologic evidence on the structure of oceanic bottoms below the so-called second or basaltic layer that is only partially known from drilling. Doubts still exist as to the scale of the abyssal thermal convection and its role as a principal mechanism in the movement of lithospheric plates, as to the part and scale of subduction and the possibility of plate overgliding, especially in the case of the asthenosphere being absent under the continents down to a depth of 400–500 km, as is now suggested. The plate concept does not explain the fact of the rather long existence of large structures of a platform, synclise, and anteclise. These critical remarks and others referred to the tectonics of the lithospheric plates, some aspects of which are pointed out and discussed in the works of many authors (Belousov 1975, Milanovsky 1978, Dobretsov 1980, Kropotkin 1980, etc.). Here only some aspects of plate tectonics will be discussed, those related to Precambrian geology.

**Paleomagnetic Studies.** As a result of paleomagnetic studies mainly the mobility concepts are being revived and are becoming quite popular after a long period of oblivion. Not long ago paleomagnetic data for the Precambrian formations were supposed to be evidence of a continental drift at the earliest stage in geologic evolution of the Earth. However, new studies by Embleton and Schmidt (Embleton and Schmidt 1979, Schmidt and Embleton 1981) gave quite unexpected data. The analysis of paleomagnetic data done on the Precambrian of Africa, Australia, North America, and Greenland determined the situation of wandering poles in the range of 2.3–1.6 b.y. and established that the poles of all these continents were situated exactly in their common apparent paths of polar wandering during different periods of the Protozoic (Proterozoic) (Fig. 78). This indicates that the continents in the Mesoprotozoic and Early Neoprotozoic had the same relative position (were of the same orientation) as they are now. In other words, there was no continental drift in the time interval mentioned (reliable data for the still earlier period are not available).

Schmidt and Embleton suggest that the Earth was half as large in size up to 1600 m.y. and all continents were closely packed; then in the range of 1600–1000 m.y., a general expansion of the planet had taken place up to its modern size, and as a result the continental blocks became separated and the space between them became the oceanic crust covered with oceans. In their opinion the tectonic regime of the Earth in the Phanerozoic changed, and the continental (lithospheric plate) drift became its dominant global feature. These concepts seem debatable and hardly realistic, presuming in particular the idea of a sharp turning change of tectonic regime in the Phanero-

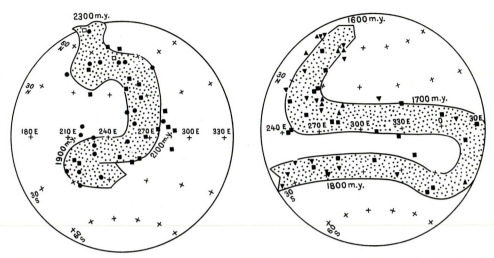

**Fig. 78.** Common apparent polar-wandering path from 2300 m.y. to 1600 m.y. (After Schmidt and Embleton 1981.) *Circles* are poles from Africa, *inverted triangles* from Australia, *upright triangles* from Greenland and *squares* from North America; *open symbols* refer to the obscured hemisphere

zoic. The data analyzed in this book evidently show that different changes in tectonic processes that were registered during almost the whole geologic history had a definite trend and at the same time were of inherited nature. The change of regime suggested by the authors had no relation with the preceding events and thus seems unnatural. Then it is quite improbable that the lithospheric plates subjected to significant horizontal movements and rotations returned to the positions they had been known to be in during the Early Precambrian.

**Problem of Origin of Oceans.** It was shown that the oldest formations in continents are ubiquitously represented by stratified rocks that make up the Katarchean gneiss-granulite complexes, whose known thickness exceeds 20 km. They are built of highly altered and recurrently granitized volcanics, mostly of basic and ultrabasic composition, alternating as a rule with subordinate metasedimentary crystalline rocks. It is important that the Katarchean complexes all over the world are of similar composition and structure. The continuous sequence of strata, and the absence of structural-facial zonation and psephytic rocks indicate that the oldest supracrustals were formed in the initial ocean that covered almost the whole planet with the exception of short periods (see Chap. 2).

It is well known that one of the important items of plate tectonics and of modern conceptions of the structure of the planet interior is the theory of a great difference in composition of the continental and oceanic crust based on the interpretation of geophysical (mainly seismic) data and partly on data of underwater drilling, dredging of sea bottom and studies of the products of oceanic volcanoes. It is believed that the oceanic crust generally consists of the upper ("first") thin sedimentary layer made up of unaltered Meso-Cenozoic rocks, of the underlying ("second") basaltic layer, drilled to a shallow depth (up to 500 m) and finally of the lower ("third"), quite thick layer that is of presumed gabbroid composition. A question arises: where are the metamorphic strata distributed in the continents and in particular the Katarchean gneiss-granulite complexes? These latter are known from every continent and the whole world: they are exposed in many places along oceanic coasts and marginal sea shores, where the exposures are reported not only in the older shields but also in the Phanerozoic foldbelts, in the areas surrounding oceans, for instance, in the Chukotka Peninsula, in Kamchatka and Taygonos, in Northern Korea, in the North American Cordilleras (British Columbia, Coast Ranges, Sierra-Madre), in the Andes (Peru), in the Appalachians and in other places, though strongly rejuvenated. Important is the fact that the Katarchean fold structures go down under the sea level in many localities as if being cut by the coast line or by the continental slope. Near the shore of Northern Korea, at approximately some 50 km from exposures of the oldest rocks on land, the Sea of Japan bottom becomes of oceanic nature according to geophysical data.

Different hypotheses exist to explain this paradox. One states that the absence of a sialic layer in oceanic bottoms is related to its metasomatic basification, but that process on the scale supposed is not possible from the point of view of thermodynamic or petrologic laws. Another states that the continuous initial continental crust became split due to expansion of the Earth's crust and crept in different directions on its surface, and the parts between these crust portions were filled in with abyssal substratum of basic composition ("sima"). In this case the expansion of the Earth must have been so great that it is hardly possible to imagine the physical process that could have provided it. A widely accepted circulation of this expansion is based on the supposition of Dirac on the decrease of the gravity constant (G) with time; physicists, however, oppose this hypothesis, for if it were true, then in the Early Precambrian the insolation of the Sun must have been so great that no hydrosphere was possible on the Earth, and consequently, no life either (insolation of the Sun depends on $G^7$; Zel'dovich and Novikov 1975). Geologic and paleontologic data argue against it. The only hypothesis of *Hydride Earth* of Larin (1981) that states the possibility of a great expansion of the planet is not substantially founded and does not accord with certain aspects of composition and inner structure of the Earth (see Dobretsov 1980 etc). Thus it must be stated that the concept of a great expansion of the Earth has no physical grounds and cannot be useful in explaining the problem discussed.

Geologists advocating tectonics of lithospheric plates do not consider as a rule the earliest stages in the formation of the Earth's crust, and apply their conception to its late stage when it became separated into two types. Only some of them (see Windley, 1976 1977) try to reconstruct an earlier history in the crust formation. For instance, Burke et al. (1976) accept the two-stage model of crust formation: at first the crust of basic composition was formed, then in some places certain areas of sialic matter became isolated over this initial crust, these representing the framework of the oldest continents. At some later stage (3.5–2.5 b.y.) of evolution, the crust of oceanic type occupied approximately 2/3 of the planet's surface and only 1/3 of it became covered with the continental type crust. Small massifs of this latter were being accreted during the junction of island arcs and partly as a result of collision. Data of the Precambrian rock record, as well as paleomagnetic studies mentioned above, however, contradict this hypothesis. It is known that intensive processes of destruction of a vast rigid continental crust were already operating at the time to which the authors referred. The main argument, however, lies in the following; it is true that local sial was formed at the very beginning of the initial (pre-geologic) stage of the Earth's evolution, but it is incredible that Katarchean supracrustal complexes of exceptionally great thickness and homogeneity all over continents were formed only on the "sialic islands". It is reasonable to suggest that the oldest sedimentary-volcanogenic strata were developed ubiquitously and then transformed into gneiss-granulite complexes during the Godthaabian and Saamian diastrophic cycles.

After analyzing geologic data it is possible to state that during the Paleoprotozoic and even the Katarchean (during the deposition of the Isuan lithostratigraphic complex in particular) there were periods of extensive regression that became expressed in many places in breaks and other types of unconformities. These regressions could appear only as a result of withdrawal of oceanic water into vast and deep depressions. We do not know any depressions of this kind in the continents, and so it is natural to assume that those depressions originated somewhere within the limits of modern oceans. We witness the same situation in the Mesoprotozoic, but it is more evident and of recurrent character. It is thus concluded that the oceanic depressions are extremely old structures. Then it was shown that the Pacific Ocean with contours close to the modern ones (Pre-Pacific) appeared at the beginning of the Neoprotozoic, and that the Atlantic Ocean (Pre-Atlantic) initiated to form at the beginning of the Epiprotozoic.

Thus the subsidence of large portions of continental crust is an indisputable fact. What had meanwhile been occurring with the crust and particularly with its gneiss-granulite layer in the areas of oceans? In my opinion, a satisfactory answer to this questions is given in the work of Artyushkov (1979). According to this author, basalt can be transformed into eclogite as a result of strong heating of a basaltic (or gabbroic) layer and phase transitions, and it can then plunge into a lighter matter of anomalous mantle possessing low viscosity. In our case it can be pictured thus: at the early stages of the Earth's evolution with rising of a heat front during a total expansion of the planet, the newly formed eclogite lithosphere, in company with the Lower Katarchean supracrustals (basic and ultrabasic granulites mainly) similar to it in physical properties and composition would have plunged into the anomalous mantle and the material of the latter would have reached the surface along fractures between the submerged blocks and would have formed extensive basaltic flows. This process could have been most extensive prior to the Saamian diastrophism when the Katarchean supracrustals (altered under granulite facies during the Godthaabian cycle or even earlier) were not strongly granitized. It is quite probable that the portions of crust with abundant granite material (formed during the pre-geologic stage and during the Katarchean) must have retained their floatability and have been added to the continental blocks or remained on the surface of oceanic bottom without being covered by basalts (an extensive granulite field in Northern Atlantic in the areas of the Rockall Plateau is an example, see Figs. 7 and 75). Probably this is the explanation of the initial heterogeneity in the Earth's structure that is expressed in an irregular distribution of continents and oceans on its surface.

It should be noted that granulites comprising oceanic bottoms are known not only from the Northern Atlantic; they are reported from some localities in the Indian Ocean, within the limits of the underwater Arabian-Indian Ridge (at a depth of 6400 m), that represents a typical mid-oceanic rise with rift zones, and also in the fracture zone in the underwater East Indian Ridge (at a depth of about 5000 m).

Granulites were also encountered while dredging in the Eltanin fracture zone in the south of the East Pacific Rise, where they were dredged together with blocks of stratified amphibolites with small-scale folds, from the lower part of a steep slope of rift, from a depth of 5600–4000 m. At a depth of 2000–1800 m gabbroid blocks were situated, and still higher (1000 m) dolerites and basalts (Kashintsev 1979). (The Eltanin fracture is situated outside the limits of drift ice in the period of its greatest extension; thus modern glacial scattering of rocks is excluded.) This unique profile thus demonstrates a major portion of the oceanic crust that accords well with a theoretical model.

It should be emphasized that some tectonists suggest that the basic granulites dredged from the oceanic bottom are the mantle rocks metamorphosed during Meso-Cenozoic spreading. The lack of traces of metamorphism in oceanic sediments of corresponding age and also certain petrographic features of granulites indicating P-T conditions that are absolutely improbable in the presumed environment, however, plainly confute this idea. At the same time, the oceanic granulites do not differ in anything from the rocks forming the Katarchean gneiss-granulite complexes (the same refers to some extent to the oceanic amphibolites that are very similar to the analogous old rocks of the continental crust).

It may thus prove to be true that large isolated blocks of Katarchean granulites are present in the oceanic crust and upper mantle, and that the oceanic depressions were formed as a result of subsidence of the lower part of the crust into the upper mantle. It is therefore not necessary to assume great expansion of the Earth or lithospheric plate drift to explain this paradox.

If the oceanic depressions had appeared at an early stage of the Earth's evolution and that is substantially evidenced by data on the Precambrian geology, then the depressions had to emplace the rocks of different age and not the Meso-Cenozoic rocks only as it is accepted now. A question arises though: what do we really know about the structure of the oceanic crust besides its uppermost sedimentary layer and the underlying basaltic layers known from drilling holes down to a depth of not more than 500 m? To what extent is the geophysical (seismic) model accepted now is valid? Bulin (1982) critically revised the data available from deep seismic sounding and came to a conclusion that the studies done in oceans cannot serve the base of the model accepted, for they are of low accuracy (compared with those done in the continents). Moreover, they are rather indicative of its similarity with the continental crust in certain aspects: its thickness is 3–4 times greater than is commonly supposed and down to a depth of 15 km below the oceanic bottom it is specified by alternation of layers of relatively low (5,5–6 km/sec) and high ($\geq$ 7 km/sec) velocity of longitudinal waves, in other words it is of a layered structure as is the case of the continental crust. Besides, detailed seismic surveys and deep drilling in continents proved that the high-velocity layers ($v_p \geq$ 7 km/sec) that were considered to be characteristic only of the oceanic crust are common in the continental crust at a depth from 2 to 15 km. Plagiogneisses and granites (with rare amphibolite inclusions) below 7,6 km depth from the superdeep drilling-hole in the Kola Peninsula are characterised by $v_p =$ 6,5 km/sec, this value was suggested to correspond to the rocks of the "basaltic layer" before these drilling data became available. Thus, now it is not excluded that the so-called third layer of the oceanic crust might contain quite thick sedimentary-volcanogenic piles of rocks of different Precambrian and Paleozoic units. The oceanic depressions are quite distant from source areas and besides it is known that commonly the areas of subsidence are characterised by occurrence of basic lavas, so it might be that these strata are principally built up of volcanics of basaltic.

All oceans have a long and complicated history. It was already stated that large depressions, at the site of the present Pacific Ocean, originated in the Early Precambrian. Later these depressions could have joined and separated again, bottoms of oceanic basins might have been drained from time to time, but the crust (and the entire lithosphere) suffered a continuous subsidence. During the Precambrian the most significant events occurred in the Early Neoprotozoic when a new, chemically isolated asthenosphere appeared under the oceanic crust in the mantle (see Chap. 5). Probably at that time a significant subsidence of lithosphere occurred and a unification of depressions into a single oceanic basin became possible, this basin being in contours and shape close to the Pacific Ocean. Somewhat later, in the Riphean, it was surrounded on all sides by paraliageosynclines (see Fig. 65). The modern appearance of the ocean, of fold structures surrounding it, of island arcs and marginal seas and also of intra-oceanic rises is the result of various events and processes at different periods of time. The most significant, however, were probably the events of the Earth's expansion during the Meso-Cenozoic that were generally located in the periphery of the ocean; in the East Pacific Rise it was the result of sea spreading, and also in the areas of marginal seas and island arcs surrounding them where detachment of the continental crust blocks was likely to be caused by the presumed euduction (Chudinov 1981) – withdrawal of the crustal or (and) mantle matter from under the continental margins toward the ocean.

No direct evidence is known as to existence of any oceanic depressions in the present site of the Atlantic Ocean during the Early Precambrian, but some geologic data favor the existence of the Pre-Atlantic Ocean (or the Iapetus as it is sometimes called) at the beginning of Epiprotozoic and its existence at least during the Vendian and Cambrian. The problem of formation of the oceanic depression proper still, however, remains complicated and debatable. According to the concept of plate tectonics it originated during a gradual drawing apart of lithospheric plates, that previously, till the beginning of the Jurassic period, had been enclosed in two supercontinents: Gondwana and Laurasia; the axial part of the Middle Atlantic Ridge where the rift depressions are situated represents a modern zone of spreading, and the linear magnetic anomalies in the slopes of the ridge are interpreted as traces or the earlier divisions caused by alternation of oceanic basalts magnetized during their outflows at different periods of inversion of a general magnetic field of the planet.

This at first sight harmonious and inviting conception has, however, some questionable moments and contradictions. Firstly, only an exceptionally great width (more than 5000 km) of the ocean could need a corresponding scale of subduction for compensation of sea spreading (with the same Earth's radius); this subduction would have had to occur in the opposite margins of lithospheric plates or, to be more exact, in their Pacific margin, for in the eastern margin of the African plate there is a zone of spreading. There are no geologic data in evidence of such gigantic underthrusts. This conception also disagress with the existence of the Pre-Atlantic Ocean somewhere on the site of the modern ocean during Late Precambrian-Cambrian periods. If this had been the case, it would have to be accepted that the old Atlantic Ocean closed and then opened again (and some suggest it to have happened several times), which seems hardly realistic, as there is no satisfactory explanation of the disappearance of new lithified oceanic crust (lithosphere) that would have had to be formed in its place. Geologic data do not permit the assumption that it wholly overthrusted or underthrusted the continents, for it is impossible to imagine a process of that kind on such a scale. The concept of a huge division of continents fails to explain the existence of an extensive area built up of Katarchean granulites on the bottom of the North Atlantic, where not only the Rokall Plateau, but also the regions adjoining North Norway and Scotland belong. Moreover, granite-gneiss, diorite, granite, and different metamorphic schists, amphibolite, and metagabbroid were dredged from the bottom in some sites of the mid-oceanic ridge, and crop out in the islands (Kashintsev 1979).

Statements on the similarity of geologic structure of the Precambrian formations on the opposite coasts of the Atlantic Ocean, in Africa, and South America can be found in many geologic publications, and are true to a very rough approximation, as this kind of similarity is known from many other very distant areas of the world. Some definite Precambrian structures on the opposite coasts of these continents, however, do not match; in any case, there are no reference objects (indicators) there that could make their correlation possible. The examples usually given (see, for instance, Porada 1979, Torquato and Gordani 1981) are not convincing and can be interpreted in different ways. A good matching of the areas of Baikal activation in both continents (see Fig. 75) can be discounted, as their boundaries are only rough and, additionally, such areas are commonly situated in the margins of the continents (see Chap. 7).

In my opinion, the modern contours of the Atlantic Ocean developed during several periods of a general but relatively minor expansion of the Earth, the intensity of which increased with time so that different modes of this process made the principal impact at different periods. The Pre-Atlantic Ocean is very likely to have been formed as a result of subsidence of large blocks of the lithosphere (together with the crust made up of a granulite "layer") into the anomalous mantle or asthenosphere due to the same "Artyushkov's mechanism" that was given for the Pacific Ocean. The subsiding lithospheric blocks were probably limited, with numerous and large vertical abyssal fractures that constituted a regmatic system. This is a probable explanation for the similarity in contours of the southern continents and also for the block structure of the oceanic bottom that is seen from numerous transversal and cross-fractures and from submarine ridges present, which create a coarse-celled morphology of the oceanic bottom. Modern contours of the Atlantic Ocean are very likely to have inherited the original ones to a certain extent, but the area of the ocean increased greatly at the expense of intensive stretching of lithosphere in a wide (up to 1000 km) belt occupied by the Mid-Atlantic Ridge, which is characterized by a sharply expressed linear structure of magnetic field. It seems possible to regard this belt as a zone of the last – Meso-Cenozoic – expansion of the Earth. It is probably the only belt within the limits of which the newly formed oceanic crust, resulting from a vast spreading, is exposed. However, in the light of petrophysical data obtained in studying similar magnetic anomaly strips in volcanogenic areas of continents, the "conveyor mechanism" idea of the opening of the oceanic sublithospheric bed admitted by the tectonic plate concept seems debatable. According to Dortman (1981), the linear magnetic anomalies in the zones of spreading could not be the result of successive emplacement of basic rocks directly and reversely magnetized in different paleomagnetic epochs, but are to be explained only by the alternation of rocks revealing different grades of magnetization. In this case there is no necessity to regard the mid-oceanic ridges as zones of continuous separation of the lithosphere, and it is more reasonable to regard them as zones of high permeability that appeared as a result of constraint stretching of the lithosphere where the parallel thick dykes or dyke-like bodies of basic rocks (the incurrent canals for basalt outflows in fractures) alternate with bands of enclosing rocks. Such dykes are known from margins of the Atlantic Ocean: on the south-eastern shore of Greenland and on the Hebrides Islands. In Greenland the boundary, of the oceanic area is fixed by a "shore belt" of thick (up to 2000 m) and extensive (tens and some hundreds of kilometers) dykes of basic rocks of Early Tertiary age that cut the older gneisses; the oceanic basalt is commonly associated with these dykes (Larsen 1978). Rather rare linear anomalies observed in flanks of mid-oceanic ridge and locally in the ocean near the continent are also likely to be fractures filled in with basic rocks. This kind of interpretation of geophysical data greatly decreases the scale of oceanic spreading.

All these statements refer not only to the Atlantic Ocean; it is assumed that the global system of the mid-oceanic rises is of the same nature, and is the main result of intensive global expansion of the Earth during the Meso-Cenozoic period. The situation of the mid-oceanic rises, rift structures, and transform fractures is possibly governed by the regmatic system initiated at much earlier stages of geologic evolution. At least it can be taken for granted in some Cenozoic continental rift systems occurring in the arch rises, for instance, in the East African and Baikal systems. Incidentally this latter, with an extension of more than 2000 km, greatly resembles the rift system of the mid-oceanic rises in its morphology and tectonic structure, in localization of earthquakes and basaltic volcanism within its limits and also in the modern high heat flow. It is possible that the long life of the older regmatic system in the oceanic areas is due to the abyssal "erosion" of lithosphere under the influence of hot "mantle flows" rising along the fractures.

If these suppositions on the origin of oceans are reasonable, they will help in solving the problem of space and the related subduction problem. In this case the continental drift could have been provided by a relatively small-scale separation of lithosphere plates that was possible accompanied by the Earth's expansion on the same scale.

**Problem of Geosynclines.** According to the plate tectonics concept, geosynclines and also foldbelts were formed in the collision zones of lithosphere plates or, in other words, in the subduction zones, and the "true" geosynclines were initiated in the oceanic bed. Advocates of this concept suggest that the foldbelts represent scars that remained after the closing of the lithosphere plates, and that the ophiolites within their limits are the remains of oceanic crust and mantle.

These ideas disagree with of studies the results of the Precambrian foldbelts of different age. First of all, the Precambrian geosynclinal formations starting from the Paleoprotozoic ("Upper Archean") greenstone strata rest either on the Katarchean gneiss-granulite basement or on the younger metamorphic and plutonic complexes overlying this latter. In many Precambrian geosynclinal belts the roots of fold structures squeezed among mobilized crystalline and metamorphic rocks (diaphthorites) are known to be exposed, but in continents there is not a single case that could serve as an illustration of the occurrence of geosynclinal piles on the basalt or mantle substratum rocks.

The same situation is typical of many Phanerozoic geosynclinal belts, including those of supposed ensimatic origin, for instance, the Hercynian belt of the Urals, the Caledonian belts of the Atlai-Sayan fold system, the Hercynian and Alpine belts of Central Asia, Minor Caucasus, North-East of the U.S.S.R. and others that are characterized by the presence of numerous intrusions of basite and hyperbasite in some zones. Detailed studies revealed that exposures of the Precambrian granite-metamorphic basement can also be registered there, including the Katarchean gneiss-granulite complex. In these geosynclines the ophiolite geosynclinal formations distributed in the zones of abyssal fractures are usually assigned to the oceanic-mantle or basaltoid sockle, but in many cases it is quite evident that they were deposited on the older metamorphic rocks. It is absolutely impossible to regard the discrete lens-like bodies of hyperbasite that represent intrusions or protrusions of abyssal matter in fracture zones as remains of a bed of the "shut-in" ocean. This type of body is also widespread in many Precambrian foldbelts of different age, where they are situated in the zones of dislocations among rocks of the continental crust. It is not excluded, however, that relatively large sheets of ultrabasic rocks in certain Phanerozoic foldbelts consist of mantle matter located in zones of spreading, but the original width of such zones rarely exceeded 60–80 km judging from palinspastic reconstructions.

Analysis of the Precambrian foldbelts shows that all the older geosynclines were initiated on the crushed crystalline basement mainly made up of the Katarchean gneiss and granulite. The situation and orientation of geosynclinal belts was under the control of regmatic system of abyssal fractures. In the Paleoprotozoic, greenstone strata occurred in fractures that separated blocks of the Katarchean basement. The Mesoprotozoic geosynclines are belts of different orientation among a mosaic of angular platform blocks made up of older Precambrian rocks; these latter also crop out ubiquitously inside geosynclines. It is important that typical platform strata (see Chap. 4) underlie many Mesoprotozoic geosynclinal complexes (in the Baltic shield, in Southern Greenland, in Hindustan, and in other regions). Large Neoprotozoic platforms were formed as a result of unification of the older platform blocks after closing of geosynclines. At the same time, intensive destruction is also registered in the Neoprotozoic. The geosynclines that originated at that time continued to exist up to the end of the Precambrian or even in the Phanerozoic, exhibiting different modes of occurrence. In the Epiprotozoic new geosynclinal belts appeared in zones of abyssal fractures, and continued to evolve later, at least during the Early Paleozoic.

The paleotectonic schemes of different continents included in this book clearly illustrate the formation of geosynclines under destruction processes. They indicate that geosynclines initiated on the sialic crust inside continents as well as in their margins. It becomes specially evident when examining the evolution of Africa (Kröner 1977, Salop 1977b). Paraliageosynclines occurring in the margins of modern continents have appeared only since the Riphean, but their basement, if it is exposed, is composed only of the older Precambrian rocks.

In the Precambrian, at least in the Early Precambrian (that is prior to the Neoprotozoic), the spreading was not great and resulted mainly in the formation and opening of fractures and in the emplacement of densely packed dykes of basic rocks along fissures. An exceptionally striking example of lithosphere stretching of this type is the

Mesoprotozoic Nagssugtoqidian mobile belt of Greenland, whose whole width of Katarchean gneiss basement (up to 250 km) is filled in with deformed parallel dykes of basic rocks that injected 2300–1950 m.y. ago (Escher and Watt 1976, Kalsbeek and Zeck 1978).

Thus, data obtained on the principal problem of location and mode of origin of geosynclines while applying facts from Precambrian studies contradict the concept of plate tectonics.

**Problem of Inheritance.** Inheritance and trend in evolution of large tectonic elements of the Earth, geosynclines and platforms, is a well established regularity. This aspect of geologic evolution is especially evident in the Precambrian. Many geosynclinal systems and belts are known to have developed during the Precambrian for many hundreds of millions of years, and in spite of common rebuilding they kept to their general tectonic plan, certain geosynclinal formations being successively followed by other geosynclinal formations and representing a single evolutionary series from the initial spilite-keratophyre formation to the orogenic (molasse) formation.This is exemplified by the Baixal geosynclinal system, which appeared on the crushed Katarchean basement in the Paleoprotozoic and continued to evolve up to the Upper Precambrian (Salop 1964–1967). It has already been stated that many geosynclinal belts that appeared at the beginning of the Riphean continued to exist in similar limits during the major part of the Phanerozoic. Many large platform elements such a shields, syneclises, and aulacogens in part are also known to have existed for many millions of years. For instance, the Baltic shield, which appeared as a large positive morphostructure soon after the Karelian folding 1900 m.y. ago (in the west after 1000 m.y. ago in part), exists up to now as an area of stable rising. Many more examples of the kind may be given. These are certainly arguments against a significant horizontal drift of continents.

**Problem of Tectonic Regime.** According to the plate tectonics concept, folding and other orogenic processes occur in the zones of subduction simultaneously with separation of lithospheric plates in the zones of spreading. Meanwhile, the data presented in this book show that the geologic phenomena recorded may find a satisfactory explanation only if alternation in time in the processes of expansion and compression of the Earth is accepted. The same conclusion is reached by many geologists studying Phanerozoic geology; it is specially emphasized in the works of Milanovsky (1978, etc.).

From the above it may be concluded that the concept of lithospheric plates does not explain the whole diversity of geologic phenomena, though it is possible that some of its elements will make a contribution as a part of a future more general tectonic theory. They will probably be useful in explaining certain tectonic events of the Phanerozoic (of the Meso-Cenozoic in particular) and in the Late Precambrian in part.

I certainly realize that not all of the ideas proposed here are substantially based. Some do not give the answer to certain facts. This is particularly true for paleomagnetic data on the Late Precambrian and Phanerozoic rocks, that seem to disagree with my idea of a limited role for the horizontal drift of continents (lithospheric plates). It is to be hoped, however, that these data will at some time receive a new interpretation. Certain proposals, however, may be discussed even now. If we examine the paleomagnetic data of Mesoprotozoic and Neoprotozoic generalized by Schmidt and Embleton (Fig. 78), they seem to have a realistic explanation in wandering of the magnetic pole relative to fixed lithosphere (continents) or in the lithosphere movement relative to fixed pole, the latter appears to be more appropriate. It is quite remarkable that the most abrupt bends of the pole-wandering path fall within the time at about 2000 m.y., which is the time of the Ladogan orogeny; and specially at the time of about 1900 m.y., which corresponds to the Karelian orogeny of the first order and at the interval of 1800–1750 m.y., when intensive uplifts and subsidence of lithosphere accompanied by intense lava outflows (the Akitkan epoch) are recorded. Probably it is related to the fact that a thick asthenosphere generated during the diastrophic epochs due to a general uplift of the thermal front and very significant movements of mass took place inside the Earth as well as on its surface, they had to call out corresponding changes in rotation regime of the planet and as a consequence an abrupt wandering of its geographic or(and) magnetic pole. In the Phanerozoic the rate of diastrophic events and the amplitude of morphostructures increased greatly and the disturbances in pole positions must have become more frequent and sharp than in the Precambrian. If this were so, even slight errors in dating (synchronism) of objects of paleomagnetic studies (in particular situated in different continents) would result in misinterpretation of paleomagnetic data and would create a false impression of the separate movement of each continent.

Naturally, it was not possible to examine in detail many aspects of global tectonics in a book devoted to geologic evolution in the Precambrian. I have paid attention only to some of these aspects that are in some manner related to Precambrian geology. However, when analyzing all data available on the early tectonic evolution of continents, it seems more reasonable to proceed from such a "pulsation hypothesis" that provides an alternation of periods of expansion and compression of the Earth against a background of a general, relatively small expansion of the planet, located, however, in certain zones; an expansion which, moreover encreases with time, reaching its maximal value at the end of the Phanerozoic. In this case, interpretation of all the facts and data at our disposal could make sense.

## III. On the Actualism Principle in the Light of Precambrian Geology

In conclusion, it seems reasonable to dwell on the principle of actualism, or more exactly on the uniformitarian principle, in our interpretation of Precambrian geology.

It has been shown that a clearly expressed global rhythm of geologic events is typical of the Precambrian. Meanwhile, as the pattern of distribution of sedimentary facied varies and the tectonic regime is nowadays complex, it was considered for a long time that in the Phanerozoic quite different tectonic conditions co-existed and simultaneously, though in different places, sea transpressions and regressions took place and that generally all geologic events were mostly of local nature. However, recent studies have shown that the Phanerozoic was also typified by a rhythmical alternation of different geologic events on a global scale, the analysis was done using data of every continent and in some cases quantitative (statistical) methods. It was noted above that the works of Pronin and other investigators confirmed the ideas of Stille, Bubnov and many other well-known tectonists as to the simultaneous manifestion of diastrophic phases (cycles) which separate the periods of relative tectonic quietness. The results of the work done by Naydin (1971), by Ronov (1980) in particular and by Khain (1977) revealed that geocratic and thalassocratic epochs of the Phanerozoic were also characteristic of the entire planet, and that periodical changes in sea extensions occurred simultaneously in platforms and in geosynclines. Ronov and his co-authors (1979) state that "the existence of this global rhythm is confirmed in spite of different modes of behavior of separate large blocks of the lithosphere... this suggests a dominance of global trends over regional ones" (p. 12). The essential difference in the tectonic history of the Phanerozoic and that of the Precambrian lies in an abrupt higher frequency of diastrophic events in the Phanerozoic.

One more question arises in this connection: why are different Precambrian units so special all over the world and the co-eval units so similar, when the Phanerozoic formations do not seem to be characterized in the same manner? Several causes may be proposed to explain this. Firstly, it should be kept in mind that the stages of evolution with the formation of the supracrustals during the Precambrian were much longer that during the Phanerozoic. It is obvious that the major Precambrian stratigraphic units enclosed between the diastrophic cycles of the first order, or in other words, the Precambrian Erathems are of a greater volume than those of the Phanerozoic. The lithostratigraphic complexes that comprise certain portions of the Precambrian Erathems were formed during the first hundreds of millions of years (with the exception of short glacial complexes) and so the length of time of their formation is much longer that of the Phanerozoic periods and even of the Mesozoic or Cenozoic Eras. Naturally, the Precambrian and Phanerozoic units are not comparable in this respect.

The climatic zonation moreover did not exist in the Precambrian, the structural-facial zonation was much less clear than in the Phanerozoic, the amplitude of tectonic movements was lower and as a consequence the contrast of tectonic and erosional forms of relief was also of lower grade. The physical environments on the Earth's surface were also different. Finally, it is very important that the geochemical evolution of atmosphere and hydrosphere mainly occurred in the Precambrian, prior to the beginning of the Neoprotozoic Era, as this affected the formation of many specific irreversible piles in the Earth's evolution (of different oxysensor formations in particular).

These special characteristics of the Precambrian and many more peculiarities of this era indicate that the actualism principle (not the method) may appear to be of limited application in interpretation of events and processes during the early stage of geologic history.

# References

Abramovich II, Klushin IG (1978) Petrochemistry and abyssal structure of the Earth. Nedra, Leningrad, 375 pp

Afanas'ev GD, Bagdasar'yan GP, Borovikov LI (1964) Geochronological scale in absolute chronology based on data of the U.S.S.R. In: Absol Vozrast Geol Formatsyi. Nauka, Moscow, pp 287–324

Aksenov EM, Keller BM, Sokolov BS, Solontsov LF, Shul'ga NL (1978) General scheme of the Upper Precambrian stratigraphy of the Russian platform. Izv Acad Sci U.S.S.R Geol Ser 12:17–34

Aldrich LT, Davis GL, James HL (1965) Ages of minerals from metamorphic and igneous rocks near Iron Mountain, Michigan. J Petrol 6(3):445–472

Al-Khatib R, Touret J (1973) Fluides corboniques dans les rochés du facies granulite du Sud de la Norvégé. Utilisation semi-quantitative de la surplatine à écrasement. Bull Soc Geol Fr 15(3–4):321–325

Allaart JH (1976) The pre–3700 m.y. old supracrustal rocks of the Isua area, central West Greenland and the associated occurrence of quartz-banded ironstone. In: Windley BF (ed) The early history of the Earth. Wiley, London, pp 177–190

Allègre CJ, Caby R (1972) Chronologie absolue du Précabrien de l'Ahaggar occidental. CR Acad Sci Ser D 275, 19:2095–2098

Allsopp HL, Roberts HR, Schreiner GD, Hunter DR (1962) Rb-Sr age measurements on various Swaziland granites. J Geophys Res 67(13):5307–5313

Allsopp HL, Ulrych TJ, Nicolaysen LO (1968) Dating some significant events in the history of the Swaziland System by the Rb-Sr isochron method. Part 2. Can J Earth Sci 5(3):605–619

Allsopp HL, Davies RD, Le Gasparis AA, Nicolaysen LO (1969) Review of Rb-Sr age measurements from the early Precambrian terrain in the south-eastern Transvaal and Swasiland. Geol Soc S Afr Spec Publ 2:433–444

Almeida FFM de (1968) Tectonic evolution of Brazils Central-Western Region in Late Proterozoic time. Anais Acad Bras Cienc (Suppl) 40:285–286

Almeida de FFM, Amaral G, Cordani UG, Kawashita K (1973) The Precambrian evolution of the South American cratonic margin south of the Amazon River. In: Nairn AE, Stehli FG (eds) The ocean basins and margins 1. Plenum Press, New York pp 411–446

Almeida FFM de, Hasui I, Brito Neves BB de (1976) The Upper Precambrian of South America. Bol Inst Geosci Univ Sao Paulo 7:45–80

Amaral G (1969) Nota prévia sobre o reconhecimento do Pré-Cambriano da região Amazônica. 23 Congr Bras Geol Salvador, Resumo, Bol Esp 1:81–82

Anderson MM (1977) A possible time span for the Late Precambrian of the Avalon Peninsula, Southeastern Newfoundland in the light of worldwide correlation of fossils, tillites, and rock units within the succession. Can J Earth Sci 9(12):1710–1726

Andrade Ramos JR de, Fraenkel MO (1974) Uranium occurrences in Brazil. In: Formation of uranium ore deposits. Int Atom Energy Agency, Vienna, pp 637–658

Anhaeusser CR (1971) Cyclic volcanicity and sedimentation in the evolutionary development of Archaean greenstone belts of Shield areas. Geol Soc Australia Sp Publ 3:57–70

Anhaeusser CR (1973) The evolution of the early Precambrian crust of Southern Africa. Philos Trans R Soc London Ser A 273:359–388

Anhaeusser CR, Roering C, Viljoen MJ, Viljoen RP (1968) The Barberton Mountain Land. A model of the elements and evolution of Archean fold belt. Geol Soc S Afr Annex 71:225–254

Anhaeusser CR, Mason R, Viljoen MJ, Viljoen RP (1969) A reappaisal of some aspects of Precambrian Shield geology. Geol Soc Am Bull 80(11):2175–2200

Archibald NJ, Bettenay LF (1977) Indirect evidence for tectonic reactivation of a pre-greenstone sialic basement in Western Australia. Earth Planet Sci Lett 33(3):370–378

Archibald NJ, Bettenay LF, Binns RA, Groves DI, Grunthorp RJ (1978) The evolution of Archean greenstone terrains, Eastern Goldfields province Western Australia. Precambrian Res 6(2):129–131

Arndt NT, Naldrett AJ, Pyke DR (1977) Komatiitic and ironrich tholeiitic lavas of Munro Township, northeast Ontario. J Petrol 18(2):319–369

Artyushkov EV (1979) Geodynamics. Moscow, Nauka, 327 pp

Aswathanarayana U (1964) Isotopic ages from the Eastern Ghats and Cuddapahs of India. J Geophys Res 69(16):3479–3486

Awramik SM (1976) Gunflint stromatolites: paleomicrobial content and significance. 25 Ses IGC, Sydney, Abstr., p. 29

Awramik SM, Barghoorn ES (1977) The Gunflint microbiota. Precambrian Res 5(2):121–142

Baadsgaard H, Lambert RSt, Krupicka J (1976) Mineral isotopic age relationships in the polymetamorphic Amitsoq gneisses, Godthaab District, West Greenland. Geochim Cosmochim Acta 40(5):513–527

Badham JPN (1978) The early history and tectonic significance of the East Arm Graben, Great Slave Lake, Canada. Tectonophysics 45(2–3):201–215

Bailey EB (1926) Domes in Scotland and South Africa: Arram and Vredefort. Geol Mag 43:481–494

Balitsky VS, Orlova VP, Ostapenko GM (1971) Solubility of quartz in hydrothermal solutions of potassium sulphide and potassium, sodium and ammonium fluoride. In: Trudy 8 Soveshchaniya Eksper. Tekhn Miner Petrogr (Novosibirsk, 1968). Nauka, Moscow, pp 220–224

Banks PO, Cain JA (1969) Zircon ages of Precambrian granitic rocks, Northeastern Wisconsin. J Geol 77(2):208–220

Baragar WRA, Donaldson JA (1973) Coppermine and Dismal Lake map-areas. Geol Surv Can Pap 71–39:20

Baragar WRA, McGlynn (1976) Early Archean basement in the Canadian Shield: a review of the evidence. Geol Surv Can Pap 76–14:21

Barr MWC, Cahen L, Ledent D (1978) Geochronology of syntectonic granites from central Zambia: Lusaka granite and granite NE of Rufunsa. Ann Soc Geol Belg 100(1977):47–54

Barton JM (1975) Rb-Sr isotopic characteristics and chemistry of the 3.6 b.y. Hebron gneiss, Labrador. Earth Planet Sci Lett 27(3):427–435

Barton JM, Fripp RP, Ryan B (1977) Rb/Sr ages and geological setting of acient dykes in the Sand River area, Limpopo Mobile Belt, South Africa. Nature (London) 267(5611):487–490

Barton JM, Ryan B, Fripp RP (1978) The relationship between Rb-Sr and U-Th-Pb whole-rock and zircon system in the 3790 m.y. old Sand River gneisses, Limpopo Mobile Belt Southern Africa. Geol Surv S Afr Open-File Rep 701:27–78

Becker RH, Clayton RN (1976) Oxygen isotope study of Precambrian banded iron formation, Hamersley Range, W. Australia. Geochim Cosmochim Acta 40(9):1153–1165

Bekker YuR, Keller BM, Kozlov VI, Rotar AF (1979) The Riphean stratotypical section. In: Stratigr Verkhnego Proterozoya U.S.S.R, Riphean and Vendian. Nauka, Leningrad, pp 71–85

Belevtsev YaN (ed) (1957) Geologic structure and iron ores of the Krivoy Rog basin. Gosgeoltekhizdat, Moscow, 280 pp

Belevtsev YaN, Sukhinin AN (1974) Certain mineral, geochemical and genetic aspects of granitoids in the central part of the Ukrainian Shield. Geol Zh 34(1):16–30

Belevtsev YaN, Rudnitsky LM, Sukhinin AN (1971) Stratigraphy and structure of the central part of the Ukrainian shield (contribution to the discussion). Geol Zh 31(2):119–134

Belokrys LS, Mordovets LF (1968) Precambrian plant remains from the Krivoy Rog area. Dokl Acad Sci U.S.S.R 183(1):196–199

Belousov VV (1975) Principles of geotectonics. Nedra, Moscow, 262 pp

Berkner LV, Marshall LC (1967) Rise of oxygen in the Earth's atmosphere. In: Advances in geophysics, vol XII. Academic Press, London, New York, pp 309–332

Bersenev II (1973) Origin and evolution of depression of the Sea of Japan. In: Voprosy Geologii Dna Yaponskogo Morya, Vladivostok, pp 15–35

Bertrand-Sarfati J (1972) Stromatolites columnaires du precambrien superieur du Sahara nord-occidental. Publ Cent Rech Zones Arides Ser Geol Paris 14:300

Bertrand-Sarfati J, Caby R (1974) Precisions sur l'age précambrien terminal (vendien) de la serie carbonatée à stromatolites du groupe d'Eleonore Bay (Groenland oriental). CR Acad Sci Ser D 288, 18:2267–2270

Bertrand-Sarfati J, Moussine-Pouchkine A (1978) Mise en evidence d'une discordance du groupe de Bandiagara sur les formations sédimentaires du Précambrien supérieur (Gourma, Mali). Bull Soc Geol Fr Suppl: CR Somm Séances 20 (2):59–61

Besairie H (1973) Précis de geologie Malgache. Ann Geol Madagascar 36, 141 pp

Beurlen K (1970) Geologie von Brasilien. (Beiträge zur regionalen Geologie der Erde, Bd 9) Gebr Borntraeger, Berlin-Stuttgart, 444 S

Bibikova EV, Makarov VA, Gracheva TV, Seslavinsky KB (1978) Age of the oldest rocks of the Omolon massif. Dokl Acad Sci U.S.S.R, 241 (2):434–436

Bickford ME, Mose DG (1972) Chronology of igneous events in the Precambrian of the St. Francois Mountains, Southern Missouri: U-Pb age of zircons and Rb-Sr ages of whole rocks and mineral separates. Geol Soc Am Annu Meet Abstr Programs 4 (7):451–452

Bickford ME, Mose DG (1975) Geochronology of Precambrian rocks in the St. Francois Mountain, Southeastern Missouri. Geology 3 (9):537–540

Bickle MJ, Martin A, Nesbet EG (1975) Basaltic and peridotitic komatiites and stromatolites above a basal unconformity in the Belingwe greenstone belt, Rhodesia. Earth Planet Sci Lett 27 (2):155–162

Birch F (1965) Energetics of core formation. J Geophys Res 70 (24):6217–6221

Bird JM (1975) Late Precambrian graben facies of the northern Appalachians. Geol Soc Am 10th Annu Meet Abstr Programs 7 (1):27

Bjørlykke K, Elvsborg A, Hoy T (1976) Late Precambrian sedimentation in central sparagmite basin of South Norway. Nor Geol Tidsskr 56 (3):233–290

Black LP, Gale NH, Moorbath S, Pankhurst RJ, McGregor VR (1971) Isotopic dating of very Early Precambrian amphibolite facies gneisses from Goodthaab district of west Greenland. Earth Planet Sci Lett 12 (3):245–259

Blais S, Auvray B, Bertrand JM, Capdevila R, Hameurt J, Vidal Ph (1977) Les grand traits géologiques de la ceinture archéenne de roches vertes de Suomussalmi-Kuhmo (Finlande orientale). Bull Soc Geol Fr (7), 19 (5):1033–1039

Blaxland AB, Breemen Van O, Emeleus CH, Anderson JG (1978) Age and origin of the major syenite centres in the Gardar province of south Greenland: Rb-Sr studies. Geol Soc Am Bull 89:231–244

Bliss NW, Stidolph PA (1969) A review of the Rodesian basement complex. Geol Soc S Afr Spec Publ 2:305–330

Blusson SL (1971) Sekwi Mountain map-area, Yukon Territory and district of Mackenzie. Geol Surv Can Pap 71–22:17

Bogdanov AA (1967) Contribution to the tectonic division of the Precambrian formations in the basement of the East European platform. Vestn Mosk Univ 1:8–26

Bogdanov YuB, Negrutza VZ, Suslova SN (1971) Precambrian stratigraphy of the eastern part of the Baltic shield. In: Stratigrafiya i Izotopnaya Geokhronologiya Dokembriya Vost Chasti Baltiyskogo Shchita. Nauka, Leningrad, pp 160–170

Bond G, Wilson JF, Winnall NJ (1973) Age of the Huntsman Limestone (Bulawayan) stromatolites. Nature (London) 244 (5414):275–276

Bondarenko LP, Dagelaysky VB (1968) Geology and metamorphism of Archaean rocks in the Central part of the Kola Peninsula. Nauka, Moscow, 168 pp

Bondesen E, Pedersen KR, Jorgensen O (1967) Precambrian organisms and the isotopic composition of organic remains in the Ketilidian of S.-W. Greenland. Medd Grønl 164 (4):10–32

Boon JD, Albritton CC (1937) Meteorite scars in acient rocks. Field Lab 5:53–64

Bowes DR, Hopgood AM, Pidgeon RT (1976) Source ages of zircon in an Archaean quartzite, Rona, Inner Hebrides, Scotland. Geol Mag 113 (6):545–552

Bozhko NA (1975) Tilloid conglomerates of the Volta and Akwapim syneclise in the Togo (Atacor) block-fold zone. In: Geol Pol Iskop Dokembrija. Tr Nauchno Issled Lab. Zarubezhgeol 29:90–97

Bozhko NA, Pykhova NG, Raaben ME (1974) On the Upper Precambrian biostratigraphy of Africa. Dokl Acad Sci U.S.S.R 214 (3):643–646

Brakel AT, Muhling PC (1976) Stratigraphy, sedimentation, and structure in western and central parts of the Bangemall Basin, Western Australia. Geol Surv W Aust Annu Rep 1975; 70–79

Bridgwater D (1974) West Greenland conglomerate. Geotimes 179 (1) (Photo on journal cover)

Bridgwater D, Allaart JH, Baadsgaard H, Collerson KD, Ermanovics I, Gorman B, Griffin W, Hanson VR, McGregor VR, Moorbath S, Nutman AP, Taylor P, Tvetin E, Watson J (1979) International field work on Archaean gneisses in the Godthåbsfjord-Isua area, southern West Greenland. Geol Surv Greenl Rep Activ 1978: 1-10

Bridgwater D, McGregor VR (1974) Field work on the very early Precambrian of the Isua area, Southern West Greenland. Geol Surv Greenl 65:49–53

Bridgwater D, Keto L, McGregor VR, Myers JS (1976) Archaean gneiss complex of Greenland. In: Esher A, Watt WS (eds) Geology of Greenland. Geol Surv Greenl, Copenhagen, pp 18–75

Bridgwater D, Allaart J, Schopf J, Klein C, Walter MR, Barghoorn ES, Strother R, Knoll AH, Gorman BE (1981) Microfossil-like objects from the Archaean of Greenland: a cautionary note. Nature (London) 289 (5793): 51–53

Brookins DG (1968) Rb-Sr and K-Ar age determinations from the Precambrian rocks of the Jardine-Grevice Mountain area, Southwestern Montana. Earth Sci Bull 1 (2):5–9

Brooks C, Hart SR (1978) Rb-Sr mantle isochrons and variations in the chemistry of Gondwanaland's lithosphere. Nature (London) 271 (5642):220–223

Brooks C, Hart SB, Hofmann A, James DE (1976) Rb-Sr mantle isochrons from oceanic regions. Earth Planet Sci Lett 32 (1):51–61

Brown JS (1973) Sulfur isotopes of precambrian sulfides in the Grenville of New York and Ontario. Econ Geol 68 (3):362–370

Brummer JJ, Mann EL (1961) Geology of the Seal Lake area, Labrador. Geol Soc Am Bull 72 (9):1361–1381

Bulin NK (1982) The oceanic Earth's crust from seismic data. Byull Mosk Ova Ispyt Prir Otd Geol, 57 (1):3–17

Burger AJ, Coertze FJ (1973) Radiometric age measurement on rocks from Southern Africa to the end of 1971. Geol Surv S Afr Bull 58, 46 pp

Burke K, Dewly JF, Kidd WS (1976) Dominance of horizontal movements, arc and microcontinental collisions during the later permobile regime. In: Windley BF (ed) The early history of the Earth. Wiley, London, pp 113–129

Caby R (1967) Une nouveau fragment du craton de l'Ouest africain dans le Nord-Ouest de l'Ahaggar (Sahara algérien): ses relations avec la serie à stromatolites, sa place dans l'orogenie du Precambrien superieur. CR Acad Sci Ser D 265:1452–1455

Caby R (1969) Une nouvelle interpretation structure et chronologique des séries à "facies suggarien" et à "facies Pharusien" dans l'Ahaggar. CR Acad Sci Ser D 268:1248–1251

Caby R (1971) Niveaux et imprégnations cuprifères du Précambrien superieur et de la Série Pourprée au Tanerzouft oriental (Sahara algérien). Publ Serv Geol Algérie 41:129–137

Caby R (1972) Evolution pré-orogénique, site et agencement de la chaine pharusienne dans le Nord-Ouest de l'Ahaggar (Sahara algérien): sa place dans l'orogenèse pan-africaine en Afrique occidentale. In: Corrél Précambrien Coll Int 1970, Rabat-Paris, pp 65–80

Caby R, Moussu H (1968) Une grande serie detritique du Sahara. Bull Soc Geol Fr (1967), 9 (6):876–882

Cahen L (1954) Geologie du Congo Belge. Vaillant-Carmanne, Liege, 577 pp

Cahen L, Lepersonne J (1967) The Precambrian of Congo, Rwanda and Burundi. In: Rankama K (ed) The Precambrian, vol III. Intersci Publ, London New York, pp 143–190

Cahen L, Snelling NJ (1966) The geochronology of Equatorial Africa. North Holland Publ, Amsterdam, 195 pp

Cahen L, Delhal J, Ledent D (1970) On the age and petrogenesis of the microcline-bearing pegmatite veins at Roan-Antelope and at Musoshi (Copperbelt of Zambia and South-East Katanga). Mus R Afr Cent Sci Geol 65:43–68

Cahen L, Ledent D, Villeneuve M (1979) Existence d'une chaine plissee proterozoique superieur au Kivu oriental (Zaire). Donnees geochronologiques relatives au supergroupe de l'Itombwe. Bull Soc Belg Geol 88 (1–2):71–83

Carey SW (1954) The rheid concept in geotectonics. J Geol Soc Aust (for 1953) 1:61–117

Carter EK, Brooks JH, Walker KR (1961) The Precambrian mineral belt of northwerstern Queensland. BMR Aust Bull 51:344

Catanzaro EI (1963) Zircon ages in Southwestern Minnesota. J Geophys Res 68:2045–2048

Catenzaro EI, Kulp JL (1964) Discordant zircon from little Belt (Montana), Beartooth (Montana) and Santa Catalina (Arizona) Mountains. Geochim Cosmochin Acta 28(1):87–124

Cawthorn RG, Strong DF (1974) The petrogenesis of komatiites and related rocks as evidence for a layered upper mantle. Earth Planet Sci Lett 23(3):369–375

Chamberlain VE, Lambert RS, Baadsgaard H, Gafe NH (1979) Geochronology of the Malton Gneiss Complex of British Columbia. Geol Surv Can Pap 1B:45–50

Chase CG, Perry EC (1972) The oceans: grows and oxygen isotope evolution. Science 1972 (4053):992–994

Choubert B (1966) Etat actuel de nos connaissanses sur la géologie de la Gayane francaise. Bull Soc Géol Fr 7(1):129–135

Choubert B (1974) Le Precambrian des Guyanes. BRGM Mem 81:204

Choubert G, Faure-Muret A (1972) Au sujet des rajeunissement des ages isotopiques. In: Corrél Précambrien Coll Int 1970, Rabat-Paris, pp 145–170

Choubert G, Faure-Muret A (eds) (1973) Tectonics of Africa. Mir, Moscow, 542 pp (in Russian)

Chudinov YuV (1981) Expansion of the Earth and tectonic movements: on direction of movements in the marginal-oceanic zones. Geotektonika 1:19–37

Chumakov NM (1978a) The Precambrian tillites and tilloids. Nauka, Moscow, 202 pp

Chumakov NM (1978b) On stratigraphy of the Precambrian upper horizons in the Southern Urals. Izv Acad Sci U.S.S.R. Geol Ser 12:35–48

Chung FuTao (1977) On the Sinian geochronological scale of China based on isotopic ages for the sinian strata in the Genshan Region, North China. Sci Sin 20(6):13–42

Church WR, Young GM (1970) Discussion of the progress report of Federal-Provincial Committee of Huronian Stratigraphy. Can J Earth Sci 7(3):62–72

Cloud PE, Semikhatov MA (1969) Proterozoic stromatolite zonation. Am J Sci 267(9):1017–1061

Coats RP, Preiss WV (in press) Revised lithostratigraphic correlations of Late Adelaidean sequences in the Kimberley Region, Western Australia. Precambrian Res

Compston W, Arriens PA (1968) The Precambrian geochronology of Australia. Can J Earth Sci 5(3):561–583

Compston W, Grawford A, Bofinger V (1966) A radiometric estimate of the duration of sedimentation in the Adelaide geosyncline, S Australia. J Geol Soc Aust 13(1):229–276

Condie KC (1973) Archean magmatism and crustal thickening. Geol Soc Am Bull 84(9):2981–2992

Condie KC (1975) A mantle plume model for the origin of Archaean greenstone belt based on trace element distribution. Nature (London) 258:413–414

Condie KC, Macke JE, Reimer TO (1970) Petrology and geochemistry of Early Precambrian greywackes from the Fig Tree Group, South Africa. Geol Soc Am Bull 81(9):2759–2776

Cowie JW, Glaessner MF (1975) The Precambrian-Cambrian boundary: a symposium. Earth Sci Rev 11:209–251

Crampton DA (1974) A note on the age of the Matsap Formation of the Northern Cape province. Trans Geol Soc S Afr 77(1):71–72

Crawford AR (1969) Reconnaissance Rb-Sr dating of the Precambrian rocks of Southern Peninsular India. J Geol Soc India 10(2):117–166

Crittenden MD, Peterman ZE (1975) Provisional Rb/Sr age of the Precambrian Uinta Mountain Group, northeastern Utah. Utah Geol 2(1):75–77

Daily B (1976) New data on the Precambrian basement in Southern Australia. Izv Acad Sci U.S.S.R. Geol Ser 3:45–52

Daly RA (1947) The Vredefort ring-structure of South Africa. J Geol Chicago 55(3):125–145

Dana J (1872) Notice of the address of Professor T. Sterry-Hunt. Am. J Sci 3rd Ser 3:338–340

Daniels JL (1966) Revised stratigraphy, paleocurrent system and palaeogeography of Proterozoic Bangemall Group. Geol Surv West Aust Annu Rep 1965:48–56

Daniels JL (1971) Talbot, Western Australia. Sheet SG/52–9. 1:250000. Geol Ser Explan Notes. Geol Surv West Aust, 28 pp

Daniels JL (1974) The geology of the Blackstone Region Western Australia. Geol Surv West Aust Bull 123:257

Darby DG (1974) Reproductive modes of *Huroniospora microreticulata* from cherts of the Precambrian iron-formation. Geol Soc Am Bull 85 (10):1595–1656

Dardenne MA, Costa C (1975) E stromatolites culunares na Série Minas (MG). Rev Bras Geoci 2 (2):99–105

Davies RD, Allsopp HL (1976) Strontium isotopic evidence relating to the evolution of the lower Precambrian crust in Swaziland. Geology 4 (9):553–556

Défosser M (1963) Contribution à l'étude geologique et hydrogéologique de la Boucle du Niger. MBR Geol Mineral 13: 174

De Laeter JR, Blockley JG (1972) Granite ages within the Archaean Pilbara block, Western Australia. J Geol Soc Aust 19:363–370

De Laeter JR, Fletcher IR, Rosman KJR, Williams IR, Gee RD, Libby WG (1981) Early Archean gneisses from the Yilgarn Block, Western Australia. Nature 292 (5821):322–323

Delhal J, Ledent D (1971) Ages U/Pb et Rb/Sr et rapports initiaux du strontium du complexe gabbro-noritique et charnockitique du bouclier du Kasai (Republique democratique du Kongo et Angola) Ann Soc Geol Belg 94 (3):211–221

Delhal J, Ledent D (1975) Données géochronologiques sur le complexe calcomagnésien du Sud-Cameroun. Rapp Annu 1974 Dép Geol Mineral Mus R Afr Cent Tervuren, pp 71–75

De Vries W (1969) Stratigraphy of the Waterberg system of the southern Waterberg area, north-western Transvaal. Annu Geol Surv S Afr 7:48–56

De Waard D, Walton M (1967) Precambrian geology of the Adirondack highlands, a reinterpretation. Geol Rundsch 2:596–610

Dietz PS (1961) Vredefort Ring Structure: meteorite impact scar? J Geol 69:499–516

Dimroth E (1970) Evolution of the Labrador Geosyncline. Geol Soc Am Bull 81 (9):2717–2742

Dimroth E (1972) The Labrador geosyncline revisited. Am J Sci 272 (6):487–506

Dimroth E, Lichtblau A (1978) Oxygen in the Archean ocean. Comparison of ferric ocide crust on archean and cainozoic pillow basalts. Neues Jahrb Mineral Abh 133 (1):1–22

Divakara RaoV, Aswathanarayana U, Qureshy MN (1972) Petrological studies in parts of the Closepet granite pluton, Mysore State. J Geol Soc India 13 (1):1–12

Dobretsov NL (1980) Global petrology. Introduction. Nauka, Novosibirsk, 200 pp

Donaldson JA (1967) Two Proterozoic clastic sequences: a sedimentological comparison. Geol Assoc Can Proc 18:33–54

Dorr JVN (1969) Physiographic, stratigraphic and structural development of the Quadrilatere Ferrifero Mines Gerais, Brazil. US Geol Surv Prof Pap 641 A: 104

Dortman NB (1981) On the petrophysical basis of interpretation of linear magnetic anomalies. Byull LOE Leningr Univ 2:154–159

Dougan TW (1976) Origin of trondhjemitic biotite-quartz-oligoclase gneisses from the Venezuelan Guyana Shield. Precambrian Res 3 (4):317–342

Dow DB, Gemuts I (1969) Geology of the Kimberley region, Western Australia: the East Kimberley. BMR Aust Bull 106:135

Drevin AYa (1967) The results of Precambrian studies in the Middle Near-Bug area by the lithologic-structural method. In: Probl Osad Geol Dokembr, 2 Nauka, Moscow, pp 88–96

Drysdall AR, Johnson RL, Moore TA, Thieme JG (1972). Outline of the geology of Zambia. Geol Mijnbouw 51 (3):265–275

Dunlop JSR, Muir MD, Milne VA, Groves DI (1978) A new microfossil asemblage from Archaean of Western Australia. Nature (London) 274 (5673):676–678

Dunn PR, Plumb KA, Roberts HG (1967) A proposal for time-stratigraphic subdivision of the Australian Precambrian. J Geol Soc Aust 13 (2):593–608

Durney DW (1972) A major unconformity in the Archaean Jones Creek, Western Australia. J Geol Soc Aust 19 (2):251–259

Du Toit AL (1954) The geology of South Africa. Oliver and Boyd, Edinburg, London, 611 pp

Edhorn AS (1973) Further investigations of fossils from the Animikie, Thunder Bay, Ontario. Proc Geol Assoc Can Spec Pap 25:37–66

Eisbacher GH (1978) Re-definition and subdivision of the Rapitan Group, Mackenzie Mountains. Geol Surv Can Pap 77–35: 21 pp

Embleton BJ, Schmidt PW (1979) Recognition of common Precambrian polar wandering: a conflict with plate tectonics. Nature (London) 282:705–707

Emslie RF (1978) Anorthosite massifs, rapakivi granites, and Late Proterozoic rifting of North America. Precambrian Res 7 (1):61–98

Escher A, Watt WS (eds) (1976) Geology of Greenland. Geol Surv Greenl, Copenhagen, 603 pp

Eskola P (1949) The problem of mantled gneiss domes. Q J Geol Soc London 104 (4):461–476

Fawler W (1972) Solar neutrinos-where are they? Nature (London) 238 (5558):24

Fedonkin MA (1980) New representatives of the Precambrian coelenterates in the north of the Russian platform. Paleontol Zh 2:7–15

Fedorov LB, Yegorov LI, Savitsky BE (1979) First occurrence of the oldest trilobites in the lower part of the Tommotian stage of the Lower Cambrian. Dokl Acad Sci U.S.S.R. 249 (5):1188–1190

Fitton MJ, Horwitz RC, Silvester G (1975) Stratigraphy of the early Precambrian in the West Pilbara, W. Australia. Perth Austral. CSIRO Min Res Lab EPI, 30 pp

Folinsbee R (1972) The Precambrian metallogenic epochs- are they atmospheric or centrospheric? In: Ocherki Sovr Geokh Analit Khimii. Nauka, Moscow, pp 253–262

Ford TD (1979) The history of the study of the Precambrian rocks of Charnwood Forest, England. Geol Assoc Can Spec Pap 19:65–80

Ford TD, Breed WJ (1973) Late Precambrian Chuar Group, Grand Canyon, Arisona. Geol Soc Am Bull 84 (4):1243–1260

Fraser JA, Donaldson JA, Fahring WF, Tremblay LP (1970) Helikian basins and geosynclines of the Northwestern Canadian Shield. Geol Surv Can Pap 70–40:213–238

French BM (1968) Sudbery structure, Ontario: some petrographic evidence for an origin by meteorite impact. In: Shock metamorphism of natural material, Mono Press Baltimore, pp 1094–1098

Frith RA, Doig R (1975) Pre-Kenoran tonalitic gneisses in the Grenville Province. Can J Earth Sci 12:844–849

Frith RA, Frith R, Helmstaedt H, Hill J, Leatherbarrow R (1974) Geology of the Indin Lake area (86 B), District of Mac Kenzie. Geol Surv Can Pap 74–1A: 167–171

Fritz WH (1980) International Precambrian-Cambrian Boundary Working Group 1979. Field Study to Mackenzie Mountains, Northwest Territories, Canada. Geol Surv Can Pap 1A:41–45

Furday RS (1968) Tillites in the Late Precambrian of Pri-Kolyma area. Dokl Acad Sci U.S.S.R. 180 (4):948–951

Fyfe WS (1973) The granulite facies, partial melting and Archaean crust. Philos Trans R Soc London Ser A 273:457–462

Gamaleya YuN (1968) Absolute age of granitoids from the Ulkan pluton. Izv Acad Sci U.S.S.R. Geol Ser 2:35–40

Gamaleya YuN, Losev AG, Popov MYa (1969) The oldest cover formations in the southeastern part of the Siberian platform. Sov Geol 4:137–144

Garan' MI (1946) The age and environment of the older formations in the western slopes of the Southern Urals. Gosgeoltekhizdat, Moscow, 49 pp

Garan' MI (1963) Western slope and central zone of the Southern Urals. In: Stratigraphiya U.S.S.R. Verkhny Dokembry. Gosgeoltekhizdat, Moscow, pp 114–161

Gaudette HE, Mendoza V, Hurley PM, Fairbairn HW (1978) Geology and age of the Parguaza rapakivi granite, Venezuela. Geol Soc Am Bull 89 (9):1335–1340

Gemuts I, Theron AC (1975) The Archaean between Coolgardie and Norseman-stratigraphy and mineralization. In: Economic Geology of Australia and Papua New Guinea Knight CL (ed) Aust Inst Min Metall Melbourne p. 66–74

Geologic map of the Republic of South Africa and Kingdoms of Lesoto and Swaziland 1:1,000,000 (1970). Geol Surv S Afr, Pretoria

Gerling EK, Lobach-Zhuchenko SB (1967) The radiological methods: their status and application in Precambrian mapping with reference to Karelia. In: Problemy Izucheniya Geologii Dokembriya. Nauka, Leningrad, pp 36–47

Gerling EK, Lobach-Zhuchenko SB, Borisenko NF (1966) New data on Jotnian absolute age determinations in the Baltic Shield. Dokl Acad Sci U.S.S.R. 166 (3):674–677

Gerling EK, Iskanderova AD, Levchenkov OA, Mikhaylov DA (1970) On age of marble from the Dzheltula and Iyengra Groups of Aldan according to U-Pb isochron method. Dokl Acad Sci U.S.S.R. 194 (6):1397–1400

Germs GJ (1972) Trace fossils from the Nama Group, South-West Africa. J Paleontol 46 (6): 864–870

Germs GJ (1974) The Nama group in South-West Africa and its relationship to the Pan-African geosyncline. J Geol 82(3):301–317

Gibbins WA, Adams CY, McNutt RH (1972) Rb-Sr isotopic studies of the Murray granite. Geol Assoc Can Spec Pap 10:61–66

Gilluly J (1949) Distribution of mountain building in geological time. Geol Soc Am Bull 60(4):561–590

Gilyarova MA (1964) Crusts of weathering and the Lammas conglomerates in the Pechenga area. Vestn Leningr Univ 6:22–30

Glaessner MF (1961) Pre-Cambrian animals. Sci Am 204(3):72–78

Glaessner MF (1978) Re-examination of *Archaeichnium*, a fossil from the Nama Group. Ann S Afr Mus 74(13):335–342

Glikson AY (1971) Archaean geosynclinal sedimentation near Kalgoorlie, Western Australia. Geol Soc Aust Spec Publ 3:443–460

Glikson AY (1976a) Earliest Precambrian ultramafic-mafic volcanic rocks: ancient oceanic crust or relic terrestrian maria? Geology 4:201–205

Glikson AY (1976b) Stratigraphy and evolution of primary and secondary greenstones; significance of data from shields of Southern Hemisphere. In: Windley BF (ed) The early history of the Earth. Wiley, London, pp 257–277

Glikson AY (1978) On the basement of Canadian greenstone belts. Geosci Can 5(1):3–12

Glikson AY (1979) Early Precambrian tonalite-trondjemite sialic nuclei. Earth-Sci Rev 15(1):1–74

Goldich SS, Hedge CE (1974) 3800-Myr granitic gneiss in south-western Minnesota. Nature (London) 252(5483):467–468

Goldich SS, Nier AO, Baadsgaard H, Hoffman IH, Krueger HW (1961) The Precambrian geology and geochronology of Minnesota. Minn Geol Surv Bull 41:18–34

Golivkin NI (1967) Precambrian stratigraphy of Starooskol and Novooskol iron-ore deposits in the KMA area. In: Geol Pol Iskop KMA. Nedra, Moscow, pp 60–75

Golubic S, Hofmann H (1976) Comparison of Holocene and mid-Precambrian *Entophysalidaceae* (Cyanophyta) in stromatolitic algal mat: cell division and degradacion. J Paleontol 50(6):1047–1082

Goodwin AM (1966) The relationship of mineralization to stratigraphy in the Michipicoten area, Ontario. Geol Assoc Can Spec Pap 3:57–73

Goodwin AM (1976) Giant impacting and the development of continental crust. In: Windley BF (ed) The early history of the Earth. Wiley, London, pp 77–95

Goodwin AM (1977) Archean volcanism in Superior Province, Canadian Shield. In: Baragar WR (ed) Volcanic regimes in Canada. Geol Assoc Can Spec Pap 16:205–241

Gordani UG, Iyer SS (1979) Geochronological investigation on the Precambrian granulitic terrain of Bahia, Brazil. Precambrian Res 9(3–4):255–274

Gor'kovets VYa, Inina KA, Rayevskaya MB (1976) On stratigraphy of the Gimola Group in the area of the Kostomuksha iron – ore deposit. In: Mater Inst Geol Karel Filial Acad Sci U.S.S.R., Petrozavodsk, pp 10–17

Grandstaff DE (1980) Origin of uraniferous conglomerates at Elliot Lake, Canada and Witwatersrand, South Africa: implications for oxygen in the Precambrian atmosphere. Precambrian Res 80(1):1-26

Gravelle M, Letolle R (1963) Sur l'age apparent de deux galenes du précambrien de la region de Silet (Ahaggar occidental, Sahara central). C R Somm Soc Geol Fr 2:45–46

Green DH (1972) Archean greenstone belt may include terrestrial equivalent of lunar area? Earth Planet Sci Lett 15(2):263–270

Greene B, Williams H (1974) New fossil localities and the base of the Cambrian in Southeastern Newfoundland. Can J Earth Sci 11(2):319–323

Grieve RAF (1980) Impact bombardment and its role in proto-continental growth of the Early Earth. Precambrian Res 10:217–247

Grieve RAF, Robertson PB (1979) The terrestrial cratering record-I. Current status of observations. Ikarus 38:212–229

Griffin WL, Taylor PN, Hakkinen JW, Heier KS (1978) Archaean and Proterozoic crustal evolution in Lofoten-Vesterolen, N. Norway. J Geol Soc London 135(6):629–647

Griffin WL, McGregor VR, Nutman A, Taylor PN, Bridgwater D (1980) Early Archaean gra-
    nulite facies metamorphism south of Ameralik, West Greenland. Earth Planet Sci Lett 50 (1):
    59–74
Grinenko VA, Grinenko LN (1974) Geochemistry of sulphur isotopes. Nauka, Moscow, 271 pp
Grout FF, Gruner IW, Schwartz GM, Thiel GA (1951) Precambrian stratigraphy of Minnesota.
    Geol Soc Am Bull 62 (107):1017–1078
Gunning HC, Ambrose JW (1939) The Timiskaming-Keewatin problem in the Rouyn-Harri-
    kanaw region, northwestern Quebec. Trans R Soc Can 33 (Sect IV):19–47
Gunning HC, Ambrose JW (1940) Malartic area, Quebec. Geol Surv Can Mem 1940, N 222:
    129 pp
Hahn-Weinheimer P (1971) Die Verwendung von stabilen Kohlenstoffisotopen-verhältnissen
    zur Diskussion der Entstehung Kohlenstoffverbindungen. Geol Rundsch 60 (4):1384–1392
Halligan R (1963) The Proterozoic rock of Western Tanganyika. Geol Surv Tanganyika Bull
    34: 34
Hamblin WK (1961) Paleogeographic evolution of the Lake Superior region from Late Keewee-
    nawan to Late Cambrian time. Geol Soc Am Bull 72 (1):1–18
Hanson GN, Himmelberg GR (1967) Ages of mafic dikes near Granite Falls, Minnesota. Geol
    Soc Am Bull 78 (11):1429–1432
Hanson GN, Goldich SS, Arth JG, Yardley DH (1971) Age of the Early Precambrian rocks of
    the Saganaga Lake-Northern Light Lake area, Ontario-Minnesota. Can J Earth Sci
    8 (9):1100–1121
Harrison JE (1972) Precambrian Belt basin of Northwestern United States: its geometry, sedi-
    mentation and copper occurrences. Geol Soc Am Bull 83 (5):1215–1240
Haughton SH (1969) Geological history of southern Africa. Geol Soc S Afr, Cape Town, 535 pp
Hawkesworth CJ, Moorbath S, O'Nions RK, Wilson JF (1975) Age relationships between
    greenstone belt and "granites" in the Rhodesian Archaean craton. Earth Planet Sci Lett
    25:251–262
Hedberg RM (1979) Stratigraphy of the Ovamboland basin South West Africa. Univ Cape
    Town Chamber Mines Precambrian Res Unit Bull 24: 325 pp
Heier KS (1964) Rubidium/strontium and strontium-87/strontium-86 ratio in deep crustal ma-
    terial. Nature (London) 202 (4931): 477–478
Henderson JB (1975) Archean stromatolites in the Northern Slave Province, Northwest Territo-
    ries, Canada. Can J Earth Sci 12 (9):1619–1630
Henderson JB, Easton RM (1977) Archean supracrustal-basement rock relationships in the
    Kesharrah Bay map-area, Slave structural province, District of Mackenzie. Geol Surv Can
    Pap Rep Activ 77–1A:217–221
Hickman AH, Laeter de JR (1977) The depositional environment and age of a shale within the
    Hardey Sandstone of the Fortescue group. Geol Surv West Aust Annu Rep 1976: 62–68
Hickman MH (1974) 3500 Myr old granite in Southern Africa. Nature (London) 251 (5473):
    295–296
Hickman MH, Lipple SL (1975) Explanatory notes on the Marble Bar, 1:250,000 geological
    sheet, Western Australia. Geol Surv West Aust Rep 1974: 20
Hietanen A (1967) Scapolite in the Belt series in the St. Joe-Clearwater region, Idaho. US Geol
    Surv Spec Pap 86: 56
Higgins AK (1970) The stratigraphy and structure of the Ketilidian rocks of Midternaes, South-
    West Greenland. Groenl. Unders Bull 87: 96
Hjelmqvist S (1966) Beskrivning till berggrundskartta över Kopparbergs län (English sum-
    mary). Sver Geol Unders 40: 26
Hoffman PF (1968) Precambrian stratigraphy, sedimentology, palaeocurrents and palaeoeco-
    logy in the East Arm of Great Slave Lake, district of Mackenzie (75L). Geol Surv Can Pap
    68–1A:140–142
Hoffman PF (1973) Aphebian supracrustal rocks of the Athapuscow aulacogen, East arm of
    Great Slave Lake, district of Mackenzie. Geol Surv Can Pap 73–1A:151–156
Hoffman PF (1980) On the relative age of the Muskox intrusion and the Coppermine River
    basalts, District of Mackenzie. Geol Surv Can Pap 80–1A:223–225
Hoffman PF, Bell IR, Hildebrand RS, Thorstad L (1977) Geology of the Athapuscow aulaco-
    gen, East Arm of Great Slave Lake, District of Mackenzie. Geol Surv Can Rep Activ 77–1A:
    117–129

Hofmann HJ (1969) Stromatolites from the Proterozoic Amikimie and Sibley Groups, Ontario. Geol Surv Can Pap 68–69: 77

Hofmann HJ (1971) Precambrian fossils, pseudofossils and problematica in Canada. Geol Surv Can Bull 189: 146

Hofmann HJ (1976) Precambrian microflora of Belcher Islands, Canada: significance and systematics. J Paleontol 50 (6):1040–1073

Hofmann HJ (1977) Aphebian stromatolites and Riphean stromatolite stratigraphy. Precambrian Res 5 (2):175–205

Holland HD (1972) The geologic history of see water- an attempt to solve the problem. Geochim Cosmochim Acta 36 (6):637–651

Holubec J (1972) Lithostratigraphy, structure and deep crustal relations of Archean rocks of the Canadian Shield, Rouyn-Noranda area, Quebec. Krustalinikum 9:63–89

Horodyski RJ, Donaldson A (1980) Microfossils from the Middle Proterozoic Dismal Lakes Group, Arctic Canada. Precambrian Res 11 (2):125–159

Hottin G (1972) Geochronologie et stratigraphie Malgaches. Essai d'interpretation. Doc Bur Geol Rep Malagasy 182: 16

Hottin G (1976) Presentation et essai d'interpretation du Precambrien de Madagascar. Bull BRGM 4 (2):117–153

Houston RS (1968) A regional study of rocks of Precambrian age in the part of the Medicine Bow Mountains in southwestern Wyoming. Wyo Geol Surv Mem 1: 167

Houzay JP (1979) Empreintes attribuables à des méduses dans la série de base de Adoudounien (Précambrien terminal de l'Anti-Atlas, Maroc) Geol. Mediter 6 (3):379–384

Hunter DR (1974 a) Crustal development in the Kaapvaal craton. I. The Archean. Precambrian Res 1 (4):259–294

Hunter DR (1974 b) Crustal development in the Kaapvaal craton. II. The Proterozoic. Precambrian Res 1 (4):295–326

Hunter DR, Barker F, Millard HT (1978) The geochemical nature of the Archean Ancient Gneiss Complex and Granodiorite Suite, Swaziland: a preliminary study. Precambrian Res 7 (2):105–127

Hurley PM, Leo GW, White RW, Fairbairn HW (1970) The Liberian age province (ca 2700 my) and adjacent provinces in Liberia and Sierra Leone. In: Hurley PM (ed) Variation in isotopic abundances of strontium, calcium and argon and related topics. Mass Inst Technol 18th Annu Prog Rep Atomic Energy Comm 1381–18: 17–37

Hurley PM, Kalliokoski J, Fairbairn HW, Pinson HW (1972 a) Progress report in the age of granulite facies rocks in the Itamaka Complex, Venezuela. Abstr Proc 9 Interguayanas Geol Conf Ciudad Bolivar 10: 82

Hurley PM, Pinson WHJr, Nagy B, Teska TM (1972 b) Ancient age of the Middle Marker Horizon: Onverwacht Group, Swaziland Sequence, South Africa. Earth Planet Sci Lett 14:360–366

Hurley PM, Fairbairn HW, Gaudette HE (1976) Progress report on the early Archaean rocks in Liberia, Sierra Leone and Guyana. In: Windley BF (ed) Early history of the Earth. London, Wiley, pp 511–521

Hurst RW, Bridgwater D, Collerson KD, Wetherill GW (1975) 3600 my Rb-Sr ages from Saglek Bay, Labrador. Earth Planet Sci Lett 27 (3):393–403

Ingram PAJ (1977) A summary of the geology of a portion of the Pilbara Goldfield, Western Australia. In: McCall GJ (ed) Archean. Search for the beginning. Dowden, Stroudsburg, pp 208–216

Iskanderova AD, Mirkina SL, Neimark LA (1977) New data on studies of metamorphic rocks and gneiss-granites in the Stanovoy area of the Aldan shield by Pb-method. In: Geol Interpret Dannych Geokhronol (summaries), Irkutsk, pp 24–25

Iskanderova AD, Neymark LA, Polevaya NI (1978) Correlation of certain Proterozoic carbonate strata of the East-European platform according to data of Pb-isochron method. In: Geokhronol Vost Eur Platformy Sochleneniya Kavkaz-Karpat Sist. Nauka, Moscow, pp 190–194

Ivantishin MN, Orsa VI (1965) The gneiss-migmatite formations and granites in Zaporozh'ye-Mishurin Rog region. In: Geokhronologiya Dokembriya Ukrainy. Naukova Dumka, Kiev, pp 26–38

Jacobson KE (1966) The problem of the Proterozoic-Paleozoic boundary in the western part of the Russian platform. Izv Acad Sci U.S.S.R. Geol Ser 7: 88–106

Jacobson KE, Krylov NS (1977) The Vendian lower boundary in its stratotypical locality. Sov Geol 7:59–70

Jahn B, Shih C (1974) On the age of the Onverwacht Group, Swaziland Sequence, South Africa. Geochim. Cosmochim Acta 38 (6):873–886

James HL (1958) Stratigraphy of pre-Keweenawan rocks in parts of Northern Michigan. US Geol Surv Prof Pap 314–C:27–44

James HL (1972) Subdivision of Precambrian: an interim scheme to be used by U.S. Geological Survey. Am Assoc Petrol Geol Bull 56 (2):1128–1133

James HL (1978) Subdivision of the Precambrian – a brief review and a report on recent decisions by the Subcommission on Precambrian Stratigraphy. Precambrian Res 7(3): 193–204

James PR (1976) Deformation of the Isua block, West Greenland: a remnant of the earliest stable continental crust. Can J Earth Sci 13(6):814–823

Jayaram S, Vekatasubramanian VS, Radhakrishna BP (1976) Rb-Sr ages of cordierite-gneisses of southern Karnataka. J Geol Soc India 17(4):557–561

Johnson RC, Hills FA (1976) Precambrian geochronology and geology of the Boxelder Canyon area, northern Laramie Range, Wyoming. Geol Soc Am Bull 87(5):809–817

Jolliffe AW (1966) Stratigraphy of the Steeprock Group, Steep Rock Lake, Ontario. Geol Assoc Can Spec Pap 3:75–98 (Precambrian Symp)

Jolly WT (1978) Metamorphic history of the Archean Abitibi Belt. Geol Surv Can Pap 78–10: 63–77

Jourde G (1972) Essai de synthese structurale et stratigraphique du Precambrien Malgache. CR Geol Madagascar (1971), p 59–69

Kalsbeek F, Zeck HP (1978) Nagssugtoqidian deformation and Kangamiut dyke intrusion in Søndre Strømfiord area, West Greenland. Geol Surv Greenl Rep (Rep Activ 1977) 90: 42

Kalyaev GI (1965) The Precambrian tectonics of the Ukrainian iron-ore formation province. Naukova Dumka, Kiev, 190 pp

Kao Chen-hsi (1962) Preliminary studies of the Sinian stratigraphy in Northern China. In: The Oldest Rocks in China. Inostranaya Literatura, Moskow, pp 39–69 (translated from the Chinese)

Kargat'yev VA (1970) Anhydrite in diopside rocks of Central Aldan area. In: Miner Syr'e 22: 65–74

Karig DE (1971) Origin and development of marginal basins in the western Pacific. J Geophys Res 76(11):2542–2561

Karlstrom KE, Houston RS (1979) Stratigraphy of the Phantom Lake Metamorphic Suite and Deep Lake Group and a review of the Precambrian tectonic history of the Medicine Bow Mountains. Contrib Geol 2:111–133

Kashintsev GL (1979) Magmatic and metamorphic rocks of the ocean bottom. Metamorphic rocks. In: Okeanologiya, geologiya okeana. Osadkoobrazovaniye i magmatism okeana. Nauka, Moscow, pp 60–69

Kay M (1956) Sediments and subsidence through time. Geol Soc Am Bull Spec Pap 62: 665–684

Kazansky YuP (1978) Remarks on the paper of V.I. Vinogradov, T.O. Reymer, A.M. Leytes, S.B. Smelov "The oldest sulphate in the Archean formations of South-African and Aldan shields and evolution of oxygenic atmosphere of the Earth". Litol Polezn Iskop 1:169–172

Kazansky YuP, Katayeva VN, Shugurova NA (1973) Composition of the older atmospheres according to studies of gas inclusions in quartz rocks. In: Geokhimiya Dokembriyskikh i Paleozoyskikh Otlozheniy Sibiri, Novosibirsk, pp 5–12

Kaźmierczak J (1979) The eukaryotic nature of *Eosphaera*-like ferriferous structures from the Precambrian Gunflint Iron Formation, Canada: a comparative study. Precambrian Res 9(1–2):1–22

Keller BM (1979) Puzzles of the Upper Precambrian. Priroda 1:66–75

Keller BM, Shul'ga PL (eds) (1978) Explanatory note to the scheme of the Upper Precambrian of the Russian platform. Acad Sci Ukrainian U.S.S.R., Kiev, 36 pp

Keller BM, Aksenov EM, Korolev BG (1974) Vendomian (Terminal Riphean) and its regional subdivisions. VINITI, Moscow, 126 pp

Khain VE (1977) Cyclicity and tectonics. In: Osnovn. Teoret. Voprosy Tsiklichn. Sedimentoge-
neza. Nauka, Moscow, pp 213–221

Khain VE (1978) From plate tectonics towards a more general global tectonogenesis. Geotektonika 3:3–25

Kharitonov LYa (1966) Structure and stratigraphy of the Karelides in the eastern part of the
Baltic shield. Nedra, Moscow, 358 pp

King AF (1977) Subdivision and paleogeography of late Precambrian and early Paleozoic rocks
in the Avalon Peninsula. Newfoundland. Geol Soc Am 12th Annu Meet Abstr Progr
9(3):284

Kirichenko GI (1967) Precambrian stratigraphy of the western margin of the Siberian platform
and surrounding foldbelts. Tr VSEGEI 112:3–48

Kiselev AS (1977) New data on stratigraphic relationships of the Krivoy Rog and Ingulets
Groups. Geokhim Rudoobras 6:46–54

Kloosterman JB (1975) Roraima, Tafelberg and Uatama formations of the Guiana shield. A
correlation. Geol Mijnbouw 54(1–2):55–60

Knaut LP, Epstein S (1976) Hydrogen and oxygen isotope ratios in modular and bedded cherts.
Geochim Cosmochim Acta 40(9):1095–1108

Knaut LP, Lowe DR (1978) Oxygen isotope geochemistry of cherts from the Onwerwacht
Group (3,4 b.y.), Transvaal, South Africa, with implications for secular variations in the
isotopic composition of cherts. Earth Planet Sci Lett 41(2):209–222

Knoll AH, Barghorn ES (1977) Archean microfossils showing cell diversion from Swaziland
System of South Africa. Science 198(4315):396–398

Knoll AH, Barghorn ES, Awramik SM (1978) New microorganisms from the Aphebian Gunflint Iron Formation, Ontario. J Paleontol 52(5):976–992

Knorre KT, Nikolayev SD, Kats AK (1970) On geologic structure and absolute age of the
northern part of surrounding areas of the Aldan shield. Byull Kom Opred Absol Vozrasta
9:104–110

Komar VA, Semikhatov MA, Serebryakov SN, Voronov BG (1970) New data on the Riphean
stratigraphy and evolution in the southeastern Siberia and North-East U.S.S.R. Sov Geol
3:37–53

Konikov AZ, Travin LV, Shalek EA (1974) Petrochemistry of metabasites and basaltoid volcanism in the Archean of Eastern Siberia. In: Problemy Dokembriyskogo Magmatisma.
Nauka, Leningrad, pp 225–231

Konikov AZ, Travin LV, Shalek EA (1975) Lithology and formational peculiarities of the Archean strata in the south of Eastern Siberia. Probl Osad Geol Dokembr 1:265–271

Krasil'shchikov AA (1973) Stratigraphy and paleotectonics in Precambrian-Early Paleozoic in
Spitsbergen. Nedra, Leningrad, 120 pp

Krasnobaev AA (1980) Results and problems of geochronologic studies in the Urals. In: Door-
dovikskaya Istoriya Urala. Izd Urals Sci Cent Acad Sci U.S.S.R. (Sverdlovsk), pp 28–39

Krasnobaev AA, Bibikova EV, Stepanov AI (1979) Geochronology of the Berdyaush massif.
In: Izotophaya Geokhronologiya Dokembriya (Summaries). Ufa, Bashkir Filial Acad Sci
U.S.S.R., pp 7–8

Krats KO (1963) Karelides geology in Karelia. Izd Acad Sci U.S.S.R., Moscow, 210 pp

Krats KO, Levchenkov OA, Ovchinnikova GV (1976) Age boundaries of the Jatulian complex.
Dokl Acad Sci U.S.S.R. 231(5):1191–1195

Kreuzer H, Harre W, Kürsten M, Schnitzer WA, Murti KS, Srivastava NK (1977) K/Ar dates
of two glauconites from the Chandarpur-series (Chhattisgarh/India). Geol Jahrb 28:23–36

Krishnan MS (1960a) Geology of India and Burma. Higginbotham, Madras, 604 pp

Krishnan MS (1960b) Precambrian stratigraphy of India. Int Geol Congr Copenhagen 1960
Rep 21:95–107

Krogh TE, Davies GL, Harris NBW, Ermanovics IF (1975) Isotopic ages in the eastern Lac
Seul region of the English River gneiss belt. Carnegie Inst Annu Rep Dir Geophys Lab
1974–1975, Washington, pp 623–625

Kröner A (1971) Late-Precambrian correlation and the relationship between the Damara and
Nama Systems of South West Africa. Geol Rundsch 60(4):1513–1523

Kröner A (1976) Proterozoic crustal evolution in parts of southern Africa and evidence for
extensive sialic crust since the end of the Archaean. Philos Trans R Soc London
280(1296):541–553

Kröner A (1977) The Precambrian geotectonic evolution of Africa: plate accretion versus plate destruction. Precambrian Res 4 (2):163–213

Kropotkin PN (1964) Relationships of the surficial and abyssal structures and a general characteristic of movement of the Earth's crust. In: Stroyeniye i Razvitiye Zemnoy Kory. Nauka, Moscow, pp 72–96

Kropotkin PN (1970) A possible role of cosmogenic factors in geotectonics. Geotectonika 2:30–46

Kropotkin PN (1972) Tectonic processes in island-arches of the Far East and their age. In: Zemnaya Kora Ostrovnykh Dug i Dal'nevostochnykh Morey. Nauka, Moscow, pp 51–68

Kropotkin PN (1980) Problems of geodynamics. In: Tektonika v Issledovaniyakh Geol. Instituta Acad Sci U.S.S.R. Nauka, Moscow, pp 176–247

Krylov IN (1963) Riphean columnar branching stromatolites of the Southern Urals and their stratigraphic value in the Upper Precambrian. Tr GIN Acad Sci U.S.S.R. 69: 133 pp

Krylov IN (1966) On the columnar stromatolites of Karelia. In: Ostatki Organizmov i Problematiki v Proterozoyskikh Obrazovaniyakh Karelii. Petrozavodsk, Karel'sk. Knizhn. Izdatel'stvo, pp 97–100

Kulp JL, Engels J (1963) Discordances in K-Ar and Rb-Sr isotopic ages. In: Radioactive dating. Publ by Int Atomic Energy Agency, Vienna, pp 219–238

Kunin NYa, Sardonnikov NM (1976) Global cyclicity of tectonic movements. Byull Mosk Ova Ispyt Prir Otd Geol 3:5–27

Ladieva VD (1965) The sedimentary-volcanogenic formations in Konsk-Belozersk zone. In: Geokhronologiya Dokembriya Ukrainy. Naukova Dumka, Kiev, pp 16–25

Lanphere MA (1968) Geochronology of the Yavapai Series of central Arizona. Can J Earth Sci 5 (3):757–762

Larin VN (1980) Hypothesis of the initial hydride composition of the Earth, 2nd edn. Nedra, Moscow, 216 pp

Larsen HC (1978) Offshore continuation of East Greenland dyke swarm and North Atlantic ocean formation. Nature (London) 274 (5668):220–223

Latulippe M (1966) The relationship of mineralization to Precambrian stratigraphy in the Matagami Lake and Val d'Or districts of Quebec. Geol Assoc Can Spec Pap 3:21–42 (Precambrian Symp)

Lauren L (1970) An interpretation of the negative gravity anomalies associated with the rapakivi granites and the Jotnian sandstone in Southern Finland. Geol Foeren Stockholm Foerh 92 Pt 1 (540):21–34

Lawson AC (1913) A standart scale for the pre-Cambrian rock of North America. In: Proc Int Geol Congr 12 Sess, pp 349–370

Leake BE, Farrow CM, Townend R (1979) A pre-2000 Myr old granulite facies metamorphosed evaporite from Caraiba, Brazil? Nature (London) 277 (5691):49–50

Leggo PJ, Compston W, Trendal AF (1965) Radiometric ages of some Precambrian rocks from the Northwest division of Western Australia. J Geol Soc Aust 12 (1):53–65

Leith CK, Lund RJ, Leith A (1935) Precambrian rocks of the Lake Superior region. US Geol Surv Prof Pap 184: 34

Lelubre M (1969) Chronologic du Précambrien au Sahara Central. Geol Assoc Can Spec Pap 5:27–32

Leo GW (1967) Age investigation in Liberia. 15th Annu Rep 1967 Dep Geol Geophys. Mass Inst Technol, Cambridge USA, pp 1–5

Leo GW, Cox DP, Carvalho JPP (1964) Geology of the southern part of Serra de Jacobina. Bol Div Geol Miner Dip Nat Prod Mineral 209:9–120

Lipsky YuN (1969) A complete map of the Moon. Nauka, Moscow

Litsarev MA, Vinogradov VI, Kuleshov VN (1977) On salinity of the Early Precambrian strata of the Vakhan Group (South-Western Pamir). Dokl Acad Sci U.S.S.R. 234 (6):1425–1428

Livingston DE, Damon PE (1968) The ages of stratified Precambrian rocks sequences in Central Arizona and northern Sonora. Can J Earth Sci 5 (3):763–772

Lobach-Zhuchenko SB, Krats KO, Gerling EK (1972) Geochronologic boundaries and geologic evolution of the Baltic shield. Nauka, Leningrad, 194 pp

Lovering JF (1979) The evidence for ∼ 4000 m.y. crustal material in Archaean times. J Geol Soc Aust 26 (5–6):268

Lowell GR, Sides JR (1973) The occurrence and origin of rapakivi granite in the St. Francois Mountain batholith of southeast Missouri. Geol Soc Am Annu Meet Abstr Progr (North-Central Sect) 5(4):332–333

Lowenstam HA (1980) What, if anything, happened at the transition from the Precambrian to the Phanerozoic? Precambrian Res 11(2):89–91

Lü Hung-yün, Sha Tzu-an (1965) Boundaries, classification, and paleogeography of Sinian strata in Southern China. Dichzhi Kesyue 4:12–60 (translated from the Chinese)

Lyubimova EA (1960) The Earth heating. In: Geologich. Resultaty Prikladn. Geokhimii i Geofisiki. Gosgeoltekhizdat, Moscow, pp 14–19

Macgregor AM (1937) The geology of the country around Hunters Road, Gwelo District. Bull Geol. Surv S Rhodesia 31: 78

Macgregor AM (1951) Some milestones in the pre-Cambrian of Southern Rhodesia. Geol Soc S Afr Trans 54:27–74

Machado F (1967) Geological evidence for a pulsating gravitation. Nature (London) 214(5095): 317–318

Machens E (1973) Contribution à l'etude des formations du socle cristallin et de la couverture sedimentaire de l'Ouest de Republique du Niger. Mem BRGM 82: 168

Mac Mannis WS (1964) La Hood formation – a coarse facies of the Belt series in Southwestern Montana. Geol Soc Am Bull 74(4):707–736

Manton MJ (1965) The orientation and origin of shatter cones in the Vredefort ring. Ann NY Acad Sci 123(2):1017–1049

Manuylova MM (ed) (1968) Geochronology of Precambrian strata in the Siberian platform and surrounding fold-belts. Nauka, Leningrad, 331 pp

Manuylova MM, Sryvtsev NA (1974) The Primorian complex of rapakivi-granites (Western Baikal region). In: Problemy Dokembriyskogo Magmatisma. Nauka, Leningrad, p 174–180

Markhinin EK, Podkletnov PE (1978) Hydrocarbons and other composite organic compounds in volcanic products. Geol Geofis 12:21–32

Martin FC (1974) Paleotectonica del Esendo de Guayana. Venezuela Minist Minas e Hydrocarburos Spec Publ p 12–18

Martin H (1965) The Precambrian geology of South West Africa and Namaqualand. Univ Cape Town, 159 pp

Martini JEJ (1978) Coesite and stishovite in the Vredefort Dome, South Africa. Nature (London) 272(5655):715–717

Masaytis VL (1981) Vredefort and Sudbury astroblems. In: Geologiya astroblem. Nedra, Leningrad, pp 149–154

Maslenikov VA (1968) Precambrian absolute geochronology in the eastern part of the Baltic shield. In: Geologiya i Glub. Stroyeniye Vost. Chasti Balt. Shchita. Nauka, Moscow-Leningrad, pp 60–77

Mason TR, Brunn von C (1977) 3-Gyr old stromatolites from South Africa. Nature (London) 266(5596):47–49

Matthews PE, Scharrer RH (1968) A graded unconformity at the base of the Early Pre-Cambrian Pongola System. Trans Geol Soc S Afr 71(3):257–272

McCartney WD (1967) Whitbourne map-area, Newfoundland. Geol Surv Can Mem 341: 135

McDougall J, Dunn PR, Compston W, Webb AW, Richards JR, Bofinger VM (1965) Isotopic age determinations on Precambrian rocks of the Carpentaria region, Northern territory, Australia. J Geol Soc Aust 12(1):69–80

McGregor VR, Mason B (1977) Petrogenesis and geochemistry of metabasaltic and sedimentary rocks enclaved in the Amitsoq gneisses, West Greenland. Am Mineral 62:887–904

Mel'nik YuP (1973) Genesis of the Precambrian ferruginous quartzites (an accumulative-biochemical variant of the volcanogenic-sedimentary hypothesis). Geol Zh 33(4):3–17

Mendelsohn F (ed) (1961) The geology of the Northern Rhodesian Copperbelt. McDonaly, London, 523 pp

Mendoza V (1974) Geologia del area Rio Suapure, parte noroccidental del Escudo de Guayana, estado Bolivar, Venezuela. Bol Geol Publ Espec 6:306–338

Michard-Vitrac A, Lancelot J, Allegre CJ, Moorbath S (1977) U-Pb ages on single zircons from the Early Precambrian rocks of West Greenland and the Minnesota River Valley. Earth Planet Sci Lett 35(1):449–453

Milanovsky EE (1978) Certain regularities in tectonic evolution and volcanism of the Earth during Phanerozoic (pulsations and expansion of the Earth). Geotektonika 6:3–16

Miller FK, McKee EH, Yates RG (1973) Age and correlation of the Windermere Group in northeastern Washington. Geol Soc Am Bull 84(11):3723–3729

Miller RMG (1980) Geology of a portion of Central Damaraland, South West Africa (Namibia) Geol Surv Rep S Africa 6: 78

Misharev DT, Amelandov AS, Zakharchenko AI, Smirnova VS (1960) Stratigraphy, tectonics and pegmatite occurrences in the north-western part of the White Sea region. Tr VSEGEI 31: 109 pp

Mladshikh SV, Ablizin BD (1967) Stratigraphy of the Upper Precambrian in the western slopes of the Middle Urals. Izv Acad Sci U.S.S.R. Geol Ser 2:67–80

Monod T (1952) L'Adrar mauritanien (Sahara occidental). Bul Dir Mines Afrique Occidental (France) 1(15); 284

Moorbath S, O'Nions RK, Pankhurst RJ (1972) Further rubidium-strontium age determinations on the very Early Precambrian rocks of the Godthaab district, West Greenland. Nature (London) Phys Sci 240(100):78–82

Moorbath S, O'Nions RK, Pankhurst RJ (1973) Early Archaean age for the Isua Iron Formation, West Greenland. Nature (London) 245(5421):38–139

Moorbath S, O'Nions RK, Pankhurst RJ (1975) The evolution of Early Precambrian crustal rocks at Isua, West Greenland – geochemical and isotopic evidence. Earth Planet Sci Lett 27(2):229–239

Moorbath S, Wilson JF, Cotterill P (1976) Early Archaean age for the Sebakwian group at Selukwe, Rhodesia. Nature (London) 264(5586):536–538

Moorbath S, Allaart JH, Bridgwater D, McGregor VR (1977a) Rb-Sr ages of early Archean supracrustal rocks and Amitsoq gneisses at Isua. Nature (London) 270(5632):43–45

Moorbath S, Wilson JF, Goodwin R, Humm M (1977b) Further Rb-Sr age and isotope data on Early and Late Archaean rocks from the Rhodesian craton. Precambrian Res 5(3):229–239

Moralev VM, Roshkova GP, Cheshikhina KG (1979) The Middle Proterozoic continental volcanic belts of the Gondwana platforms. Geol Razvedka 2:15–24

Morey GB (1973) Stratigraphic framework of Middle Precambrian rocks in Minnesota. Geol Assoc Can Spec Pap 12:211–249

Morton RD (1974) Sandstone-type uranium deposits in the Proterozoic strata of Northwestern Canada. In: Formation of uranium ore deposits. Int Atomic Energy Agency, Vienna, pp 255–273

Moshkin VN, Dagelayskaya IN, Sobotovich EV (1977) Pb-isochron age of anorthosites of the Sekhtag-Dzhugdzhur Massifs of the Aldan-Stanovoy shield. In: Geol Interpret Dannykh Geokhronol (summaries), Irkutsk, pp 8–9

Murthy VR (1976) Composition of the core and the early chemical history of the Earth. In: Windley BF (ed) The early history of the Earth. Wiley, London, pp 21–31

Nagy LA (1974) Transvaal stromatolite: first evidence for the diversification of cells about 2 b.y. years ago. Science 183(4124):514–516

Naqvi SM, Divakara RaoV, Narain H (1978) The primitive crust: evidence from the Indian Shield. Precambrian Res 6(3–4):323–345

Naydin DP (1971) Changes in ocean levels in Mesozoic and Cenozoic. Byull Mosk Ova Ispyt Prir Otd Geol 3:10–18

Negrutza VZ (1968) The Middle Proterozoic of Karelia. In: Geologicheskoye Stroyeniye U.S.S.R., 1. Nedra, Moscow, pp 118–131

Negrutza VZ, Negrutza TF (1968) The Jatulian geology problem. Tr VSEGEI 143:81–96

Neimark LA (1981) U-Th-Pb isotopic-geochemical systems in Early Precambrian high-grade metamorphic rocks of the Aldan-Stanovoy Shield. Thesis, VSEGEI, Leningrad 22 pp

Neruchev SG (1977) An attempt of a quantitative estimation of parameters of the older atmospheres of the Earth. Izv Acad Sci U.S.S.R. Geol Ser 10:5–22

Newell ND (1967) Revolutions in the history of life. Geol Soc Am Spec Pap 89:63–91

Nuzhnov SV (1967) The Riphean strata in the south-eastern part of the Siberian platform. Nauka, Moscow, 160 pp

Obradovich JD, Peterman ZE (1968) Geochronology of the Belt Series, Montana. Can J Earth Sci 5(3/2):737–747

Ohmoto H (1972) Systematics of sulfur and carbon isotopes in hydrothermal ore deposits. Econ Geol 67:551–578

Old RA, Rex DC (1971) Rubidium and strontium age determination of some Pre-Cambrian granitic rocks S. E. Uganda. Geol Mag 108(5):353–360

Orpen JL, Wilson JF (1981) Stromatolites at ~ 3500 Myr and a greenstone-granite unconformity in the Zimbabwean Archaean. Nature (London), 291 (5812):218–220

Oskwarek J, Perry E (1977) Temperature limits of early Archaean ocean from oxygen isotope variations in the Isua supracrustal sequence, West Greenland. Nature (London) 259(5540): 192–194

Pedersen K, Lam Y (1968) Precambrian organic compounds from the Ketilidian of South-West Greenland. Medd Grønl 185(5):16

Perry EC, Monster J, Reimer T (1971) Sulfur isotopes in Swaziland system barite and evolution of the Earth's atmosphere. Science 171(3975):1015–1016

Perry EC, Tan FC, Morey GB (1973) Geology and stable isotope geochemistry of the Biwabik iron formation, northern Minnesota, Econ Geol 68(7):1110–1125

Persson L (1974) Precambrian rocks and tectonic structures of an area in north-eastern Småland, Southern Sweden. Sver Geol Unders Ser C 703:1–55

Peterman ZE, Zartman RE, Sims PK (1978) Gneiss of Early Archean age in Northern Michigan. U.S.A. Geol Surv Open-File 701:332–334

Pflug HD, Jaeschke-Boyer H (1979) Combined structural and chemical analysis of 3800-Myr old microfossils. Nature (London) 280(5722):485–486

Pidgeon RT (1978a) Geochronological investigation of granite batolith of the Archaean granite-greenstone terrain of the Pilbara block, Western Australia. In: Proc 1978 Archaean Geochem Conf, Toronto, pp 360–362

Pidgeon RT (1978b) 3450 m.y.-old volcanics in the Archaean layered greenstone succession of the Pilbara Block, Western Australian. Earth Planet Sci Lett 37(3):421–428

Piirainen T (1968) Die Petrologie und die Uranlagerstätten des Koli-Kaltimogebietes in finnischen Nordkarelien. Bull Comm Geol Finl 237, 99 pp

Piirainen T (1978) General geology and metallogenetic features of Finland. In: Metallogeny of Baltic Shield Helsinki Symposium, pp 2–20

Playford PhE (1979) Stromatolite research in Western Australia. J R Soc West Aust 62(1–4):13–20

Plumb KA (1979) The tectobic evolution of Australia. Earth-Sci Rev 14(3):205–249

Plumb KA, Gemuts I (1975) Precambrian geology of the Kimberley region, Western Australia. 25 Int Geol Congr Excurs Guide 44 C, Canberra, 69 pp

Polishchuk VD (ed) (1970) Geology, hydrogeology and iron-ores in Kursk Magnetic Anomaly basin, 1. Geology, 1. Precambrian. Nedra, Moscow, 439 pp

Polkanov AA, Gerling EK (1961) Geochronology and geologic evolution of the Baltic shield and surrounding foldbelts. Tr LAGED Acad Sci U.S.S.R. 12:7–102

Polovinkina YuI (1960) Stratigraphic subdivision of the older gneissic strata in the Ukraine. Dokl Acad Sci U.S.S.R. 134(4):909–912

Polunovsky RM (1969) Characters of gneissic rocks in Central Pri-Azov area and the problems of their stratigraphy. Dokl Acad Sci U.S.S.R. 187(6):1360–1363

Porada H (1979) The Damara-Ribeira orogen of the Pan-African-Brasiliano cycle in Namibia (Southwest Africa) and Brasil as interpreted in terms of continental collision. Tectonophysics 57(2–4):237–266

Posadas VG, Kalliokoski J (1967) Rb-Sr of the Enerucijada granite intrusive in the Itamaca Venesuela. Earth Planet Sci Lett 2(3):210–214

Preiss WV (1977) The biostratigraphic potential of Precambrian stromatolites. Precambrian Res 5(2):207–219

Preiss WV, Krylov IN (1980) The Precambrian stratigraphy and plant remains in Southern Australia. Izv Acad Sci U.S.S.R. Geol Ser 7:61–74

Preiss WV, Walter MR, Coats RP, Wells AT (1978) Lithological correlations of Adelaidean glaciogenic rocks in part of the Amadeus, Ngalia, and Georgina Basins. BMR J Aust Geol Geophys 3:45–53

Pretorius DA (1975) The nature of the Witwatersrand gold-uranium deposits. Univ Witwatersrand Johannesburg Econ Geol Res Unit Annu Rep 1974, 16: 26

Priem HN, Boelrijk NA, Hebeda EH (1971) Isotopic age of the trans-Amazonian acidic magmatism and the Nickerie Metamorphic Episode in the Precambrian basement of Suriname, South America Geol Soc Am Bull 82(6):1667–1679

Priem HN, Boelrijk NA, Hebeda EH, Verdurmen AA, Verschure RH (1973) Age of the Precambrian Roraima formation in northeastern South America: evidence from isotopic dating of Suriname pyroclastic volcanic rocks in Suriname. Geol Soc Am Bull 84(5):1677–1684

Pringle JR (1973) Rb-Sr age determinations on shales associated with Varanger Ice age. Geol Mag 109(6):465–472

Prokof'yev VA (1971) Archean stratigraphy and tectonics in the Kitoy and Irkut Rivers basins (the South-Eastern Sayan region). In: Mater Geol Sibirsk. Platform Smezhnykh Oblastey 5:99–111

Pronin AA (1969a) Caledonian cycle in the tectonic evolution of the Earth. Nauka, Leningrad, 231 pp

Pronin AA (1969b) Hercynian cycle in the tectonic evolution of the Earth. Nauka, Leningrad, 195 pp

Pronin AA (1973a) Alpine cycle in the tectonic evolution of the Earth. Mesozoic. Chronology of tectonic movements. Nauka, Leningrad, 222 pp

Pronin AA (1973b). Alpine cycle in the tectonic evolution of the Earth. Cenozoic. Nauka, Leningrad, 200 pp

Pushcharovsky YuM (1980) Tectonic problems in oceans. In: Tektonika v Issledovaniyakh Geol. Inst Acad Sci U.S.S.R. Nauka, Moscow, pp 123–175

Raaben ME (ed) (1978) The Riphean lower boundary and the Aphebian stromatolites. Nauka, Moscow, 198 pp

Rabkin MI, Lopatin BG (1966) Metamorphic and magmatic formations of the Anabar Shield. In: Magmat. i Metamorf. Obrazovanya Sibiri. Nedra, Moscow, pp 156–168

Ramdohr P (1961) The Witwatersrand controversy: a final comment on the review by Professor C.F. Davidson. Min Mag 105(1):18–21

Ramsay CR, Davidson LR (1970) The origin of scapolite in the regionally metamorphosed rocks of Mary Kathleen, Queensland, Aust Contrib Miner Petrol 25:41–51

Randall MA (1963) Katherine, N.T. Sheet SD/53–9. 1:250,000 Geol Ser Explan Notes. BMR, Australia. 26 pp

Rankama K (1970) Proterozoic, Archean and other weeds in Precambrian rock garden. Bull Geol Soc Finl 42:211–222

Ravich MG, Grikurov GE (1978) Geologic Map of Antarctic. Izd Mingeo U.S.S.R., Leningrad

Reichelt R (1972) Géologie du Gourma (Afrique occidentale). Un "seuil" et un basin du Precambrien supérieur. Stratigraphie, tectonique, metamorphisme. Mem BRGM 53: 213

Reid A (1974) Stratigraphy of the type area of the Roraima Group, Venezuela. Bol Geol Publ Espec 6:343–353

Reid RR, Greenwood WR, Morrison DA (1970) Precambrian metamorphism of the Belt Supergroup in Idaho. Geol Soc Am Bull 81(3):915–917

Richards R (1978) Lead isotopes and ages of galenas from the Pilbara region, Western Australia. J Geol Soc Aust 24(7):465–473

Rix P (1965) Milinqimbi, N.T. Sheet SD 53–2, 1:250,000 Geol Ser Explan Notes. BMR, Aust, 28 pp

Roberts DC, Ardus DA, Dearnley R (1973) Precambrian rocks on the Rockall Bank. Nature (London) Phys Sci 244:21–23

Robertson DS (1974) Basal proterozoic units as fossil time markers and their use in uranium prospection. In: Formation uranium ore deposits. Int Atom Energy Agency, Vienna, pp 495–512

Robonen VI, Rybakov SI, Ruchkin GV (1978) Pyrite deposits of Karelia. Tr Inst Geol Karel Filial Acad Sci U.S.S.R., 37. Nauka, Leningrad, 192 pp

Roddick JC, Compston W, Durney DW (1976) The radiometric age of the Mount Keith Granodiorite, a maximum age estimate for an Archaen greenstone sequence in the Yilgarn block, Western Australia. Precambrian Res 3(1):55–78

Ronov AB (1976) Volcanism, carbonate origin, Life (regularities in globale geochemistry of carbon). Geokhimiya 8:1252–1257

Ronov AB (1980) Sedimentary covers of the Earth (quantitative regularities in composition, structure and evolution). Nauka, Moscow, 80 pp

Ronov AB, Khain VE, Balukhovsky AN, Seslavinsky KB (1976) Change in distribution, volume and rate of accumulation of sedimentary and volcanogenic strata during Phanerozoic (within limits of modern continents). Izv Acad Sci U.S.S.R. Geol Ser 2:44–57

Ronov AB, Khain VE, Balukhovsky AN (1979) Comparative evaluation of volcanism intensity in the continents and oceans. Izv Acad Sci U.S.S.R. Geol Ser 5:5–12

Ronov AB, Khain VE, Seslavinsky KB (1980) The Lower and Middle Riphean lithologic complexes of the world. Sov Geol 5:59–79

Roscoe SM (1957) Stratigraphy of Quirke Lake-Elliot Lake sector, Blind River, Ontario. Geol Surv Can Spec Publ 2:54–58

Roscoe SM (1969) Huronian rocks and uraniferous conglomerates in the Canadian Shield. Geol Surv Can Pap 68–40: 205

Ross CP (1970) The Precambrian of the United States of America: Northwestern United States – the Belt series. In: Rankama K (ed) The Precambrian, vol IV. Intersci Publ, New York London, pp 145–252

Rotman VK (1975) Paleovolcanism of island-arches in the north-western part of the Pacific Ocean and some aspects of the new global tectonics. Tr VSEGEI 234:138–148

Rozanov AYu, Sokolov BS (1980) The Precambrian-Cambrian boundary: up-to-date status. In: Dokl Sov Geol Dokembry. (MGK, 26 Sess). Nauka, Moscow, pp 159–164

Rubey WW (1951) Geologic history of sea water. Geol Soc Am Bull 62:1111–1173

Rubinshtein MM (1967) Orogenic phases and periodicity in folding in the light of new data of absolute geochronology. Geotektonika 2:21–30

Rudnik VA, Sobotovich EV, Terent'yev VM (1969) About the Archean age of rocks of the Aldan complex. Dokl Acad Sci U.S.S.R. 188 (4):897–900

Rutten MG (1971) The origin of Life by natural causes. Elsevier, Amsterdam, 420 pp

Ryan BD, Blenkinsop J (1971) Geology and geochronology of the Hellroaring Creek Stock, British Columbia. Can J Earth Sci 8 (1):85–95

Sagan C, Mullen G (1972) Earth and Mars: evolution of atmospheres. Science 177 (4043):52–56

Sakko M (1971) Varhais-Karjalaisten metadiabaasien radiometrisia zirconikiä. Geology 23 (9–10):117–119

Sakko M, Laajoki K (1975) Whole-rock Pb-Pb isochron age for the Pääkkö iron formation in Väyryiänkylä, South Poulanka area, Finland. Bull Geol Soc Finl 47 (1–2); 113–116

Salmo, British Columbia (1965) Map 1145A, 1:63,360. Geol Surv Can, Dep Mines, Techn Surv

Salop LJ (1963) Geologic interpretation of K-Ar absolute age data for rocks. Geol. Geofiz 1:3–21

Salop LJ (1964) Precambrian geochronology and some peculiarities of the early stage of geologic evolution of the Earth. Proc Int Geol Congr 22 Sess, New Delhi, Pt 10 Archean and Precambrian Geology, pp 131–149

Salop LJ (1964–1967) Geology of Baikal Mountain area. Nedra, Moscow, 1 (1964), 515 pp; 2 (1967), 699 pp

Salop LJ (1966) Contribution to the stratigraphy of the Lower Precambrian in Southern India. In: Probl Geol 22 Sess Int Geol Congr. Nauka, Moscow, pp 59–70

Salop LJ (1968) Precambrian of the U.S.S.R. Proc Int Geol Congr 23 Sess, Prague, Geology of the Precambrian, pp 61–73

Salop LJ (1970) Revision of the geochronological scale of the Precambrian. Byull Mosk Ova Ispyt Prir Otd Geol 45 (4):115–135; 45 (5):5–26

Salop LJ (1971a) Basic features of the stratigraphy and tectonics of the Precambrian in the Baltic shield. In: Problemy Geologii Dokembriya Baltiyskogo Schita i Pokrova Russkoy platformy. Tr VSEGEI 175:6–87

Salop LJ (1971b) Two types of Precambrian structures: gneiss folded ovals and domes. Byull Mosk Ova Ispyt Prir Otd Geol 46 (4):5–30 (English transl. in: Int Geol Rev Am Geol Inst 1972, 1:1209–1228)

Salop LJ (1972a) The problem of gold-uraniferous conglomerates: its geological aspects. Tr VSEGEI 178:150–174

Salop LJ (1972b) A unified stratigraphic scale of the Precambrian. Proc Int Geol Congr 24 Sess, Montreal, Precambrian Geology, pp 253–259

Salop LJ (1973a) A unified stratigraphic scale of the Precambrian. Nedra, Leningrad, 310 pp

Salop LJ (1973b) The Precambrian tillites and the Great glaciations. Byull Mosk Ova Ispyt Prir Qtd Geol 6:74–80

Salop LJ (1974a) On the Precambrian of the Great Lakes (Canada). Sov Geol 1:97–112
Salop LJ (1974b) On the stratigraphy and tectonics of the Mama-Chuya mica-bearing area in the Precambrian. Tr VSEGEI 199:83–143
Salop LJ (1977a) Precambrian of the Northern Hemisphere. Amsterdam, New York, Elsevier, 378 pp
Salop LJ (1977b) The Precambrian of Africa. Nedra, Leningrad, 304 pp
Salop LJ (1977c) Glaciations, rapid changes in organic evolution and their relationships with cosmic phenomena. Byull Mosk Ova Ispyt Prir Otd Geol 1:5–32 (English transl. in: Int Geol Review, 1977, 19:1271–1291
Salop LJ (1979a) Subdivision of the Precambrian on the geohistorical basis. In: Obshchiye Voprosy Raschlenenia Dokembria U.S.S.R. Nauka, Leningrad, pp 10–52
Salop LJ (1979b) The gneiss-granulite complex – a Karelides basement in the Ladoga agea and a Svecofenides basement in the south of Finland. Byull Mosk Ova Ispyt Prir Otd Geol 5:3–17
Salop LJ (1980) A turning-point in geologic evolution of the Earth at the Middle-Late Precambrian boundary (1900–1600 m.y.) Dokl Sov Geol, Dokembriy. Nauka, Moscow, pp 138–144 (Int Geol Congr 25 Sess)
Salop LJ, Murina GA (1970) An age of the Berdyaush rapakivi pluton and the problem of geochronological boundaries of the Lower Riphean. Sov Geol 6:15–27
Salop LJ, Scheinmann YuM (1969) Tectonic history and structures of platform and shields. Tectonophysics 7(5–6):565–597
Salop LJ, Travin LV (1974) New data on stratigraphy and tectonics of the central part of the Aldan shield. Tr VSEGEI 199:5–82
Salop LJ, Travin LV, Shalek EA (1974) Stratigraphy and tectonics of the southern part of Baikal Ridge in the Precambrian. Tr VSEGEI 199:144–172
Salotti Ch, Heinrich EW, Giardini AA (1971) Abiotic carbon and the formation of graphite deposits. Econ Geol 66(6):929–932
Sandimirova GP, Plyusnin GS, Petrova ZI (1979) Rb-Sr age of the Sharyzhalgay Group (Southern Baikal Region). In: Izotopnaya Geochronologiya Dokembriya (summaries). Bashkir Filial Acad Sci U.S.S.R., Ufa, pp 160–161
Sarkar SN (1972) Present status of Precambrian geochronology of Peninsular India. 24 Sess Int Geol Congr Sect 1, Precambrian geology, Montreal, pp 260–272
Sarkar SN, Saha AK, Miller JA (1967) Potassium argon ages from oldest metamorphic belt in India. Nature (London) 215(5104):946–948
Sarkar SN, Saha AK, Boelryk NA, Hebeda EH (1979) New data on the geochronology of the Older Metamorphic Group and the Singbhum granite of Singbhum-Keonjhar-Mayurbhanj region, Eastern India. Indian J Earth Sci 6(1):32–51
Savitsky VE, Zhuravleva IT, Kir'yanov VV (1980) The Nemakit-Daldyn facial stratotype of the Precambrian-Cambrian boundary in Siberia. In: Dokembry Dokl Sov Geol 26 Sess MGK, Nauka, Moscow, pp 164–170
Saxena SK (1977) The charnockite geotherm. Science 198(4317):614–617
Sayyah TA (1965) Geochronological studies of the Kinsley stock, Nevada, and the Raft River Range, Utah. Ph D thesis Utah Univ, Salt Lake City, Utah, 112 pp (unpublished)
Schidlowski M (1972) Probleme der atmospherischen Evolution in Präkambrium. Geol Rundsch 60(4):1351–1384
Schidlowski M (1975) Archean atmosphere and evolution of the terrestrial oxygen budget. In: Windley BF (ed) The early history of the Earth. Wiley, London, pp 525–534
Schmidt PW, Embleton BJ (1981) A geotectonic paradox. Has the Earth expanded? J Geophys 49(1):20–26
Schnitzer WA (1969) Die jung-algonkischen Sedimentationsräume Peninsula-Indiens. Neues Jahrb Geol Palaeontol Abh 133(2):191–198
Schopf JW (1974) The development and diversification of Precambrian life. Orig Life 5(1–2):119–135
Schopf JW (1977) Biostratigraphic usefulness of stromatolitic microbiotas: a preliminary analysis. Precambrian Res 5(2):143–174
Schopf JW, Prasad KN (1978) Microfossils in Collenia-like stromatolites from the Proterozoic Vempalle Formation of the Cuddapah Basin, India. Precambrian Res 6(3–4):347–366

Schopf JW, Ford TD, Breed W (1973) Microorganisms from the Late Precambrian of the Grand Canyon, Arizona. Science 179 (4080):1319–1321

Schwab FL (1977) Grandfather Mountain Formation, depositional environment, provenance, and tectonic setting of late Precambrian alluvium in the Blue Ridge of North Carolina. J Sediment Petrol 47 (2):800–810

Schwab FL (1978) Secular trends in the composition of sedimentary rock assemblages. Archean through Phanerozoic time. Geology 6 (9):532–536

Sdzuy K (1978) The Precambrian-Cambrian boundary beds in Morocco. Geol Mag 115 (2): 83–94

Sedgwick A (1838) A synopsis of the English series of stratified rocks interior to the old red sandstone; with an attempt to determine the successive natural groups and formations. Proc Geol Soc London 21 (58):684

Semenenko NP, Shcherbak NP, Bartnitsky EN, Sobotovich EV (1974) Geochronologic basis of the lower age boundary of the Krivoy Rog group. Izv Acad Sci U.S.S.R. Geol Ser 11:18–29

Semikhatov MA (1974) Stratigraphy and geochronology of the Proterozoic. Nauka, Moscow, 300 pp

Semikhatov MA (1978) Certain Aphebian carbonate stromatolites in the Canadian shield. In: Raaben ME (ed) Nizhnyaya Granitsa Rifeya i Stromat. Afebiya. Tr GIN Acad Sci U.S.S.R., pp 111–148

Semikhatov MA (1979) New Precambrian stratigraphie scheme: analysis and conclusions. Izv Acad Sci U.S.S.R. Geol Ser 11:5–22

Semikhatov MA, Aksenov EM, Bekker YuR (1979) Subdivision and correlation of the Riphean in the U.S.S.R. In: Stratigrafiya Verkhnego Proterozoya U.S.S.R. (Riphean and Vendian). Nauka, Leningrad, p 6–42

Shafeyev AA (1970) The Precambrian of southwestern Pri-Baikal area and of Khamar-Daban. Nauka, Moscow, 179 pp

Shatsky NS (1945) Outlines of tectonics in the Volga-Urals oil-bearing area and in the adjacent western slope of the Southern Urals. Mater Poznan Geol Stroyeniya U.S.S.R. N Ser 2 (6):129

Shenfil' VYu (1978) Correlation of the Riphean strata of Siberia according to stromatolites. In: Novoye v Strat. i Paleont. Pozdnego Dokembr. Sibirsk. Platformy. Acad Sci U.S.S.R., Novosibirsk, pp 22–37

Shride AF (1967) Younger Precambrian geology in Southern Arizona. US Geol Surv Prof Pap 566, 89 pp

Shul'diner VI, Ozersky AF (1967) The Lower Precambrian geology of the Shilka-Olekma inter-stream area. Izv Acad Sci U.S.S.R. Geol Ser 8:102–113

Simmons GC, Maxwell CH (1961) Grupo Tamanduá da Série Rio das Velhas. Braz Div Geol Miner Bol 211: 30

Sinha AK (1972) U-Th-Pb systematics of the age of the Onverwacht Series, South Africa. Earth Planet Sci Lett 16 (2):219–227

Skiöld T (1976) The interpretation of the Rb-Sr and K-Ar ages of Late Precambrian in south-western Sweden. Geol Foeren Stockholm Foerh 98–1 (564):3–29

Slawson WF (1976) Vredefort core: a crossection of the upper crust? Geochim Cosmochim Acta 40 (1):117–121

Slettene RL, Wilcox LE, Blouse RS, Sunders JR (1973) A Bougeur gravity anomaly map of Africa. Defense Mapping Agence Aerospace Center St. Louis AFS, Missouri, Tech Pap 73–3:1–13

Smith AG, Barnes WC (1966) Correlation of and facies changes in the carbonaceous, calcareous, and dolomitic formations of the Precambrian Belt-Purcell supergroup. Geol Soc Am Bull 77 (12):1399:1426

Smith IEM, Williams JG (1980) Geochemical variety among Archean granitoids in Northwestern Ontario. Geol Assoc Canada Spec Pap 20:181–192

Smith JV (1976) Development of the Earth-Moon system with implications for the geology of the early Earth. In: Windley BF (ed) The early history of the Earth, Wiley, London, pp 3–19

Smith JW (1964) Bauhinia Downs, N.T. Sheet SE/53–3. 1:250,000. Geol Ser Explan Notes. BMR, Australia, 20 pp

Smyth WR, Marten BE, Ryan AB (1978) A major Aphebian-Helikian unconformity within the Central Mineral Belt of Labrador: definition of new groups and metallogenic implication. Can J Earth Sci 15(12):1954–1966

Snelling NY, McConnell RB (1969) The geochronology of Guyana. Geol Mijnbouw 48(2): 201–213

Snizhko AM (1974) Occurrences of oncolites and catagraphs in the upper formation of the Krivoy Rog Group. Dokl Acad Sci Ukraininan S.S.R. 5(7):595–599

Sobotovich EV, Grashchenko SM, Aleksandruk VM (1963) Age of the oldest rocks determined by Pb-isochron and Sr-isotope spectral methods. In: Dokl Sov Geol (MGK, 22 Sess) Probl 3, Moscow, pp 429–430

Sobotovich EV, Iskanderova AD, Kamenev EN (1973) New data on the Azoic age of the oldest strata of the Earth (Enderby Land and Okhotsk Massif). In: Opred. Abs. Vozrasta Rudnykh Mestor. i Molodykh Magmat. Protsessov (summaries). Acad Sci U.S.S.R., Moscow, pp 87–89

Sobotovich EV, Iskanderova AD, Korol'kov VP (1977) The Early Archean age of rocks of the Taygonos and Omolon Massifs in the Pacific mobile belt. In: Geol Interpret Dannykh Geokhronol (summaries), Irkutsk, pp 10–11

Sokolov BS (1952) On the age of the oldest sedimentary cover of the Russian platform. Izv Acad Sci U.S.S.R. Geol Ser 5:12–20

Sokolov BS (1974) Problems of the Precambrian-Cambrian boundary. Geol Geofiz 2:3–29

Sokolov BS (1979) Precambrian paleontology. In: Paleontol. Dokembriya i Rannego Kembriya. Nauka, Leningrad, pp 5–16

Sokolov BS (1980) The Vendian System; pre-Cambrian geobiological environment. Paleontology and Stratigraphy. Nauka, Moscow, pp 9–21 (MGK, 25 Sess)

Sokolov VA (1963) Geology and lithology of carbonate rocks of the Middle Proterozoic in Karelia. Acad Sci U.S.S.R., Moscow Leningrad, 185 pp

Solontsov LF, Aksenov EM (1970) Stratigraphy of the Valday Group of the East-European platform. Izv Vuzov Geol Razved 6:3–13

Solov'yev DS, Halpern M (1975) First Archean isotope ages obtained in Antarctic on rocks of the crystalline basement. Inform Byull Sov Antarct Eksped 90:23–25

Sougy J (1972) Etat des connaissances géologique sur la partie mauritanienne de la Dorsale réguibat précambrienne. In: Corrél Précambrien Coll Int 1970, Rabat Paris, pp 95–103

Srinivasan R, Sreenivas BL (1972) Dharwar stratigraphy. J Geol Soc India 13(1):75–85

Srinivasan R, Sreenivas BL (1976) Greenstone-granite pluton and gneiss-granulite belts of the type Dharwar craton, Karnataka, India. Abstr 25 Sess Int Geol Congr, Sydney, p 19

Steward JH (1970) Upper Precambrian and Lower Cambrian strata in the Southern Great Basin California and Nevada. US Geol Surv Prof Pap 620: 206

Stockley GM (1943) The pre-Karroo stratigraphy in Tanganyika. Geol Mag 80(5):161–170

Stockwell CH (1964) Fourth report on structural provinces, orogenesis, and classification of rocks of the Canadian Precambrian Shield. Geol Surv Can Pap 64–17:1–121

Stockwell CH (1973) Revised Precambrian time scale for the Canadian Shield. Geol Surv Can Pap 72–52: 4

Stockwell CH, McGlynn JC, Emslie RF, Sanford BV, Norris AW, Donaldson JA, Fahrig WF, Currie KL (1970) Geology of the Canadian Shield. In: Douglas RJ (ed) Geology and economic minerals. Publ Geol Surv Can, p 45–150

Strakhov NM (1963) Lithogenesis, its types and evolution during the Earth's history. Gosgeoltekhizdat, Moscow, 535 pp

Suslova SN (1976) Komatiites from the Lower Precambrian metamorphosed volcanogenic strata of the Kola Peninsula. Dokl Acad Sci U.S.S.R. 228(3):697–700

Svetov AP, Sokolov VA (1976) Proterozoic orogenic and platform basaltic volcanism in the southern part of the Baltic shield (with reference to Southern and Central Karelia). In: Geol Petrol Metallogen Kristallich Obrazov Vost Eur Platformy, 2. Nedra, Moscow, pp 125–129

Sweet IP, Mendum JR, Morgan CM, Pontifex IR (1974) The geology of the northern Victoria River region, Northern Territory. BMR Aust Rep 166: 117

Tatsumoto M (1978) Isotopic composition of lead in oceanic basalt and its implication to mantle evolution. Earth Planet Sci Lett 38(1):63–87

Taylor PN (1975) An early Precambrian age for migmatitic gneisses from Vikan BO, Vesterålen, North Norway. Earth Planet Sci Lett 27(1):35–42

Thomson BP (1966) The lower boundary of the Adelaide system and older basement relationships in South Australia. J Geol Soc Aust 13:203–228

Thomson BP, Coats RP, Mirams RC, Forbes BG, Dalgarno CR, Johnson JE (1964) Precambrian rock groups in the Adelaide Geosyncline: a new subdivision. Q Geol Notes Geol Surv S Aust 9: 16

Thomson BP, Daily B, Coats RP, Forbes BG (1976) Late Precambrian and Cambrian geology of the Adelaide "Geosyncline" and Stuart Shelf, South Australia. 25th Int Geol Congr Sydney, Excurs Guide 33a: 53

Timofeev BV, German TN, Gnilovskaya MB (1980) The Precambrian biota of Eurasia. Dokl Sov Geol Dokembriy. Nauka, Moscow, p 170–176 (MGK, 25 Sess)

Tkachenko BV (ed) (1970) Reference section for the Upper Precambrian strata of the Western slopes of the Anabar Uplift. Izd NIIGA, Leningrad, 166 pp

Torquato JR, Gordani UG (1981) Brazil-Africa geological links. Earth-Sci Rev 17(1–2): 155–176

Torquato JR, Tanner M, Oliveira de MAF (1978) Idaqe radiometrica do Campo Formosa, da uma idade minima para o Grupo Jacobina. Rev Bras Geosci 8(3):171–179

Touret J (1971) Le facies granulite en Norvege meridionale. II. Les inclusions-fluides. Lithos 4(4):423–436

Travin LV (1974) On certain stratified structures of the Archean rocks of the Aldan shield. Tr VSEGEI 199:199–206

Trendall AF (1975a) Precambrian. Main areas of Proterozoic sedimentary rocks. Hamersley Basin. Geol Surv West Aust Mem 2:119–143

Trendall AF (1975b) Geology of the Hamersley Basin. 25th Int Geol Congr Sydney, Excurs Guide 43A: 44

Trendall AF (1976) Striated and faceted boulders from the Turee Creek Formation – evidence for a possible Huronian Glaciation on the Australian continent. Geol Surv W Aust Annu Rep 1975:88–92

Trendall AF, Bockley JG (1970) The iron formation of the Precambrian Hamersley Group, Western Australia. Geol Surv W Aust Bull 119: 366

Trimble DE (1976) Geology of the Michard and Pocatello Quadrangles, Bannock and Power Counties, Idaho. US Geol Surv Bull 1400: 88

Tugarinov AI, Bibikova EV (1971) On geochronology of Brazil. Geokhimiya 7:799–808

Tugarinov AI, Voytkevich GV (1970) Geochronology of the Precambrian of the continents, 2nd ed. Nedra, Moscow, 431 pp

Tugarinov AI, Bibikova EV, Zykov SI (1963) Absolute age of sedimentary rocks by Pb-U-method. Geokhimiya 3:266–283

Tugarinov AI, Bibikova EV, Krasnobayev AA, Makarov VA (1970) Precambrian geochronology in the Urals. Geokhimiya 4:501–509

Tugarinov AI, Bibikova EV, Gracheva TV (1973) Zircon chronology of the eastern part of the Baltic shield. In: Opred. Absol. Vozrasta Rudnykh Mestorozhd. i Molodykh Magmat. Protsessov (summaries). Acad Sci U.S.S.R. Moscow, pp 31–32

Tugarinov AI, Zykov SI, Stupnikova NI (1977) Age problem of the oldest formations of the Stanovoy Range. In: Geol Interpret Dannykh Geokhronol (summaries), Irkutsk, p 8

Urey HC (1973) Cometary collisions and geological periods. Nature (London) 242(5392): 32–33

Van Breemen O, Dodson MH (1972) Metamorphic chronology of the Limpopo Belt, Southern Africa. Geol Soc Am Bull 83(7):2005–2018

Van Breemen O, Allaart JA, Aftalion M (1972) Rb-Sr whole rock and U-Pb zircon age studies on granites of the Early Proterozoic mobile belt of South Greenland. Grønl Geol Unders Rep Activ 1971, 45: 45–48

Van Breemen O, Dodson MH, Vail JR (1966) Isotopic age measurements on the Limpopo orogenic belt, Southern Africa. Earth Planet Sci Lett 1(6):401–406

Van Niekerk CB, Burger AJ (1964) The age of Ventersdorp System. Geol Surv S Afr Ann 3:75–86

Van Niekerk CB, Burger AJ (1969a) A note on the minimum age of the acid lava of the Onverwacht Series of the Swaziland System. Trans Geol Soc S Afr 72–I:9–23

Van Niekerk CB, Burger AJ (1969b) Lead isotopic data relating to age of the Dominion Reef Lava. Trans Geol Soc S Afr 72–II:37–47

Van Niekerk CB, Burger AJ (1978) A new age for the Ventersdorp acidic lava. Trans Geol Soc S Afr 81 (2):155–163

Van Schmus WR (1978) Geochronology of the Southern Wisconsin rhyolites and granites. Geosci Wis 2:19–24

Van Schmus WR, Medaris LC, Banks PO (1975) Geology and age of the Wolf River Batholith, Wisconsin. Geol Soc Am Bull 86 (7):907–914

Väyrynen H (1954) Suomen Kalliopera. Helsinki, 300 pp

Velikoslavinsky SD (1978) Petrology and geochemistry of crystalline schist of basic composition in the central part of the Aldan shield. Thesis. VSEGEI, Leningrad, 25 pp

Venkatasubramanian VS, Narayanaswamy R (1974) The age of some gneissic pebbles in Kalduga conglomerate, Karnataka, South India. J Geol Soc India 15:318–319

Verbeek T (1970) Géologie et lithologie du Lindien (Précambrien supererieur du Nord de la Republique democratique du Congo). Ann Mus R Afr Cent Ser 8 Sci Geol 66: 311

Verhofstad J (1970) The geology of the Wilhelmina mountains in Suriname, with special reference to the occurrence of Precambrian ash-flow tuffs. Diss Graad doct wiskunde en naturwetens, Univ Amsterdam, 97 pp

Verhoogen J, Turner FJ, Weiss LE, Wahrhafting C, Fyfe WS (1970) The Earth. An introduction to physical geology. Holt-Rinehart, New York, 748 pp

Viljoen MJ, Viljoen RP (1969) An introduction to the geology of the Barberton granito-greenstone terrain. Geol Soc S Afr Spec Publ 2:9–28

Vinogradov AP (1967) Geochemistry of the Ocean. Introduction. Nauka, Moscow, 214 pp

Vinogradov VI (1977) On traces of the Early Precambrian evaporites (according to data of sulphur isotope composition). In: Problemy Osadkonakoplenija, 1. Novosibirsk, p 105–108

Vinogradov VI, Reymer TO, Leytes AM, Smelov SB (1976a) The oldest sulphate in the Archean formations of South-African and Aldan shields and evolution of oxygenic atmosphere of the Earth. Litol Polezn Iskop 4:12–27

Vinogradov VI, Smelov SB, Litsarev MA (1976b) Older K-Ar age of the Dzheltula Group of the Aldan shield. Dokl Acad Sci U.S.S.R. 230 (1):164–166

Vogt PR, Avery OE (1974) Detailed magnetic surveys in the northeast Atlantic and Labrador Sea. J Geophys Res 79 (2):363–389

Volkova NA, Kir'yanov VV, Pyatiletov VG (1980) Microfossils from the Upper Precambrian of the Siberian platform. Izv Acad Sci U.S.S.R. Geol Ser 1:23–29

Volobuev MI, Zykov SI, Stupnikova NI (1964) Geochronology of the Yenisei Ridge. In: Absol. Vosrast Geol. Formatsyi, Nauka, Moscow, pp 108–127

Volobuev MI, Zykov SI, Stupnikova NI (1973) Geochronology of Grenvillides basement and geosynclinal formations of the Yenisei Ridge. In: Opred. Abs. Vozrasta Rudnykh Mestorozhd. i Molodykh Magmat. Porod. Nauka, Moscow, pp 39–47

Volobuev MI, Zykov SI, Stupnikova NI (1977a) Geochronology of the Precambrian granitoids of the area surrounding the Siberian platform from the south-west. In: Geol Interpret Dannykh Geokhronol (summaries), Irkutsk, p 33

Volobuev MI, Zykov SI, Stupnikova NI (1977b) Isotope age of the Precambrian metamorphic complexes of South-Western Balkhash region, Eastern Sayan and Yenisei Ridge. In: Geol Interpret Dannykh Geokhronol (summaries), Irkutsk, p 32

Walker RG (1978) A critical appraisal of Archean basin-craton complexes. Can J Earth Sci 7:1213–1218

Wallace RM (1965) Geology and mineral resources of the Pico de Itabirito District, Minas Gerais, Brasil. US Geol Surv Prof Pap 341–F: 68

Walpole BP, Crohn PW, Dunn PR, Randal MA (1968) Geology of the Katharine-Darwin Region, Northern Territory. BMR Aust Bull 82: 304

Walsh J (1966) Geology of the Karasuk area, Republic of Kenya. Geol Surv Kenya Rep 72: 34

Walter MR (1972) Stromatolites and biostratigraphy of the Australian Precambrian and Cambrian. Paleontol Assoc (London) Spec Pap Paleontol 11: 190

Walter MR, Awramik SM (1979) Frutexites from stromatolites of the Gunflint Iron Formation of Canada, and its biologial affinities. Precambrian Res 9 (1):23–33

Walter MR, Buick R, Dunlop ISR (1980) Stromatolites 3400–3500 Myr old from the North Pole area, Western Australia. Nature (London) 248 (5755):443–445

Watters RR (1977) The Sinclair Group: definition and regional correlation. Trans Geol Soc S Afr 80 (1):9–16

Weber F (1968) Une série précambrienne du Gabon: le Francevillien, sedimentologie, géoche-mie, relations avec les gites minéraux associés. Mem Ser Carte Geol Alsace-Lorraine 28: 330

Welin E, Lundqvist T (1970) New Rb-Sr age data for the Sub-Jotnian volcanics (Dala porphy-ries) in the Los-Hamra region, Central Sweden. Geol Foeren Stockholm Foerh 92 (540): 35–39

Wells AT, Forman DJ, Randford LC, Cook PJ (1970) Geology of the Amadeus Basin Central Australia. BMR Aust Bull 100: 229

Williams H, King AF (1979) Trepassey map area, Newfoundland. Geol Surv Can Mem 389: 24

Williams IR (1969) Structural layering in the Archean of Kurnalpi 1:250,000 Sheet area, Kal-goorlie region. Geol Surv W Aust Annu Rep 1968: 40–41

Wilshire HG (1971) Pseudotachylite from Vredefort Ring, South Africa. J Geol 79 (2):193–206

Wilson IH (1978) Volcanism on Proterozoic continental margin in northwestern Queensland. Precambrian Res 7 (3):205–235

Wilson IH, Derrick GM (1976) Precambrian geology of the Mount Isa region, northwest Queensland. 25th Int Geol Congr Sydney, Excurs Guide 5A, 5C: 44

Wilson JF, Bickle MJ, Hawkewworth CJ, Martin A, Nisbet EG, Orpen JL (1978) Granite-greenstone terrains of the Rhodesian Archaean craton. Nature (London), 271 (5640):23–27

Windley BF (ed) (1976) The Early history of the Earth. Wiley, London, 619 pp

Windley BF (1977) The evolving continents. Wiley, London, 385 pp

Winter H (1976) A lithostratigraphic classification of the Ventersdorp succession. Trans Geol Soc S Afr 79 (1):31–48

Wood DA, Gibson JL, Thompson RN (1976) Elemental mobility during zeolite facies metamor-phism of the tertiary basalt of Eastern Iceland. Contrib Mineral Petrol 55 (3):241–256

Wynne-Edwards HR (1967) Westport map-area, Ontario, with special emphasis on the Precam-brian rocks. Geol Surv Can Mem 346: 142 pp

Yankauskas TV (1978) Plant microfossils from the Riphean strata of the Southern Urals. Dokl Acad Sci U.S.S.R. 242 (4):913–915

Yeliseeva GD, Shcherbak NP, Kazantseva NP (1973) Pb-isochron age of carbonate rocks of the Near-Bug area. In: Opred. Absol. Vozrasta Rudn. Mestorozhd. i Molodykh Magmat. Prot-sessov (summaries). Acad Sci U.S.S.R., Moscow, pp 25–26

Yershov VM, Markov SN, Khayritdinov RK (1969) Absolute age of rocks of the Zigalga For-mation of the Urals. Geokhimiya 5:623–627

Young GM (1966) Huronian stratigraphy of the Mac Gregor Bay area, Ontario: relevance to the paleogeography of the Lake Superior region. Can J Earth Sci 3 (2):203–210

Young GM (ed) (1973a) Tillites and aluminous quartzites as possible time markers for Middle Precambrian (Aphebian) rocks of North America. In: Huronian stratigraphy and sedimen-tation. Geol Assoc Can Spec Pap 12: 97–128

Young GM (ed) (1973b) Huronian stratigraphy and sedimentation. Geol Assoc Can Spec Pap 12: 276

Young GM (1978) Some aspects of the evolution of the Archean crust. Geosci Can 5 (3): 140–149

Young GM, Jefferson CW, Delaney GD, Yeo GM (1979) Middle and Late Proterozoic evolu-tion of the northern Canadian Cordillera and Shield. Geology 7:125–128

Zagorodny VG, Mirskaya DD, Suslova SN (1964) Geologic structure of the Pechenga sedi-mentary-volcanogenic group. Nauka, Moscow Leningrad, 207 pp

Zaytsev YuA, Filatova LN (1971) New data on the Precambrian structures in Ulu-Tau. In: Voprosy Geol. Centr. Kazakhstana, 10. Moscow Univ, pp 21–92

Zel'dovich YaB, Novikov ID (1975) Structure and evolution of the Universe. Nauka, Moscow, 735 pp

Zhuravleva ZA (1964) The Riphean and Lower Cambrian oncolites and catagraphs in Siberia and their stratigraphic importance. Tr Geol Inst Acad Sci U.S.S.R. 114: 73

Zubtsov EI (1972) Precambrian tillites of Tien-Shan and their stratigraphic value. Byull Mosk Ova Ispyt Prir Otd Geol 1:42–56

Zykov SI, Stupnikova NI, Lebedeva LM (1979) Standard values of isotope ratios and decay constants in geochronometry. In: Izotopnaya Geokhronologiya Dokembriya (summaries). Bashkir Filial Acad Sci U.S.S.R., Ufa, pp 45–48

# Subject Index

Abitibi belt 76, 80, 81, 92–96, 108
Acrytarchs 3, 232, 235, 248, 261, 310, 332, 335, 336, 358, 364
Activation, see thermal-tectonic activation
Activation zones of Baikalian cycle 349, 376–378, 404
Actualism principle 408–410
Actualistic method 52, 410
Adelaide belt (geosyncline) 301, 350, 351, 370, 374
Adelaidean era (= Epiprotozoic or Katangan era) 12, 308
Adrar-Iforas zone 267
Africa (Precambrian of) 27–30, 43, 45, 57, 74, 78, 82–91, 151–158, 217–221, 262–272, 298–301, 367–370
Ahaggar 29, 269, 319
Algeria 29, 132, 221, 266–270
Angola 74, 218
Belingwe area 89, 90
Benin 30, 57, 132, 317
Cameroun 28, 219, 314
Congo 218
Copper belt 310–313
Eastern Sahara 123
Equatorial Africa 43, 45, 91, 96, 123, 218–220, 262, 266, 310, 315
Gabon 132, 218, 313
Ghana 30, 220, 318
Guinea 74
Ivory Coast 220
Katanga 42, 311–313
Kenya 30, 74, 91
Komati River area 108
Liberia 28–30, 74

Madagascar 28, 42, 43, 45, 61
Malavi 28, 30
Mali 266–268, 319
Mauritania 132, 220, 318
Morocco 367, 368
Murchison Range 122
Namibia 131, 218, 270, 315–317, 368–370
Natal 78, 87
Niger 320
Nigeria 320
Nile basin 123
North-Western Africa 30, 43, 220, 266–270, 317–320
Republic of S. Africa 82, 83, 122, 131, 151–158, 317
Ruanda-Burundi 131, 218
Selukwe area 89
Senegal 319
Sierra-Leone 29, 30, 74
Southern Africa 74, 82–91, 95, 151–157, 217, 218, 270–272, 315–317
Swaziland 27, 30, 74, 78, 82–87, 108–122
Tanganyika Lake area 264
Tanzania 30, 91, 131, 218, 263, 264, 314
Togo 132, 317
Transvaal 127, 151–158
Uganda 30, 74, 91, 131, 218, 219, 230, 314
Upper Volta 220
Victoria Lake area 91
Vort Victoria area 90
Western Africa 132, 195, 220, 317–320
Zaire 28, 74, 218, 263, 310
Zambia 47, 131, 218, 310
Zimbabwe 28, 30, 74, 78, 81, 89–91, 98, 121, 317
African platform 300, 301, 376, 377

Akitkan fracture zone 207–210, 287
Akitkanian Suberathem, see Neoprotozoic
Albany-Fraser zone 302
Alberton zone 253, 254
Aldanian (= Katarchean) Era 11, 22
Aldan platform (craton) 146, 192
Aldan Shield 24, 33, 34, 36–42, 44, 146, 192, 211, 336, 362
Aldanian stage 364
Algonkian 260
Amadeus aulacogen 324, 350, 351, 370
Amphibole-granulite subfacies 32, 47
Anabar (Anabara) shield, massif 22, 42–44, 192, 249, 363
Angara craton 192
Anorthosite 7, 32, 64, 68, 205, 213–215, 226, 288, 289, 306, 387
Ansongo zone 266, 267
Antarctic Continent (Precambrian of) 29, 30, 305
Anti-Atlas belt 294, 295, 300, 317, 348, 349
Aphebian 10, 130
Appalachian belt (geosynkline) 302, 326, 327, 350, 375
Archean 9, 10, 73
Archeoprotozoic (= Paleoprotozoic) 10, 72
Arunta block (craton) 103, 187, 190, 222
Ashburton zone 169, 190
Asia (Precambrian of) 24, 25, 30, 33, 34, 36–42, 44, 46, 74, 75, 101, 102, 145–151, 207–212,

# Index of Local Stratigraphic Units and of Some Intrusive Formations

Age is given in brackets (Cm – Cambrian, EC – Eocambrian, EP – Epiprotozoic, KA – Katarchean, MP – Mesoprotozoic, NP – Neoprotozoic, PP – Paleoprotozoic)

V. V. Beloussov

# Geotectonics

1980. 134 figures. X, 330 pages
ISBN 3-540-09173-4
Distribution rights for all socialist countries:
Mir Publishers, Moscow

**Contents:** Introduction. – Tectonics of
Continents: General Tectonic Movements of
the Earth's Crust. Tectonic Movements
Within the Crust. Patterns of the Evolution
of Continents. – Tectonics of Oceans. – The
Earth's Internal Structure, Composition, and
Deep Processes. – Literature. – Index.

Geotectonics occupies a special, interdisci-
plinary position among the various branches
of geology. In this book, V. V. Beloussov, head
of the Department of Planetary and Marine
Geophysics of the USSR Academy of Sciences'
Institute of the Physics of the Earth, and of
Moscow University's Laboratory of Tectonophysics, guides the reader through the complex
geological, geophysical and geochemical data
underlying geotectonic theory. The author
concentrates primarily on observable processes responsible for the formation of the crustal
structure while providing a critical review of the
major hypothetical conjectures advanced to
explain them. His efforts will be of inestimable
value to students and researchers dealing with
general tectonic movements of the Earth's
crust, types of tectonic processes, categories of
oscillatory movement, tectonics of continents,
tectonics of oceans, endogenous continental
regimes, and magmatic and metamorphic
processes.

Springer-Verlag
Berlin
Heidelberg
New York

# Geology of the Northwest African Continental Margin

Editors: **U. v. Rad, K. Hinz, M. Sarnthein, E. Seibold**

1982. 325 figures. XI, 703 pages.
ISBN 3-540-11257-X

**Contents:** Introduction. – Structure and Geodynamic Evolution of the Continental Margin. – Comparison of the Northwest African and Northeast American Margins. – Evolution of Volcanism. – Cretaceous Stratigraphy, Sedimentation and Paleoenvironment. – Cenozoic Stratigraphy, Sedimentation and Paleoenvironment. – Inorganic/Organic Geochemistry.

The Northwest African continental margin provides an excellent example of a mature passive continental margin. In this thick sedimentcover it bears a clue to the evolution of the Atlantic Ocean Basins during the past 200 million years. A dense net of single- and multichannel seismic reflection lines, numerous Deep Sea Drilling Project sites and petroleum exploration wells, and the wellknown onshore geology make the Northwest African margin one of the best-documented margins of the world.
This publication provides a topical synthesis of geological, paleontological, geochemical, and geophysical studies of the Northwest African continental margin. Concentrating primarily on DSDP Legs 14, 41, 47a, 50 and 79, it also includes deep crustal physical surveys, the results of Neogene and Quaternary cores, and field work on volcanic islands and onshore coastal basins conducted by various government agencies, industry, and academic institutions.

Springer-Verlag
Berlin
Heidelberg
New York